Frequently Used Formulas

$n(A \cup B) = n(A) + n(B)$, when A and B are disjoint

$n(A \cup B) = n(A) + n(B) - n(A \cap B)$, for all finite sets

$n(A \cup B \cup C) = n(A) + n(B) + n(C) - n(A \cap B) - n(B \cap C) - n(A \cap C) + n(A \cap B \cap C)$

Slope of line $m = \dfrac{y_2 - y_1}{x_2 - x_1}$

point slope form: $y - y_1 = m(x - x_1)$ slope–intercept form: $y = mx + b$

Probability $= \dfrac{f}{n}$ ${}_nP_r = \dfrac{n!}{(n-r)!}$ ${}_nC_r = \dfrac{n!}{r!(n-r)!}$

Addition rule (for mutually exclusive events): Prob (A or B) = Prob (A) + Prob (B)

Addition rule (general case): Prob(A or B) = Prob(A) + Prob(B) − Prob(A and B)

Conditional probability formula: $\text{Prob}(A|B) = \dfrac{\text{Prob}(A \text{ and } B)}{\text{Prob}(B)}$

Multiplication rule: Prob(A and B) = Prob($A|B$) · Prob(B)

Multiplication rule (for independent events): Prob(A and B) = Prob(A) · Prob(B)

Bayes' rule:

$$\frac{\text{Prob}(B|A_n) \cdot \text{Prob}(A_n)}{\text{Prob}(B|A_1) \cdot \text{Prob}(A_1) + \text{Prob}(B|A_2) \cdot \text{Prob}(A_2) + \text{Prob}(B|A_3) \cdot \text{Prob}(A_3) + \cdots + \text{Prob}(B|A_n) \cdot \text{Prob}(A_n)}$$

Prob(x successes out of n trials) $= \dfrac{n!}{x!(n-x)!}\, p^x q^{n-x}$

If payoff matrix is of form $\begin{array}{|c|c|} \hline a & b \\ \hline c & d \\ \hline \end{array}$, optimal strategies for nonstrictly determined games are:

$$R_1 = \frac{d-c}{a+d-b-c} \qquad R_2 = \frac{a-b}{a+d-b-c}$$

$$C_1 = \frac{d-b}{a+d-b-c} \qquad C_2 = \frac{a-c}{a+d-b-c}$$

Value of game $v = \dfrac{ad - bc}{a+d-b-c}$

The z-score or z-value of any number x in a normal distribution: $z = \dfrac{x - \mu}{\sigma}$

Coefficient of linear correlation: $r = \dfrac{n(\Sigma\, xy) - (\Sigma\, x)(\Sigma\, y)}{\sqrt{n(\Sigma\, x^2) - (\Sigma\, x)^2}\,\sqrt{n(\Sigma\, y^2) - (\Sigma\, y)^2}}$

Regression equation: $y = m_y + b(x - m_x)$

where $m_y = \dfrac{\Sigma\, y}{n}$, $m_x = \dfrac{\Sigma\, x}{n}$

and $b = \dfrac{n(\Sigma\, xy) - (\Sigma\, x)(\Sigma\, y)}{n(\Sigma\, x^2) - (\Sigma\, x)^2}$

Standard error of the estimate: $\sqrt{\dfrac{\Sigma(Y - Y_p)^2}{n-2}}$

Mathematical expectation: $m_1p_1 + m_2p_2 + m_3p_3 + \cdots$

Using Finite Mathematics

Using Finite Mathematics

Joseph Newmark

The College of Staten Island

1817

HARPER & ROW, PUBLISHERS, New York

Cambridge, Hagerstown, Philadelphia, San Francisco, London, Mexico City, São Paulo, Sydney

Sponsoring Editor: Sharmon Hilfinger/Don Shauger
Project Editor: Molly Scully
Designer: Dick Kharibian
Assistant Production Manager: Marian Hartsough
Senior Production Manager: Kewal K. Sharma
Photo Researcher: Myra Schachne
Compositor: Syntax International Pte. Ltd.
Printer and Binder: R. R. Donnelley & Sons Company
Art Studio: Syntax International Pte. Ltd.
Cover Photo: Don Mansa, DeWys

Using Finite Mathematics

Library of Congress Cataloging in Publication Data

Newmark, Joseph.
Using finite mathematics.

Includes bibliographies and index.
1. Mathematics—1961– I. Title.
QA39.2.N49 510 80-25187
ISBN 0-06-385752-9

To Trudy, Sharon, Rochelle, and Stephen

Special Features

- Newspaper and magazine clippings at the beginning of each chapter, which illustrate ideas studied in the chapter in applied context.
- Chapter objectives section at the beginning of each chapter, which pinpoint the main ideas of the chapter.
- Review of algebra in several chapters as needed rather than in an appendix.
- Historical notes for each chapter.
- Two forms of mastery tests for each chapter including both short-answer and long-answer questions.
- Summary, study guide, and list of formulas for each chapter.
- Calculator problems in many exercises and examples have been designed so that they can be solved by means of a calculator.
- Well over 1000 examples, exercises, and case studies, which cover many different areas.
- Suggested course outlines for both one-semester and two-semester courses.

Contents

Preface

This student-oriented text has been specifically designed so that it can be used by students from a variety of disciplines and at various educational levels. Such students will find the nontechnical writing style easy to read and understand. Each new idea is introduced in an intuitive manner and then applied to realistic settings.

Many exercises and examples illustrate how the ideas of finite mathematics can be applied to a variety of situations. A wide range of topics is included, thereby allowing greater flexibility in the ordering of topics. The wide range of topics makes this text suitable for a one-semester or two-semester course in finite mathematics designed for liberal arts, life science, and managerial science students. A suggested course outline is also provided.

Numerous teaching and learning aids have been incorporated in this text. Among these are:

1. *Newspaper and magazine clippings.* Each chapter starts with a newspaper or magazine article that presents the ideas discussed in the chapter in applied context. This is intended to motivate the reader by showing how the ideas mentioned in the chapter are applied in everyday situations.

2. *Chapter objectives.* Each chapter is introduced by chapter objectives, which pinpoint the main ideas of the chapter.

3. *Algebra review.* Numerous ideas from algebra are reviewed in several chapters as they are needed, rather than in an appendix where they are often overlooked.

4. *Comments.* In frequent comments various points that are often overlooked or misunderstood by students are carefully analyzed.

5. *Calculator problems.* Since hand-held calculators are becoming an integral part of the classroom, many exercises and examples have been designed so that they can be solved easily with a calculator. Such exercises are clearly marked by a color box.

6. *Concluding remarks.* Each chapter concludes with a summary, a study guide, and a list of formulas to remember.

7. *Mastery tests.* Each chapter contains two sets of mastery tests designed to measure the reader's comprehension of the material covered in the chapter, as well as to present a few challenging problems to the more demanding student.

8. *Historical notes.* Most chapters contain brief historical notes of interest on mathematicians who made contributions to the field of finite mathematics.

9. *Course outlines.* A suggested course outline is provided for both one-semester and two-semester courses.

There are more than 1000 exercises and about 300 illustrative examples with step-by-step explanations. Answers to most exercises are provided at the end of the text. A solutions manual providing detailed solutions to all problems is available from the publisher.

Suggested Course Outline

The following course outlines indicate how this text can be used in schools that offer finite mathematics in either one-semester or two-semester courses. It is assumed that each course meets about 40 times during the semester for 40 minutes each session. The instructor should have no difficulty in organizing other course outlines to suit individual needs from the wide range of topics covered.

One-Semester Course

Text material	Approximate amount of time
Chapter 1	3 lessons
Chapter 2 (skip 2.6, 2.8)	3 lessons
Chapter 3	6 lessons
Chapter 4	4 lessons
Chapter 6 (skip 6.7, 6.8)	6 lessons
Chapter 7 (skip 7.5)	4 lessons
Chapter 8	3 lessons
Chapter 9 (skip 9.6)	5 lessons
Chapter 10 (skip 10.5)	4 lessons
Chapter 12 (up to 12.5)	2 lessons
	38*

* The remaining time can be devoted to exams and review.

Two-Semester Course

Semester I

Text material	Approximate amount of time
Chapter 1	4 lessons
Chapter 2	5 lessons
Chapter 3	6 lessons
Chapter 4	4 lessons
Chapter 5	6 lessons
Chapter 6	7 lessons
Chapter 7	5 lessons
	37*

* The remaining time can be devoted to exams and review.

Semester II

Text material	Approximate amount of time
Chapter 8	3 lessons
Chapter 9	6 lessons
Chapter 10	5 lessons
Chapter 11	3 lessons
Chapter 12	6 lessons
	23*

* The remaining time can be devoted to exams and review.

Acknowledgments

I wish to thank the many students who used the text in the form of mimeographed class notes and who made valuable suggestions for its improvement. I also wish to thank the many people who reviewed the manuscript for their valuable suggestions. These include Dan Anderson (Parkland College), Ignacio Bello (Hillsborough Community College), Daniel W. Dodway (Broome Community College), and Dr. Kenneth A. Rager (Metropolitan State College).

Finally, and most important, I wish to thank my wife, Trudy, and our children Sharon, Rochelle, and Stephen, for their understanding and patience as they endured the enormous strain associated with completing this project. Without their encouragement this project could not have been undertaken.

JOSEPH NEWMARK

Using Finite Mathematics

CHAPTER 1 Sets

Objectives to be Achieved in this Chapter

- *To analyze* what a set is, the notation used for it, and several ways of indicating the elements or members of a set.
- *To distinguish* between finite and infinite sets. Much of the recent work in set theory involves infinite sets.
- *To study* subsets in general and proper subsets in particular, as well as the notation used for each.
- *To discuss* ways of forming new sets by performing the elementary set operations of union and intersection.
- *To study* pictorial ways of representing sets and set operations.
- *To learn* rules that can be used to determine the number of elements in sets under different conditions.
- *To apply* the ideas of sets to a variety of problems in the business, social science, and natural science areas.

Government to Investigate Summer Lunch Program

DAILY REPORTER AUGUST 3, 1978

Washington (Aug. 2)—Investigators for the United States Department of Agriculture announced yesterday that they were subpoenaing the business records of the Gartel Food Distribution Corporation for allegedly submitting fraudulent claims for reimbursement. According to these investigators, the summer lunch program provides free breakfasts, lunches, and snacks for the needy at selected sites. Although this program has been in existence for several years, it has been plagued by charges of mismanagement and inefficiency since its inception.

During the summer of 1975 Congresswoman Elizabeth Holtzman of New York City conducted an investigation which exposed much waste and fraud in the program. Since then the government has tightened its controls. Nevertheless, the Gartel Food Distribution Corp. is accused of consistently submitting false claims. Specifically, for the month of July the following claims for reimbursement involving 100 sites were submitted: 40 sites were served lunch, 43 sites were served breakfast, 25 sites were served snacks, 19 sites were served breakfast and lunch, 9 were served snacks and lunch, 23 were served breakfast and snacks, and 7 sites were served all three.

Government officials claim that these figures are inconsistent. Officials of the Gartel Corp. were unavailable for comment.

The newspaper article indicates that, according to government investigators, the claims for reimbursement submitted by the Gartel Food Distribution Corporation were inconsistent. How can we determine that these figures are inconsistent? Must we apply some advanced accounting techniques? Is the inconsistency due to the fact that the numbers do not add up to 100? As a matter of fact, the sum of these numbers is greater than 100, so it appears that the Gartel Corp. is losing money by submitting claims for only 100 sites.[1]

1.1 Introduction

In recent years great emphasis has been placed on the ideas of sets and the "new math." Many educators believe that the concepts of sets simplify the discussion of mathematical ideas and make them easier to comprehend. While this may or may not be true, sets do play an important role in mathematics today.

In this chapter we analyze some basic ideas of sets and set relations and discuss several applications in the life and management sciences. In later chapters, particularly when we discuss probability, we will apply the ideas of sets to a variety of important problems.

Historical Note

The ideas of set theory can be traced back to the great German mathematician, Georg Cantor. Although born in St. Petersburg, Russia, on March 3, 1845, he spent most of his adult life (both as a student and as a teacher) in Germany.

[1] This article is discussed further in the Form B Mastery Test questions at the end of this chapter. See question 1.

In an attempt to define numbers in an "abstract manner," Cantor developed the theory of infinite sets. His ideas about infinite sets were so radically different and disturbing that some of his fellow mathematicians conspired to keep his works from being published. In particular, one of his former professors, Leopold Kronecker (1823–1891), attacked Cantor with such vehemence that the controversy is said to have contributed to Cantor's several nervous breakdowns. Kronecker is also alleged to have used his influence to prevent Cantor from being appointed as a professor at the University of Berlin.

These ongoing controversies were too much for Cantor to withstand, and eventually he had to be committed to a mental asylum. He died in a mental hospital in 1918.

1.2 Sets and Subsets

We can think of a *set* as any collection of objects. Thus, we can speak of the set of all bones in the human body, the set of all courses offered by a college, the set of all endangered species of animals, or the set of all products sold by a nationwide supermarket chain. In each case the objects that belong to the set are called the *elements* or *members* of the set.

One way of describing a set is simply to list all the elements of the set and enclose them within set braces, { }. Thus, the set of all presidents of the United States who were born in October is written as

{John Adams, Rutherford Hayes, Chester Arthur, Theodore Roosevelt, Dwight Eisenhower, Jimmy Carter}

Similarly the set of all deodorants available in Rick's Pharmacy is written as

{Right Guard, Ban, Brut, Arrid, . . . , Sure}

Capital letters such as A, B, C, . . . are used to indicate particular sets, whereas lowercase letters such as a, b, c, . . . are used to indicate members of a set. For example, the set A of all the vowels in the English alphabet is written as

$$A = \{\text{a, e, i, o, u}\}$$

We can express the fact that e is a member of set A by writing $\text{e} \in A$. Similarly we can express the fact that x does not belong to set A by writing $\text{x} \notin A$. The symbol $\in$ (the Greek letter epsilon) is read "is an element of," whereas the symbol $\notin$ is read "is *not* an element of" or "does not belong to."

Notation If an object a is an element of set A, we write $a \in A$. If an object a is *not* an element of set A, we write $a \notin A$.

Example 1 If $B = \{$the set of all baseball teams in the National League East$\}$, then

$B = \{$Chicago Cubs, Philadelphia Phillies, Pittsburgh Pirates, Montreal Expos, New York Mets, St. Louis Cardinals$\}$

In this case

New York Mets $\in B$ and New York Yankees $\notin B$

Two sets may contain exactly the same elements. For example, the sets $\{a, b, c\}$ and $\{c, a, b\}$ contain exactly the same elements. These sets are said to be equal.

Definition 1.1 Two sets A and B that contain exactly the same elements are called *equal sets*. We indicate equal sets by writing $A = B$. Two sets that do not contain exactly the same elements are not equal.

Comment In determining whether two sets are equal or not, the order in which the elements appear within the braces does not matter. Thus, the sets $\{a, b, c\}$ and $\{c, b, a\}$ are equal, since both contain the same elements. Similarly, the transcripts of the two students shown in Figure 1.1 indicate that the sets of courses completed by both are equal.

Comment Notice that Mary took Math 100 twice. This is because she failed it the first time and had to retake it in a later semester. Nevertheless, in determining the different courses that Mary has taken, we write Math 100 only once, since all members of a set must be distinct.

In many situations, it may be inconvenient or impossible to list all the elements of a set. In such cases sets may be specified by means of a verbal statement describing a property that the elements of the set have in common. This is frequently accomplished by using what is called *set-builder notation*. For example, consider the set

$A = \{$autumn, winter, spring, summer$\}$

In this case we have actually listed the elements of set A. Since this set consists of the four seasons of the year, we can write set A using set-builder notation as

$A = \{x \mid x$ is a season of the year$\}$

This is read "A equals the set of all x such that x is a season of the year." In this notation x represents a typical element of the set, the vertical line stands for the words "such that," and following the vertical line are the conditions that must be satisfied for x to be a member of set A.

In some cases, it may be helpful to write a complete list of all the elements of the set. In other situations, it may be difficult to see an identifying property of the elements. For example, consider the set

$A = \{$William Harrison, Zachary Taylor, Abraham Lincoln, James Garfield, William McKinley, Warren Harding, Franklin Roosevelt, John Kennedy$\}$

THE COLLEGE OF STATEN ISLAND

A FEDERATION OF STATEN ISLAND COMMUNITY AND RICHMOND COLLEGES
STATEN ISLAND, N.Y. 10301

DATE OF ISSUE: 03/11/79

PAGE: 1

CURRICULUM: Electro-Mechanical Technology

CANDIDATE FOR: AAS

IDENTIFICATION NUMBER: 423-66-2379

STUDENT NAME AND ADDRESS:
PETER QUINTAS
120 Styyvesant Place
Staten Island, N.Y. 10301

DATE OF BIRTH: 03/09/55

DATE OF H.S. GRAD.: 06/75

HIGH SCHOOL OF GRADUATION: Port Richmond H.S.

END OF SESSION OR DATE OF ACTION MO. DAY YR.	COURSE NUMBER	COURSE TITLE	GRADE	CREDITS	QUALITY POINTS
01/31/77		FALL SEMESTER 1976			
	MTH 120	Algebra and Trigonometry	B	3	9
	PHY 110	College Physics I	B	3	9
	ELT 120	D. C. Fundamentals Theory	B	4	12
	ELT 121	D. C. Fundamentals Lab	B	1	3
	MET 121	Engineering Drawing I	A	2	8
	ECO 101	Economics	A	3	12
05/30/77		SPRING SEMESTER 1977			
	MTH 224	Technical Calculus I	B	3	9
	MET 100	Intro To Techniques in Engineering Technology	B	2	6
	MET 230	Mechanics	B	3	9
	ELT 220	A.C. Fundamentals Theory	A	3	12
	ELT 221	A.C. Fundamentals Lab	A	1	4
	MET 221	Engineering Drawing II	C	1	2
	PED 100	Body Conditionning	C	1	2
01/31/78		FALL SEMESTER 1977			
	ENG 110	English/Speech Workshop	B	5	15
	HST 100	Mankind in Time	B	3	9
	MET 340	Machine Design I	B	3	9
	ELT 360	Electric Machines I	B	3	9
	ELT 361	Electric Machines II Lab	C	1	2
	PED 157	Horse Back Riding	C	1	2
06/12/78		SPRING SEMESTER 1978			
	MET 342	Machine Design II	C	3	6
	PHY 150	College Physics II	B	3	9
	PHY 151	College Physics II Lab	B	1	3
	MTH 100	Math Concepts I	B	3	9
	MET 113	Engineering Processes	B	1	3
	MET 330	Mechanics of Materials	B	3	9
01/31/79		FALL SEMESTER 1978			
	MTH 225	Technical Calculus II	A	3	12
	ENG 111	Communications Workshop	C	3	6
	ELT 460	Electric Machines II	A	3	12
	ELT 461	Electric Machines II Lab	C	1	6

GPA CREDITS ATTEMPTED	TOTAL CREDITS COMPLETED	QUALITY POINTS	GRADE POINT AVERAGE
70	70	218	3.11

THIS TRANSCRIPT IS NOT OFFICIAL UNLESS SIGNED AND EMBOSSED WITH THE SEAL OF THE COLLEGE.

OFFICE OF THE REGISTRAR
AUTHORIZED SIGNATURE

DATE
JAN 31, 1979

54087

SOMERSET DATA FORMS, BELLE MEAD, N.J. (201) 359-8254

C-48377

Figure 1.1

THE COLLEGE OF STATEN ISLAND

A FEDERATION OF STATEN ISLAND COMMUNITY AND RICHMOND COLLEGES

STATEN ISLAND, N.Y. 10301

DATE OF ISSUE: 03/11/79

PAGE: 1

CURRICULUM: Electro-Mechanical Technology

CANDIDATE FOR: A A S

IDENTIFICATION NUMBER: 131-46-5794

STUDENT NAME AND ADDRESS:
MARY RYAN
715 Ocean Terrace
Staten Island, N.Y. 10301

DATE OF BIRTH: 02/04/54

DATE OF H.S. GRAD.: 06/75

HIGH SCHOOL OF GRADUATION: Port Richmond H.S.

END OF SESSION OR DATE OF ACTION MO. DAY YR.	COURSE NUMBER	COURSE TITLE	GRADE	CREDITS	QUALITY POINTS
01/31/77		FALL SEMESTER 1976			
	MTH 100	Math Concepts I	F	3	0
	PHY 110	College Physics I	C	3	6
	ELT 120	D.C. Fundamentals Theory	B	4	12
	ELT 121	D.C. Fundamentals Lab	B	1	3
	MET 121	Engineering Drawing I	A	2	8
	MTH 120	Algebra & Trigonometry	C	3	6
05/30/77		SPRING SEMESTER 1977			
	MET 221	Engineering Drawing II	B	1	3
	MET 100	Introduction To Techniques in Engineering Technology	B	2	6
	MTH 224	Technical Calculus I	C	3	6
	ELT 220	A.C. Fundamentals Theory	A	3	12
	ELT 221	A.C. Fundamentals Lab	B	1	3
	MET 230	Mechanics	B	3	9
01/31/78		FALL SEMESTER 1977			
	MET 118	Engineering Processes	C	1	2
	MET 330	Mechanics of Materials	A	3	12
	MET 340	Machine Design I	B	3	9
	PHY 150	College Physics II	B	3	9
	PHY 151	College Physics II Lab	A	1	4
	MTH 100	Math Concepts I	C	3	6
06/12/78		SPRING SEMESTER 1978			
	ENG 110	English / Speech Workshop	B	5	15
	ELT 360	Electric Machines I	C	3	6
	ELT 361	Electric Machines I Lab	C	1	2
	HST 100	Mankind in Time	A	3	12
	MET 342	Machine Design II	B	3	9
	PED 100	Body Conditionning	C	1	2
01/31/79		FALL SEMESTER 1978			
	ELT 460	Electric Machines II	B	3	9
	ELT 461	Electric Machines II Lab	B	1	3
	PED 157	Horseback Riding	C	1	2
	MTH 225	Technical Calculus II	B	3	9
	ECO 101	Economics	A	3	12
	ENG 111	Communications Workshop	B	3	9

GPA CREDITS ATTEMPTED	TOTAL CREDITS COMPLETED	QUALITY POINTS	GRADE POINT AVERAGE
73	70	206	2.82

54085

THIS TRANSCRIPT IS NOT OFFICIAL UNLESS SIGNED AND EMBOSSED WITH THE SEAL OF THE COLLEGE.

OFFICE OF THE REGISTRAR

AUTHORIZED SIGNATURE

DATE: JAN 31, 1979

Figure 1.1 (*continued*)

Do you notice any common property of the elements of set A?

On the other hand, if we express the same set as

$$A = \{x \mid x \text{ is a U.S. president who died while in office}\}$$

then the common property that is intended is clear.

The following examples further illustrate this idea.

Example 2 Rewrite the following sets using set-builder notation.

a) {April, June, September, November}
b) $\{1, 3, 5, 7, 9, \ldots\}$
c) {Connecticut, Rhode Island, Massachusetts, Vermont, New Hampshire, Maine}

Solution Using set-builder notation, we have

a) $\{x \mid x \text{ is a month with 30 days}\}$
b) $\{x \mid x \text{ is an odd number greater than } 0\}$
c) $\{x \mid x \text{ is one of the six eastern New England states of the United States}\}$

Example 3 List the elements of the sets described.

a) $M = \{x \mid x \text{ is the blood type of an individual}\}$
b) $C = \{y \mid y \text{ is a country in which capital punishment has been abolished}\}$

Solution The elements that belong to each of the sets described are:

a) $M =$ {type A, type B, type AB, type O}
b) $C =$ {Argentina, Australia, Brazil, Denmark, Switzerland, Italy, West Germany, . . . , Sweden}

Comment When using set-builder notation, great care must be exercised to make sure that the set is *well defined.* In other words, it must be clear whether or not any given object is an element of the set. Thus, set $B = \{x \mid x \text{ is an honest person}\}$ is not a well-defined set, since it is not clear whether Richard Millhouse, for example, is an element of that set or not. The set does not specify what constitutes an honest person.

Some sets contain a specific number of elements; that is, the elements in the set can be counted. Such sets are called *finite sets.* On the other hand, some sets have an unlimited number of elements; that is, the elements in the sets cannot be counted. These sets are called *infinite sets.* For example, the set of different U.S. coins in circulation is a finite set with six elements, namely {penny, nickel, dime, quarter, half-dollar, dollar piece}, and the set of positive even numbers $\{0, 2, 4, 6, 8, \ldots\}$ is an infinite set.

A finite set may have no elements at all and is then called a null set or empty set.

Definition 1.2 A set that has no elements is called the *null set* or the *empty set* and is denoted by $\varnothing$ or $\{\quad\}$.

For example, the set of all secretaries who can type more than 2000 words per minute and the set of submarines that carry attack bomber planes are examples of null sets.

Now consider Table 1.1, which lists the birthdays of the presidents of the United States. Let A be the set of all presidents who were born in June. How many elements are in set A?

Table 1.1 Birthdays of the Presidents of the United States

Name	Birthday	Name	Birthday
George Washington	Feb. 22, 1732	James Garfield	Nov. 19, 1831
John Adams	Oct. 30, 1735	Chester Arthur	Oct. 5, 1830
Thomas Jefferson	Apr. 13, 1743	Grover Cleveland	Mar. 18, 1837
James Madison	Mar. 16, 1751	Benjamin Harrison	Aug. 20, 1833
James Monroe	Apr. 28, 1758	William McKinley	Jan. 29, 1843
John Quincy Adams	July 11, 1767	Theodore Roosevelt	Oct. 27, 1858
Andrew Jackson	Mar. 15, 1767	William Taft	Sept. 15, 1857
Martin Van Buren	Dec. 5, 1782	Woodrow Wilson	Dec. 28, 1857
William H. Harrison	Feb. 9, 1773	Warren Harding	Nov. 2, 1865
John Tyler	Mar. 29, 1790	Calvin Coolidge	July 4, 1872
James Knox Polk	Nov. 2, 1795	Herbert Hoover	Aug. 10, 1874
Zachary Taylor	Nov. 24, 1784	Franklin D. Roosevelt	Jan. 30, 1882
Millard Fillmore	Jan. 7, 1800	Harry S. Truman	May 8, 1884
Franklin Pierce	Nov. 23, 1804	Dwight D. Eisenhower	Oct. 14, 1890
James Buchanan	Apr. 23, 1791	John F. Kennedy	May 29, 1917
Abraham Lincoln	Feb. 12, 1809	Lyndon B. Johnson	Aug. 27, 1908
Andrew Johnson	Dec. 29, 1808	Richard M. Nixon	Jan. 9, 1913
Ulysses S. Grant	Apr. 27, 1822	Gerald Ford	July 14, 1913
Rutherford Hayes	Oct. 4, 1822	Jimmy Carter	Oct. 1, 1924

Example 4 Determine which of the following sets are finite, which are infinite, and which are null sets.

a) The set of transatlantic telephone cables
b) The set of manned space flights to Mars
c) The set of all major stock exchanges in Rhode Island
d) The set of all numbers
e) The quantity of food available to feed the world population
f) The set of women who have been president of the following companies: General Motors, American Telephone & Telegraph (ATT), U.S. Steel Corp., Exxon Corp., International Business Machines (IBM)
g) The world population
h) The set of all U.S. states that begin with the letter Z

Solution The sets described in parts a, e, and g are finite sets. The sets described in parts b, c, f, and h are finite and null sets. The set described in part d is an infinite set.

Employees in savings plan	Employees in dental plan
Bill Mary Joe George	Bill Mary

Consider the following situation: The personnel director of a company compiles a list of employees who are enrolled in the payroll savings bond plan and of those who are enrolled in the company optional dental plan.

You will notice that every employee who is enrolled in the dental plan is also enrolled in the savings plan. If we let A represent the set of employees enrolled in the dental plan and let B represent the set of employees enrolled in the savings plan, then every element of set A is also an element of set B. In this case we say that set A is a subset of set B, or that set A is contained in set B. On the other hand, not every element of set B is an element of set A. Joe, for example, is enrolled in the savings plan but not in the dental plan. Thus, set B is *not* a subset of set A.

Definition 1.3 Set A is said to be a *subset* of set B if every element of set A is also an element of set B. If set A is a subset of set B, we write $A \subseteq B$.

Definition 1.4 If A is a subset of B and B contains one or more additional elements not in A, then A is called a *proper subset* of B. This is denoted as $A \subset B$.

Example 5 Determine which of the following sets are subsets, proper subsets, or neither.

a) $A = \{$carburator, radiator, tires$\}$; $B = \{$all parts of a car$\}$
b) $M = \{$all antibiotics$\}$; $N = \{$Aureomycin, penicillin, erythrocin, streptomycin$\}$

Solution a) A is a subset and a proper subset of B.
b) M is neither a subset nor a proper subset of N.

Comment It should be apparent that every set is a subset of itself; that is, for any set A, we have $A \subseteq A$.

Note The null set, $\varnothing$, is assumed to be a subset of every set. Thus, if set A is any nonnull set, then $\varnothing \subset A$.

How many subsets does a set have? To answer this question, consider a set $A = \{x\}$, which has only one element in it. There are two subsets possible:

$$\{x\} \quad \text{and} \quad \varnothing$$

Suppose we are now given a set with two elements. For example, let $A = \{e, x\}$. Then we can list all the subsets:

$$\{e\} \qquad \{x\} \qquad \{e, x\} \qquad \varnothing$$

You will notice that there are four possible subsets. If a set has three elements, there are eight possible subsets. Thus, if $A = \{a, b, c\}$, the eight possible subsets are

$$\{a\} \quad \{b\} \quad \{c\} \quad \{a, b\} \quad \{a, c\} \quad \{b, c\} \quad \{a, b, c\} \quad \emptyset$$

If a set has four elements, how many possible subsets are there?

Example 6 *Investigating discrimination charges.* Four female employees of the Amtex Corporation have complained to the board of directors that the company discriminates against women and minorities when making appointments to top-level management positions. The women are Manya White, Mabel Szeto, Sherry Blackman, and Trudy Rodriguez. The board decides to appoint a committee from among these four women to present the facts. How many different committees are possible?

Solution Since we are not told how many people are to be on the committee, we can determine the total number of possible committees by computing the number of possible subsets of a set with four elements. There are 16 possible subsets: {Manya}, {Mabel}, {Sherry}, {Trudy}, {Manya, Mabel}, {Manya, Sherry}, {Manya, Trudy}, {Mabel, Sherry}, {Mabel, Trudy}, {Sherry, Trudy}, {Manya, Mabel, Sherry}, {Manya, Mabel, Trudy}, {Manya, Sherry, Trudy}, {Mabel, Sherry, Trudy}, {Manya, Mabel, Sherry, Trudy}, and $\emptyset$. Because the null set corresponds to a set with no members in it, we get a total of 16 − 1 or 15 possible committees to investigate the discrimination charges.

Past experience in studying sets has shown that it is helpful to identify a set U such that all sets in the discussion are subsets of U. Such a set is called the *universal set*, or simply the *universe*. For example, if a discussion centers around Johnny Cash, the Carpenters, Elvis Presley, Pat Boone, and Frank Sinatra, then an appropriate universal set would be "the set of contemporary singers or singing groups." Similarly, in discussing which items a company should advertise during the coming fiscal year, the discussion would be limited to the products produced by the company. When determining which students will graduate from Staten College with honors, the universal set would be the set of all students eligible to graduate. Those students who have not completed their required courses and who are therefore not eligible to graduate would not enter into the discussion and would not even be considered.

In many cases the universal set is clear from the nature of the problem and is not explicitly stated. Furthermore, this set may vary from situation to situation but remains the same throughout a given problem. For example, in a sociology class the universal set may be the set of all people in the world, whereas in an ecology class the universal set may be the set of all people who do not have enough food to eat.

From your past experience with mathematics, you may have found that a diagram often helps to clarify things. The same is true

Figure 1.2 The universal set U.

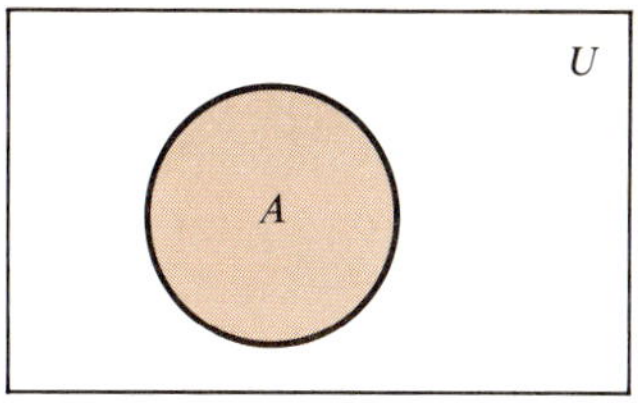

Figure 1.3 Set A.

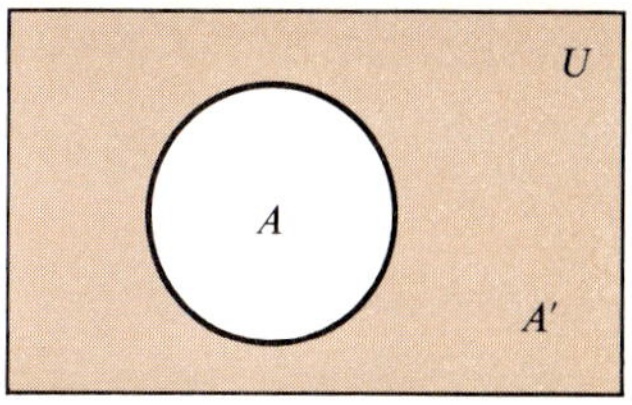

Figure 1.4 The complement of set A.

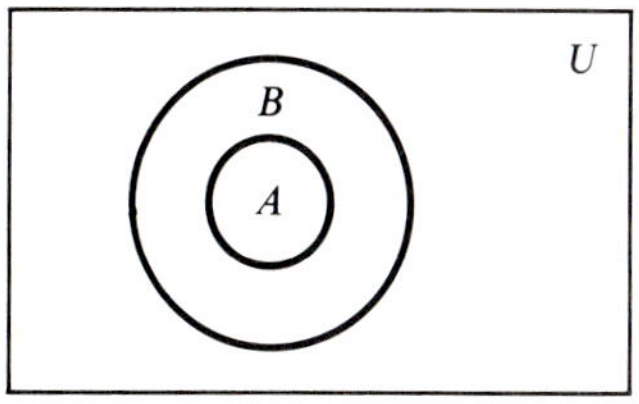

Figure 1.5 Set A is a proper subset of set B.

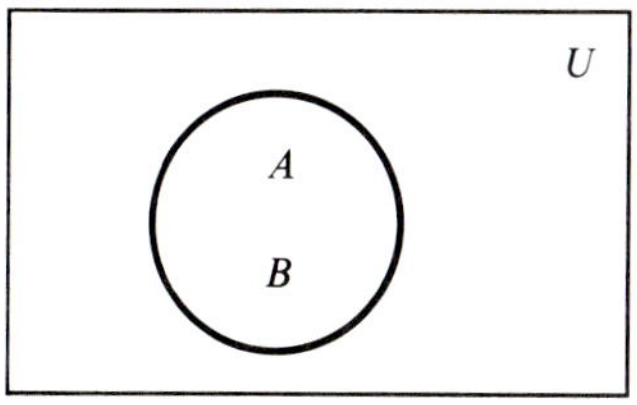

Figure 1.6 Sets A and B are equal.

for sets, where we find it helpful to draw graphical representations of sets. Such diagrams are called *Venn diagrams* in honor of the British logician, John Venn (1834–1923).

When using Venn diagrams, the universal set can be represented by our drawing a rectangle as shown in Figure 1.2. We usually put the letter U in the upper right-hand corner of the rectangle to designate the universal set.

We represent a subset A of U by drawing a circle within the rectangle. We shade the interior of the circle and designate it as set A. This is shown in Figure 1.3.

Those remaining points that are in the universal set but not in set A are shaded in Figure 1.4. These points are called the complement of set A.

Definition 1.5 The *complement* of set A is the set of elements in the universal set U that are not in set A. The complement of set A is denoted as $\bar{A}$ or A'.

We can also picture subsets with Venn diagrams. If A is a proper subset of B, we draw circle A completely enclosed within circle B as shown in Figure 1.5. If A and B are equal sets, we draw one circle and label it as both A and B as shown in Figure 1.6.

Example 7 Let the universal set U be $\{1, 2, 3, 4, 5, 6\}$. Find the complement of the indicated sets.

a) $A = \{2, 4, 6\}$
b) $B = \{x \mid x \text{ is an odd number between 0 and 6}\}$
c) $U = \{1, 2, 3, 4, 5, 6\}$

Solution

a) The complement of set $A = \{2, 4, 6\}$ consists of those elements that are in the universal set but not in set A. Thus, $A' = \{1, 3, 5\}$.
b) Since $B = \{x \mid x \text{ is an odd number between 0 and 6}\}$, we have $B = \{1, 3, 5\}$. Thus, $B' = \{2, 4, 6\}$, or A.
c) The complement of the universal set is the null set. Thus, $U' = \varnothing$.

Example 8 *Artificial sweeteners.* Let A be the set of artificial sweeteners that have been banned by the Food and Drug Administration because they have resulted in cancer when fed to laboratory mice. Can you list the elements in the complement of set A?

Example 9 *Paying no taxes.* Let B be the set of all people who earned more than \$100,000 during 1979 and who paid no taxes to the Internal Revenue Service. Do you know the number of elements in set B, or is it the null set? (Surprisingly enough, set B is not the null set.)

Exercises

1. Are the following sets well defined? Explain your answer.
   a) The set of all honest politicans
   b) The set of all good-looking secretaries
   c) The set of all talented business executives
   d) The set of all dogs that can fly
   e) The set of easy courses offered at your college
   f) The set of all square triangles
   g) The set of all small numbers
   h) The set of all rational numbers less than one.

2. List the elements in the following sets.
   a) $\{x | x$ is a consonant in the English alphabet$\}$
   b) $\{x | x$ is a planet in our solar system$\}$
   c) $\{x | x$ is a leap year in the twentieth century$\}$
   d) $\{x | x$ is a positive integer less than 7$\}$
   e) $\{y | y$ is a number such that $y + 5 = 7\}$
   f) $\{y | y$ is the Celsius temperature at which water freezes$\}$
   g) $\{x | x$ is a state in the United States that begins with the letter B$\}$

3. Rewrite each of the following sets using set-builder notation.
   a) {Ford, General Motors, Chrysler, American Motors}
   b) {a, e, i, o, u}
   c) {2, 3, 5, 7, 11, 13, 17, 23, 29, . . .}
   d) {TWA, Pan Am, United, American, Delta, . . . , Eastern}
   e) {Nebraska, Nevada, New Hampshire, New Jersey, New York, New Mexico, North Carolina, North Dakota}
   f) {basketball, baseball, football, soccer, hockey, . . .}
   g) {pitcher, first baseman, second baseman, third baseman, shortstop, catcher, left fielder, right fielder, center fielder}
   h) {the bald eagle}
   i) {O, T, T, F, F, S, S, E, N, . . .}

4. Determine which of the following are finite sets, infinite sets, or null sets.
   a) The set of all the food in the world
   b) The set of all stars in the universe
   c) The set of months with 32 days
   d) The set of U.S. banks that are forced to close each year because of financial difficulties
   e) The set of all trees in the world
   f) The set of all fractions with denominators of 3

g) The set of all people who weigh a ton
h) The set of all possible genetic makeups of an individual person
i) $\{x \mid x$ is an even prime number greater than $4\}$

For Exercises 5–10, let $A = \{1, 2, 3, 4, 5\}$, $B = \{1, 3, 4\}$, and $C = \{4, 5, 6\}$. Determine which of the following are true.

5. $A \subseteq B$ **6.** $B \subset A$ **7.** $C \subset A$

8. $B \subset C$ **9.** $A \subset C$ **10.** $A \subset B$

For Exercises 11–18, indicate whether the statement is true or false.

11. The sets $A = \{$Erie, Michigan, Superior, Huron, Ontario$\}$ and $B = \{$the five Great Lakes$\}$ are equal.

12. The sets $\{x \mid x$ is a letter in the word "circle"$\}$ and $\{y \mid y$ is a letter in the word "relic"$\}$ are equal.

13. $\{a, b\} \in \{a, b, c\}$

14. $\{\varnothing\} = \varnothing$

15. $\{0\} = \varnothing$

16. $\{a\} \subset \{a, b, c\}$

17. $a \subset \{a, b, c\}$

18. If $A \subseteq B$ and $B \subseteq A$, then $A = B$.

19. For each set in the left column, select a set from the right column that is a subset of it.

a) $\{x, y, z\}$	1. $\{x\}$
b) $\{$even counting number$\}$	2. $\{\frac{1}{2}, \frac{1}{5}, \frac{1}{4}, \frac{1}{3}\}$
c) $\{$odd number between 2 and 8$\}$	3. $\{6, 8, 10, 12\}$
d) $\{\frac{1}{2}, \frac{1}{3}, \frac{1}{4}, \frac{1}{5}\}$	4. $\{3, 5\}$
e) $\{$an odd number between 2 and 3$\}$	5. $\{$x, m$\}$
f) $\{$the consonants in the word "axiom"$\}$	6. $\varnothing$
	7. $\{3, 5, 7, 9\}$

20. In each of the following replace the ? by the symbol $\in$ or $\notin$ to obtain a true statement.

a) 웃 ? {웃, □, ○}

b) ▱ ? {□, ○, ▭}

c) 1 ? {4, 2, 7, 1}

d) △ ? {☼, 웃, ▱, □, ○}

21. Let the universal set $U = \{1, 2, 3, 4, 5, 6\}$, $A = \{1, 4, 6\}$, and $B = \{2, 5\}$. Find:

a) A' b) B' c) U' d) $\varnothing'$

22. For each of the following, find *all* the subsets.

a) $\{a\}$ b) $\{a, b\}$ c) $\{a, b, c\}$
d) $\{a, b, c, d\}$ e) $\{a, b, c, d, e\}$ f) $\varnothing$

★ 23. Using the results of Exercise 22, complete the following chart:[2]

Number of elements in set	1	2	3	4	5	6	7	8	9	10	25	50	n
Total number of possible subsets													

24. *Choosing teams.* Jennifer, Heather, George, and Jimmy have challenged several tennis players for the tennis court. In how many different ways can Jennifer, Heather, George, and Jimmy choose two players from among themselves to compete for the tennis court?

★ 25. *Selecting a menu.* Steve enters the kitchen and finds the following items on the table: bacon, eggs, milk, juice, and cereal. His doctor has recommended that he eat a *minimum* of three of these items. In how many different ways can Stephen select his menu?

26. In each of the following, find the smallest set that contains the given sets as subsets.
   a) {*Playboy, Penthouse, Fortune*} and {*Sports Illustrated, Time, Cosmopolitan, Reader's Digest, Playboy*}
   b) {golf, soccer, tennis}, {shuffleboard, handball}, {swimming, baseball, tennis}, and {shuffleboard}

Photo: Mimi Forsyth, Monkmeyer

★ 27. Assume that the U.S. Congress is composed of only Democrats, Republicans, and Conservatives. Draw a Venn diagram picturing the political makeup of Congress.

28. How many different committees can be formed from each of the following?
   a) {Sharon, Rochelle}
   b) {Morris, Fluffy, Pussy}
   c) {Sam, Helen, Hilda, Ike}

★ 29. *Loading mail trucks.* A postal worker is loading a truck with identical mail bags. There are five more identical bags to be loaded; however, there is room for only two of them in the truck. In how many different ways can the driver select the three bags *not* to be loaded on the truck?

[2] Exercises marked with a star are slightly more difficult and may require some additional time and thought.

★ 30. *Taking pictures.* Gertrude has just arrived at the photographer's with her three children to take family portraits. In how many different ways can the children arrange themselves in groups of three to have their pictures taken?

1.3 Elementary Set Operations

Now that we are familiar with sets and their relationships, we will study several elementary operations that can be performed with sets.

Suppose company A markets a product in New York, Los Angeles, San Francisco, Chicago, and Miami, and company B markets the same product in New York, Houston, Los Angeles, San Francisco, St. Louis, and Boston. You will notice that *both* companies market their product in New York, Los Angeles, and San Francisco. This illustrates the idea of the intersection of two sets.

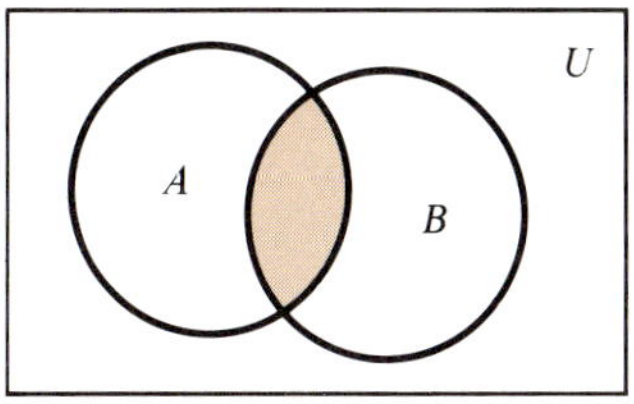

Figure 1.7 The shaded portion represents the intersection of sets A and B.

Definition 1.6 If A and B are two given sets, then the set of all elements that belong to *both* set A and set B is called the *intersection* of A and B and is denoted by the symbol $A \cap B$.

In the previous case

$$A = \{\text{New York, Los Angeles, San Francisco, Chicago, Miami}\}$$

$$B = \{\text{New York, Houston, Los Angeles, San Francisco, St. Louis, Boston}\}$$

so that

$$A \cap B = \{\text{New York, Los Angeles, San Francisco}\}$$

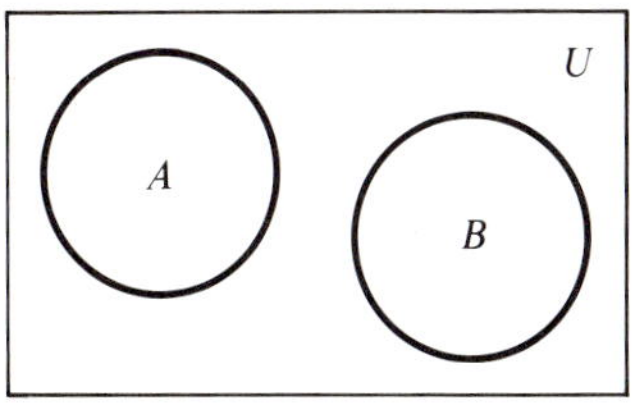

Figure 1.8 The intersection of sets A and B when they are disjoint. Nothing is shaded.

Company C markets the same product in Los Angeles, Denver, and Reno, and company D markets the product in Trenton, Baltimore, Atlantic City, and Philadelphia. In this case companies A, B, and C all market the product in Los Angeles, so that $A \cap B \cap C = \{$Los Angeles$\}$. On the other hand, companies C and D do not market the product in the same cities. Thus, $C \cap D = \varnothing$.

Definition 1.7 Two sets that have no elements in common are called *disjoint sets.*

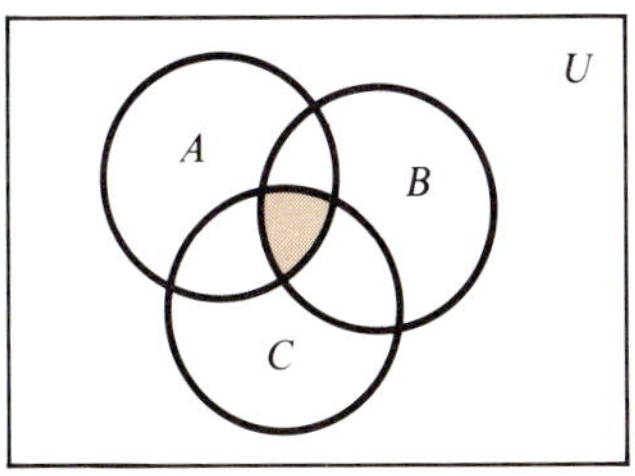

Figure 1.9 The shaded portion represents the intersection of sets A, B, and C.

In the previous case sets A, B, C, and D are disjoint, since $A \cap B \cap C \cap D = \varnothing$.

We can picture the intersection of two sets using Venn diagrams. The shaded portion in Figure 1.7 represents the intersection of sets A and B, since this region is common to both circles. On the other hand, when two sets have nothing in common (that is, they are disjoint), then nothing is shaded, as shown in Figure 1.8.

Using Venn diagrams, we can also picture the intersection of three sets. The shaded portion in Figure 1.9 represents the intersection of sets A, B, and C.

Example 10 *Stockholder proposals.* At a recent stockholders' meeting, proposals presented by the company's shareholders were grouped in sets in the following way:

$A = \{$shareholders in favor of raising the dividend$\}$

$B = \{$shareholders in favor of the new company pension plan$\}$

$C = \{$shareholders in favor of retaining the present board of directors$\}$

Describe in words each of the following sets:

a) $A \cap B$ b) $A \cap B \cap C$ c) $(A \cap B)'$
d) $A \cap B' \cap C'$ e) $A' \cap B' \cap C'$

Solution

a) $A \cap B$ represents shareholders who are in favor of both raising the dividend and approving the new pension plan.
b) $A \cap B \cap C$ represents shareholders who are in favor of raising the dividend, approving the new pension plan, and retaining the present board of directors.
c) $(A \cap B)'$ represents shareholders who are *not* in favor of raising the dividend or in favor of approving the new pension plan.
d) $A \cap B' \cap C'$ represents shareholders who favor raising the dividend, oppose the new pension plan, and oppose retaining the present board of directors.
e) $A' \cap B' \cap C'$ represents shareholders who oppose all three proposals.

Example 11 *Voter opinion.* In a recent poll voters were asked their opinion on a number of issues. Their responses were grouped in sets in the following way:

$A = \{$voters who approve of President Carter's foreign policy$\}$

$B = \{$voters who want the death penalty reinstated$\}$

$C = \{$voters who want taxes to be lowered$\}$

$D = \{$voters who believe that inflation is America's main enemy$\}$

Using set notation, describe each of the following:

a) {voters who approve of President Carter's foreign policy but do not want the death penalty reinstated or taxes lowered}
b) {voters who want taxes to be lowered, believe that inflation is America's main enemy, and oppose President Carter's foreign policy}

Solution a) $A \cap B' \cap C'$ b) $C \cap D \cap A'$

In some cases we are interested in those elements that are *either* in set A *or* in set B *or in both.* We have the following:

Definition 1.8 If A and B are two given sets, then the set of all elements that belong to *either* set A *or* set B *or both* is called the *union* of sets A and B and is denoted by the symbol $A \cup B$.

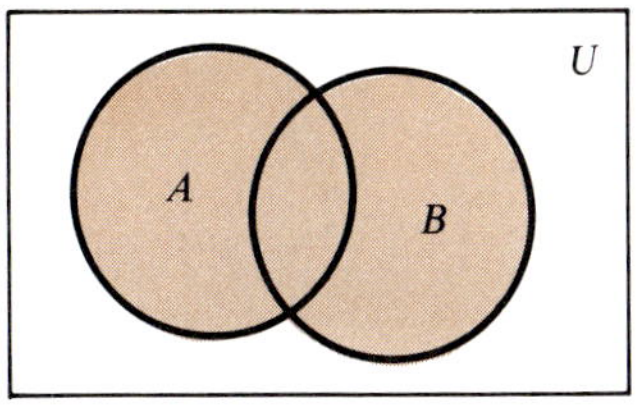

Figure 1.10 The union of sets A and B.

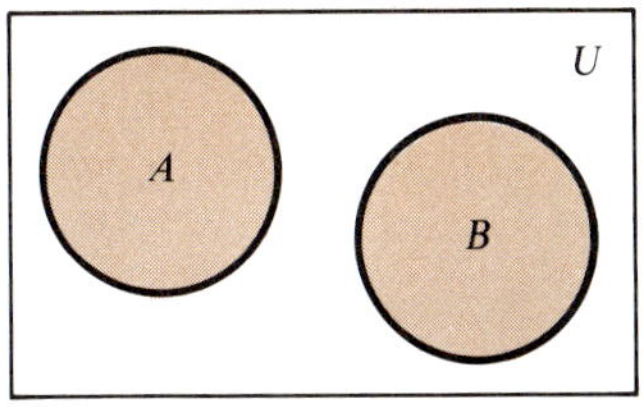

Figure 1.11 The union of sets A and B when the sets are disjoint.

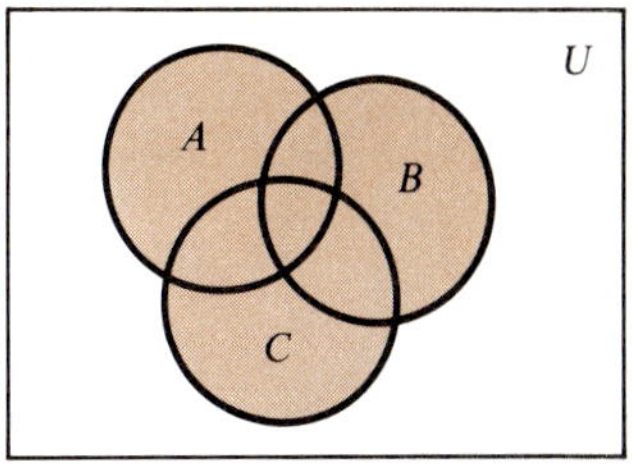

Figure 1.12 The union of three sets, A, B, and C.

Example 12 If $A = \{1, 2, 3, 4, 5\}$ and $B = \{4, 5, 6, 7\}$, then $A \cup B = \{1, 2, 3, 4, 5, 6, 7\}$.

Comment In Example 12 the elements 4 and 5 are in both sets A and B. When we form the union of both sets, however, we write these elements only once.

We can also picture the union of two sets by using Venn diagrams. The shaded portion in Figure 1.10 represents the union, since this region is in either set. It does not make a difference whether the sets are disjoint or not, as shown in Figure 1.11.

When dealing with three sets A, B, and C, the union represents those elements that are either in set A or set B or set C or all of them as shown in Figure 1.12.

Example 13 If $A = \{a, b, c, d\}$, $B = \{b, d, e\}$, and $C = \{a, e, f\}$, find

a) $A \cup B \cup C$
b) $(A \cup B) \cap C$
c) $(A \cup B) \cap (A \cup C)$
d) $(A \cap B) \cup C$
e) $(A \cup B)' \cap C$

Solution

a) $A \cup B \cup C$ means those elements that are either in set A or in set B or in set C. Thus, $A \cup B \cup C = \{a, b, c, d, e, f\}$.
b) To find $(A \cup B) \cap C$, we first find $A \cup B$ and then take the intersection of the results with set C. We have $A \cup B = \{a, b, c, d, e\}$, so that $(A \cup B) \cap C = \{a, e\}$.
c) $A \cup B = \{a, b, c, d, e\}$ and $A \cup C = \{a, b, c, d, e, f\}$, so that $(A \cup B) \cap (A \cup C) = \{a, b, c, d, e\}$.
d) $A \cap B = \{b, d\}$, so that $(A \cap B) \cup C = \{a, b, d, e, f\}$.
e) $A \cup B = \{a, b, c, d, e\}$, so that $(A \cup B)' = \{f\}$. Thus, $(A \cup B)' \cap C = \{f\}$.

Example 14 Using Venn diagrams, show that $(A \cup B)' = A' \cap B'$.

Solution We draw two diagrams, one to represent $(A \cup B)'$ and one to represent $A' \cap B'$. To draw the diagram for $(A \cup B)'$, we first draw $A \cup B$ and then shade everything *outside* $A \cup B$ as shown in Figure 1.13.

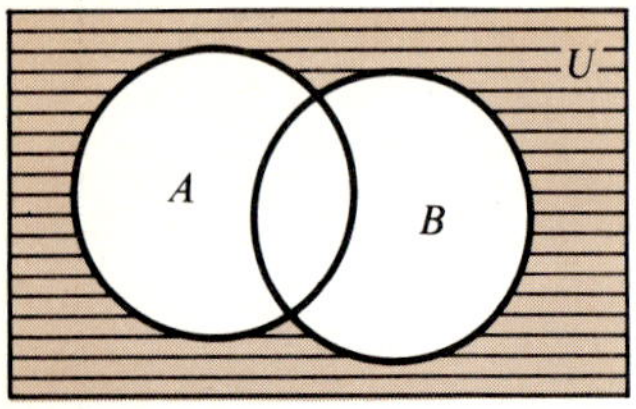

Figure 1.13 The shaded portion represents $(A \cup B)'$.

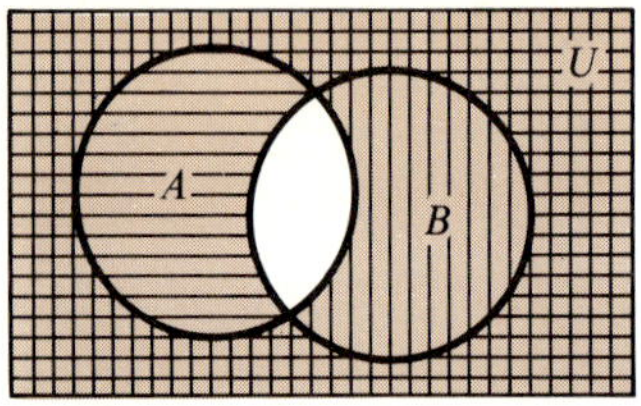

Figure 1.14 The portion that has been shaded twice represents $A \cap B'$.

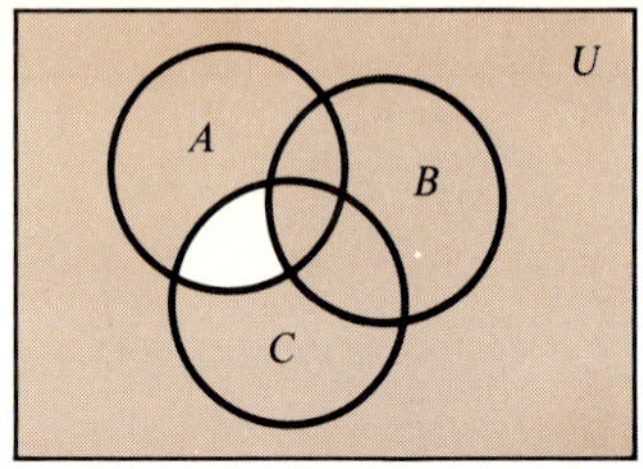

Figure 1.15 The shaded portion represents $(A' \cup B) \cup C'$.

To draw the diagram for $A' \cap B'$, we first shade in *vertically* everything outside A. This is A'. We then shade in *horizontally* everything outside B. This is B'. The part that has been shaded twice represents $A' \cap B'$. This is shown in Figure 1.14.

You will notice that in both diagrams the end results of the shadings are the same. This means that $(A \cup B)'$ and $A' \cap B'$ are equal.

Example 15 Using Venn diagrams, shade the region represented by $(A' \cup B) \cup C'$.

Solution We draw three circles and shade $A' \cup B$ followed by C'. The part that has any shading is $(A' \cup B) \cup C'$. This is shown in Figure 1.15.

Example 16 *Promotions.* The promotions committee of the mathematics department is reviewing the records of eight teachers it is considering for promotion. The following chart has been prepared by the department secretary.

Candidate's name	Serves on many school committees	Has good student evaluations	Does research and has scholarly publications
Richard Gonzales	Yes	Yes	Yes
Charles Protter	No	Yes	Yes
Mary White	Yes	No	Yes
Alice Goodman	Yes	No	Yes
Arlene Vitale	No	No	Yes
Sue Wheelbarger	No	Yes	No
Orlando Ruiz	Yes	Yes	Yes
Spencer Wayne	Yes	Yes	No

Let

$A = \{$all teachers who serve on many school committees$\}$

$B = \{$all teachers who have good student evaluations$\}$

$C = \{$all teachers who do research and have scholarly publications$\}$

List the names of those faculty members who are in the following sets:

a) $A \cup B \cup C$ b) $(A \cap B) \cup C$
c) $A \cap B \cap C$ d) $(A \cup B) \cap (A \cup C)$

Solution a) $A \cup B \cup C = \{$Richard Gonzales, Charles Protter, Mary White, Alice Goodman, Arlene Vitale, Sue Wheelbarger, Orlando Ruiz, Spencer Wayne$\}$
b) $(A \cap B) \cup C = \{$Richard Gonzales, Charles Protter, Mary White, Alice Goodman, Arlene Vitale, Orlando Ruiz, Spencer Wayne$\}$
c) $A \cap B \cap C = \{$Richard Gonzales, Orlando Ruiz$\}$
d) $(A \cup B) \cap (A \cup C) = \{$Richard Gonzales, Charles Protter, Mary White, Alice Goodman, Orlando Ruiz, Spencer Wayne$\}$

Example 17 In Example 16 the committee has decided to promote only those members who are in the set $A' \cap (B \cup C')'$. Which members (if any) will be promoted?

Solution If the committee will promote only those members who are in the set $A' \cap (B \cup C')'$, then only Charles Protter and Arlene Vitale will be promoted.

Exercises

31. If $A = \{a, b, c, d, e\}$, $B = \{c, d, e, f, g\}$, and $C = \{a, e, g\}$, find each of the following:

a) $A \cap B$ b) $B \cap C$ c) $A \cap B \cap C$
d) $A \cup B \cup C$ e) $A \cup (B \cap C)$ f) $(A \cup B) \cap C$
g) $(A \cup B') \cap C$ h) $A' \cap B' \cap C'$ i) $(A' \cup B) \cap C'$
j) $(A \cup B)' \cap C$

32. The following sets list items that are manufactured by four companies in their various divisions:

$A = \{$cosmetics, televisions, paper plates, paper cups$\}$
$B = \{$toasters, tires, televisions$\}$
$C = \{$tires, toys, processed meats, billboard signs$\}$
$D = \{$toys, paints, sailboats$\}$

List the products in each of the following sets:

a) A' b) $A' \cup B$ c) $(B' \cap A') \cup C$
d) $B' \cup A'$ e) $(A \cup B)'$ f) $A' \cap B' \cap C$
g) $A \cup B \cup C$ h) $(A \cap B)'$ i) $B \cap (C' \cup A)$
j) $A \cup (B \cup C)'$

33. Using Venn diagrams, shade the region represented by each of the following:

a) $A \cap B'$ b) $A \cup (B \cap C')$ c) $A \cap (B' \cup C')$
d) $A \cap (B \cup C)$ e) $(A \cup C)' \cap B$ f) $[(A \cup C) \cap B]'$

34. Which of the following pairs of sets are disjoint?

a) $\varnothing$ and $\{2, 4\}$
b) $\{1, 2, 3\}$ and $\{4, 5, 6\}$
c) The set of all people who weigh a ton and the set of all cats that can read Russian

35. A psychiatrist has classified all her patients into the following categories:

$M =$ the set of all male patients

$F =$ the set of all female patients

$T =$ the set of all tall patients (taller than 5 feet 5 inches)

$S =$ the set of all short patients (shorter than or equal to 5 feet 5 inches)

$R =$ the set of all people who write with their right hand

$L =$ the set of all people who write with their left hand

Describe in words the members of the following sets:

a) $M \cap R$ b) $M \cap R \cap T$ c) $(M \cup S) \cap R'$
d) $M \cup F$ e) $F \cap T \cap L$

36. Refer to Exercise 35. Ann, who is tall and writes with her left hand, is a patient of the psychiatrist. Which of the following statements is true?

a) Ann $\in M \cap T \cap R$ b) Ann $\in F \cap S' \cap L'$
c) Ann $\in M' \cap T \cap L$ d) Ann $\in (M \cup S \cup L)'$

37. *Classifying loans.* A large bank in New York City classifies all applicants for loans into categories before coding them. The categories are:

$M =$ the set of all male applicants

$F =$ the set of all female applicants

$A =$ the set of all applicants whose salary is greater than \$20,000

$B =$ the set of all applicants whose salary is \$20,000 or less

$H =$ the set of all applicants for home improvement loans

$P =$ the set of all applicants for personal finance loans

$C =$ the set of all applicants for car purchase loans

Describe in words the members of the following sets:

a) $M \cap A \cap H$ b) $(F \cup M) \cap C$
c) $F \cap B \cap P'$ d) $M \cap A \cap H \cap C$
e) $A \cap F \cap H$ f) $H' \cap B' \cap F'$

38. Using Venn diagrams, show that

a) $(A \cap B) \cap C = A \cap (B \cap C)$
b) $(A \cup B) \cap C = (A \cap C) \cup (B \cap C)$
c) $A' \cup B' = (A \cap B)'$
d) $A \cup (B \cup C) = (A \cup B) \cup C$

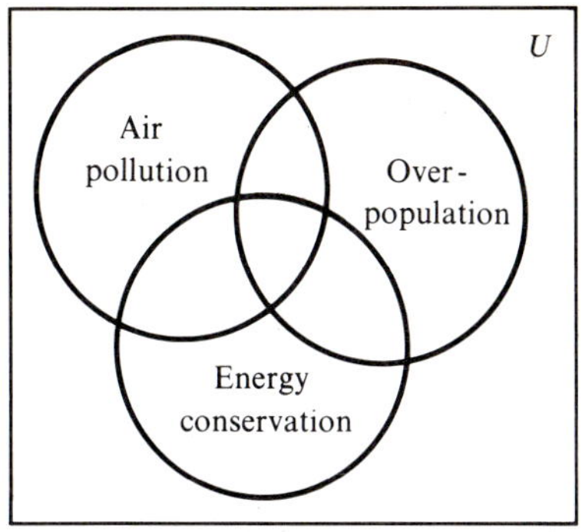

Figure 1.16 Environmental issues of concern to students in an ecology class.

Environmental Issues. In Exercises 39–45, draw a Venn diagram similar to the one in Figure 1.16 to indicate which environmental issue most concerned students and then shade the region described by each.

39. Students who were concerned about air pollution, overpopulation, and energy conservation.

40. Students not concerned about any of these issues.

41. Students concerned about overpopulation and energy conservation but not air pollution.

42. Students concerned about overpopulation only.

43. Students concerned about any of these issues.

44. Students concerned about overpopulation and energy conservation.

45. Students not concerned about air pollution.

1.4 Counting Elements in Sets: Application of Sets

In the previous section we studied the union and intersection of sets. Suppose we are given two sets A and B. (We will concern ourselves with only finite sets.) We can obviously count the number of elements in each set. Then what can we say about the number of elements in $A \cup B$?

To get us started, we introduce a new symbol. Let A be any set with a finite number of elements. Then we denote the number of elements in set A by the symbol $n(A)$. This symbol is read "the number of elements in set A." Thus, if $A = \{1, 2, 7, 10\}$, then $n(A) = 4$, and if $B = \{1, 2, 3, 4, 5, \ldots\}$, then $n(B)$ is undefined since B is an infinite set. Also, if $C = \phi$, then $n(C) = 0$ since C contains no elements.

Example 18 If $A = \{\text{a, e, i, o, u}\}$ and $B = \{\triangle, \square, 🯅\}$, then $n(A) = 5$ and $n(B) = 3$. What about $n(A \cup B)$? We can find $n(A \cup B)$ by first finding $A \cup B$ and then counting the number of elements. We have

$$A \cup B = \{\text{a, e, i, o, u}, \triangle, \square, 🯅\}$$

so that

$$n(A \cup B) = 8$$

Notice that sets A and B are disjoint. Furthermore,

$$n(A \cup B) = n(A) + n(B)$$

since

$$8 = 5 + 3$$

We can generalize the results of Example 18 in the form of a rule.

Rule 1 If A and B are disjoint sets then

$$n(A \cup B) = n(A) + n(B)$$

Now consider the sets $A = \{1, 4, 7, 9\}$ and $B = \{7, 9, 11, 13, 15\}$. You will notice that $A \cup B = \{1, 4, 7, 9, 11, 13, 15\}$ and that

$$n(A) = 4 \qquad n(B) = 5 \qquad n(A \cup B) = 7$$

Obviously, $n(A \cup B) \neq n(A) + n(B)$. Can you see why?

In this case sets A and B are not disjoint, so we have to revise Rule 1. Observe that $A \cap B = \{7, 9\}$ and $n(A \cap B) = 2$. Thus, 7 and 9 have been counted twice, so we have

$$n(A \cup B) = n(A) + n(B) - n(A \cap B)$$
$$7 = 4 + 5 - 2$$

Therefore, when sets A and B are not disjoint, we have the following rule.

Rule 2 If A and B are any finite sets (not necessarily disjoint), then

$$n(A \cup B) = n(A) + n(B) - n(A \cap B)$$

Comment When sets A and B are disjoint, Rule 2 is the same as Rule 1, since in that case $n(A \cap B) = 0$.

Comment The reason we subtract $n(A \cap B)$ should be obvious. In the previous case we counted the elements 7 and 9 twice, once in computing $n(A)$ and once in computing $n(B)$. To avoid double counting any element, we subtract $n(A \cap B)$.

Example 19 Let $A =$ {i, Δ, 🯅, ☼, θ, m} and $B =$ {γ, 🯅, a, θ, w}. Then $A \cup B =$ {i, Δ, 🯅, ☼, θ, m, γ, a, w} and $A \cap B =$ {🯅, θ}. Also, $n(A) = 6$, $n(B) = 5$, and $n(A \cap B) = 2$, so that

$$\begin{aligned} n(A \cup B) &= n(A) + n(B) - n(A \cap B) \\ &= 6 + 5 - 2 \\ &= 9 \end{aligned}$$

Example 20 *Strikes.* Recently both the sanitation workers and the fire fighters of a large city went out on strike demanding higher pay. The mayor refused to give in to the strikers' demands. Nevertheless, he decided to sample public opinion. In a survey of 1500 taxpayers who favored raises, the following information was obtained:

400 taxpayers favored giving the sanitation workers a raise

225 taxpayers favored giving *both* the sanitation workers and the fire fighters a raise

How many of the taxpayers sampled favored giving the fire fighters a raise?

Solution Let A be the set of taxpayers who favored giving the sanitation workers a raise, and let B be the set of taxpayers who favored giving the fire fighters a raise. From the given information, $n(A) = 400$ and $n(A \cap B) = 225$. Since the survey consisted of 1500 taxpayers who favored raises, $n(A \cup B) = 1500$. Applying Rule 2, we have

$$\begin{aligned} n(A \cup B) &= n(A) + n(B) - n(A \cap B) \\ 1500 &= 400 + n(B) - 225 \end{aligned}$$

so that

$$1500 = 175 + n(B)$$

and

$$1325 = n(B)$$

Thus, 1325 of the taxpayers surveyed favored giving the fire fighters a raise.

Example 21 *Quality control.* A quality control engineer inspected 100 cars and found that 13 of them had paint defects, 17 had mechanical defects, and 5 had both paint *and* mechanical defects. How many cars had no defects at all?

Solution Let A be the set of all cars with paint defects, and let B be the set of all cars with mechanical defects. From the given information, $n(A) = 13$, $n(B) = 17$, and $n(A \cap B) = 5$. When we apply Rule 2, we get

$$\begin{aligned} n(A \cup B) &= n(A) + n(B) - n(A \cap B) \\ &= 13 + 17 - 5 \\ &= 25 \end{aligned}$$

Thus, 25 cars had either of the two defects mentioned. Since the engineer inspected 100 cars, the number of cars with no defects at all is $100 - 25$, or 75 cars.

Rule 2 can be extended to the union of three sets. We generalize this in Rule 3.

Rule 3 If A, B, and C are finite sets, then

$$\begin{aligned} n(A \cup B \cup C) = n(A) + n(B) + n(C) - n(A \cap B) - n(A \cap C) \\ - n(B \cap C) + n(A \cap B \cap C) \end{aligned}$$

Example 22 Let $A = \{\text{x, y, z, ✲}\}$, $B = \{\text{웃, z, □}, \theta, \alpha, \beta\}$, and $C = \{\text{x, 웃}, \theta, \phi\}$. Then

$$\begin{aligned} A \cap B &= \{\text{z}\} \\ A \cap C &= \{\text{x}\} \\ B \cap C &= \{\text{웃}, \theta\} \\ A \cap B \cap C &= \phi \end{aligned}$$

so that

$$\begin{aligned} n(A) &= 4 & n(A \cap B) &= 1 \\ n(B) &= 6 & n(A \cap C) &= 1 \\ n(C) &= 4 & n(B \cap C) &= 2 \\ & & n(A \cap B \cap C) &= 0 \end{aligned}$$

Applying Rule 3 gives

$$\begin{aligned} n(A \cup B \cup C) &= n(A) + n(B) + n(C) - n(A \cap B) - n(A \cap C) \\ &\quad - n(B \cap C) + n(A \cap B \cap C) \\ &= 4 + 6 + 4 - 1 - 1 - 2 + 0 \\ &= 10 \end{aligned}$$

We can verify the results by finding $A \cup B \cup C$, which is

$$A \cup B \cup C = \{\text{x, y, z, ✲, 웃, □}, \theta, \alpha, \beta, \phi\}$$

and

$$n(A \cup B \cup C) = 10$$

Example 23 *False advertising claims.* A new medicine was recently marketed. The manufacturer claims that this product is a miracle drug since it is able to eliminate much human suffering. To support this claim, the manufacturer submits the following test results for the drug when administered to 10,000 patients suffering from one of several conditions.

4495 were relieved of their hay fever discomfort

6120 were relieved of their arthritis (and muscular) aches

3755 were relieved of their headaches

1165 were relieved of their hay fever discomfort and headaches

1690 were relieved of their arthritis aches and headaches

1730 were relieved of their arthritis aches and hay fever discomfort

900 were relieved of all three symptoms

A consumer group decides to test this "fantastic miracle" drug. After analyzing the manufacturer's data, the consumer group labels the report as a fraud. Explain why.

Solution Let A be the set of all patients who were relieved of their hay fever discomfort after taking the drug, let B be the set of all patients who were relieved of their arthritis aches after taking the drug, and let C be the set of all patients who were relieved of their headaches after taking the drug. Then $A \cup B \cup C$ is the set of all people with at least onc of the conditions mentioned for whom the drug had an effect. We have

$$n(A) = 4495 \qquad n(A \cap B) = 1730$$
$$n(B) = 6120 \qquad n(A \cap C) = 1165$$
$$n(C) = 3755 \qquad n(B \cap C) = 1690$$
$$n(A \cap B \cap C) = 900$$

Applying Rule 3 gives

$$\begin{aligned} n(A \cup B \cup C) &= n(A) + n(B) + n(C) - n(A \cap B) - n(A \cap C) \\ &\quad - n(B \cap C) + n(A \cap B \cap C) \\ &= 4495 + 6120 + 3755 - 1730 - 1165 - 1690 \\ &\quad + 900 \\ &= 10{,}685 \end{aligned}$$

Thus, there were 10,685 people who had at least one of the conditions mentioned and for whom the drug had an effect. Since the manufacturer claimed that the drug was administered to 10,000 suffering patients, the data are inconsistent. Therefore, the consumer group labeled the report a fraud.

The examples discussed so far in this section can also be analyzed by means of Venn diagrams. This will be done in the remainder of this section.

Example 24 *Rising postage costs.* A large company in Detroit is concerned about its rising postage costs. The manager of the mailroom submits the following report to management about the 4500 pieces of mail handled during the week of February 1–5.

2100 were mailed to addresses within the city

1425 weighed more than 2 ounces

4000 were sent first class

2000 were sent first class to addresses within the city

1300 weighed more than 2 ounces and were sent first class

425 weighed more than 2 ounces and were sent to addresses within the city

400 weighed more than 2 ounces and were sent first class to addresses within the city.

a) How many pieces of first-class mail weighing less than 2 ounces were mailed to addresses within the city?
b) How many pieces of mail weighing less than 2 ounces were mailed not by first class to addresses outside the city?

Solution We can easily solve this problem using Venn diagrams. We draw three circles to represent the different categories of mail handled. This is shown in Figure 1.17. On this diagram we first enter the number of pieces of mail weighing more than 2 ounces that were sent first class to addresses within the city. From the given information, this number is 400. Since we are told that 425 pieces of mail weighed more than 2 ounces and were mailed to addresses within the city, there are 425 − 400 or 25 that weighed more than 2 ounces and that were sent to addresses within the city but *not* by first class. This number is entered in Figure 1.17.

Similarly we know that 1300 pieces of mail weighing more than 2 ounces were sent first class. Thus, 1300 − 400 or 900 weighed more than 2 ounces and were sent first class *but not* to addresses within the city. This number is entered in Figure 1.17.

Also we know that 2000 pieces of mail were sent first class to addresses within the city. Thus, 2000 − 400 or 1600 first-class pieces of mail weighing *not* more than 2 ounces were sent to addresses within the city. This number is entered in Figure 1.17.

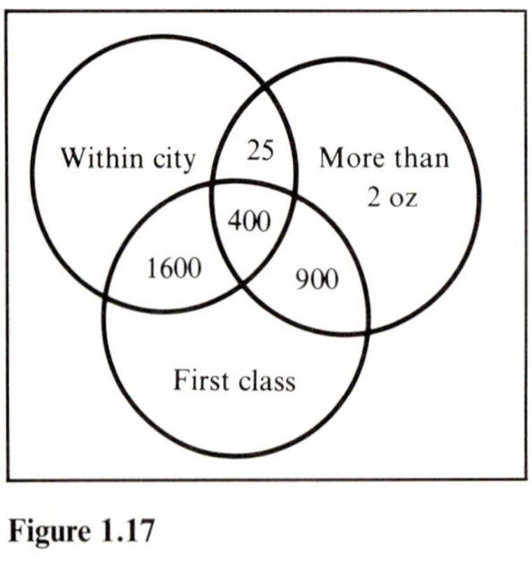

Figure 1.17

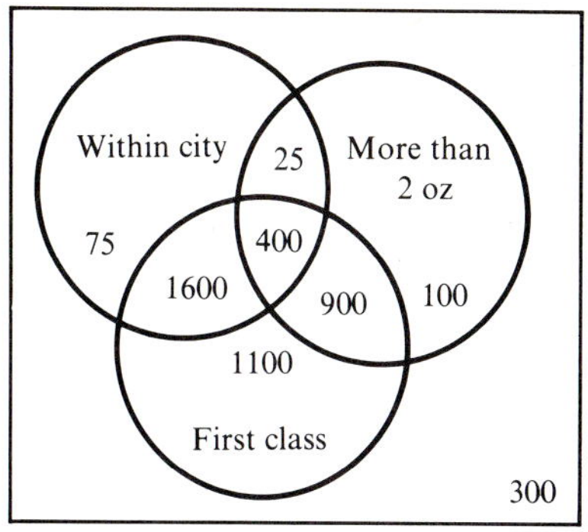

Figure 1.18

According to the given information, 4000 pieces of mail were sent first class. If we look at the first-class circle of Figure 1.17, we notice that we have so far accounted for 900 + 400 + 1600, or 2900. This leaves 4000 − 2900 or 1100 to be put in the remainder of the first-class circle. This is shown in Figure 1.18.

Similarly, 1425 pieces of mail weighed more than 2 ounces. In the circle for more than 2 ounces in Figure 1.17, we have so far accounted for 25 + 400 + 900, or 1325. This leaves 1425 − 1325 or 100 to be in the remainder of the circle for more than 2 ounces. This is shown in Figure 1.18.

Also there are 1600 + 400 + 25 or 2025 already in the circle for within the city. This leaves 2100 − 2025 or 75 for the remainder of this circle.

Now we add all the numbers appearing in the diagram:

$$75 + 25 + 400 + 1600 + 100 + 900 + 1100 = 4200$$

Thus, 4200 pieces of mail have been accounted for. Since the report covered 4500 pieces of mail, we are left with 4500 − 4200 or 300 that were not sent first class, did not weigh more than 2 ounces, and were not mailed to addresses within the city.

We can now answer the questions.

a) From the diagram, 1600 pieces of mail weighing less than 2 ounces were mailed first class to addresses within the city.
b) There were 300 pieces of mail that weighed less than 2 ounces and were mailed not by first class to addresses outside the city.

Example 25 *Physical fitness.* The Atlantic Health Club recently polled 220 of its members to determine which activity its members preferred to maintain their physical fitness. The following results were obtained:

95 liked jogging

115 liked weight lifting

55 liked push-ups

30 liked jogging and weight lifting

25 liked jogging and push-ups

35 liked weight lifting and push-ups

10 liked all three activities

a) How many did not like any of these activities?
b) How many liked jogging but not weight lifting or push-ups?
c) How many liked weight lifting but not jogging or push-ups?
d) How many liked push-ups and weight lifting but not jogging?

Solution From the given information we draw a Venn diagram and label it as shown in Figure 1.19. Remember to first use the fact that 10 liked all three activities. From the diagram we can now answer the questions.

a) 35 club members did not like any of these activities.
b) 50 members liked jogging but not weight lifting or push-ups.
c) 60 members liked weight lifting but not jogging or push-ups.
d) 25 liked push-ups and weight lifting but not jogging.

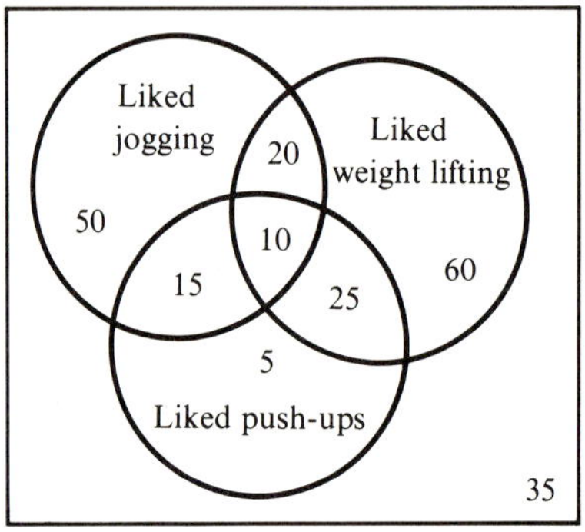

Figure 1.19

Exercises

46. *Drugs on campus.* In an effort to control the use of drugs on campus, the administration of State University polled 200 students to determine which plan, A or B, to adopt. The following results were obtained:

75 approved of plan A

25 approved of both plans A and B

How many students approved of plan B only?

47. *Drug use.* The 200 students discussed in Exercise 46 were then asked to indicate whether they used drugs. The following results were obtained:

28 used either marijuana or heroin

14 used marijuana only

18 used heroin only

a) How many students used *both* marijuana and heroin?
b) How many students did *not* use marijuana or heroin?

48. *Registering your car.* The district office of the Motor Vehicle Bureau has 195 clerks who process applications as described below:

75 process driver's licenses only

30 process auto registrations only

15 process both driver's licenses and auto registrations

a) How many clerks process either driver's licenses or auto registrations?
b) How many clerks do not process driver's licenses or auto registrations?

49. *Cash or charge.* Rochelle's Department Store has 90 checkout counters, which accept cash, checks, or charge cards as shown here:

70 accept cash

35 accept checks (with proper identification)

32 accept charge cards

20 accept cash and charge cards

25 accept cash and checks

7 accept checks and charge cards

5 accept cash, checks, and charge cards

a) How many checkout counters accept cash only?
b) How many accept charge cards only?
c) How many accept checks only?

50. *Parks.* A parks department supervisor submits the following report to the mayor on the condition of the city's 90 parks:

44 need new benches

35 need new fences

30 need new swings

19 need new swings and benches

6 need new swings and fences

21 need new fences and benches

6 need new benches, fences, and swings

a) How many parks need new benches and fences but *not* new swings?
b) How many parks need new swings and fences but *not* new benches?
c) How many parks need only new benches?
d) How many parks need only new swings?
e) How many parks do not need new swings, benches, or fences?

51. *Sporty cars.* A sociology student collected the following information on the cars of the 50 students in the class:

32 had tape decks

25 had bucket seats

31 had AM-FM radios

11 had tape decks and bucket seats

14 had tape decks and AM-FM radios

12 had bucket seats and AM-FM radios

4 had all three

The report was rejected. Can you see why?

52. *Polluting our lakes.* An ecology class conducted a survey of a state's 175 polluted lakes and submitted the following report on the cause of pollution:

75 were polluted by raw sewage

86 were polluted by industrial wastes

73 were polluted by phosphates

19 were polluted by raw sewage and phosphates

24 were polluted by raw sewage and industrial wastes

23 were polluted by industrial wastes and phosphates

7 were polluted by all three

a) How many lakes were polluted by *only one* of the three causes mentioned?
b) How many lakes were polluted by *at least one* of the causes mentioned?

★53. *Classifying your blood.* A person's blood is classified in two ways: It can contain the A antigen, the B antigen, or both the A and B antigens. Of course, it may not contain any of these antigens. Additionally the blood may contain the Rh antigen, in which case the blood is Rh-positive. If the blood does not contain the Rh antigen, it is Rh-negative. Thus

Blood is type A positive if it contains both the A and Rh antigens but not the B antigen.

Blood is type A negative if it contains the A antigen only.

Blood is type B positive if it contains both the B and Rh antigens only.

Blood is type B negative if it contains the B antigen only.

Blood is type AB positive if it contains the A, B, and Rh antigens.

Blood is type AB negative if it contains the A and B antigens only.

Blood is type O positive if it contains only the Rh antigen.

Blood is type O negative if it does not contain the A, B, or Rh antigen.

The blood samples of 150 nursing students were analyzed and the following results were obtained:

47 had the A antigen

76 had the B antigen

34 had both the A and B antigens

107 had the Rh factor

33 had the A and Rh factors

59 had the B and Rh factors

28 had the A, B, and Rh factors

a) How many students had type A negative blood?
b) How many students had type O negative blood?
c) How many students had type AB positive blood?
d) How many students had type AB negative blood?
e) For these students what is the most common type of blood?
f) For these students what is the rarest type of blood?

54. *Servicing your car.* A large car dealership employs 100 mechanics who perform the functions indicated:

60 tune the car engines

30 service the air conditioning

25 change the oil and grease

13 tune the car engines and service the air conditioning

8 tune the car engines and change the oil and grease

15 service the air conditioning and change the oil and grease

6 perform all three functions

a) How many mechanics tune the car and service the air conditioning only?
b) How many mechanics change the oil and grease only?
c) How many mechanics do none of the three functions mentioned, that is, do nothing?

55. *Selecting a magazine.* A flight stewardess submits the following report to the captain on the magazines requested by 200 passengers:

56 asked for *Fortune*

141 asked for *Sports Illustrated*

74 asked for *Newsweek*

17 asked for *Fortune* and *Sports Illustrated*

40 asked for *Fortune* and *Newsweek*

36 asked for *Newsweek* and *Sports Illustrated*

15 asked for all three magazines

a) How many passengers requested at least one of these magazines?
b) How many passengers requested exactly one of these magazines?
c) How many passengers requested exactly two of these magazines?
d) How many passengers requested none of these magazines?

56. *Lung cancer.* A survey of 200 patients with lung cancer revealed the following information:

95 were cigarette smokers

115 worked with dangerous chemicals (for example, asbestos)

55 lived in an area with large amounts of air pollution

30 were cigarette smokers who lived in an area with large amounts of air pollution

25 were cigarette smokers who worked with dangerous chemicals

35 worked with dangerous chemicals and lived in an area with large amounts of air pollution

10 were cigarette smokers who worked with dangerous chemicals and lived in an area with large amounts of air pollution

a) How many patients lived in an area with large amounts of air pollution and worked with dangerous chemicals but did not smoke?
b) How many patients smoked only?
c) How many patients did not smoke, did not work with dangerous chemicals, and did not live in an area with large amounts of air pollution?

57. *Music preference.* A disc jockey surveyed 500 randomly selected people in the audience to determine their musical preference. He submitted the following report:

278 preferred rock

302 preferred classical

233 preferred jazz

79 preferred rock and classical

243 preferred rock and jazz

119 preferred classical and jazz

73 preferred all three types of music

The disc jockey was fired. Can you see why?

1.5 Summary

In this chapter we discussed some ideas about sets. After describing a set, we analyzed the different types of sets. We distinguished between finite and infinite sets. Cantor's radical work on sets involved infinite sets. We indicated how subsets are formed and how to determine how many of them there are. These results will be used in later chapters.

We then discussed how we can form new sets by performing elementary set operations. Although we did not study all possible set operations, we analyzed those that will prove helpful later. Finally we presented several formulas that can be used to determine the number of elements in a set. These formulas can be applied to solve a variety of survey problems.

We also indicated that sets can be analyzed by means of Venn diagrams, which are graphical representations of sets. Venn diagrams are quite helpful when analyzing survey problems. Although we discussed only some applications of sets, the range of uses is considerable. The interested reader can find such additional applications by consulting the references given.[3]

Study Guide

You should be able to demonstrate your knowledge of the following ideas presented in this chapter by giving definitions, descriptions, or specific examples. Page references (in parentheses) are given for each term so that you can check your answer.

Set (p. 3)
Members of a set (p. 3)
Set-builder notation (p. 4)
Finite sets (p. 7)
Null or empty sets (p. 7)
Proper subsets (p. 9)
Venn diagrams (p. 11)
Intersection of sets (p. 15)
Union of sets (p. 16)
Solving survey problems using sets (p. 25)
Elements of a set (p. 3)
Equal sets (p. 4)
Well-defined sets (p. 7)
Infinite sets (p. 7)
Subsets (p. 9)
Universal set (p. 10)
Complement of a set (p. 11)
Disjoint sets (p. 15)
Counting elements in sets (p. 21)

Formulas to Remember

The following list summarizes the formulas given in this chapter.

Rule 1: $n(A \cup B) = n(A) + n(B)$ when A and B are disjoint sets

Rule 2: $n(A \cup B) = n(A) + n(B) - n(A \cap B)$ for any finite sets A and B

Rule 3: $n(A \cup B \cup C) = n(A) + n(B) + n(C) - n(A \cap B) - n(A \cap C) - n(B \cap C) + n(A \cap B \cap C)$ for any finite sets A, B, and C

[3] For a more thorough discussion of sets, consult *Mathematics As a Second Language* by Joseph Newmark and Frances Lake (see "Suggested Further Reading" at the end of this chapter).

FORM A

For questions 1–6 let the universal set $U = \{1, 2, 3, 4, 5, 6, 7, 8, 9, 10\}$, $A = \{2, 4, 6, 8\}$, $B = \{3, 5, 9\}$, and $C = \{4, 5, 6, 8, 9\}$.

1. Find $A \cap B$.
a) C b) U c) $\{2, 3, 4, 5, 6, 8, 9\}$
d) $\varnothing$ e) A

2. Find $A \cup B$.
a) C b) U c) $\{2, 3, 4, 5, 6, 8, 9\}$
d) $\varnothing$ e) A

3. Find $A \cap C$.
a) A b) $\varnothing$ c) $\{2, 4, 5, 6, 8, 9\}$
d) $\{4, 6, 8\}$ e) U

4. Find A'.
a) $\{1, 3, 5, 7, 9, 10\}$ b) U
c) $\varnothing$ d) $\{1, 7, 10\}$
e) None of these

5. Find $(A \cup B)'$.
a) $\{1, 3, 5, 7, 9, 10\}$ b) U
c) $\varnothing$ d) $\{1, 7, 10\}$
e) None of these

6. Find $[(A \cup B) \cap C]'$.
a) U b) $\varnothing$ c) C'
d) $A' \cup B'$ e) None of these

7. Which of the following sets is not well defined?
a) All men over 60 feet tall b) All books
c) All lovable children d) All stars in the universe

8. If A is *any* set, what is $A \cup \varnothing$?
a) U b) A' c) A d) $\varnothing$

9. The diagram below represents

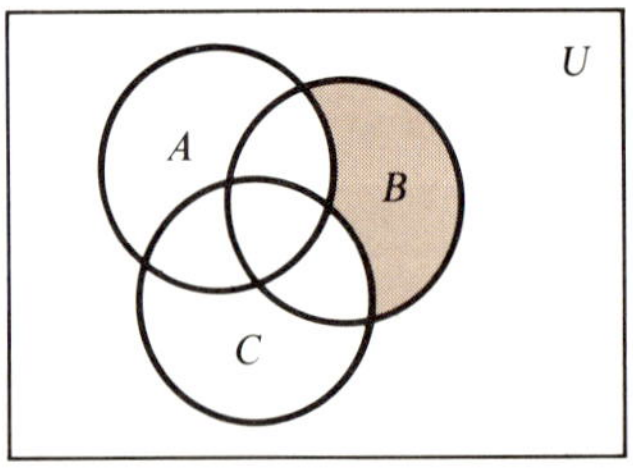

a) $B' \cap C$ b) $B \cap C'$ c) $B \cap A'$
d) $B \cap A' \cap C'$

10. If $A = \{\text{all cats}\}$ and $B = \{\text{all mammals}\}$, then

a) $A \subset B$ b) $B \subset A$
c) $B \subseteq A$ d) None of these

11. If a set has 12 elements, it has

a) 12^2 subsets b) 2^{12} subsets
c) 12 subsets d) 24 subsets

12. Disjoint sets are

a) Sets with a different number of elements
b) Sets that are not well defined
c) Unequal sets
d) Sets with no common elements

13. If $A \cup B = A$, then

a) $A \subseteq B$ b) $B \subseteq A$ c) $A = U$ d) $A \cup B = B$

14. In a survey of 200 people, the following information was obtained: 70 drank Coke, 55 drank 7-Up, 60 drank Pepsi, 50 drank 7-Up and Coke, 40 drank Coke and Pepsi, 35 drank 7-Up and Pepsi, and 30 drank all three. How many drank only Coke?

a) 10 b) 70 c) 90 d) 110

15. In the survey of question 14, how many drank none of the three?

a) 90 b) 110 c) 30 d) 80

Use the diagram to answer questions 16–18.

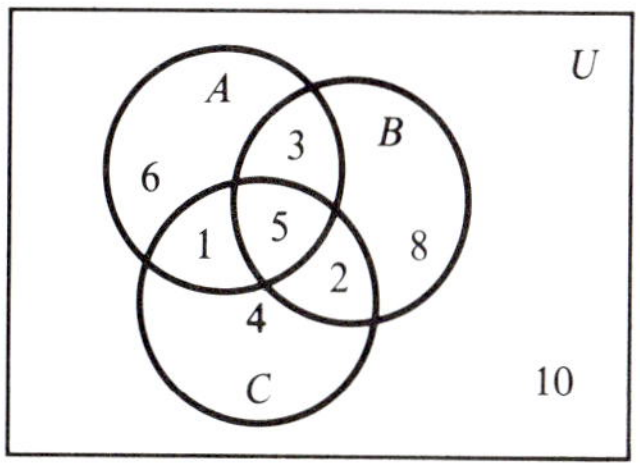

16. Find $n(A \cap B')$.

a) 8 b) 15 c) 6 d) 7

17. Find $n(A \cup B \cup C)$.

a) 29 b) 40 c) 24 d) 5

18. Find $n(A \cap B' \cap C')$.

a) 5 b) 3 c) 6 d) 10

FORM B

1. Refer to the newspaper article at the beginning of this chapter. Explain why the business records of the Gartel Food Distribution Corporation were subpoenaed.

2. *The missing hot dogs.* The owner of a concession stand at a baseball park accuses one of her workers, Paul, of eating three hot dogs without paying for them. Paul denies this and claims that at yesterday's game he sold all 45 hot dogs that were given to him as follows:

23 hot dogs with mustard
20 hot dogs with sauerkraut
23 hot dogs with relish
9 hot dogs with mustard and sauerkraut
13 hot dogs with relish and sauerkraut
8 hot dogs with mustard and relish
6 hot dogs with mustard, relish, and sauerkraut

Is the owner of the concession stand correct in her accusations, that is, what happened to the three missing hot dogs?

★3. Suppose A and B are any sets such that $(B \cap A') \cup (B' \cap A) = A$. What can be said about set B?

★4. Let the universal set $U = \{a, b, c, d, e, f\}$, $A = \{b, e\}$, and B and C be any two nonempty sets. Find the elements in set C such that $B \cap C = \varnothing$, $B \cup C = \{a, b, c, d, f\}$, and $A \cup B = \{a, b, e\}$.

■★5. *Classifying girls.*[4] The girls in the Alpha Beta Sorority were classified as in the following table. Using the letters given in the table, find the

	Redhead (R)	Blonde (B)	Brunette (N)	Black hair (L)
Tall (T)	6	4	18	7
Average (A)	3	9	12	15
Short (S)	8	14	17	16

number of people in each of the following categories:

a) $T \cap B$
b) $A \cap B' \cap N' \cap L'$
c) $N \cup (T \cap R)$
d) $(T' \cap A') \cup L$
e) $(A' \cap N')' \cup B$
f) $(T \cup A \cup S) \cup (R \cup B \cup N \cup L)$

■★6. *Hiring guards.* The First National Bank is interested in hiring 40 additional security guards. An analysis of the résumés submitted by 300 applicants indicates that 147 were men, 152 had no prior experience, 178 had police records, 82 were men with no prior experience, 56 were women with no prior experience but with police records, and 63 were men with no prior experience but with police records. How many applicants had no police record but had some prior experience?

[4] Problems preceded by a ■ may be solved by using a hand-held calculator.

■★7. *Bees.* A scientist is conducting experiments with bees and classifies the bees (for security reasons) as in the table. Using the codes given in the table, find the number of bees that the scientist has in each of the following categories:

	Type of Bee				
	Wasp (W)	Killer (K)	Bumble (B)	Stingless (S)	Honey (H)
Female or queen (Q)	40	78	63	171	103
Male or drone (D)	1600	1247	1982	1206	1748

a) $Q' \cap H$
b) $D \cup (Q' \cap S)$
c) $D \cup Q$
d) $(Q \cap D) \cup (W \cap H)$
e) $(B' \cup S') \cup D'$
f) $(Q' \cup B') \cap (S' \cup H')$

8. *No vacancy?* As described in the Historical Note at the beginning of this chapter, Georg Cantor's ideas about infinite sets were quite astounding. For example, consider the following situation raised by Cantor: A mathematician arrives at the Cantor Hotel and requests a room. He is told that every room is taken. After a moment's hesitation, the manager of the hotel devises a scheme whereby the new guest can be accommodated and there will be no shortage of rooms. He tells the mathematician, "You will take room 1, the guest that is in room 1 will be moved to room 2, the guest that is in room 2 will be moved to room 3, and so on. The guest that is in room n will be moved to room $n + 1$." After this switch is accomplished every guest has a room.

How many rooms does the Cantor Hotel have?

Suggested Further Reading

Breuer, Joseph, *Introduction to the Theory of Sets*, Englewood Cliffs, N.J.: Prentice-Hall, 1958.

Dinkines, F., *Elementary Theory of Sets*, New York: Appleton-Century-Crofts, 1964.

Fraenkel, A., *Abstract Set Theory*, 3d ed., New York: Humanities Press, 1966.

Hahn, H., "Infinity," in James R. Newman, ed., *The World of Mathematics*, vol. III, pp. 1593–1611, New York: Simon & Schuster, 1956.

Kline, M., *Mathematical Thought from Ancient to Modern Times*, New York: Oxford University Press, 1972.

Mathematics in the Modern World (readings from *Scientific American*), articles 28 and 30. San Francisco: Freeman, 1968.

Newmark, J., and F. Lake, *Mathematics As a Second Language*, 2d ed., Reading, Mass.: Addison-Wesley, 1977.

CHAPTER 2 Functions and Graphs

Objectives to be Achieved in this Chapter

- *To discuss* what a function is, how it is used, and the notation for it.
- *To analyze* a rectangular coordinate system that can be used to draw the graphs of functions.
- *To study* the straight line, including its graph, its slope, its intercepts when graphed, and several alternate forms of the equation of a line.
- *To learn* how to draw the graph of a second-degree equation.
- *To discuss* how to find the point(s) of intersection of two graphs (if there are any).
- *To apply* linear equations to business problems such as calculating depreciation, paying real estate taxes, and determining break-even and equilibrium costs.

U.S. Dollar Sinks to New Low

DAILY REPORTER JULY 20, 1978

New York (July 20)—The U.S. dollar hit an all-time low yesterday on the foreign currency market. In Tokyo on the Japanese Foreign Exchange the dollar dropped to an all-time low. Japanese yen were trading at 190 to the American dollar.

Many government economists believe that the value of the American dollar is a function of our unfavorable balance of trade. Recently our foreign trade deficit has been increasing largely due to rising oil imports, and as a result the value of our dollar has been dropping. To counteract this functional relationship, President Carter has been urging Americans to conserve energy.

This newspaper article indicates that many economists believe the value of the American dollar is a function of and directly related to our unfavorable balance of trade. How can we determine whether and under what conditions there is such a functional relationship? Can we find an equation that will predict the value of the American dollar when we are given the balance of trade figures?

2.1 Introduction

Many of us have encountered charts of numbers indicating that the value of one thing depends on some other quantity. For example, the price of a long-distance telephone call between two cities depends on the length of the conversation as shown in Table 2.1. Similarly the sticker price of a particular new car depends on the number of options ordered as shown in Table 2.2.

These tables illustrate what the mathematician calls a *function.* In this chapter we discuss the ideas of functions and their graphs. Our discussion is not intended to be a complete treatment of functions and graphs, as the reader is undoubtedly familiar with some (or all) of the

Table 2.1

Length of conversation (minutes)	Cost
3	$1.00
4	1.30
5	1.60
6	1.90
7	2.00
8	2.10

Table 2.2

Number of options	Sticker price
0	$5200
1	5500
2	5600
3	5740
4	5800
5	5875
6	6132

ideas mentioned. It is intended merely as a review of some basic ideas of functions and graphs that the reader will need as a common starting point in studying finite math.

Historical Note

The idea of graphs can be traced back to the ancient Egyptians and Greeks. Many Greek mathematicians, including Euclid and Apollonius, studied graphs. The *Elements*, written by Euclid, contains a detailed treatment of figures formed by straight lines and circles. *Conic Sections*, written by Apollonius, presents a thorough discussion of the graphs of conic sections, which he named ellipses, parabolas, and hyperbolas.

In the seventeenth century the French mathematicians Pierre de Fermat (1601–1665) and René Descartes (1596–1650) developed a new approach to geometry, specifically an algebraic method of representing and analyzing curves. This is the basis of the coordinate geometry we discuss in this chapter.

In 1637 Descartes published his major work on philosophy, *Discours de la méthode pour bien conduire sa raison et chercher la vérité dans les sciences* (Discourse on the method of rightly conducting the reason and seeking truth in the sciences). This book had three appendixes. In one of these, *La Géométrie*, Descartes explained how the "method" can be applied to coordinate geometry. He analyzed geometric ideas by means of algebraic techniques. His analysis paved the way for the development of a new branch of mathematics called analytic geometry. Many algebraic symbols that we use today are contained in this work.

So successful was Descartes' publication that he soon was invited to the courts of many kings and queens. He routinely declined most of these invitations. In 1649, however, he was invited to the court of Queen Christina of Sweden to give the queen instruction in philosophy, and he accepted this invitation. Queen Christina preferred to work in the mornings in the unheated castle, whereas Descartes was accustomed to staying in bed until noon when it warmed up a bit. Several months after arriving in Stockholm, Descartes contracted pneumonia. He died at the age of 54. His death had apparently been brought on by the rigors of the Swedish winter, to which Descartes could not adapt.

2.2 Functions and Function Notation

To understand the idea of a function, let us consider the charts of numbers given in the following examples.

Example 1 *Selling gas.* A gas station dealer finds that the number of gallons of gas sold per week is dependent on the price charged per gallon as shown in the table.

Price per gallon of gas	\$0.95	\$0.98	\$1.02	\$1.05	\$1.08	\$1.10	\$1.11	\$1.13	\$1.15
Number of gallons sold per week	4000	3500	3200	3000	2700	2000	1800	1500	900

Example 2 *Mass inoculation.* The Calville health department finds that the number of people inoculated per day with smallpox vaccine depends on the number of doctors available to administer the shot as shown in the table.

Number of doctors available	1	2	3	4	5
Number of people inoculated	25	38	61	76	89

Example 3 *Fish breeding and pollution.* According to state environmental officials, the number of fish that will breed in a state fish hatchery depends on the percentage of pollutants in the water as shown in the table.

Percentage of pollutants	5%	8%	10%	15%	20%
Number of fish that will breed	1000	775	600	450	100

Example 4 *Radar and speeding tickets.* Police department records indicate that the number of speeding tickets issued by the Melville police department on any given day is dependent on the number of radar-equipped speed traps in operation as shown in the table.

Number of speed traps in operation	0	2	3	4	5
Number of speeding tickets issued	0	20	30	40	60

In each of the preceding tables, there is a correspondence between the numbers in the top row and the numbers in the bottom row. For example, from the table in Example 1 we see that the gas station dealer will sell 3500 gallons of gas when the price per gallon is \$0.98 and 900 gallons when the price is \$1.15. Further examination of this table indicates that each number in the top row is associated with one and only one number (that is, a unique number) in the bottom row. Mathematicians say that the number of gallons of gas sold *is a function of* or *depends on* the price of a gallon of gas.

Definition 2.1 A *function* is a rule that assigns to each number in a set of numbers a *unique* second number.

Using Definition 2.1, we say:

1. In Example 1 the number of gallons of gas sold is a function of the price per gallon.
2. In Example 2 the number of people inoculated with the vaccine is a function of the number of doctors available.
3. In Example 3 the number of fish that will breed is a function of the percentage of pollutants in the water.
4. In Example 4 the number of speeding tickets issued is a function of the number of radar traps in operation.

Functions are not always expressed by means of tables of numbers. Sometimes letters are used. To see this, consider the function given in Table 2.3. In this example the Hillview Medical Center pays out a certain amount of money daily depending on the number of people who donate blood. Notice that we have used a letter, rather than a number, as the first entry in the top row. This is because the number of people that may want to donate blood is zero or any counting number. Since we cannot enter all the possible values in the first column, we use the letter x as the first entry. We then write some sample values for that letter as the remaining entries in the left-hand column of the table. The letter x used in this manner to represent any one of many possible values is called a *variable*. Thus, if the variable x equals 5, this means that 5 people will donate blood. The amount of money paid out, the y value, will then be \$125. Therefore, we can define a function as a relationship between two variables, x and y, such that for each value substituted for x, there is obtained a *unique* value of y.

Table 2.3

Number of people donating blood, x	Amount of money paid out, y
x	\$ $25x$
2	50
5	125
8	200
20	500
50	1250
100	2500

Since we will be studying functions a great deal in this text, we introduce some special notation.

Notation We abbreviate the words "y is a function of x" as $y = f(x)$. This is read "y equals f of x."

Example 5 The equation $y = x^2$ represents a function. If we replace x by 2, we get $y = 2^2 = 4$. If we replace x by -6, we get $y = (-6)^2 = 36$. For each value of x, we get a unique value of y. In this case y is a function of x, and we write $y = f(x) = x^2$.

Example 6 $y = 2x - 6$ is a function, since if we replace x by any number, we get a unique value of y. We write $y = f(x) = 2x - 6$.

Example 7 The equation $y = \pm\sqrt{x}$ is *not* a function. The symbol $\sqrt{x}$ is read "the square root of x." This means that we are looking for a number that gives x when we multiply it by itself. Thus, $\sqrt{36} = 6$ since $6 \cdot 6 = 36$. The symbol $\pm\sqrt{x}$ means either the positive square root or the negative square root of x. Therefore, if x is 36, y may be either $+6$ or -6. For any given x, we do *not* get a unique value for y.

Example 8 $y = -\sqrt{x}$ is a function because now there is a unique value of y for each value of x. Thus, if $x = 36$, then $y = -6$ only.

Example 9 If $f(x) = 3x + 1$, find the following:

a) $f(1)$ b) $f(2)$ c) $f(-1)$
d) $f(\frac{1}{3})$ e) $f(B)$ f) $f(x + h)$

Solution a) The symbol $f(1)$ means the value of y when $x = 1$. Thus

$$f(1) = 3(1) + 1 = 4$$

b) When $x = 2$, the y value is

$$f(2) = 3(2) + 1 = 7$$

c) When $x = -1$, the y value is

$$f(-1) = 3(-1) + 1 = -2$$

d) When $x = \frac{1}{3}$, the y value is

$$f\left(\frac{1}{3}\right) = 3\left(\frac{1}{3}\right) + 1 = 2$$

e) When $x = B$, the y value is

$$f(B) = 3B + 1$$

f) The symbol $f(x + h)$ means the value of y when x is replaced by $x + h$. Thus,

$$f(x + h) = 3(x + h) + 1 = 3x + 3h + 1$$

Example 10 If $f(x) = (5x + 2)/(3x + 1)$, then

$$f(1) = \frac{5(1) + 2}{3(1) + 1} = \frac{7}{4}$$

$$f(0) = \frac{5(0) + 2}{3(0) + 1} = \frac{2}{1} = 2$$

$$f(x + h) = \frac{5(x + h) + 2}{3(x + h) + 1} = \frac{5x + 5h + 2}{3x + 3h + 1}$$

Example 11[1] If $f(x) = |2x - 1|$, then

$$f(1) = |2(1) - 1| = |1| = 1$$
$$f(4) = |2(4) - 1| = |7| = 7$$
$$f(-2) = |2(-2) - 1| = |-5| = 5$$
$$f(0) = |2(0) - 1| = |-1| = 1$$

Definition 2.2 In using the function notation, the set of all values of x that can be substituted in the function is called the *domain* of the function. The set of all y values that these substitutions create is called the *range* of the function.

Example 12 gives several illustrations of the domain and range of a function.

Example 12

Equation	Domain	Range
$y = 3x + 2$	All real numbers	All real numbers
$y = x^2$	All real numbers	All nonnegative numbers
$y = \|2x + 1\|$	All real numbers	All nonnegative numbers
$y = \frac{2x + 7}{x^2 - 9}$	All real numbers except $x = \pm 3$	All real numbers
$y = \sqrt{x + 1}$	All real numbers that are -1 or greater	All nonnegative numbers

Comment Throughout this text we will always assume that the domain is the set of real numbers.

Comment Note that $f(x)$ does not mean f times x. It is read "f of x" and represents the range, that is, the y values. Thus, $y = f(x)$.

[1] The symbol | | represents the *absolute value* of a number. The absolute value of a number means the number itself disregarding its sign. For example, $|+4| = 4$, $|-4| = 4$, and $|-3.25| = 3.25$. In general, $|x| = x$ if x is positive or 0 and $|x| = -x$ if x is negative. Thus, $|x|$ is always positive or 0.

1. For each of the following, determine whether y is a function of x. If not, explain why.

a) *The cost of mailing a first-class letter in 1979.*

Weight of envelope, x (ounces)	Postage required, y
1	15¢
2	28
3	41
4	54
5	67

b) *New York State 1978 tax rate.*

If taxable income, x, is Over	But not over	New York State tax, y, is
$0	$1,000	2% of amount
1,000	3,000	$20 plus 3% of excess over $1,000
3,000	5,000	80 plus 4% of excess over 3,000
5,000	7,000	160 plus 5% of excess over 5,000
7,000	9,000	260 plus 6% of excess over 7,000
9,000	11,000	380 plus 7% of excess over 9,000
11,000	13,000	520 plus 8% of excess over 11,000
13,000	15,000	680 plus 9% of excess over 13,000
15,000	17,000	860 plus 10% of excess over 15,000
17,000	19,000	1,060 plus 11% of excess over 17,000
19,000	21,000	1,280 plus 12% of excess over 19,000
21,000	23,000	1,520 plus 13% of excess over 21,000
23,000	25,000	1,780 plus 14% of excess over 23,000
25,000		2,060 plus 15% of excess over 25,000

c) *Listings in the telephone book.*

Number of listings in the yellow pages of a telephone directory for a city, x	Number of telephones in the city, y

(For the sake of this problem, assume that no one has an unlisted telephone number.)

d) *Buying whole life insurance policies from a particular company.*

Age nearest birthday, x		Payments, y, ordinary life		
Male	Female	Annual	Semi-annual	Quarterly
20	25	$11.13	$6.27	$3.74
21	26	11.50	6.46	3.84
22	27	11.86	6.65	3.93
23	28	12.26	6.86	4.04
24	29	12.68	7.07	4.15
25	30	13.07	7.28	4.26
26	31	13.47	7.48	4.37
27	32	13.90	7.71	4.48
28	33	13.34	7.94	4.60
29	34	13.81	8.18	4.73
30	35	15.30	8.44	4.86
31	36	15.82	8.71	5.00
32	37	16.37	8.99	5.15
33	38	16.94	9.29	5.30
34	39	17.54	9.60	5.47
35	40	18.18	9.93	5.64
36	41	18.86	10.29	5.82
37	42	19.58	10.66	6.02
38	43	20.32	11.05	6.22
39	44	21.11	11.46	6.43
40	45	21.94	11.89	6.65
41	46	22.80	12.34	6.89
42	47	23.72	12.81	7.13
43	48	24.69	13.32	7.40
44	49	25.71	13.85	7.67
45	50	26.91	14.47	8.00
46	51	28.17	15.13	8.34
47	52	29.52	15.83	8.70
48	53	30.95	16.57	9.09
49	54	32.47	17.36	9.50
50	55	34.10	18.21	9.94
51	56	35.82	19.11	10.40
52	57	37.65	20.06	10.90
53	58	39.61	21.08	11.42
54	59	41.70	22.16	11.99
55	60	43.95	23.33	12.60
56	61	43.79	23.25	12.55
57	62	46.01	24.41	13.15
58	63	48.36	25.63	13.79
59	64	50.85	26.92	14.46
60	65	53.50	28.30	15.18

e) *United States customs.*

Number of hourly plane arrivals, x, at a large U.S. airport	Number of customs agents, y, needed
1	1
2	2
3	2
4	3
5	3
6	5
7	6

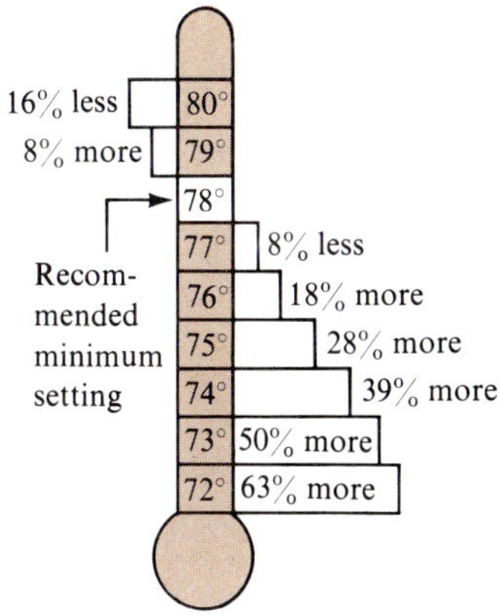

f) *How to set your air conditioner.* According to government energy officials, the recommended setting for your air conditioner is no lower than 78°F. The amount of energy consumed, y, is related to the temperature (thermostat) setting as shown in the accompanying diagram.

2. For each of the accompanying tables,

a) Determine whether y is a function of x.

b) If y is a function of x, can you find an equation expressing this relationship?

c) Find the missing value of y.

Number of exemptions claimed on federal tax return, x	1	2	3	4	5	11
Amount to be subtracted from adjusted gross income, y	\$750	\$1500	\$2250	\$3000	\$3750	?

Number of packs of cigarettes purchased, x	1	2	3	4	5	15
Cost, y	75¢	150¢	225¢	300¢	375¢	?

3. *Selling cars.* A car salesman receives a flat weekly salary of \$150 plus \$50 extra for each used car he sells. Thus, if the salesman sells seven used cars his weekly salary will be \$500, and if he sells ten used cars his weekly salary will be \$650. Let x be the number of cars the salesman sells and let y be his salary.

a) Is y a function of x?

b) In one week the salesman sells 25 cars. Find his salary for the week.

c) Can you find an equation expressing the relationship between the salesman's salary, y, and the number of cars sold, x?

4. A group of sociologists recently concluded a survey on family income and average family size in the city of Pleasant Springs. Their results are shown in the table. Based on this survey, can we conclude that family size is a function of family income? Explain your answer.

Family income	Average number of children in family
Under \$9,000	6
Between \$9,000 and \$13,000	5
Between \$13,000 and \$16,000	4
Between \$16,000 and \$18,000	3
Between \$18,000 and \$22,000	2
Over \$22,000	1

5. For each of the following, find $f(3)$, $f(-5)$, $f(0)$, $f(\frac{1}{3})$, $f(c)$, and $f(x + h)$.

a) $f(x) = 2x + 9$ b) $f(x) = 3x^2 - 1$

c) $f(x) = \dfrac{5x + 3}{2x - 1}$ d) $f(x) = \sqrt{2x + 3}$

e) $f(x) = \sqrt{\dfrac{2x + 3}{x + 4}}$ f) $f(x) = x^3$

g) $f(x) = x^3 + 7x^2 - 5x + 1$ h) $f(x) = |2x - 3|$

★6. Barbara moved into her four-room flat in Greenwich Village in New York City on January 1, 1970. The rent was \$200 per month. Her apartment is rent controlled and, as a result, rent increases have been limited to 5% a year. Let x be the number of years Barbara has lived in the apartment and let y be her monthly rent.

a) Is y a function of x?

b) Assuming the apartment will remain rent controlled, what will Barbara's rent be in 1982?

★7. *Conserving energy.* In an effort to conserve energy (see the newspaper article at the beginning of this chapter), many states require air conditioners to carry a label indicating their EER, energy efficiency ratio. It is always advisable to buy a unit with a high EER since such units have a greater efficiency. To find the EER for any air conditioner, simply divide the cooling capacity of the unit, which is given in terms of British thermal units of BTUs (this usually can be read from the nameplate), by the number of watts consumed. For example, for an air conditioner with a BTU rating of 8400 and a wattage rating of 1200, the

$$\text{EER} = \frac{8400}{1200} = 7$$

a) Is the EER a function of the number of BTUs?

b) Find the EER of an air conditioner that has a BTU rating of 14,000 and a wattage rating of 1530.

2.3 The Rectangular Coordinate System

Most of us are familiar with the concept of a graph, although we might not use this name specifically. For example, consider the map of San Francisco. Chinatown has a map location of approximately I7. This means that we move across the bottom scale until we arrive at the I position. We draw a vertical line through this location. Next, 7 is found on the side of the map, and we draw a horizontal line through this point. The place where the horizontal and vertical lines meet (intersect) is the approximate location of Chinatown. What is the map location of Alcatraz? Alcatraz has a map location of N3. What is the map location of Fisherman's Wharf?

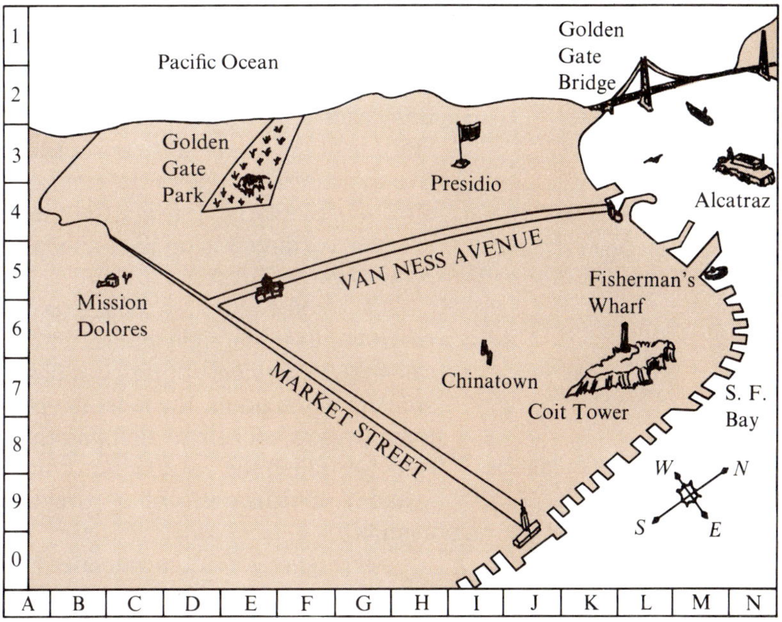

The same ideas are used when we graph functions. Coordinate geometry allows us to draw the graphs of the functions we will discuss in this book. These functions usually will be expressed or can be expressed by means of an equation.

To get started, let us first discuss a convenient way of representing numbers, that is, by means of a *number line*. A horizontal line is drawn and any point on it is selected as the starting point. This point is labeled O and is called the *origin* (see Figure 2.1). Since a line can be extended indefinitely in either direction, we indicate this by putting arrows at the left and at the right.

Figure 2.1

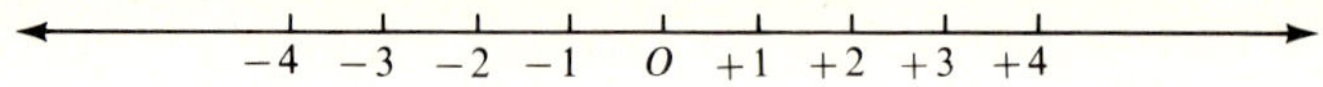

Figure 2.2 A horizontal number line.

Next we select any convenient unit of length and mark off points in succession both to the right and to the left of the origin. When we move in a direction to the right of O, we say that we are moving in the *positive direction.* When we move to the left, we say that we are moving in the *negative direction*, as shown in Figure 2.2. The number that names a point is called the *coordinate* of that point.

Number lines do not necessarily have to be drawn horizontally; they may also be drawn vertically as shown in Figure 2.3.

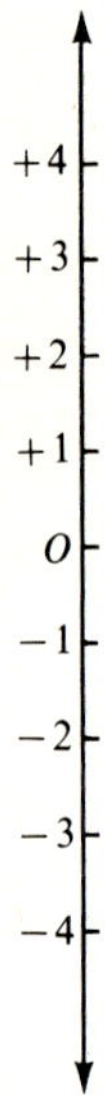

Figure 2.3 A vertical number line.

In the rectangular coordinate system we draw two lines (one vertical and one horizontal) that are perpendicular to each other. The horizontal line is called the *x axis* and the vertical line is called the *y axis.* The point where the x axis and the y axis meet is called the *origin.* Both axes are labeled with a number scale as shown in Figure 2.4. Starting at the origin, if we move to the right we are going in the positive direction, as opposed to moving to the left, which is the negative direction. Similarly, if we start at the origin and move up we are going in a positive direction on the y axis, as opposed to moving down, which is the negative direction.

Look at Figure 2.4. To get to point P, we start at the origin and move 4 units to the right and then from that point move up 3 units. We call 4 the *x coordinate* or *abscissa* of the point and 3 the *y coordinate* or *ordinate* of the point. We label the point as (4, 3), always being careful to write the x coordinate first and then the y coordinate. The numbers (4, 3) are called the *coordinates* of the point P. We always enclose the x and y coordinates within parentheses and separate them with a comma.

In Figure 2.5 the coordinates of point Q are $(-3, 1)$. This means that we start at the origin, move 3 units to the left, and then, from that

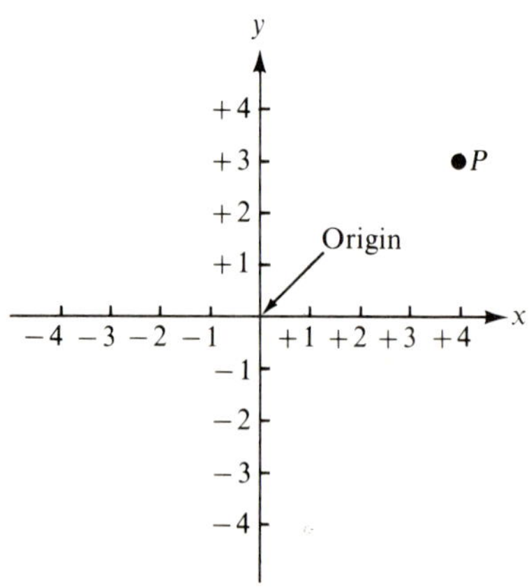

Figure 2.4 The rectangular coordinate system.

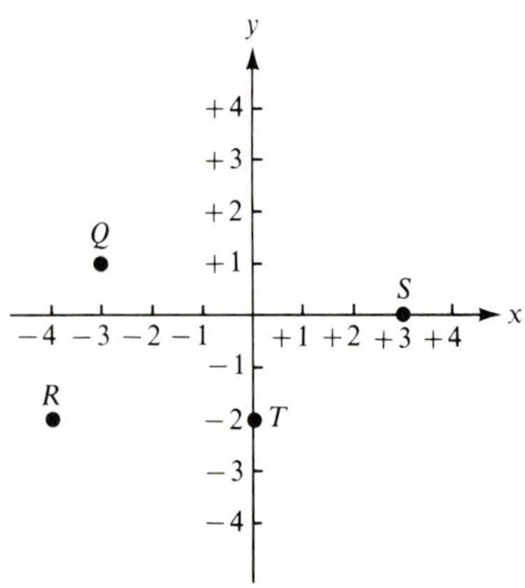

Figure 2.5

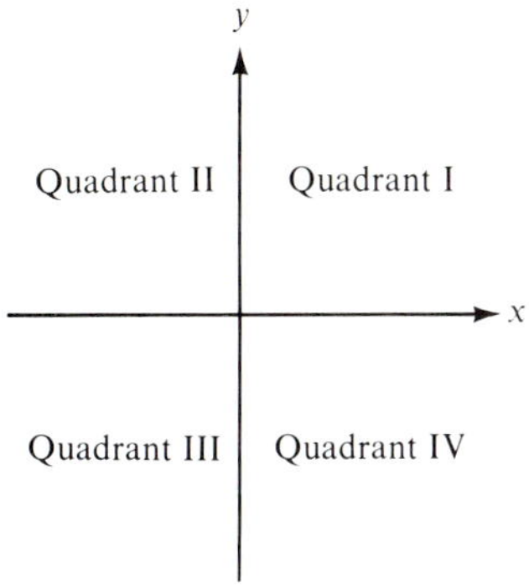

Figure 2.6 The different quadrants.

point, move up 1 unit. The coordinates of point R are $(-4, -2)$. We start at the origin, move 4 units to the left, and then, from that point, move down 2 units. The coordinates of point S are $(3, 0)$. We start at the origin, move 3 units to the right, and then move 0 units up or down. The coordinates of point T are $(0, -2)$. We start at the origin and move 0 units to the right or left. Then, from that point, we move down 2 units.

Notice that the x and y axes divide the plane (paper) into four regions. Each region is called a *quadrant*, and these are labeled counterclockwise as shown in Figure 2.6.

The values of x and y for any point may be positive or negative depending on which quadrant the point is in. The different possibilities are summarized in the following table:

If point is in quadrant	then x is	and y is
I	Positive	Positive
II	Negative	Positive
III	Negative	Negative
IV	Positive	Negative

What are the values of x and y when the point is on the line between two quadrants?

The system we have just described is called a *rectangular coordinate system*. It is the one that is most commonly used in mathematics. Other systems are possible; however, in this text we will use only a rectangular coordinate system. This system has many applications in the social sciences. One is given in the next example.

Example 13 The accompanying table summarizes some data on the U.S. Postal Service. We can picture this information on a rectangular coordinate system as shown in Figure 2.7.

Year	Number of post offices, x (includes Puerto Rico and all outlying areas)	Number of pieces of mail, y, handled (millions)
1967	33,624	71,873
1968	32,260	79,517
1969	32,064	82,005
1970	32,002	84,885
1971	31,947	86,983
1972	31,686	87,156
1973	31,385	89,683

Source: U.S. Postal Service, Annual Report of the Postmaster General, 1975.

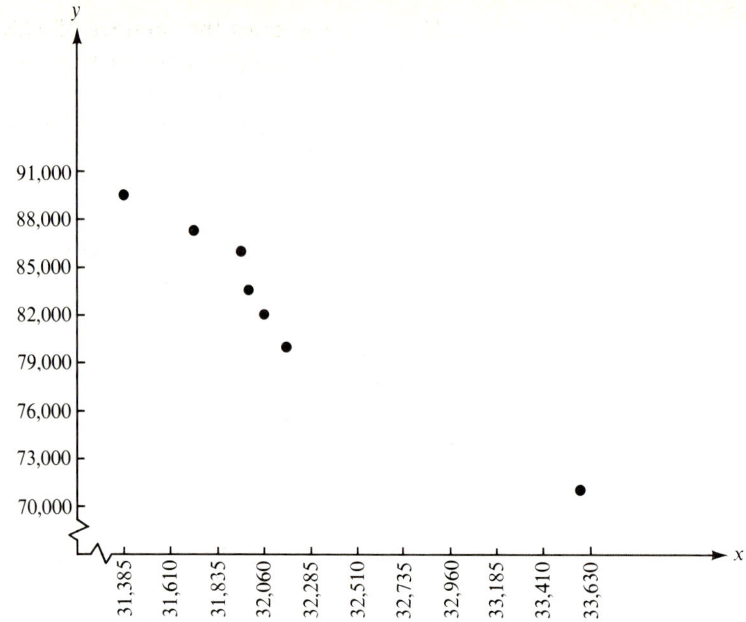

Figure 2.7

Exercises

8. Picture the information given in the accompanying table on a rectangular coordinate system.

State	Total number of reported crimes (per 100,000 population) for the calendar year 1973,[a] x	State and local police expenditures (millions of dollars) for the calendar year 1973,[b] y
Connecticut	3664	\$90
Maine	2544	18
Massachusetts	4521	184
New Hampshire	2329	16
Rhode Island	4678	25
New Jersey	4083	269
New York	4307	1051
California	6305	787
Idaho	3458	15
Oregon	5297	54
Washington	5090	84
Nebraska	2811	30

[a] *Source: Uniform Crime Reports for the United States*, U.S. Federal Bureau of Investigation, 1974.
[b] *Source: Expenditure and Employment Data for the Criminal Justice System*, U.S. Law Enforcement Assistance Administration, 1974.

9. The table indicates the number of U.S. banks that were closed permanently or temporarily because of financial difficulties and the amount of money on deposit at the time of failure. Picture the information on a rectangular coordinate system.

Year or period	Number of banks closed, x	Amount of money on deposit (millions), y
1951–1955	17	\$58
1956–1960	19	41
1961–1965	28	99
1966–1970	11	50
1971	3	5
1972	2	57
1973	3	21

Source: Federal Reserve Bulletin, Board of Governors of the Federal Reserve System, 1974.

10. The number of pints of blood in storage on any given day at Willow Hospital depends on the number of operations scheduled for that day.

Number of operations scheduled, x	Number of pints of blood in storage, y
10	320
12	327
13	329
15	335
16	339
17	345
18	348

a) Using an appropriate scale on each axis, locate each point on a coordinate system.
b) Join the points together in the order given.
c) If 20 operations are scheduled for tomorrow, find the predicted number of pints of blood in storage.

11. *Population estimates.* The accompanying table indicates the population change in the United States from 1890 to 1970. (Each symbol represents 10 million people.) Graph the data by labeling the years on the x axis

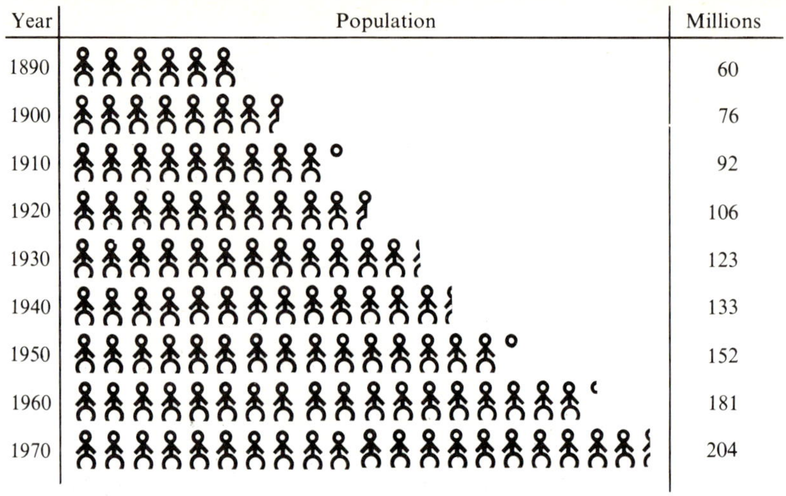

and the population on the y scale. From the graph estimate the population of United States in 1980.

12. *IQ scores and television.* A leading psychologist believes that IQ scores and the number of hours per week that a child watches television are related. In support of this claim, the following data have been collected for six students:

IQ scores, x	Number of hours spent watching TV, y
123	6
115	10
111	12
107	14
103	16
99	18

a) Using an appropriate scale on each axis, locate each point on a coordinate system.
b) Join the points in the order given.
c) What is the predicted IQ of a child who spends 20 hours a week watching television?

★13. *Energy consumption.* The accompanying table indicates how the energy consumption by an individual in the United States has changed over the years.

Year	Natural gas consumption per capita (million barrels), x	Crude oil consumption per capita (million barrels), y
1960	249	103
1965	278	111
1970	330	145
1971	338	148
1972	346	158
1973	360	165
1974	369	175
1975	374	182
1976	377	187

Source: Energy Facts, The Library of Congress, November 1976.

a) Graph the data on a rectangular coordinate system.
b) Based on the graph, can we conclude that the increase in crude oil consumption per capita is *proportional* to the increases in natural gas consumption per capita?

2.4 The Graph and Equation of a Straight Line

Consider the equation $y = 3x + 2$. One solution of this equation is $x = 2$ and $y = 8$. We can write this solution in abbreviated form as (2, 8). Another solution is $x = 1$ and $y = 5$, which we write as (1, 5). Other solutions are (3, 11), $(-1, -1)$, (0, 2), and so forth. Actually there are infinitely many such pairs of values that are solutions for this equation. To find other solutions, we let x be any convenient number and solve for y as shown in the following table:

x	$y = 3x + 2$	y	Point
4	$3(4) + 2$	14	(4, 14)
3	$3(3) + 2$	11	(3, 11)
2	$3(2) + 2$	8	(2, 8)
1	$3(1) + 2$	5	(1, 5)
0	$3(0) + 2$	2	(0, 2)
-1	$3(-1) + 2$	-1	$(-1, -1)$
-2	$3(-2) + 2$	-4	$(-2, -4)$
-3	$3(-3) + 2$	-7	$(-3, -7)$
-4	$3(-4) + 2$	-10	$(-4, -10)$

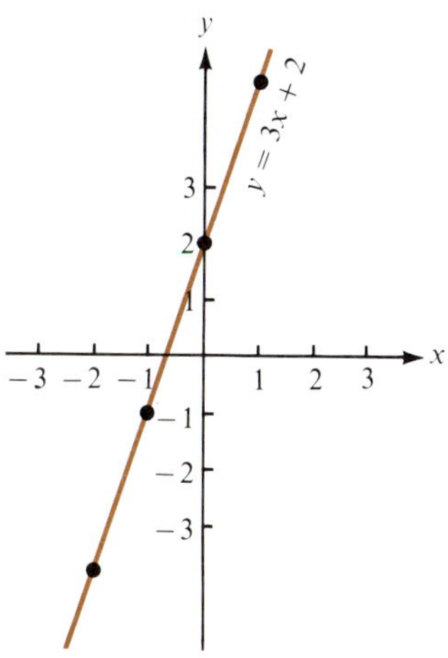

Figure 2.8 Graph of $y = 3x + 2$.

Let us plot these points on a rectangular coordinate system as shown in Figure 2.8. If we join these points together, we get a straight line, which is called the *graph* of the equation. Actually the graph of the equation consists of *all* the points (x, y) that are solutions of the equation.

Example 14 Draw the graph of $y = 4x - 1$.

Solution We set up the accompanying table of values by finding several pairs of numbers that are solutions to the equation. Then we plot the points and join them together. The resulting straight line is the graph of $y = 4x - 1$. See Figure 2.9.

x	$y = 4x - 1$	y
2	$4(2) - 1$	7
1	$4(1) - 1$	3
0	$4(0) - 1$	-1
-1	$4(-1) - 1$	-5
-2	$4(-2) - 1$	-9

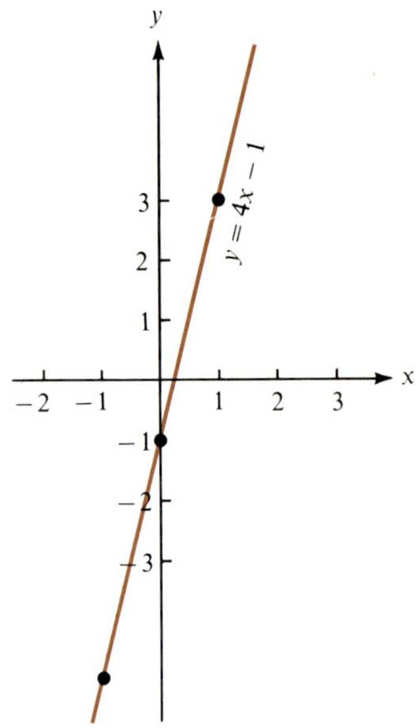

Figure 2.9 Graph of $y = 4x - 1$.

Example 15 Draw the graph of $y = -2x + 3$.

Solution We set the accompanying table of values by finding several pairs of numbers that are solutions to the equation. The graph is given in Figure 2.10.

x	$y = -2x + 3$	y
2	$-2(2) + 3$	-1
1	$-2(1) + 3$	1
0	$-2(0) + 3$	3
-1	$-2(-1) + 3$	5
-2	$-2(-2) + 3$	7

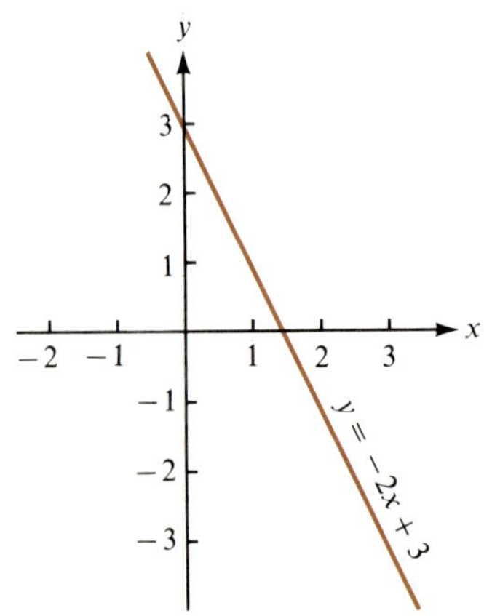

Figure 2.10 Graph of $y = -2x + 3$.

Example 16 Draw the graph of $y = 2$.

Solution We set up a table of values as we did in the previous examples. In this case we notice that y must always be 2 no matter what value x has. The graph of $y = 2$ is given in Figure 2.11. It is a line parallel to the x axis.

x	y
2	2
1	2
0	2
−1	2
−2	2

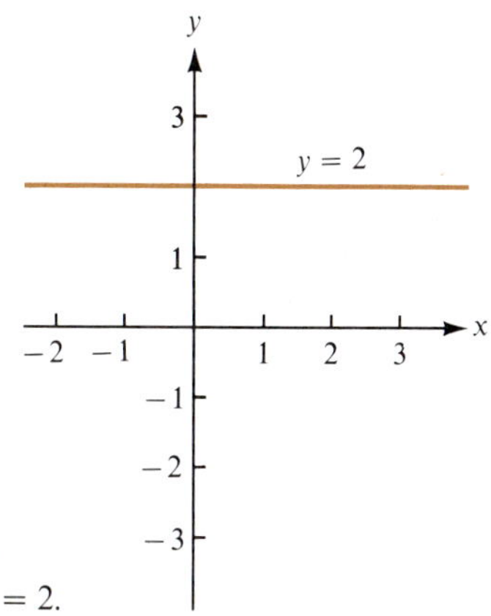

Figure 2.11
Graph of $y = 2$.

Example 17 Draw the graph of $x = 3$.

Solution We set up a table of values. In this case we notice that x must always be 3 no matter what value y has. The graph of $x = 3$ is given in Figure 2.12. It is a line parallel to the y axis.

x	y
3	4
3	7
3	0
3	−2
3	−5

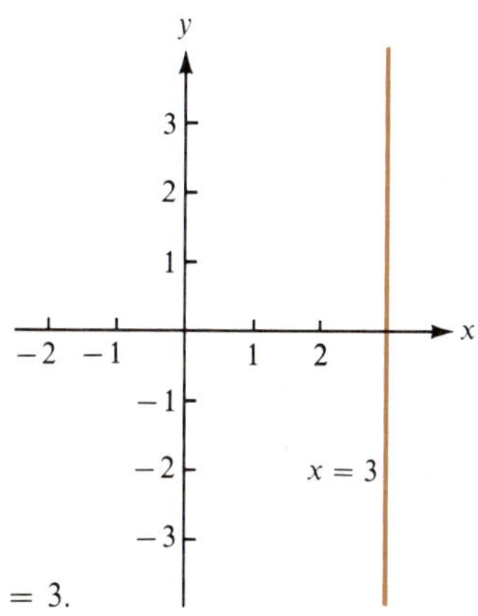

Figure 2.12
Graph of $x = 3$.

All the graphs we have drawn so far are straight lines. Actually the graph of any equation of the form $Ax + By + C = 0$ (where A, B, and C are constants and A and B are not both 0) will be a straight line. An equation whose graph is a straight line is called a *linear equation.*

The results of the previous examples can be summarized as follows:

1. The graph of $Ax + By + C = 0$ (where A, B, and C are constants and A and B are not both 0) is a straight line.
2. The graph of $x = c$ is a straight line parallel to the y axis.
3. The graph of $y = b$ is a straight line parallel to the x axis.

Example 18 Is the point (4, 5) on the graph of $7x + 8y = 68$?

Solution The point (4, 5) on the graph of $7x + 8y = 68$ only if $x = 4$, $y = 5$ is a solution of the equation. Substituting, we get

$$7(4) + 8(5) = 68$$

Thus the point (4, 5) is on the graph of $7x + 8y = 68$.

Exercises

14. Draw the graph of each of the following equations.

a) $y = 3x + 4$ b) $y = x + 5$ c) $y = -2x + 5$
d) $y = -3x - 3$ e) $y = 4x$ f) $3x = 9$
g) $4y = -8$ h) $3x + 6y = 12$
i) $7x + 14y = 21$ j) $y = (x/3) + 4$

15. Draw the graph of $y = |x - 1|$.

★16. Draw the graph of $|x - 7| = 4$.

★17. Draw the graph of $|y - 5| = 7$.

★18. Draw the graph of $|x| - |y| = 5$.

★19. Draw the graph of $|y| - |x| = 5$.

★20. Compare the graphs and equations in Exercises 18 and 19. Comment.

2.5 Intercepts and Slope of a Straight Line

Consider the graph of $2x + 3y = 6$, which is given formally in Figure 2.13. Notice that this graph crosses the x axis at the point (3, 0) and crosses the y axis at the point (0, 2). Mathematicians refer to these points as the x intercept and the y intercept, respectively.

Definition 2.3 The point at which a graph crosses the x axis is called the *x intercept.* To find the x value of this point, set $y = 0$ and solve for x.

Definition 2.4 The point at which a graph crosses the y axis is called the *y intercept.* To find the y value of this point, set $x = 0$ and solve for y.

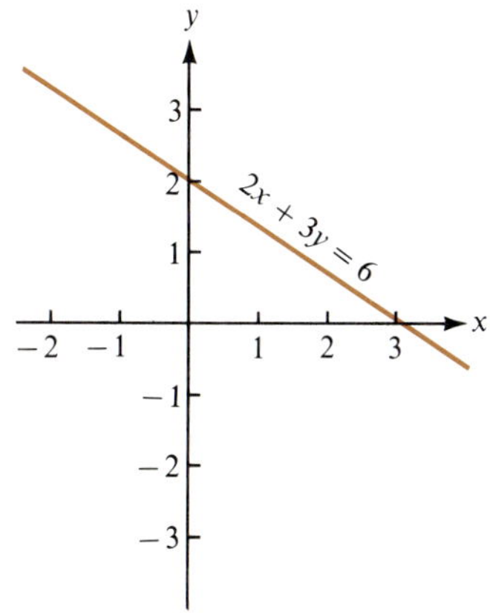

Figure 2.13 Graph of $2x + 3y = 6$.

Example 19 Find the x intercept and the y intercept of the line $4x + 3y = 12$.

Solution We apply Definitions 2.3 and 2.4. To find the x intercept, we let $y = 0$ and solve for x, getting

$$\begin{aligned} 4x + 3(0) &= 12 \\ 4x &= 12 \\ x &= 3 \end{aligned}$$

Thus, the x intercept is at (3, 0). To find the y intercept, we let $x = 0$ and solve for y, getting

$$\begin{aligned} 4(0) + 3y &= 12 \\ 3y &= 12 \\ y &= 4 \end{aligned}$$

Thus, the y intercept is at (0, 4). The graph of this equation is given in Figure 2.14.

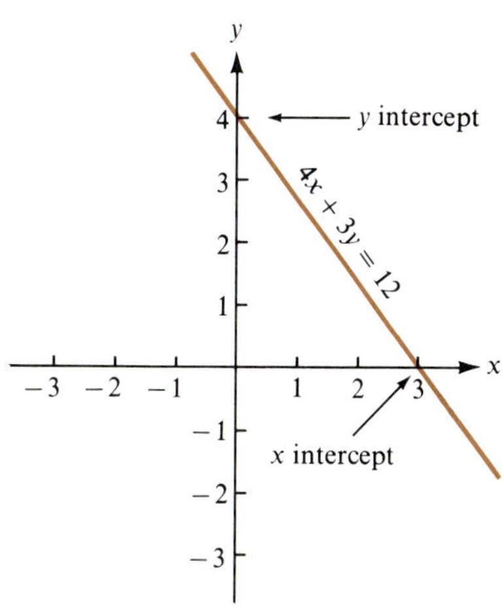

Figure 2.14 Graph of $4x + 3y = 12$.

Example 20 Draw the graph of a straight line whose x intercept is at $(-3, 0)$ and whose y intercept is at $(0, -2)$.

Solution Since we are told that the x intercept is at $(-3, 0)$, we know that the graph crosses the x axis at the point $(-3, 0)$. Also, since the y intercept is at $(0, -2)$, the graph crosses the y axis at the point $(0, -2)$. Thus, we plot these two points on a coordinate axis and join them to get the straight line shown in Figure 2.15.

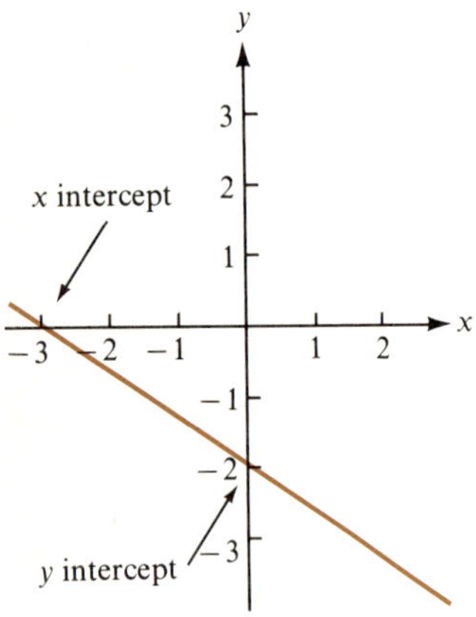

Figure 2.15 Graph of a line whose x intercept is at $(-3, 0)$ and whose y intercept is at $(0, -2)$.

Another important number associated with the graph of a straight line is its *slope*. We define the slope of a line as the change in the vertical distance divided by the change in the horizontal distance. To understand this concept, consider the roof shown in Figure 2.16. To climb to the top, a person would cover a horizontal distance of 365 centimeters and a vertical distance of 730 centimeters. Therefore, in this case the slope of the roof is $\frac{730}{365}$ or 2. The slope of the roof shown in Figure 2.17 is $\frac{365}{1095}$ or $\frac{1}{3}$. The first roof has a greater slope and hence is steeper.

Now consider the photograph shown in Figure 2.18. On August 7, 1974, Philippe Petit walked across a guy wire connecting the twin towers of the World Trade Center in New York City. In accomplishing this feat, the only change that occurred was the horizontal distance covered. There was no vertical distance covered. Such a line has zero slope.

On May 26, 1977, George Willig scaled the side of the World Trade Center in New York City. In climbing up the 110 floors, the only change that occurred was the vertical distance. There was no horizontal distance covered. Such a line has no slope, since we would have to divide by zero. See Figure 2.19.

More formally, suppose we are given two distinct points, P_1 and P_2, in a rectangular coordinate plane as shown in Figure 2.20.

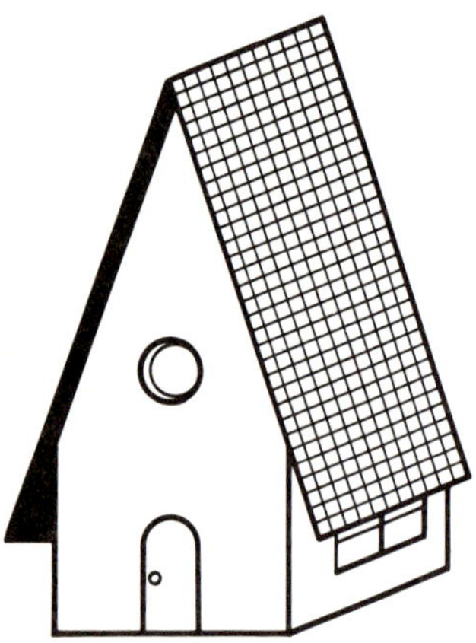

Figure 2.16

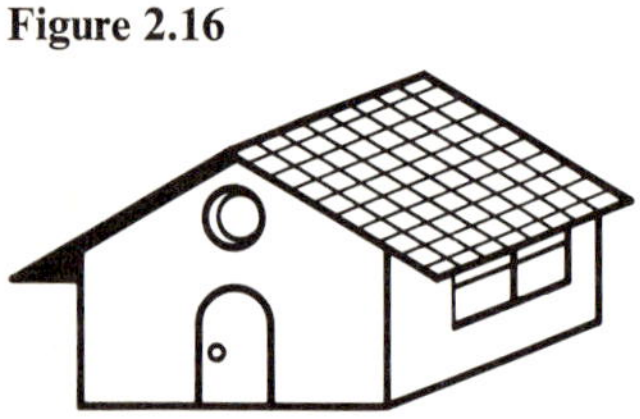

Figure 2.17

Figure 2.18 (*Source: United Press International*)

Figure 2.19
(*Source: United Press International*)

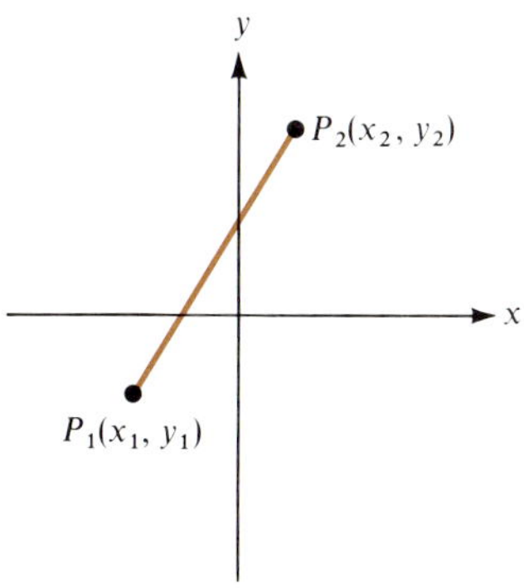

Figure 2.20

The subscripts are used merely to distinguish the points. Let the coordinates of these points be $P_1(x_1, y_1)$ and $P_2(x_2, y_2)$. These two points determine a unique number called the slope of the line.

Definition 2.5 The *slope* of the line joining the points (x_1, y_1) and (x_2, y_2) is given by

$$\frac{y_2 - y_1}{x_2 - x_1}$$

provided that $x_2 \neq x_1$. If $x_2 = x_1$, then $x_2 - x_1 = 0$ and

$$\frac{y_2 - y_1}{x_2 - x_1}$$

does not exist.[2] We usually use the letter m to represent the slope.

Example 21 Find the slope of the line joining the points (2, 5) and (7, 8).

Solution We apply Definition 2.5. Let $x_1 = 2$, $y_1 = 5$, $x_2 = 7$, and $y_2 = 8$. Then

$$x_2 - x_1 = 7 - 2 = 5$$
$$y_2 - y_1 = 8 - 5 = 3$$

so that

$$\frac{y_2 - y_1}{x_2 - x_1} = \frac{3}{5}$$

Thus, the slope of the line joining the points (2, 5) and (7, 8) is $\frac{3}{5}$.

[2] In mathematics, division by zero is not defined. For a detailed discussion of division involving zero, see pages 124–127 in *Mathematics As a Second Language*, 2d ed., by Joseph Newmark and Frances Lake (Reading, Mass.: Addison-Wesley, 1977).

Comment In applying Definition 2.5, it does not make a difference which point is called $P_1(x_1, y_1)$ and $P_2(x_2, y_2)$. In both cases we get the same answer. To illustrate, suppose in Example 21 we let $x_1 = 7$, $y_1 = 8$, $x_2 = 2$, and $y_2 = 5$. Then the slope of the line is

$$\frac{y_2 - y_1}{x_2 - x_1} = \frac{5 - 8}{2 - 7} = \frac{-3}{-5} = \frac{3}{5}$$

This is exactly the same answer we obtained in Example 21.

Example 22 Find the slope of the line joining the points.

a) (3, 4) and (5, 1)
b) (7, 6) and (9, 6)
c) (4, 5) and (4, −2)

Solution We apply Definition 2.5.

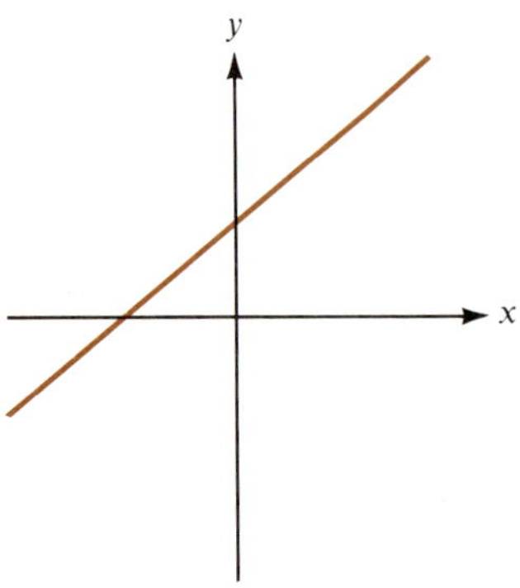

Figure 2.21 Positive slope (line rises as we move from left to right).

a) $\dfrac{y_2 - y_1}{x_2 - x_1} = \dfrac{1 - 4}{5 - 3} = \dfrac{-3}{2} = -\dfrac{3}{2}$

b) $\dfrac{y_2 - y_1}{x_2 - x_1} = \dfrac{6 - 6}{9 - 7} = \dfrac{0}{2} = 0$

c) $\dfrac{y_2 - y_1}{x_2 - x_1} = \dfrac{-2 - 5}{4 - 4} = \dfrac{-7}{0}$ This cannot be done. Therefore, there is no slope. (See footnote 2.)

It can be shown that if the slope of a line is positive, then the line rises as we go from left to right. If the slope of a line is negative, then the line falls as we go from left to right. If the slope of a line is zero, then the line is parallel to the x axis. If a line has no slope, then the line is parallel to the y axis. Thus, the slope of a line tells us the "direction" of the line. These possibilities are pictured in Figures 2.21–2.24.

Lines that are parallel have the same slope. Can you see why? See Figure 2.25.

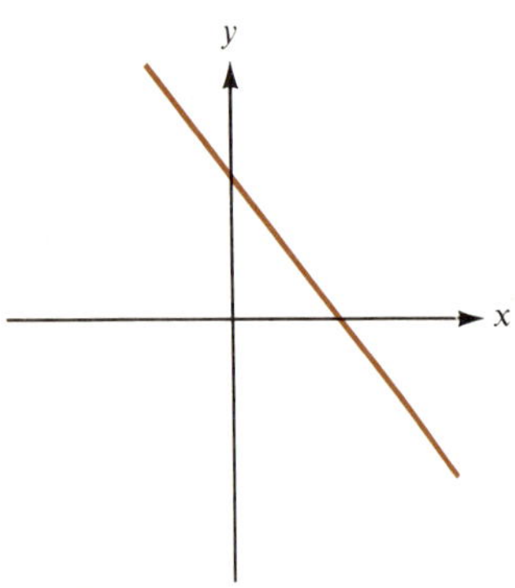

Figure 2.22 Negative slope (line falls as we move from left to right).

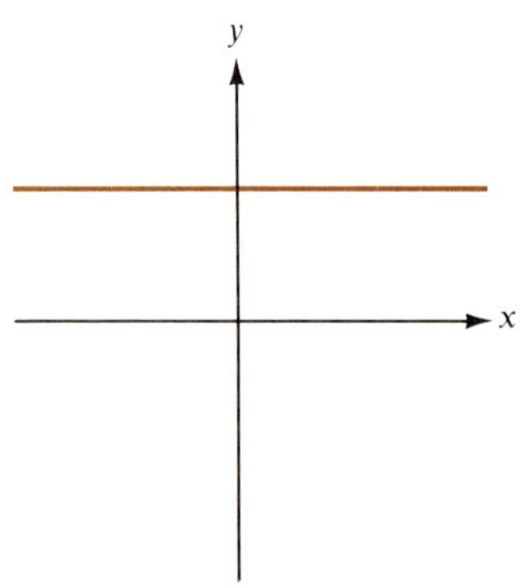

Figure 2.23 Zero slope (line is parallel to x axis).

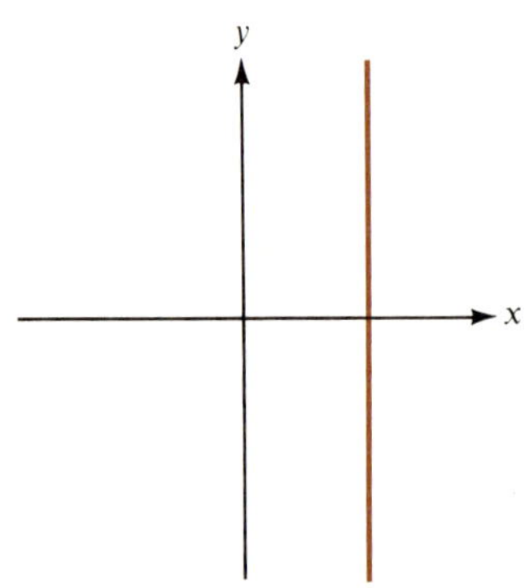

Figure 2.24 No slope (line is parallel to y axis).

Let us again analyze the graph of $y = 3x + 2$. Several pairs of values that satisfy this equation are shown in the accompanying table.

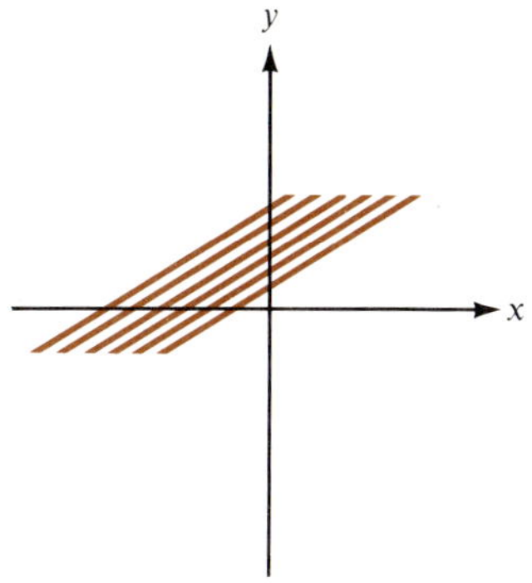

Figure 2.25 Parallel lines (all lines have the same slope).

x	y
2	8
1	5
0	2
−1	−1
−2	−4

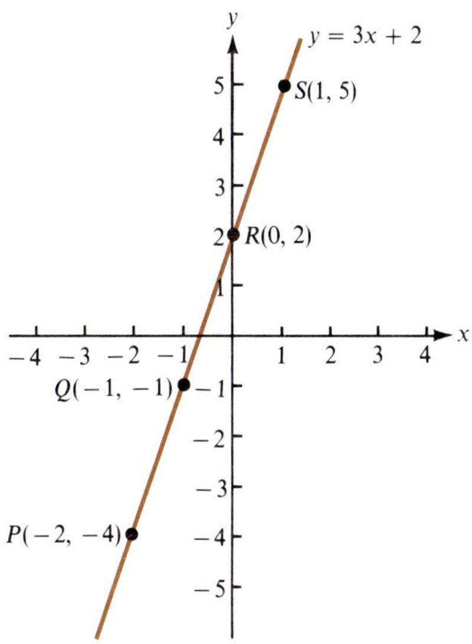

Figure 2.26 Graph of $y = 3x + 2$.

The graph is given in Figure 2.26. To calculate the slope of this line, we pick any two points on the line, say P and Q, and find the slope of the line joining these points. We have

$$\text{Slope} = m = \frac{-4-(-1)}{-2-(-1)} = \frac{-4+1}{-2+1} = \frac{-3}{-1} = +3$$

If we had selected points R and S, the slope of the line would still be 3. In both cases we get the same value for the slope; it is 3. More generally, if we select any two points on this line, the slope will always be 3. We say that *the slope of a line is constant.* Notice that 3 is the coefficient of x in the equation $y = 3x + 2$. Next we observe that the line crosses the y axis at (0, 2). In the equation $y = 3x + 2$, the 2 is the number that stands alone. We generalize our results in the following formula.

Formula 2.1 The equation $y = mx + b$ represents a straight line whose slope is m and whose y intercept is at $(0, b)$.

The accompanying table gives several examples of slope and y intercept.

Equation	Slope	y-intercept at
$y = 2x + 5$	2	(0, 5)
$y = -3x + 4$	-3	(0, 4)
$y = 5x$	5	(0, 0)
$y = -x$	-1	(0, 0)
$y = 4$	0	(0, 4)
$y = x + 1$	1	(0, 1)
$x = 2$	Undefined	There is none

Example 23 Find the equation of a line with a slope of 3 and whose y intercept is at (0, 14).

Solution Using Formula 2.1, we find that the equation is $y = 3x + 14$.

Example 24 Find the equation of a line that has a slope of -2 and passes through the origin.

Solution Since the line passes through the origin, its y intercept is at (0, 0). Thus, using Formula 2.1, the equation is $y = -2x + 0$, or $y = -2x$.

We will be interested in finding the equation of a line under different conditions, so we list several different forms of the equation, depending on the information given.

Name	Given information		Equation
Slope-intercept form	Slope $= m$	y intercept $(0, b)$	$y = mx + b$
Point-slope form	Slope $= m$	point $= (x_1, y_1)$	$y - y_1 = m(x - x_1)$

Example 25 Find the equation of a line that has a slope of 4 and passes through the point $(-2, -3)$.

Solution Since we are given the slope of the line and a point on the line, we use the point-slope form. We have $m = 4$, $x_1 = -2$, and $y_1 = -3$, so that when we substitute these values into the equation

$$y - y_1 = m(x - x_1)$$

we get

$$\begin{aligned} y - (-3) &= 4[x - (-2)] \\ y + 3 &= 4(x + 2) \\ y &= 4x + 5 \end{aligned}$$

The equation of the line then is $y = 4x + 5$.

Example 26 Find the equation of the line that passes through the point (3, 13) and is parallel to the line whose equation is $y = 2x - 5$.

Solution The line that we are interested in is parallel to the line whose slope is 2. A point on the line is (3, 13). Now that we know the slope of the line and a point on the line, we use the point-slope form with $m = 2$, $x_1 = 3$, and $y_1 = 13$. We have

$$\begin{aligned} y - y_1 &= m(x - x_1) \\ y - 13 &= 2(x - 3) \\ y - 13 &= 2x - 6 \\ y &= 2x + 7 \end{aligned}$$

Therefore, the equation of the line is $y = 2x + 7$.

Exercises

21. Find the slope of the line joining the points.

a) (2, 1) and (7, 8) b) (5, 6) and (−3, −1)
c) (−2, −3) and (−4, −5) d) (6, −2) and (−4, −8)
e) (2, 0) and (0, −5) f) (7, 7) and (0, 0)
g) (4, 5) and (6, −7) h) (−2, −7) and (−6, −9)
i) (−3, −4) and (−7, −4) j) (4, 3) and (4, 9)

22. Find the slope and the y intercept of the line whose equation is

a) $y = 5x - 9$ b) $y = (x/2) + 7$
c) $4x - 7y = 9$ d) $3x + y = 2$
e) $y = x$ f) $y = 1$

23. Is the point (−5, −6) on the graph of $4x + 3y = -39$?

24. Find the x intercept and the y intercept of each of the following equations.

a) $7x + 5y = 35$ b) $3x + 4y = 24$
c) $y = 7x$ d) $9x - 5y - 45 = 0$
e) $7y = 21$ f) $2x = 10$

25. Find the equation of the line under the indicated conditions:

a) Has slope −4 and y intercept at (0, −9)
b) Passes through the points (−3, −4) and (7, 11)
c) Has slope −7 and passes through the point (5, −7)
d) Passes through the points (−4, −7) and (−3, −9)
e) Has slope −3/2 and passes through the origin

26. Find the equation of the line that passes through the point (−2, 1) and is parallel to the line whose equation is $y = 3x - 1$.

27. Find the equation of the line that passes through the point (−5, −5) and has the same y intercept as the line whose equation is $2y = 5x + 10$.

28. Find the equation of the line that is parallel to the line whose equation is $y = 2x + 9$ and that has the same y intercept as the line with the equation $y = 7x - 1$.

29. Are the following sets of lines parallel?

a) $y = 7x - 1$, $y = 7x + 1$
b) $y = -2x$, $7y + 14x = 3$
c) $5x = 8y + 3$, $16y = 10x - 6$

30. The vertices of a parallelogram $ABCD$ have the following coordinates:

$A(-4, 8)$ $B(4, 12)$ $C(14, 4)$ $D(x, 0)$

a) Find the slope of the line joining A to B.
b) Find the slope of the line joining D to C in terms of x.
c) Using the results of parts a and b, find the value of x.

2.6 The Graph of a Second Degree Equation[3]

Until now we have been considering equations whose graphs are straight lines. Of course, graphs need not always be straight lines. For our purposes we will draw nonlinear graphs by finding values of x and y that are solutions of the second-degree equations. (An equation that can be written in the standard form as $ax^2 + bx + c = 0$ where a, b, and c are real numbers and $a \neq 0$ is an example of a second-degree equation.)

Example 27 Draw the graph of $y = x^2$.

Solution First we find several solutions of this equation. Some solutions are (0, 0), (1, 1), (2, 4), (3, 9), and (−1, 1). We put these in a table and then plot the points. The graph is given in Figure 2.27. Such a graph is called

x	y
0	0
1	1
2	4
3	9
4	16
−1	1
−2	4
−3	9
−4	16

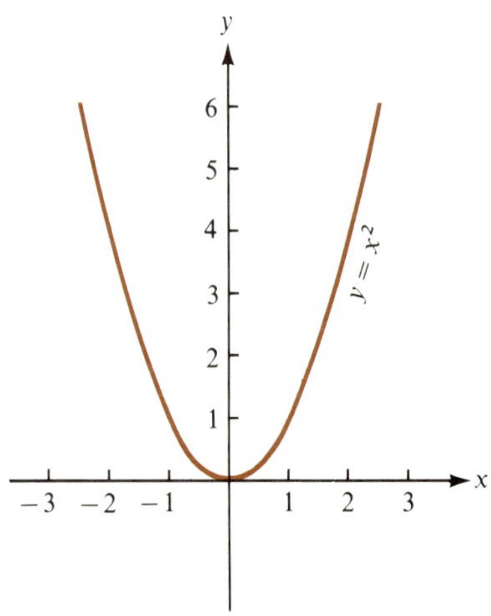

Figure 2.27 Graph of $y = x^2$.

a *parabola*. Although we need only two points to draw the graph of a straight line, the same is not true for this parabola. Why?

In this example, y is a function of x, since there is a unique value of y associated with any x.

[3] This section may be omitted without affecting the continuity of the text.

Second-degree equations and their graphs have many applications as the following examples illustrate.

Example 28 On the basis of past experience, a large milk company claims that its daily income from a certain product is fairly constant. Thus, a decrease in price produces a corresponding increase in demand. Let x represent the price (in cents) of the item and y represent the quantity (in thousands of containers) sold per day. The company's president claims that an equation expressing a relationship between x and y is $xy = 800{,}000$. Draw the graph.

Solution We set up a table of values and then draw the graph, which is given in Figure 2.28. Notice that this curve is not a straight line. Such a graph

x	y
20¢	40,000
25	32,000
30	26,666.67
35	22,857.14
40	20,000
45	17,777.78
50	16,000

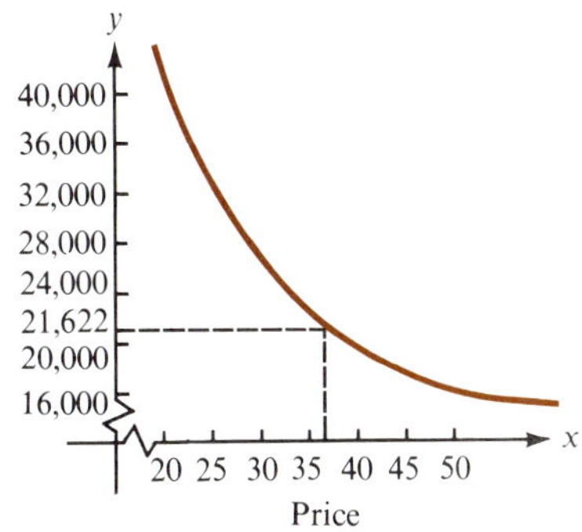

Figure 2.28 Graph of $xy = 800{,}000$.

is called a *hyperbola*. From this graph we can predict that when the price of the item is 37¢, the company will sell approximately 21,622 items.

Comment In selecting values for x in the table of values for Example 28, we avoided using negative numbers. Can you see why? Can x be 0?

Example 29 *Swimming in a polluted lake.* A resort hotel has been polluting a lake by dumping raw sewage into it. In an effort to remove the coliform bacteria that have formed, local health officials have added various chemicals. The number of dangerous bacteria, y, present x days after the chemicals have been added is given by the equation $y = 90(8 - x)^2$. The health officials will not permit anyone to swim in the lake unless the coliform bacteria count is 90 or less. Assuming no further pollution, after how many days will the lake be safe for swimming?

Solution We set up a table of values and then draw the graph, which is given in Figure 2.29. From the graph we find that when $y = 90$, $x = 7$. Thus, the lake will be safe for swimming 7 days after treatment.

x	$y = 90(8 - x)^2$	y
0	$90(8 - 0)^2$	5760
1	$90(8 - 1)^2$	4410
2	$90(8 - 2)^2$	3240
3	$90(8 - 3)^2$	2250
4	$90(8 - 4)^2$	1440
5	$90(8 - 5)^2$	810
6	$90(8 - 6)^2$	360
7	$90(8 - 7)^2$	90
8	$90(8 - 8)^2$	0

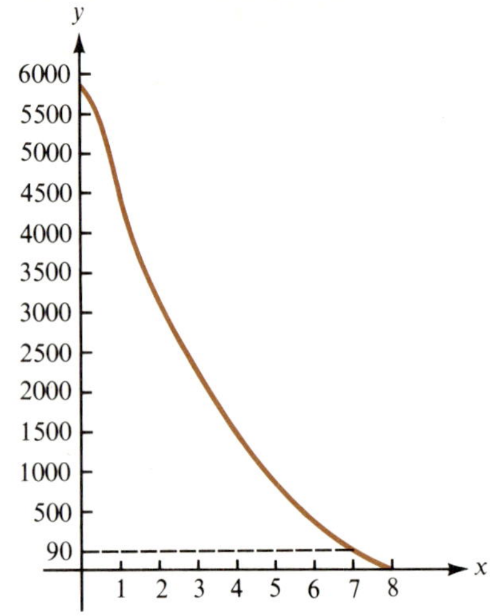

Figure 2.29

Example 30 Medical authorities at Midstate Medical Center have discovered that x days after an adult has been exposed to a certain contagious disease, the person's systolic blood pressure, y, fluctuates according to the function $f(x) = 125 + 8x - x^2$. (This equation is considered valid only when x is between 0 and 10 days, inclusive.)

a) Draw the graph of this function.
b) Assuming no complications, how many days after exposure to the disease will a person's systolic blood pressure be at a maximum? What will the person's systolic blood pressure be then?

Solution a) We set up a table of values and then draw the graph, which is given in Figure 2.30.

x	$125 + 8x - x^2$	y
0	$125 + 0 - 0$	125
1	$125 + 8 - 1$	132
2	$125 + 16 - 4$	137
3	$125 + 24 - 9$	140
4	$125 + 32 - 16$	141
5	$125 + 40 - 25$	140
6	$125 + 48 - 36$	137
7	$125 + 56 - 49$	132
8	$125 + 64 - 64$	125
9	$125 + 72 - 81$	116
10	$125 + 80 - 100$	105

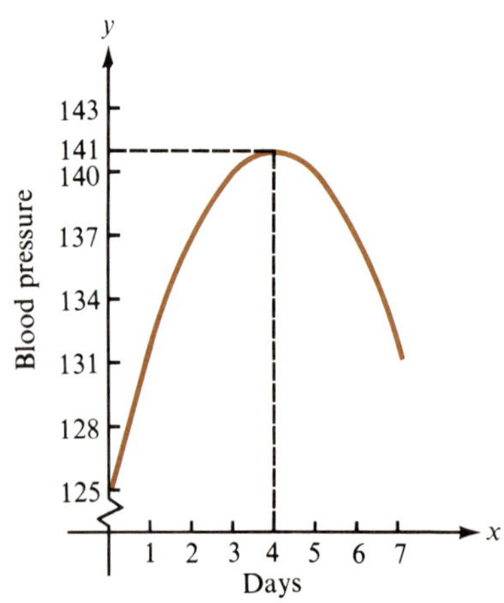

Figure 2.30

b) From the graph we find that the curve has a maximum when $x = 4$. At that point $y = 141$. Thus, the person's systolic blood pressure will be at a maximum of 141 after 4 days. Thereafter it decreases.

Draw the graph of the equations given in Exercises 31–38.

31. $y + x^2 = 4$ **32.** $x^2 + y^2 = 9$ **33.** $2x^2 + y^2 = 4$

34. $y^2 = -2x^2 + 7$ **35.** $y = x^2 + 2x + 4$ **36.** $4x^2 - 5y^2 = 20$

37. $xy = 8$ **★ 38.** $x(y - 5) = 25$

★ 39. In an effort to protect the toll collectors, a local environmental group takes hourly samples of the sulfur oxide pollutants (in parts per million) in the air at the city's toll bridges. It is discovered that the number of pollutants present in the air depends on the time of day. Let $x =$ the number of hours after midnight and $y =$ the number of pollutants in the air. Thus, $x = 1$ means 1:00 A.M., $x = 2$ means 2:00 A.M., etc. Then $y = 0.16 + 0.016x - (x^2/1000)$ expresses the relationship between x and y.

a) Draw the graph of this function.

b) At what hour is the concentration of sulfur oxide pollutants greatest?

★ 40. A farmer wishes to enclose as large an area as possible to provide space for his chickens to wander. He has 1000 meters of fencing for the job. Let $x =$ the width of the space to be enclosed and $y =$ the area of the enclosed space. Then an equation expressing the relationship between x and y is $y = x(500 - x)$.

a) Draw the graph of this function.

b) What should the dimensions of the enclosed space be?

★ 41. A large auto supply store claims that by adding x quarts of antifreeze to a typical (18-quart) cooling system of a car, it can be kept from freezing when the temperature reaches y degrees Fahrenheit. The equation expressing this relationship is $y = x^2 - 16x + 32$.

a) Draw the graph of this function.

b) What is the lowest temperature against which this solution will protect?

c) How many quarts of antifreeze must be added to achieve this protection?

2.7 Intersection of Graphs

When dealing with two or more equations, we are often interested in finding the values of x and y that the equations have in common. There are several methods we can use to find these x and y values. In this section we discuss graphical methods. In later chapters we will discuss other techniques.

When solving two or more equations for the common values, we realize that these x and y values are represented by the points where the graphs of the two equations intersect. Furthermore, for the point (x, y) to be on the graph of an equation, the x and y values of that point must satisfy the equation. Thus, when two graphs intersect, the x and y values of this *point of intersection* will satisfy both equations.

To find such common (x, y) values, we simply draw the graphs of both equations on the same set of axes. Then we find the coordinates of the point(s) where the graphs meet. Each point of intersection is said to be a *solution* to both equations.

Example 31 Find the point of intersection of the two functions.

$$y = 3x + 1$$
$$y = -x + 9$$

Solution On the same set of axes we graph both functions as shown in Figure 2.31. From the graphs we read the point of intersection. It is $x = 2$ and $y = 7$, or $(2, 7)$.

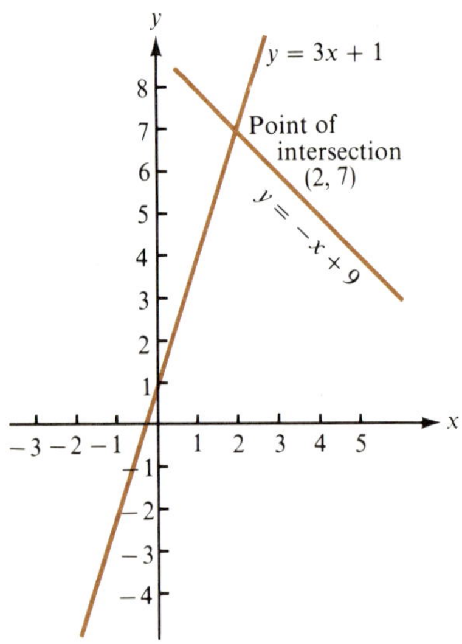

Figure 2.31

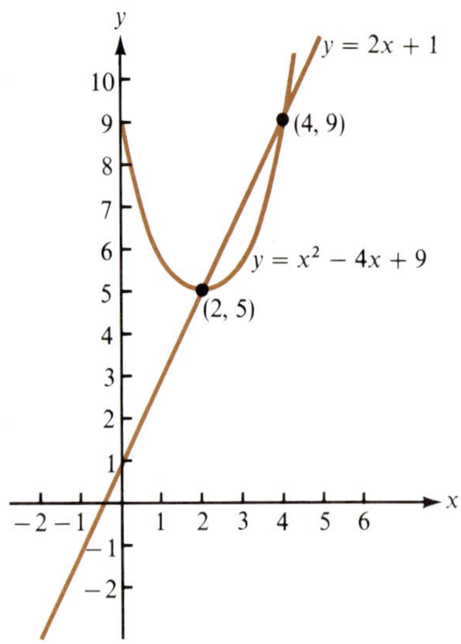

Figure 2.32

Example 32 Find the point(s) of intersection of the equations.

$$y = x^2 - 4x + 9$$
$$y = 2x + 1$$

Solution On the same set of axes we graph both equations as shown in Figure 2.32. From the graphs we read the coordinates of the points of intersection of the two graphs. These are $(2, 5)$ and $(4, 9)$.

Comment In some situations it may not be possible to find *any* points of intersection for two functions. This is true, for example, when the two functions represent parallel lines.

Example 33 Find the point of intersection of

$$y = 5x + 1$$
$$y = 5x + 3$$

Solution The graphs of these two functions are given in Figure 2.33. The slope of both lines is 5 and the lines are therefore parallel. There is no point of intersection.

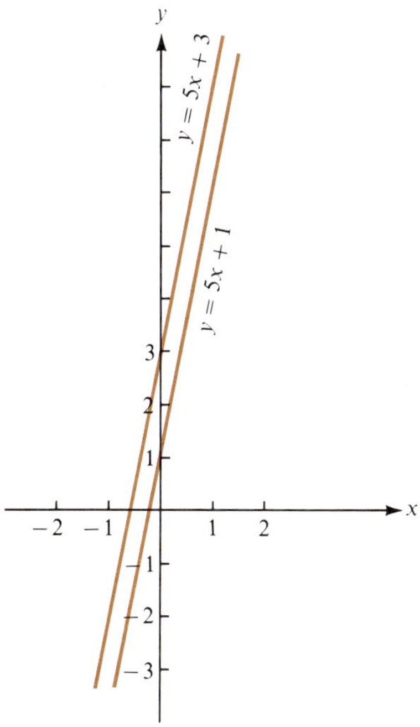

Figure 2.33

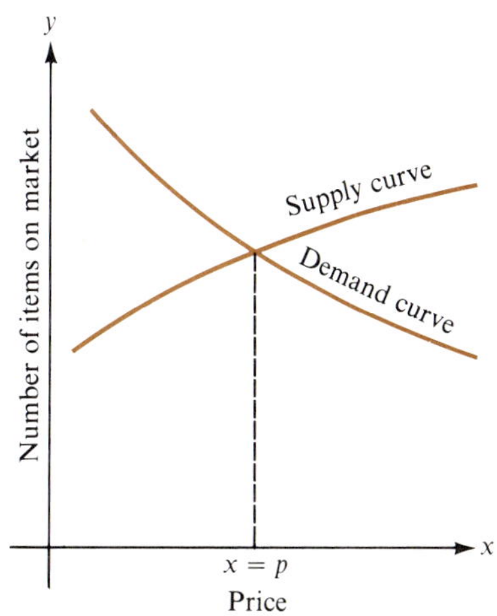

Figure 2.34

Often in economics, when discussing the *laws of supply and demand*, we are interested in determining the market price of an item. When there is pure competition, the price of an item is affected by the supply and demand for that item. If the supply is too great, the market will be overstocked with the item. This in turn will lead to a decrease in price. Similarly, if the demand is too great, a shortage of the item will result, which in turn will lead to an increase in price.

In Figure 2.34 the supply curve and the demand curve intersect when $x = p$. This point of intersection is called the *market equilibrium price*. We conclude that the market price for the item is $x = p$.

Although our discussion is an oversimplification of the many factors that have to be considered, it does indicate how we apply the idea of the intersection of graphs. In later chapters we will discuss the ideas of supply and demand in greater detail.

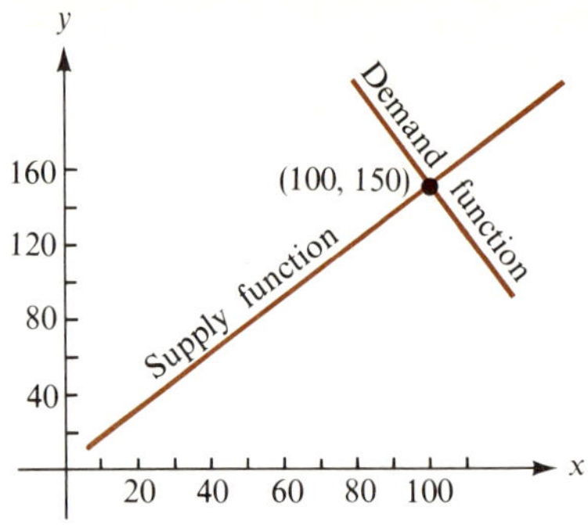

Figure 2.35

Example 34 *Supply and demand.* The supply function for a certain deodorant is given by $\frac{3}{2}x$ and the demand function is given by $-\frac{5}{2}x + 400$, where x is the price (in cents) of the deodorant.

a) Find the demand when the deodorant sells at 90¢ each.
b) Find the equilibrium price.

Solution a) The demand function for the deodorant is given by $-\frac{5}{2}x + 400$. When the deodorant sells at 90¢, then $x = 90$, so we replace x in the demand function by 90. We get

$$-\frac{5}{2}(90) + 400 \quad \text{or} \quad 175$$

Thus, 175 units will be demanded when the deodorant sells at 90¢ each.

b) To find the equilibrium price, we draw the supply function and the demand function on the same set of axes. This is shown in Figure 2.35. From the graphs we note that the point of intersection is at (100, 150). Thus, $x = 100$ is the equilibrium price. At this price the supply is $\frac{3}{2}x = \frac{3}{2}(100) = 150$. This supply is called the *equilibrium supply*. Similarly, when $x = 100$, the demand is $-\frac{5}{2}x + 400 = -\frac{5}{2}(100) + 400 = 150$. This is called the *equilibrium demand.*

Exercises

Find the point(s) of intersection (if any) for each pair of equations in Exercises 42–52.

42. $y - x = 3$
$2x + y = 6$

43. $x + 2y = 15$
$y = 2x$

44. $y - x = -2$
$x - 2y = 4$

45. $y + 2x + 6 = 0$
$y = 2x$

46. $2x + 3y = 5$
$3x + 2y = 10$

47. $4x - 3y = 7$
$8x - 6y = 7$

48. $y + 3 = 2x$
$3x^2 = 4 + 2y$

49. $y = x^2 + 3$
$y = -x + 5$

50. $x^2 + y^2 = 16$
$y = 3$

51. $x^2 + 5y^2 = 60$
$-5y^2 + x^2 = 12$

52. $x^2 - y^2 = 25$
$y = 5$

53. The supply and demand functions for a certain type of toy are given by $\frac{5}{3}x$ and $-\frac{4}{3}x + 600$, respectively.
a) Find the demand when each toy sells for 90¢.
b) Find the equilibrium price.
c) Find the equilibrium supply.

★ **54.** a) Find the equilibrium price for a certain type of tool if the supply function for the tool is given by $x^2 + 6x + 10$ and the demand function is given by $x^2 - 4x + 25$.
b) Find the number of units supplied and demanded at this price.

★ **55.** The supply and demand functions for a certain type of hairspray are given by $\frac{9}{5}x$ and $-\frac{1}{5}x + 200$, respectively.
a) Find the supply and demand when the hairspray is given away for free. (*Hint:* When the hairspray is given away for free, then $x = 0$.)
b) Find the equilibrium price, equilibrium supply, and equilibrium demand.

2.8 Applications of Linear Functions

One of the greatest difficulties that has to be overcome when working with linear functions is to arrive at an equation. There is no simple way to read a verbal problem (one expressed in words) and automatically decide what the appropriate equation is. Careful thought is required before determining the correct approach. In each case the problem must be analyzed to determine what information is given and what information is being sought. Once this is done, it will prove helpful to use symbols to represent the unknown quantities and to write an equation that can be used to solve the problem.

In the next few examples we indicate how linear equations can be solved by means of graphs.

Example 35 *Converting inches to centimeters.* A small child has a 12-inch ruler. On one side it is calibrated in inches, and on the other side it is calibrated in centimeters and goes to approximately 30.5 centimeters. Thus, using the ruler, 0 inches equals 0 centimeters and 12 inches equals approximately 30.5 centimeters.
a) Locate the given information on a rectangular coordinate system and then use the graph to find the approximate number of centimeters in 15 inches.
b) Derive an equation for converting inches to centimeters.

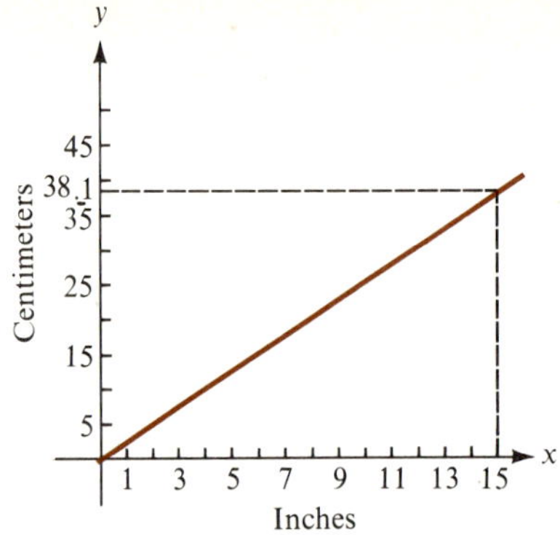

Figure 2.36

Solution a) To solve the problem graphically, we let x represent the measurement in inches and let y be the measurement in centimeters. From the given information, when $x = 0, y = 0$ and when $x = 12, y = 30.5$. We first locate these points on a rectangular coordinate system as shown in Figure 2.36. Based on this graph, we can conclude that there are approximately 38.1 centimeters in 15 inches.

b) To derive a linear equation for converting inches to centimeters, we notice that the slope of the line joining the points (0, 0) and (12, 30.5) is

$$\frac{30.5 - 0}{12 - 0} = 2.54$$

Furthermore, since the line goes through the origin, the y intercept is 0. Using the slope-intercept form for the equation of a line (page 64), we have

$$\begin{aligned} y &= mx + b \\ &= 2.54x + 0 \end{aligned}$$

or

$$y = 2.54x$$

Thus, the equation for converting inches to centimeters is $y = 2.54x$, where x is the number of inches and y is the number of centimeters. Can you derive an equation for converting centimeters to inches?

Example 36 *Real estate taxes.* The real estate tax for 1978 in New York City was \$8.60 per \$100 of assessed value.

a) Draw a graph that would enable a homeowner to determine his or her real estate taxes.
b) What is the real estate tax on a house that is assessed at \$8000?

Solution a) Let x be the assessed value of the house and let y be the real estate taxes. Based on the given information, when $x = 0$, $y = 0$ and when $x = 100$, $y = 8.6$. We locate these points on a rectangular coordinate system as shown in Figure 2.37 and then draw a straight line connecting them. The resulting line is the tax graph.

b) From the tax graph, the real estate tax on a house that is assessed at \$8000 is \$688.

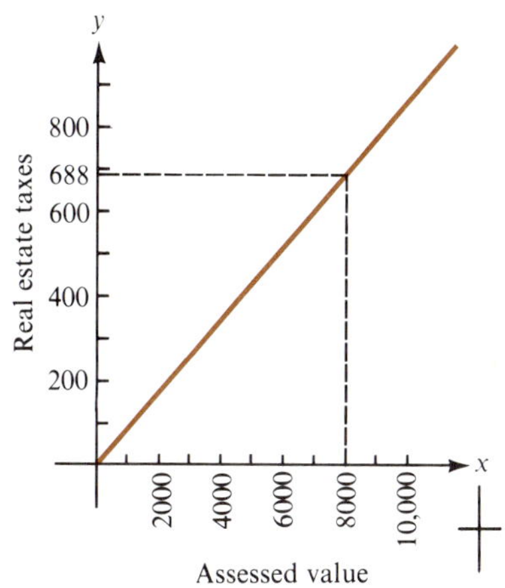

Figure 2.37

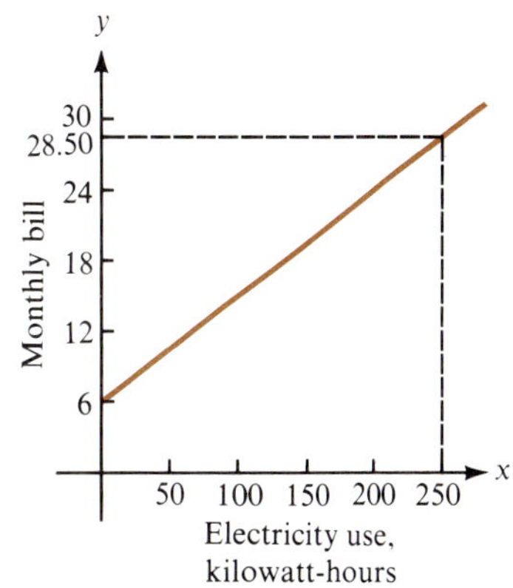

Figure 2.38

Example 37 *Electric bill.* The Thomas Edison Electric Company charges its customers a flat fee of \$6 per month plus an additional 9¢ for each kilowatt-hour of electricity used.

a) Find an equation for determining the monthly electric bill, y, when x kilowatt-hours of electricity are used.
b) Draw the graph of this equation.
c) Using the graph, estimate the monthly electric bill when 250 kilowatt-hours of electricity are used.

Solution a) According to the company rates, each customer pays 9¢ for every kilowatt-hour of electricity used. Thus, when x kilowatt-hours of electricity are used, the cost is 0.09 times x, or $0.09x$. Since there is a flat fee of \$6 per month regardless of the amount of electricity used, the total monthly bill is $y = 6 + 0.09x$.

b) The graph of this equation is given in Figure 2.38.

c) From the graph, when 250 kilowatt-hours of electricity are used, the estimated monthly electric bill is \$28.50.

The graphical solution of two simultaneous linear equations has important applications in business. We will discuss two such applications—the break-even point and straight-line depreciation.

Often the cost, C, of producing x items can be expressed as a linear function. Similarly, the revenue, R, that is obtained when x items are produced can be expressed as a linear function. Obviously, when the revenue, R, exceeds the cost, C, then the business is profitable. On the other hand, when the cost, C, exceeds the revenue, R, then the business is operating at a loss. If the revenue, R, and the cost, C, are the same, then there is no profit or loss. The business is said to be operating on a *break-even* basis.

Example 38 *Break-even point.* A certain manufacturer sells toys at \$3 apiece. It costs the manufacturer \$1.50 to produce each toy. If the manufacturer has overhead expenses (for such items as maintenance, mortgage costs, and salaries) of \$200 per day, how many toys must be sold daily to break even?

Solution If x toys are sold by the manufacturer at \$3 apiece, then the revenue derived is $R = 3x$. Since it costs the manufacturer \$1.50 to produce each toy, the production cost for x toys is $1.5x$. To this we must add the fixed overhead expenses of \$200. Thus the total daily cost of producing x toys is $C = 1.5x + 200$. In order to break even, the revenue must equal the cost. Thus, we draw the revenue function and the cost function on the same coordinate axes. This is shown in Figure 2.39. We notice that the lines meet when x is approximately 133.33.

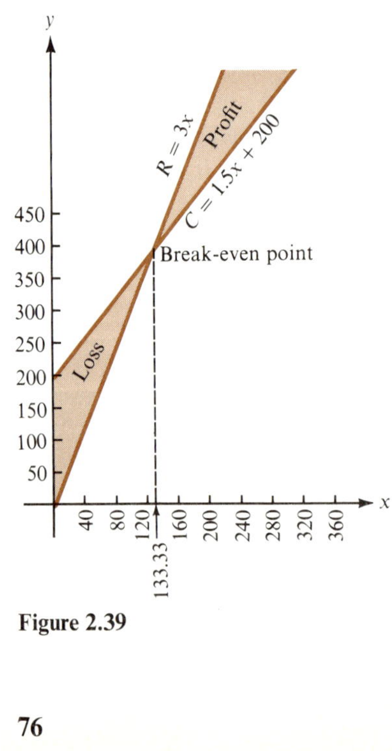

Figure 2.39

If fewer than 133.33 toys are sold per day, the manufacturer loses money. If more than 133.33 toys are sold per day, the manufacturer makes money. If exactly 133.33 toys are sold, the manufacturer breaks even. This is clearly illustrated in the diagram.

Since most businesses want to make a profit, they are interested not only in the break-even point but also in how many items must be manufactured (sold) so that the profit is realized. For this reason we will discuss inequalities in a later chapter.

When a manufacturer buys a machine or some piece of equipment, it will not last forever. After a while the equipment either breaks or becomes obsolete. This loss in value over a period of time is called *depreciation.* Accountants use several methods for computing depreciation. One such technique is known as the *straight-line method.* With this method of depreciation we first determine the useful life of the equipment. Thus, if a piece of equipment will last n years, then each year it will depreciate $1/n$ in value. For example, if a particular lift truck has a useful life of 5 years, then it would lose 1/5 of its value each year. This assumes that at the end of 5 years the lift truck will be worthless. Often this is not the case, since a piece of equipment may have a *scrap value* at the end of its useful life. In this instance depreciation is calculated on the difference between the purchase price and the scrap value. This represents the *net cost.*

Formula 2.2 *Annual straight-line depreciation.* If an item costs C dollars and has a scrap value of S dollars after n years, then the *annual allowable depreciation* is

$$\frac{C - S}{n}$$

Formula 2.3 *Accumulated depreciation.* If an item has an annual depreciation of $(C - S)/n$, then the amount of depreciation accumulated after x years is

$$\frac{C - S}{n} \cdot x$$

Definition 2.6 The *undepreciated balance* is the difference between the accumulated depreciation and the net cost.

Example 39 To reduce air pollution, a utility company installs a special antipollution device that has a useful life of 9 years. The purchase price is \$75,000 and the estimated scrap value is \$3000.

a) Find the annual allowable depreciati n.
b) Find the accumulated depreciation aft r 4 years.
c) Find the undepreciated balance after 4 years.

Solution The cost, C, of the antipollution device is \$75,000 and the estimated scrap value, S, is \$3,000. Thus, the net cost of the device is $C - S =$ \$75,000 – \$3,000, or \$72,000. The useful life is 9 years, so $n = 9$.

a) Applying Formula 2.2., the annual allowable depreciation is

$$\frac{C - S}{n} = \frac{75{,}000 - 3{,}000}{9} = \frac{72{,}000}{9} = \$8{,}000$$

The annual allowable depreciation is \$8000.

b) After 4 years the accumulated depreciation is found by Formula 2.3:

$$\frac{75{,}000 - 3{,}000}{9} \cdot 4 = \$32{,}000$$

c) The net cost is \$72,000 and the accumulated depreciation is is \$32,000. Thus, the undepreciated balance is

\$72,000 – \$32,000 or \$40,000

Exercises

---

56. *Tow trucks.* The Schlep Tow Truck Company charges the following rates:

\$10 for the first mile or less

\$1 for each additional mile or fraction thereof

a) Find an equation for the charge, y, of towing a car x miles (where $x > 1$).
b) What is the charge for towing a car 22 miles?
c) Graph the equation found in part a.

57. *Pollution.* Tests performed by environmentalists on a lake in 1974 indicated that there were 20 milligrams of pollutants per 5000 liters of water. In 1978 the level of pollutants had risen to 45 milligrams per 500 liters of water. Assuming that the level of pollutants is rising at a constant rate (that is, no antipollution policies are started), find the predicted level of pollution in 1986.

58. *Killer bees.* A researcher is experimenting with killer bees. The number of bees present in a specially controlled laboratory environment is given by the formula $N(t) = 2000 + 400t$, where t is measured in months and $t = 0$ corresponds to January 1978.

a) Find $N(0)$.
b) Find $N(5)$.

c) Determine the total number of bees present in the laboratory on January 1, 1979.

d) Assuming that this trend continues, how many killer bees will the researcher have by December 31, 1980?

★ **59.** A fire department official believes that as the temperature decreases, the number of fires increases. In support of this claim, the statistics in the accompanying table have been collected:

Outside temperature, x	Number of fires reported, y
40°	28
35	32
30	35
25	39
20	51
15	55
10	56
5	66

a) Graph this information on a rectangular coordinate system.

b) Draw a line that passes through two of these points.

c) Find an equation for predicting the number of fires, y, when given the outside temperature, x.

d) If the outside temperature drops to $-5°$, what is the predicted number of fires reported?

★ **60.** *Investing your money.* A man has determined that he can live comfortably on an annual income of $8,000. He has $100,000 that he can invest in municipal bonds that pay 8% annual interest, or he can put the money in a savings certificate of a bank that pays $7\frac{1}{2}\%$ interest a year. How much should he invest at each rate so that he earns $8,000 a year in income? (Solve by using graphical techniques.)

61. A nursing home installs a special elevator that has an estimated useful life of 20 years. The elevator costs $50,000 and has no scrap value.

a) Find the annual allowable depreciation.

b) Find the accumulated depreciation after eight years.

c) Find the undepreciated balance after eight years.

62. A service station purchased an engine compression diagnostic machine that has a useful life of seven years. The machine costs $25,000 and has a salvage value of $4,000.

a) Find the annual allowable depreciation.

b) Find the accumulated depreciation after five years.

c) Find the undepreciated balance after eight years.

63. The Teldex Corporation manufactures antique telephones. It costs the company \$14 to manufacture a telephone, which it in turn sells for \$42. The company has daily overhead expenses of \$448. How many telephones must be sold daily to break even? Graph your results.

64. A manufacturer produces Hula-Hoops at a cost of 10¢ apiece and sells them for 25¢ apiece. The weekly overhead expenses are \$210. What is the break-even point? Graph your results.

2.9 Algebra Review: Exponents and Radicals

In later chapters we will be working with expressions involving exponents and radicals. Thus, let us pause now to review some rules of exponents and radicals from algebra.

Recall that b^m means

$$\underbrace{b \cdot b \cdot b \cdot \cdots \cdot b}_{m \text{ of them}}$$

where m is a positive integer. Thus, b^3 means $b \cdot b \cdot b$, and 2^6 means $2 \cdot 2 \cdot 2 \cdot 2 \cdot 2 \cdot 2 = 64$. Accordingly,

$$2^3 \cdot 2^4 = \underbrace{2 \cdot 2 \cdot 2}_{3 \text{ of them}} \cdot \underbrace{2 \cdot 2 \cdot 2 \cdot 2}_{4 \text{ of them}}$$

which can be written as 2^7. Thus,

$$b^m \cdot b^n = b^{m+n}$$

where m and n are positive integers. (We simply add the exponents.) Also, by definition,

$$b^0 = 1 \qquad \text{if } b \neq 0$$

Thus,

$$2^0 = 1$$
$$(1000)^0 = 1$$

Similarly,

$$(b^m)^n = b^{mn}$$

where m and n are positive integers. For example,

$$(3^2)^4 = 3^{2 \cdot 4} = 3^8$$

Alternately,

$$(3^2)^4 = 3^2 \cdot 3^2 \cdot 3^2 \cdot 3^2 = 3^{2+2+2+2} = 3^8$$

In either case our answer is 3^8.

What is the product of $b^m \cdot b^{-m}$? Using the previous rule, we have

$$b^m \cdot b^{-m} = b^{m-m} = b^0 = 1$$

Therefore,

$$b^{-m} = \frac{1}{b^m}$$

where m is any positive integer, and $b \neq 0$. Thus,

$$\frac{b^m}{b^n} = b^{m-n}$$

where m and n are positive integers, and $b \neq 0$. For example,

$$\frac{7^5}{7^3} = 7^{5-3} = 7^2$$

In all the preceding examples, the exponents were integers. Of course, this need not be the case. A number can have a rational number as an exponent also. Thus, $x^{1/2}$ is defined as $\sqrt{x}$, and $x^{1/3}$ is defined as $\sqrt[3]{x}$. Similarly, $x^{1/4}$ is defined as $\sqrt[4]{x}$. More generally, $x^{1/n} = \sqrt[n]{x}$, where $n \neq 0$.

Applying the definition of $x^{1/n}$ gives

$$\sqrt[n]{x^m} = (x^m)^{1/n} = x^{m/n}$$

Also,

$$(\sqrt[n]{x})^m = (x^{1/n})^m = x^{m/n}$$

Therefore,

$$x^{m/n} = \sqrt[n]{x^m} \quad \text{or} \quad x^{m/n} = (\sqrt[n]{x})^m \qquad (n \neq 0)$$

For example, $27^{2/3}$ means

$$(\sqrt[3]{27})^2 = 3^2 = 9$$

and $32^{2/5}$ means

$$(\sqrt[5]{32})^2 = 2^2 = 4$$

So far we have discussed b^x when x is any integer (positive, negative, or zero) or when it is any rational number. How do we evaluate b^x when x is an irrational number? In other words, what does $3^{\sqrt{2}}$ or 7^{π} mean? Can we multiply 3 by itself $\sqrt{2}$ times? Can we multiply 7 by itself π times?

It turns out that any number with an irrational exponent, such as $3^{\sqrt{2}}$, can be defined by a limiting process. We can think of $3^{\sqrt{2}}$ as the limit of the sequence

$$3^1, 3^{1.4}, 3^{1.41}, 3^{1.414}, 3^{1.4142}, \ldots$$

where 1.4142 . . . is the decimal approximation of $\sqrt{2}$. Similarly, we can think of 7^π as the limit of the sequence

$$7^3, 7^{3.1}, 7^{3.14}, 7^{3.141}, 7^{3.1415}, \ldots$$

where 3.14159 . . . is the decimal approximation of π.

We can summarize our discussion of exponents in the following rules.

	Direct notation	Exponential notation	Name
Rule 1	$b^x = \underbrace{b \cdot b \cdot b \cdot \cdots \cdot b}_{x \text{ of them}}$ (where x is a positive integer)	$f(x) = b^x$	Definition of exponents
Rule 2	$b^x \cdot b^y = b^{x+y}$	$f(x) \cdot f(y) = f(x + y)$	Product rule
Rule 3	$(b^x)^y = b^{xy}$	$[f(x)]^y = f(xy)$	Power of a product
Rule 4	$b^{-x} = \dfrac{1}{b^x}$	$f(-x) = \dfrac{1}{f(x)}$	Negative exponents
Rule 5	$b^0 = 1$	$f(0) = 1$	Zero exponents
Rule 6	$b^{x/y} = (\sqrt[y]{b})^x = \sqrt[y]{b^x}$	$f\left(\dfrac{x}{y}\right) = \sqrt[y]{f(x)}$	Fractional exponents
Rule 7	$\dfrac{b^x}{b^y} = b^{x-y}$	$\dfrac{f(x)}{f(y)} = f(x - y)$	Quotient rule

Example 40 Rewrite the following expressions without negative exponents.

a) 4^{-3} b) 10^{-5} c) $(\frac{2}{3})^{-2}$
d) x^{-2} e) x^{-1}

Solution a) $4^{-3} = \dfrac{1}{4^3} = \dfrac{1}{4 \cdot 4 \cdot 4} = \dfrac{1}{64}$

b) $10^{-5} = \dfrac{1}{10^5} = \dfrac{1}{10 \cdot 10 \cdot 10 \cdot 10 \cdot 10} = \dfrac{1}{100{,}000} = 0.00001$

c) $(\frac{2}{3})^{-2} = \dfrac{1}{(\frac{2}{3})^2} = \dfrac{1}{(\frac{2}{3})(\frac{2}{3})} = \dfrac{1}{(\frac{4}{9})}$

$$= 1 \div \frac{4}{9} = 1 \cdot \frac{9}{4} = \frac{9}{4}$$

d) $x^{-2} = \dfrac{1}{x^2}$

e) $x^{-1} = \dfrac{1}{x^{+1}} = \dfrac{1}{x}$

Example 41 Solve for x in each of the following equations.

a) $2^x = 8$ b) $5^{-x} = \frac{1}{125}$ c) $4^{x+1} = 64$

Solution We rewrite each of the exponential equations so that both sides of the equation have the same base.

a) $2^x = 8$. We rewrite this as

$$2^x = 2^3 \quad \text{since} \quad 2^3 = 8$$

If the bases are the same, then the exponents must be the same. Therefore, we conclude that $x = 3$.

b) $5^{-x} = \frac{1}{125}$. We rewrite this as

$$5^{-x} = \frac{1}{5^3} = 5^{-3}$$

Therefore, $-x = -3$ and $x = 3$.

c) $4^{x+1} = 64$. We rewrite this as

$$4^{x+1} = 4^3$$

Therefore, $x + 1 = 3$, so $x = 2$.

Exercises

65. Evaluate each of the following expressions.

a) $16^{3/2}$ b) $8^{5/3}$ c) 10^0
d) $4^{-1/2}$ e) $(\frac{1}{2})^{-3}$ f) $(2^3)(2^2)(2^5)$
g) $49^{-3/2}$ h) $(4^{3/2})(4)^{7/2}$ i) $16^4/16^3$
j) $(5^2)^3$ k) $2 \cdot 3^2$ l) $(2^2 \cdot 3^3)^3$

66. Solve each of the following exponential equations for x.

a) $2^x = 32$ b) $4^x = 256$ c) $5^{2x} = 25$
d) $4^{3x+1} = 64$ e) $3^{-1} = 81^x$ f) $6^x = 36^{x+5}$
g) $3^{4-2x} = 27^{2x}$ h) $4^{-x} = \frac{1}{16}$

67. Rewrite without negative exponents.

a) 4^{-2} b) $16^{-1/4}$ c) 10^{-1}
d) $(\frac{4}{7})^{-1}$ e) $(\frac{2}{5})^{-3}$ f) $(\frac{5}{7})^{-2/3}$
g) $x^{-1}y^{-2}$ h) $2x^{-2}$ i) $4x^{-1/3}$
j) $x^{-1/2}y^{-1/3}$ k) $2x^{-1/2}y^{-3/7}$ l) $\dfrac{25}{x^{-3}y^{-4}}$

2.10 Summary

In this chapter we discussed the idea of a function and the notation we use for it. When studying a function, we are often interested in its graph. Thus, we introduced the rectangular coordinate system and the procedure used for graphing functions, both linear and nonlinear.

When we draw the graph of a linear function, the intercepts and slope of the line are quite helpful. The intercepts and slope can be read directly from the equation of the line. Different versions of the equation of a line were discussed and we indicated when each one is used.

We applied the ideas of the graph of a function to solve a variety of real-life problems. These problems were chosen from business and the life and social sciences. Specifically, we discussed depreciation and break-even analysis, which are important topics in economics.

Finally we discussed how to find the common solution for two equations (if there is one). We simply draw both graphs on the same set of axes and read off the coordinates of the point(s) of intersection. In later chapters we will discuss algebraic techniques for solving two simultaneous equations.

Study Guide

You should be able to demonstrate your knowledge of the following ideas presented in this chapter by giving definitions, descriptions, or specific examples. Page references are given for each term so that you can check your answer.

Function (p. 42)
Absolute value (p. 44)
Range of a function (p. 44)
Origin (p. 50)
Negative direction (p. 50)
x and y axes (p. 50)
Ordinate (p. 50)
Rectangular coordinate system (p. 51)
Linear equation (p. 57)
Slope (p. 61)
Hyperbola (p. 67)
Solution (p. 70)
Market equilibrium price (p. 71)
Equilibrium demand (p. 72)
Depreciation (p. 77)
Scrap value (p. 77)
Annual allowable depreciation (p. 77)
Variable (p. 42)
Domain of a function (p. 44)
Number line (p. 49)
Positive direction (p. 50)
Coordinate (p. 50)
Abscissa (p. 50)
Quadrant (p. 51)
Graph of an equation (p. 55)
x and y intercepts (p. 58)
Parabola (p. 66)
Point of intersection (p. 70)
Laws of supply and demand (p. 71)
Equilibrium supply (p. 72)
Break-even point (p. 76)
Straight-line method (p. 77)
Net cost (p. 77)
Accumulated depreciation (p. 77)
Undepreciated balance (p. 77)

Formulas to Remember

The following list summarizes all the formulas given in this chapter.

1. Slope of the line joining the points (x_1, y_1) and (x_2, y_2):

$$m = \frac{y_2 - y_1}{x_2 - x_1} \qquad (x_2 \neq x_1)$$

2. Slope-intercept form:

$$y = mx + b$$

where m = slope and the y intercept is at $(0, b)$.

3. Point-slope form:

$$y - y_1 = m(x - x_1)$$

4. Annual allowable depreciation:

$$\frac{C - S}{n}$$

5. Accumulated depreciation after x years:

$$\frac{C - S}{n} \cdot x$$

6. Undepreciated balance:

Net cost − accumulated depreciation

Mastery Tests

FORM A

1. Which of the following lines has zero slope?

   a) $x = 4$ b) $y = 3$

   c) $2x + 3y = 7$ d) $x - y = 0$

2. If (7, 9), (5, 8), (6, 4), and $(x, 3)$ represent a function, find the integers that x may *not* represent. Explain your answer.

3. Consider the situation pictured in Figure 2.40. What is the slope of the train tracks? Can you tell?

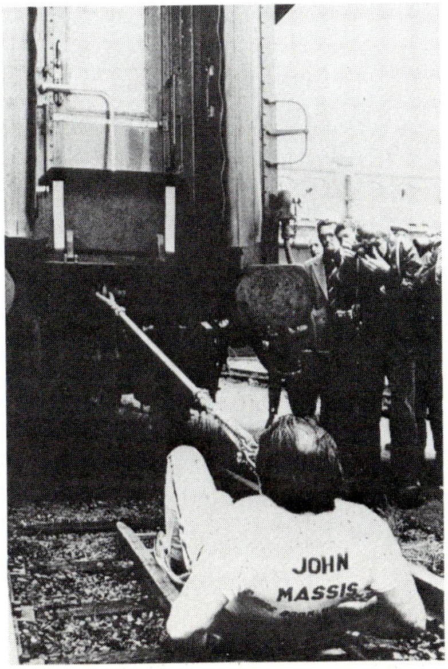

Figure 2.40 (*Source: Wide World Photos*)

4. Consider the relationship between x and y shown in the following table:

x	1	7	6	1
y	4	3	10	5

Is y a function of x? Explain your answer.

5. The graph of $Ax + By = 10$ passes through the point (0, 2). Find the value of B.

6. If the graph of $7x - 3y = k$ passes through the point $(-3, -1)$, find the value of k.

7. A scientist is performing experiments with certain bacteria. Four hours after starting the experiment, there are 70 bacteria present. Two hours later there are 105 bacteria present. Assuming a linear relationship, how many bacteria will there be after ten hours?

8. A line is drawn connecting a point in the first quadrant with a point in the third quadrant. The line has:
   a) Positive slope
   b) Negative slope
   c) Zero slope
   d) Undefined slope

9. When drawn on the same set of axes, in how many points do the graphs of these equations intersect?

   $x^2 + 4y^2 = 64$

   $4y^2 - x^2 = 16$

   a) One
   b) Two
   c) Three
   d) Four
   e) No points

10. A waitress receives a weekly salary of \$125 plus 65% of the tips she makes. If x represents her tips and y her weekly salary, write an equation for finding her weekly salary.

11. The following instructions appear in a mail-order catalog for determining the total cost for ordering merchandise.

Line 10	Cost of items ordered	\$________
Line 11	Handling charge (2% of amount on line 10)	________
Line 12	Sales tax (5% of amount on line 10)	________
Line 13	C.O.D. charges	
	\$1.00 if cost of items is less than \$6	________
	\$1.50 if cost of items is more than \$6	________
	\$0 if payment is enclosed	________
Line 14	Total charges (sum of lines 10, 11, 12, and 13)	\$________

   a) Is the total charge a function of the cost of the items ordered?
   b) If Mary orders \$8.35 worth of merchandise, find the total charge.
   c) If Mary intends to enclose a check for the total amount due, what are the total charges?

1. Water freezes at 0°C or 32°F and boils at 100°C or 212°F.
   a) Using degrees Celsius on the x axis and degrees Fahrenheit on the y axis, find an equation relating a Celsius and a Fahrenheit temperature.
   b) In an effort to conserve energy, government officials recommend that thermostats be set at 68°F during the winter. What is the equivalent reading on the Celsius scale?

2. *Depreciation.* The Handy Tool Company recently purchased a special riveting machine for $25,000. The company expects the machine to last ten years, after which the scrap value will be $5,000.
   a) Find the yearly allowable depreciation.
   b) Find the undepreciated balance after six years.

■★ 3. *Sum of the years' digits method of depreciation.* Many assets, particularly new cars and trucks, lose more of their value in the early years than at the end. For such assets there are other methods of depreciation that are used. One of these is called the *sum of the years' digits method.* To use this method, we first add all the digits less than or equal to the useful life. For example, suppose a company buys a typewriter for $800, which it estimates to have a useful life of seven years. In this case the sum of the digits is $1 + 2 + 3 + 4 + 5 + 6 + 7 = 28$. Then each year's depreciation is a fraction of the original cost. The denominator of the fraction is the sum 28, and the numerator is the remaining years in the useful life. The original cost of the item is multiplied by this fraction each year to obtain the allowable depreciation.

In our case the cost of the typewriter is $800. Assume that the scrap value is zero. We then have the allowable depreciation as shown in the accompanying table.

Year			Allowable depreciation
1	$\frac{7}{28} \cdot 800$	=	$200.00
2	$\frac{6}{28} \cdot 800$	=	171.43
3	$\frac{5}{28} \cdot 800$	=	142.86
4	$\frac{4}{28} \cdot 800$	=	114.29
5	$\frac{3}{28} \cdot 800$	=	85.71
6	$\frac{2}{28} \cdot 800$	=	57.14
7	$\frac{1}{28} \cdot 800$	=	28.57
			$800.00

   a) Use the sum of the years' digits method to find the depreciation for each of the eight years of the life of a piece of furniture that costs $640. (Assume zero scrap value.)
   b) What would the annual allowable depreciation be if the straight-line depreciation method had been used?

■ ★ 4. An insurance company purchases a photocopying machine for $30,000. The company expects the machine to last five years, after which the scrap value will be $5,000. Find the yearly allowable depreciation by the straight-line method and by the sum of the years' digits method.

■ ★ 5. *200% declining balance method of depreciation.* Another method used for depreciating an item that loses more value at the beginning of its useful life than at the end is the 200% declining balance method. In applying this method we first compute the percentage of the original price that would be depreciable yearly under the straight-line method. This figure is simply the fraction $1/n$, where n is the useful life in years. For example, if an item has a useful life of 5 years, then this figure is $1/5 = 0.20$, or 20%. We then double this figure, getting 40%. Starting with the original price in year 1, this percentage (40%) is deducted from the remaining balance in each successive year. To illustrate the technique, suppose the item costs $1000 and has zero scrap value. Then the annual allowable depreciation is as shown in the table. Using the 200% declining balance method, calculate the annual allowable depreciation on a new piece of equipment that costs $2000 and has a useful life of eight years. (Assume zero scrap value.)

Year	Yearly depreciation	Remaining undepreciated balance
1	40% of $1000 = $400 original cost	$1000 − $400 = $600
2	40% of undepreciated balance = 40% of $600 = $240	$600 − $240 = $360
3	40% of $360 = $144	$360 − $144 = $216
4	40% of $216 = $86.40	$216 − $86.40 = $129.60
5	40% of $129.60 = $51.84	$129.60 − $51.84 = $77.76

■ ★ 6. A rug company purchases a special forklift for $6300. The useful life is estimated to be five years, after which the gears wear out and the machine has to be replaced. The scrap value is $400.

a) Calculate the annual allowable depreciation using the straight-line method, the 200% declining balance method, and the sum of the years' digits method.

b) Which is the best depreciation method?

7. It costs the Scientific Precision Company $5 to produce each calculator, which it sells for $15. The daily overhead costs are $750. Find the break-even point.

8. Mrs. Johnsbury has just retired. She wishes to invest her accumulated $80,000 either in municipal bonds that pay 9% or in time deposit certificates paying 7% issued by a savings bank. She estimates that she will need exactly $6200 a year of dividend income in order to live comfortably. Using graphical techniques, find how much money she should invest in each.

9. The demand function for a certain item is given by $90 - \frac{5}{3}x$ and the supply function is given by $\frac{1}{3}x + 60$, where x is the number of items produced. Find the equilibrium price.

★10. Estimate the equation of the line through the origin that "best fits" the following points: (2, 7), (3, 8), (4, 12), and (6, 18).

★11. The board of directors of a company is reviewing the sales record of the company since Mary Southerman took over as president. It has the figures as in the accompanying table. *Note:* Year 0 corresponds to the sales figure for the year prior to the one in which Mary took over. Year 1 corresponds to the sales figure for year 1 with Mary as president.

Year	Yearly sales (thousands)
0	$75
1	79
2	80
3	84
4	89
5	90

a) Graph the data, with the years plotted on the x axis and yearly sales on the y axis.
b) Join the points (3, 84) and (5, 90) with a straight line.
c) Find the equation of the line.
d) Using the equation obtained in part c, estimate the predicted yearly sales in year 6.

12. a) Use the formula obtained in part c of question 11 to predict the sales for years 0, 1, 2, 3, 4, and 5.
b) Compare the actual yearly sales with the predicted sales. Comment.

13. The ideas in the last three questions are discussed in many textbooks in sections entitled "Method of Least Squares" and "Linear Correlation and Regression." For a more thorough discussion of these ideas, see Chapter 9 of *Statistics and Probability in Modern Life*, 2d ed., by Joseph Newmark (New York: Holt, Rinehart and Winston, 1977).

Suggested Further Reading

Bishir, J. W., and D. W. Drews, *Mathematics in the Behavioral and Social Sciences*, New York: Harcourt Brace Jovanovich, 1970.

Howell, J. E., and D. Teichroew, *Mathematical Analysis for Business Decisions*, 2d ed., Homewood, Ill.: Irwin, 1971.

Kline, Morris, *Mathematics for Liberal Arts Students*, Reading, Mass.: Addison-Wesley, 1967.

Newmark, Joseph, *Statistics and Probability in Modern Life*, 2d. ed., New York: Holt, Rinehart and Winston, 1977.

Selby, Samuel, and Leonard Sweet, *Sets, Relations and Functions*, 2d ed., New New York: McGraw-Hill, 1969.

CHAPTER 3 Matrices

Objectives to be Achieved in this Chapter

- *To discuss* what a matrix is, how it is used, and the notation for it.
- *To learn* how to add and subtract matrices and under what conditions we cannot perform these operations.
- *To study* how to multiply two matrices and under what conditions we cannot multiply matrices.
- *To apply* matrices to the solution of two and three linear equations in several unknowns. The idea is known as the Gauss-Jordan method and involves the concept of an augmented matrix.
- *To indicate* how the inverse of a matrix can be applied to the solution of simultaneous linear equations.
- *To apply* matrices to many different areas, such as business, health, communications, and cryptography.

Food Prices Edge Up Again

DAILY TIMES MAY 11, 1978

New York City (May 10)—According to statistics released yesterday by government statisticians, the consumer price index rose for the fourth consecutive month. The consumer price index is the price that a typical family of four would have to pay for food. Last month's rise was only 0.9%.

In commenting on the statistics, officials were quick to point out that the rise was not as high as had been expected and indicated that this was an encouraging sign that inflation was slowing down somewhat.

In support of this assertion, the officials pointed to the following monthly changes in the consumer price index for several cities.

This month's rise in the consumer price index was due mainly to the increase in the price of eggs and vegetables.

	Jan.	Feb.	Mar.	Apr.
NYC	0.9%	0.2%	0.7%	0.9%
Chicago	1.4	0.9	1.1	0.9
Washington, D.C.	0.7	1.0	0.6	1.3
Miami	0.6	0.8	1.5	0.3

Source: Bureau of Labor Statistics, U.S. Department of Labor.

The newspaper article compares changes in the monthly consumer price indexes for several cities over several months. Since such information is released monthly by the Bureau of Labor Statistics, it would be advisable to organize such information to make comparisons easy. To facilitate such comparative studies in this chapter, we introduce matrices, which are nothing more than arrays (tables) of numbers.

3.1 Introduction

In recent years there has been a growing need for an efficient way of organizing numerical data into two different ways at the same time. Matrices provide us with such a method. For example, a camp nutritionist is comparing the nutritional information of several breakfast cereals to serve to campers. She has collected the following information:

	1 ounce Cheerios	Cheerios plus $\frac{1}{2}$ cup vitamin D milk
Calories	110	190
Protein (grams)	4	8
Carbohydrates (grams)	20	26
Fat (grams)	2	6

If the headings or labels for the table are omitted, we can write the same information in matrix form as

$$\begin{pmatrix} 110 & 190 \\ 4 & 8 \\ 20 & 26 \\ 2 & 6 \end{pmatrix}$$

Such a rectangular array of numbers is called a *matrix*.

The nutritional content of a 1-ounce serving of Kellogg's Rice Krispies would be written in matrix form with the same organization as

$$\begin{pmatrix} 110 & 180 \\ 2 & 6 \\ 25 & 31 \\ 0 & 4 \end{pmatrix}$$

With the widespread use of computers, the many time-consuming computations that involve matrices can be done easily in a relatively short time. It is for this reason that the theory of matrices has been finding wide applications in the life, management, and social sciences.

In this chapter we provide some ideas about the nature of matrix theory and its wide range of application.

Historical Note

The theory of matrices is a little more than 100 years old and originated with the works of Arthur Cayley (1821–1895), a famous English algebraist, and James Sylvester (1814–1897).

Cayley was born at Richmond, Surrey, on August 16, 1821, and at the age of 17 enrolled in Trinity College, Cambridge. Cambridge refused to award him his degree when he completed his course of study, however, because he did not want to take the religious vows required of all students. Instead he became a successful lawyer. In 1863 Cambridge established the Sadlerian Professorship of pure mathematics and offered the job to Cayley. He eagerly accepted it.

Sylvester was denied his law degree from Cambridge because he was Jewish. Instead he began working as an actuary in London, where he met Cayley who was practicing law there. They became good friends and discussed many of their math ideas. Sylvester finally received his degree from Cambridge in 1872, 40 years after he first entered college.

Sylvester taught at many institutions including Johns Hopkins University from 1877 to 1883.

Cayley and Sylvester were influenced by the works of William Rowan Hamilton (1804–1865). Hamilton was interested in *quaternions*, which he believed to have great value in the sciences. He was particularly concerned with optics.

It is alleged that Hamilton knew 13 different languages, including Greek, Latin, Hebrew, Sanskrit, Arabic, and Persian, when he was only 13. By the age of 17, he had already begun his first work on mathematical physics. He was knighted in 1835. After 1843 he devoted most of his time to the study of quaternions, which he found very difficult to apply.

After Hamilton's death there was much controversy about using vector algebra to replace quaternions. In 1943 the Irish Academy held its Quaternion Centenary Celebration, which formally established quaternions as an important tool in mathematics and physics.

3.2 What Is a Matrix?

As indicated in the preceding section, a matrix (plural: matrices) is any rectangular array (table) of numbers enclosed within parentheses. The numbers that make up the matrix are called *entries* or *elements* of the matrix. It is important to note that a matrix is not a number.

Matrices come in all sizes and shapes. The following are all matrices.

$$\begin{pmatrix} 1 & -4 \\ 2 & 3 \end{pmatrix} \qquad \begin{pmatrix} 3 & -4 & 7 & 6 \\ 8 & -2 & 9 & 2 \end{pmatrix} \qquad \begin{pmatrix} 0 & 2 \\ -1 & 2 \\ 3 & -5 \\ 7 & -4 \\ 8 & 7 \end{pmatrix} \qquad (1)$$

(a) (b) (c) (d)

The matrix in (a) consists of 2 rows and 2 columns. We classify a matrix by its *dimension*, that is, by the number of rows and the number of columns. In this case we call the matrix a 2×2 matrix. The matrix in (b) is a 2×4 matrix, since it consists of 2 rows and 4 columns. The matrix in (c) is a 5×2 matrix, since it has 5 rows and 2 columns. The matrix in (d) is a 1×1 matrix.

Comment Do not confuse the rows of a matrix with the columns of a matrix. A row of a matrix is the numbers that appear next to one another horizontally down, whereas a column of a matrix is the numbers that appear vertically across.

Definition 3.1 If a matrix has m rows and n columns, we say that its *dimension* (size) is $m \times n$.

Definition 3.2 Any matrix that has the same number of rows and columns is called a *square matrix*.

Example 1 The following are square matrices:

$$\begin{pmatrix} 6 & 1 \\ 5 & 3 \end{pmatrix} \quad \begin{pmatrix} 7 & -1 & 6 \\ -2 & 9 & -1 \\ 3 & 2 & 4 \end{pmatrix} \quad \begin{pmatrix} 8 & 2 & 4 & 2 \\ 6 & 1 & -3 & -5 \\ 0 & -5 & 8 & 9 \\ 2 & 6 & -1 & 7 \end{pmatrix}$$

The following are not square matrices:

$$\begin{pmatrix} 2 & -3 & 5 \\ 6 & 0 & 1 \end{pmatrix} \quad (1 \quad 7) \quad \begin{pmatrix} 8 & -9 \\ -1 & 6 \\ 3 & 7 \end{pmatrix}$$

Notation Matrices are usually denoted by boldface capital letters, such as **A**, **B**, **C**, ...

Definition 3.3 Any matrix that has only one row or one column is called a *row matrix* or *column matrix*, respectively. Some books refer to these matrices as *row vectors* or *column vectors*, respectively.

Example 2 The matrix

$$\mathbf{A} = (6 \quad 2 \quad 1)$$

is a 1×3 row matrix or row vector. The matrix

$$\mathbf{B} = \begin{pmatrix} 6 \\ 5 \\ 4 \\ 3 \end{pmatrix}$$

is a 4×1 column matrix or column vector. The matrix

$$\mathbf{C} = (2 \quad 1)$$

is a 1×2 row vector. What type of matrix (vector) is (4)?

Definition 3.4 Two matrices are said to be equal if they are of the same dimension and if their corresponding entries are equal. When matrices **A** and **B** are equal, we write **A** = **B**. This is read "matrix **A** equals matrix **B**."

Example 3 If

$$\mathbf{A} = \begin{pmatrix} 1 & 7 \\ 6 & 4 \end{pmatrix} \quad \text{and} \quad \mathbf{B} = \begin{pmatrix} 1 & 7 \\ 6 & 4 \end{pmatrix}$$

then **A** = **B**.

Example 4 If

$$\mathbf{A} = \begin{pmatrix} 7 & 3 \\ 2 & x \end{pmatrix} \quad \text{and} \quad \mathbf{B} = \begin{pmatrix} y & 3 \\ 2 & 5 \end{pmatrix}$$

then matrix **A** will equal matrix **B** only if $x = 5$ and $y = 7$.

Example 5 If

$$\mathbf{A} = \begin{pmatrix} 1 & -7 \\ 3 & 2 \end{pmatrix} \quad \text{and} \quad \mathbf{B} = \begin{pmatrix} 1 & -7 & 4 \\ 3 & 2 & 5 \end{pmatrix}$$

then $\mathbf{A} \neq \mathbf{B}$, since **A** and **B** are not of the same dimension.

Example 6 Let

$$\mathbf{A} = \begin{pmatrix} 3x + 4y \\ 2x - 3y \end{pmatrix} \quad \text{and} \quad \mathbf{B} = \begin{pmatrix} 18 \\ -5 \end{pmatrix}$$

Then for matrix **A** to equal matrix **B**, we must have $3x + 4y = 18$ and $2x - 3y = -5$. This represents the two simultaneous equations

$$3x + 4y = 18$$
$$2x - 3y = -5$$

that must be solved for x and y. This will be analyzed in Sections 3.5 and 3.6.

Exercises

1. Find the dimension of each of the following matrices.

a) $\begin{pmatrix} 1 & 4 & 3 & 6 \\ 2 & 8 & -2 & 0 \\ -4 & -5 & 2 & -3 \end{pmatrix}$ b) $\begin{pmatrix} 2 & -3 & 6 \\ 0 & -2 & -3 \\ 7 & 5 & -1 \\ 2 & 0 & -5 \end{pmatrix}$

c) $\begin{pmatrix} 1 & 0 & 0 \\ 0 & 1 & 0 \\ 0 & 0 & 1 \end{pmatrix}$ d) $(1 \quad 4 \quad 3)$

e) $\begin{pmatrix} 7 & -6 & 8 \\ -2 & 9 & 0 \end{pmatrix}$ f) $\begin{pmatrix} 5 \\ 9 \\ 2 \end{pmatrix}$

g) (1)

2. Which (if any) of the matrices in Exercise 1 are square matrices?

3. Are the following matrices equal? Explain your answer.

$$\begin{pmatrix} 1 & -7 & 3 \\ 2 & 1 & 4 \end{pmatrix} \qquad \begin{pmatrix} 1 & 2 \\ -7 & 1 \\ 3 & 4 \end{pmatrix}$$

In Exercises 4–8 find a value for the unknown variable(s) so that matrices **A** and **B** will be equal. If this is not possible, explain why.

4. $\mathbf{A} = \begin{pmatrix} 2 & 1 \\ 7 & x \end{pmatrix}$ $\quad \mathbf{B} = \begin{pmatrix} 2 & 1 \\ 7 & 3 \end{pmatrix}$

5. $\mathbf{A} = \begin{pmatrix} 5 & -6 \\ -7 & y \end{pmatrix}$ $\quad \mathbf{B} = \begin{pmatrix} x & -6 \\ -7 & 8 \end{pmatrix}$

6. $\mathbf{A} = \begin{pmatrix} 7 & -3 & 9 \\ -4 & 6 & -2 \end{pmatrix}$ $\quad \mathbf{B} = \begin{pmatrix} 7 & -x & 3y \\ -4 & 6 & -2 \end{pmatrix}$

7. $\mathbf{A} = \begin{pmatrix} 1 & -7 & 6 \\ x & 4 & 3 \end{pmatrix}$ $\quad \mathbf{B} = \begin{pmatrix} 1 & -7 \\ 9 & 4 \end{pmatrix}$

8. $\mathbf{A} = \begin{pmatrix} 3 & 2x+4 & a \\ 2 & y-3 & -2 \end{pmatrix}$ $\quad \mathbf{B} = \begin{pmatrix} 3 & 3x+1 & 7 \\ 2 & 3y+7 & -2 \end{pmatrix}$

9. *Insurance.* The District Insurance Company employs six agents who sell various types of insurance policies. During one week these employees sold policies as shown in the accompanying table. Rewrite the information in matrix form. Put the type of insurance in the top row of the matrix.

	Insurance			
Employee	Health	Car	Life	Fire
Pete	7	4	8	2
Julian	6	5	9	0
Dawn	2	3	0	6
Heather	5	1	5	3
Jason	9	2	0	2
Leona	3	5	8	0

10. *Births.* The accompanying table summarizes the blood type of the mothers admitted to Maternity Hospital during the week of January 1–7.

	Type A	Type B	Type AB	Type O
Mon.	10	7	8	17
Tues.	1	6	9	19
Wed.	5	8	6	17
Thurs.	8	6	4	11
Fri.	9	0	2	13
Sat.	3	7	4	15
Sun.	9	6	8	14

a) Rewrite the information in matrix form. Put the type of blood in the top row of the matrix.
b) How many patients with type A blood were admitted during the week?
c) How many patients did not have type O blood?
d) What is the total number of patients admitted?

11. *City-dwellers' affluence.* Consider the accompanying table about city-dwellers' affluence.
a) Rewrite the information in matrix form.
b) Which city had the lowest new-home prices?

Metropolitan area	Median family income	Per capita income	Home-building permits	New-home prices	Existing-home prices
Washington	$21,282	$8,454	25,468	$72,900	$80,600
San Francisco	$19,158	$8,412	21,926	$91,600	$84,400
Seattle	$18,537	$7,680	24,125	$60,300	$58,600
Chicago	$18,502	$7,785	49,222	$78,700	$66,200
St. Louis	$17,546	$6,755	15,857	$52,100	$42,400
Boston	$17,186	$6,889	8,450	$67,800	$54,700
Dallas	$17,112	$7,096	44,340	$64,000	$57,800
Atlanta	$16,909	$6,716	17,891	$61,400	$52,000
New York	$15,166	$7,568	12,530	$78,400	$66,100
Los Angeles	$14,961	$7,529	39,322	$85,800	$81,600

Source: U.S. Departments of Commerce and Labor, 1978.

3.3 The Addition of Matrices

Adding matrices is an easy process. We simply add the corresponding elements. For example, if

$$\mathbf{A} = \begin{pmatrix} 7 & 3 \\ 9 & 2 \end{pmatrix} \quad \text{and} \quad \mathbf{B} = \begin{pmatrix} 8 & 7 \\ -2 & 9 \end{pmatrix}$$

then

$$\mathbf{A} + \mathbf{B} = \begin{pmatrix} 7 & 3 \\ 9 & 2 \end{pmatrix} + \begin{pmatrix} 8 & 7 \\ -2 & 9 \end{pmatrix} = \begin{pmatrix} 7+8 & 3+7 \\ 9+(-2) & 2+9 \end{pmatrix} = \begin{pmatrix} 15 & 10 \\ 7 & 11 \end{pmatrix}$$

Example 7 Add the matrices.

$$\mathbf{A} = \begin{pmatrix} 5 & 7 & 0 \\ -2 & 3 & 1 \end{pmatrix} \quad \text{and} \quad \mathbf{B} = \begin{pmatrix} 4 & -6 & 2 \\ -9 & -3 & -4 \end{pmatrix}$$

Solution Since the matrices are of the same dimension, we simply add the corresponding elements.

$$\mathbf{A}+\mathbf{B}=\begin{pmatrix}5 & 7 & 0\\ -2 & 3 & 1\end{pmatrix}+\begin{pmatrix}4 & -6 & 2\\ -9 & -3 & -4\end{pmatrix}$$
$$=\begin{pmatrix}5+4 & 7+(-6) & 0+2\\ -2+(-9) & 3+(-3) & 1+(-4)\end{pmatrix}$$
$$=\begin{pmatrix}9 & 1 & 2\\ -11 & 0 & -3\end{pmatrix}$$

Example 8 Add the matrices.

$$\mathbf{A}=\begin{pmatrix}1\\ -5\\ 7\end{pmatrix}\quad\text{and}\quad\mathbf{B}=\begin{pmatrix}-4\\ 8\\ 13\end{pmatrix}$$

Solution Since the matrices are of the same dimension, we simply add the corresponding elements.

$$\mathbf{A}+\mathbf{B}=\begin{pmatrix}1\\ -5\\ 7\end{pmatrix}+\begin{pmatrix}-4\\ 8\\ 13\end{pmatrix}=\begin{pmatrix}1+(-4)\\ -5+8\\ 7+13\end{pmatrix}=\begin{pmatrix}-3\\ 3\\ 20\end{pmatrix}$$

Example 9 Add the matrices.

$$\mathbf{A}=\begin{pmatrix}9 & 2 & 5 & 3\\ -2 & 8 & 7 & 1\\ 0 & 2 & 1 & 0\\ 5 & 1 & 4 & 6\end{pmatrix}\quad\text{and}\quad\mathbf{B}=\begin{pmatrix}8 & -1 & 8 & 2\\ -1 & 0 & 3 & 5\\ 7 & 1 & 6 & 2\\ 9 & 8 & 4 & 9\end{pmatrix}$$

Solution
$$\mathbf{A}+\mathbf{B}=\begin{pmatrix}9 & 2 & 5 & 3\\ -2 & 8 & 7 & 1\\ 0 & 2 & 1 & 0\\ 5 & 1 & 4 & 6\end{pmatrix}+\begin{pmatrix}8 & -1 & 8 & 2\\ -1 & 0 & 3 & 5\\ 7 & 1 & 6 & 2\\ 9 & 8 & 4 & 9\end{pmatrix}$$
$$=\begin{pmatrix}17 & 1 & 13 & 5\\ -3 & 8 & 10 & 6\\ 7 & 3 & 7 & 2\\ 14 & 9 & 8 & 15\end{pmatrix}$$

Example 10 Add the matrices.

$$\mathbf{A}=\begin{pmatrix}0 & 0\\ 0 & 0\end{pmatrix}\quad\text{and}\quad\mathbf{B}=\begin{pmatrix}5 & -2\\ 7 & 3\end{pmatrix}$$

Solution
$$\mathbf{A}+\mathbf{B}=\begin{pmatrix}0 & 0\\ 0 & 0\end{pmatrix}+\begin{pmatrix}5 & -2\\ 7 & 3\end{pmatrix}=\begin{pmatrix}5 & -2\\ 7 & 3\end{pmatrix}=\mathbf{B}+\mathbf{A}$$

Comment In Example 10 the sum is the same as the original matrix **B**. Adding matrix **A** to matrix **B** accomplished nothing. The matrix

$$\begin{pmatrix} 0 & 0 \\ 0 & 0 \end{pmatrix}$$

is called the 2×2 *identity matrix* for addition.

Example 11 Add the matrices.

$$\mathbf{A} = \begin{pmatrix} 4 & -7 & 9 \\ 3 & 0 & 1 \end{pmatrix} \quad \text{and} \quad \mathbf{B} = \begin{pmatrix} 5 & -2 & 3 & 6 \\ 6 & -1 & 9 & 0 \end{pmatrix}$$

Solution These matrices cannot be added because matrix **A** has 2 rows and 3 columns whereas matrix **B** has 2 rows and 4 columns.

Comment When the matrices are not of the same dimension, they cannot be added.

Subtraction of matrices is performed in a manner similar to addition. We simply subtract the corresponding elements.

Example 12 Perform the subtraction.

$$\begin{pmatrix} 5 & 6 \\ 3 & -1 \\ -2 & 0 \\ 7 & 9 \end{pmatrix} - \begin{pmatrix} 50 & 1 \\ 8 & -3 \\ 9 & 1 \\ 12 & 11 \end{pmatrix}$$

Solution

$$\begin{pmatrix} 5 & 6 \\ 3 & -1 \\ -2 & 0 \\ 7 & 9 \end{pmatrix} - \begin{pmatrix} 50 & 1 \\ 8 & -3 \\ 9 & 1 \\ 12 & 11 \end{pmatrix} = \begin{pmatrix} 5-50 & 6-1 \\ 3-8 & -1-(-3) \\ -2-9 & 0-1 \\ 7-12 & 9-11 \end{pmatrix} = \begin{pmatrix} -45 & 5 \\ -5 & 2 \\ -11 & -1 \\ -5 & -2 \end{pmatrix}$$

Example 13 Subtract.

$$\begin{pmatrix} 9 & -2 \\ 8 & 7 \end{pmatrix} - \begin{pmatrix} 5 & 4 \\ -2 & 7 \end{pmatrix}$$

Solution

$$\begin{pmatrix} 9 & -2 \\ 8 & 7 \end{pmatrix} - \begin{pmatrix} 5 & 4 \\ -2 & 7 \end{pmatrix} = \begin{pmatrix} 9-5 & -2-4 \\ 8-(-2) & 7-7 \end{pmatrix} = \begin{pmatrix} 4 & -6 \\ 10 & 0 \end{pmatrix}$$

Example 14 If

$$\mathbf{A} = \begin{pmatrix} 5 & 1 & 3 \\ 2 & 7 & 6 \end{pmatrix} \quad \text{and} \quad \mathbf{B} = \begin{pmatrix} 9 & 3 & 8 \\ 6 & -2 & 7 \end{pmatrix}$$

verify that $\mathbf{A} + \mathbf{B} = \mathbf{B} + \mathbf{A}$.

Solution $\mathbf{A}+\mathbf{B}=\begin{pmatrix}5 & 1 & 3\\ 2 & 7 & 6\end{pmatrix}+\begin{pmatrix}9 & 3 & 8\\ 6 & -2 & 7\end{pmatrix}$

$$=\begin{pmatrix}5+9 & 1+3 & 3+8\\ 2+6 & 7+(-2) & 6+7\end{pmatrix}=\begin{pmatrix}14 & 4 & 11\\ 8 & 5 & 13\end{pmatrix}$$

$$\mathbf{B}+\mathbf{A}=\begin{pmatrix}9 & 3 & 8\\ 6 & -2 & 7\end{pmatrix}+\begin{pmatrix}5 & 1 & 3\\ 2 & 7 & 6\end{pmatrix}$$

$$=\begin{pmatrix}9+5 & 3+1 & 8+3\\ 6+2 & -2+7 & 7+6\end{pmatrix}=\begin{pmatrix}14 & 4 & 11\\ 8 & 5 & 13\end{pmatrix}$$

Comment Since $\mathbf{A}+\mathbf{B}=\mathbf{B}+\mathbf{A}$, we say that the addition of matrices is commutative.

Exercises

12. For each of the following perform the indicated operations or indicate why the operation cannot be performed.

a) $\begin{pmatrix}4 & 9\\ 3 & -2\end{pmatrix}+\begin{pmatrix}7 & -6\\ 2 & 9\end{pmatrix}$

b) $(1 \quad 4 \quad 3)+(2 \quad 7 \quad -6)$

c) $\begin{pmatrix}2 & -1 & -3\\ 4 & 2 & 6\end{pmatrix}+\begin{pmatrix}-3 & 9 & 8\\ 2 & 0 & 1\end{pmatrix}$

d) $\begin{pmatrix}7\\ 3\\ 2\\ 1\end{pmatrix}-\begin{pmatrix}-3\\ 10\\ -3\\ 31\end{pmatrix}$

e) $\begin{pmatrix}8 & -1\\ 12 & 15\end{pmatrix}-\begin{pmatrix}3 & 0 & 2\\ 5 & 1 & 7\end{pmatrix}$

f) $\begin{pmatrix}4 & 2 & 7\\ 12 & -32 & -17\\ 61 & 14 & 13\end{pmatrix}+\begin{pmatrix}15 & 0 & 21\\ 3 & -8 & 16\\ 17 & 22 & 19\end{pmatrix}$

g) $\begin{pmatrix}2 & 14 & 61\\ 9 & -2 & 5\end{pmatrix}-\begin{pmatrix}3 & 7 & 6\\ 4 & 3 & 1\\ 0 & 1 & 5\end{pmatrix}$

h) $\begin{pmatrix}4 & 1 & 6\\ 2 & -1 & 3\end{pmatrix}-\begin{pmatrix}3 & 9 & 12\\ 17 & 25 & 7\end{pmatrix}$

i) $\begin{pmatrix}7\\ 3\\ 5\end{pmatrix}+\begin{pmatrix}2\\ 0\\ 1\end{pmatrix}$

j) $\begin{pmatrix} 7 & -3 & 4 \\ 16 & -22 & 9 \\ 10 & 1 & 5 \end{pmatrix} + \begin{pmatrix} 0 & 0 & 0 \\ 0 & 0 & 0 \\ 0 & 0 & 0 \end{pmatrix}$

k) $\begin{pmatrix} 4 \\ 2 \\ 1 \end{pmatrix} + \begin{pmatrix} 0 \\ 0 \\ 0 \end{pmatrix}$

l) $\begin{pmatrix} 32 & -1 \\ 6 & -3 \end{pmatrix} - \begin{pmatrix} 0 & 0 \\ 0 & 0 \end{pmatrix}$

m) $\begin{pmatrix} 0 & 0 \\ 0 & 0 \end{pmatrix} - \begin{pmatrix} 32 & -1 \\ 6 & -3 \end{pmatrix}$

13. *Inflation.* The accompanying chart indicates how the prices of several food items have changed in each of two years for a few cities. (A + sign indicates an increase in price; *a* – sign indicates a decrease in price.)

1978

	1 quart milk	1 dozen eggs	1 pound bread	1 quart orange juice	1 pound meat
City A	+3	+12	+4	+6	−2
City B	+8	+ 5	+2	+1	−1
City C	−1	0	−2	0	+7

1979

	1 quart milk	1 dozen eggs	1 pound bread	1 quart orange juice	1 pound meat
City A	+2	+6	−1	+7	+64
City B	+4	+1	0	+6	+47
City C	0	−2	+3	+5	+40

Note: All price changes are in cents.

a) Find the net effect of inflation over the two-year period by adding the two matrices.

b) Which item's price increased the most and in which city?

c) Which item's price increased the least or decreased the most and in which city?

14. *Warehouse storage.* A large furniture company has four warehouses in a city from which it ships various items. The inventories of each warehouse are carefully recorded for each month as indicated.

		January 1 Inventory				*February 1 Inventory*			
		Warehouse				Warehouse			
		A	B	C	D	A	B	C	D
Type of furniture	Chairs	71	38	27	53	64	32	17	53
	Sofas	12	9	16	24	10	8	16	20
	Tables	89	64	57	73	80	61	56	70

a) How many items were removed from the inventory during January?
b) Which warehouse shipped the most merchandise?
c) Which warehouse shipped the least merchandise?

15. *Nutrition.* Refer to the nutrition example given in the beginning of Section 3.1. If a camper eats 1 ounce of Cheerios and 1 ounce of Rice Krispies, each with $\frac{1}{2}$ cup of vitamin D milk, find the camper's intake of calories, protein, carbohydrates, and fat.

16. Find x and y such that

$$\begin{pmatrix} 2 & 3 \\ 7 & x \end{pmatrix} + \begin{pmatrix} y & 2 \\ 9 & 4 \end{pmatrix} = \begin{pmatrix} 7 & 5 \\ 16 & 3 \end{pmatrix}$$

17. If

$$\mathbf{A} = \begin{pmatrix} 7 & 6 & 1 \\ 4 & -3 & 1 \\ 6 & 0 & 3 \end{pmatrix} \quad \mathbf{B} = \begin{pmatrix} 5 & 9 & -1 \\ 0 & 2 & 4 \\ 7 & 3 & 8 \end{pmatrix} \quad \mathbf{C} = \begin{pmatrix} 1 & 7 & 0 \\ 2 & 9 & -3 \\ 5 & 8 & 6 \end{pmatrix}$$

verify the associative law for addition: $(\mathbf{A} + \mathbf{B}) + \mathbf{C} = \mathbf{A} + (\mathbf{B} + \mathbf{C})$.

18. An instructor is teaching several girl scouts how to swim. On Monday the instructor spent 35, 26, 18, and 19 minutes with Wendy, Gail, Juanita, and Rosalie, respectively. On Tuesday the instructor spent 19, 22, 25, and 29 minutes, and on Wednesday 16, 10, 11, and 14 minutes, respectively, with these girls.
a) Rewrite this information in matrix form.
b) Find the total amount of time spent by the instructor with each girl in these three days.
c) What is the total time spent by the instructor over the three-day period in teaching these girls how to swim?

3.4 Matrix Multiplication

In this section we discuss how to multiply matrices. The procedure is more complicated than the addition of matrices, since we *do not* simply multiply the corresponding elements. As we shall see, our somewhat involved procedure for multiplying matrices is justified by the important practical problems to which it can be applied. After analyzing several examples, you will find that the procedure is not difficult at all.

To illustrate the technique, let us multiply the two matrices

$$\mathbf{A} = \begin{pmatrix} 5 & 3 & -9 & 2 \\ 2 & 6 & 1 & 0 \end{pmatrix} \quad \text{and} \quad \mathbf{B} = \begin{pmatrix} 4 & 3 & 0 \\ 1 & -7 & 6 \\ -2 & -3 & 9 \\ 8 & 5 & 2 \end{pmatrix}$$

We first note that matrix $\mathbf{A}$ is a 2×4 matrix, since it has 2 rows and 4 columns. Matrix $\mathbf{B}$ is a 4×3 matrix, since it has 4 rows and 3 columns.

In order to multiply matrices, it is not necessary that both matrices be of the same dimension. The only requirement that must be satisfied is that the number of columns in the first matrix be the same as the number of rows in the second matrix. In the above instance this number is 4. Our result will be a 2×3 matrix. We can indicate the relationship between the dimensions of the matrices to be multiplied and the dimension of the product matrix by using a diagram similar to the following:

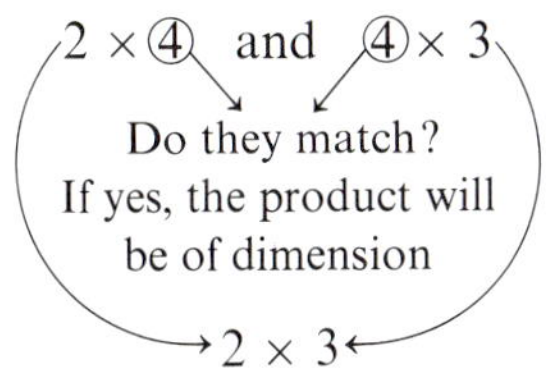

Thus, we know that our answer will be a 2×3 matrix as shown:

$$\begin{pmatrix} 5 & 3 & -9 & 2 \\ 2 & 6 & 1 & 0 \end{pmatrix} \cdot \begin{pmatrix} 4 & 3 & 0 \\ 1 & -7 & 6 \\ -2 & -3 & 9 \\ 8 & 5 & 2 \end{pmatrix} = \begin{pmatrix} ? & ? & ? \\ ? & ? & ? \end{pmatrix}$$

How do we determine the entries for this matrix? To determine the element that belongs in the first row, first column, we simply multiply each element of row 1 of the first matrix by the *corresponding* element of column 1 of the second matrix and then add. We get

$$5(4) + 3(1) + (-9)(-2) + 2(8) = 57$$

To obtain the entry for the second row, first column, we multiply each element of row 2 of the first matrix by the corresponding element of column 1 of the second matrix and then add. We get

$$2(4) + 6(1) + 1(-2) + 0(8) = 12$$

Similarly, to obtain the entry for the first row, second column, we multiply row 1 of the first matrix by column 2 of the second matrix. This gives

$$5(3) + 3(-7) + (-9)(-3) + 2(5) = 31$$

In a similar manner we obtain all the other entries. We get

$$\begin{pmatrix} 5 & 3 & -9 & 2 \\ 2 & 6 & 1 & 0 \end{pmatrix} \cdot \begin{pmatrix} 4 & 3 & 0 \\ 1 & -7 & 6 \\ -2 & -3 & 9 \\ 8 & 5 & 2 \end{pmatrix}$$

$$= \begin{pmatrix} 5(4) + 3(1) + (-9)(-2) + 2(8) & 5(3) + 3(-7) + (-9)(-3) + 2(5) & 5(0) + 3(6) + (-9)(9) + 2(2) \\ 2(4) + 6(1) + 1(-2) \quad + 0(8) & 2(3) + 6(-7) + 1(-3) \quad + 0(5) & 2(0) + 6(6) + 1(9) \quad + 0(2) \end{pmatrix}$$

$$= \begin{pmatrix} 57 & 31 & -59 \\ 12 & -39 & 45 \end{pmatrix}$$

Let us further illustrate the technique with several examples.

Example 15 If

$$\mathbf{A} = \begin{pmatrix} 7 & 1 & 5 \\ 3 & 6 & 2 \end{pmatrix} \quad \text{and} \quad \mathbf{B} = \begin{pmatrix} 4 & 6 & 2 & 7 & 1 \\ 3 & 1 & 1 & 3 & 6 \\ 7 & 4 & 0 & 5 & 2 \end{pmatrix}$$

find $\mathbf{A} \cdot \mathbf{B}$.

Solution Matrix $\mathbf{A}$ is 2×3 and matrix $\mathbf{B}$ is 3×5. We have

2×3 and 3×5

Do they match?
If yes, the product will
be of dimension

2×5

We then compute the entries for the 2×5 matrix:

$$\begin{pmatrix} 7 & 1 & 5 \\ 3 & 6 & 2 \end{pmatrix} \cdot \begin{pmatrix} 4 & 6 & 2 & 7 & 1 \\ 3 & 1 & 1 & 3 & 6 \\ 7 & 4 & 0 & 5 & 2 \end{pmatrix}$$

$$= \begin{pmatrix} 7(4) + 1(3) + 5(7) & 7(6) + 1(1) + 5(4) & 7(2) + 1(1) + 5(0) & 7(7) + 1(3) + 5(5) & 7(1) + 1(6) + 5(2) \\ 3(4) + 6(3) + 2(7) & 3(6) + 6(1) + 2(4) & 3(2) + 6(1) + 2(0) & 3(7) + 6(3) + 2(5) & 3(1) + 6(6) + 2(2) \end{pmatrix}$$

$$= \begin{pmatrix} 66 & 63 & 15 & 77 & 23 \\ 44 & 32 & 12 & 49 & 43 \end{pmatrix}$$

Example 16 If

$$\mathbf{A} = \begin{pmatrix} 5 & 2 & 6 \\ -1 & 4 & 0 \end{pmatrix} \quad \text{and} \quad \mathbf{B} = \begin{pmatrix} 5 & 1 & -3 & 8 \\ 1 & 0 & 5 & 9 \end{pmatrix}$$

find $\mathbf{A} \cdot \mathbf{B}$.

Solution Matrix $\mathbf{A}$ is 2×3 and matrix $\mathbf{B}$ is 2×4. We have

2×3 and 2×4

Do they match?

No!

Since they do not match, this multiplication cannot be performed.

Example 17 If

$$\mathbf{A} = (2 \quad 1 \quad 7 \quad 6) \quad \text{and} \quad \mathbf{B} = \begin{pmatrix} 7 \\ 3 \\ 6 \\ 2 \end{pmatrix}$$

find $\mathbf{A} \cdot \mathbf{B}$.

Solution Matrix **A** is 1×4 and matrix **B** is 4×1. We have

1×4 and 4×1

Do they match?
If yes, the product will
be of dimension

1×1

We have

$$(2 \quad 1 \quad 7 \quad 6) \cdot \begin{pmatrix} 7 \\ 3 \\ 6 \\ 2 \end{pmatrix} = (2 \cdot 7 + 1 \cdot 3 + 7 \cdot 6 + 6 \cdot 2) = (71)$$

Comment In Example 17 the answer is (71). It is necessary to put this answer in parentheses, since it is a matrix. It would be wrong to write the answer as 71.

Example 18 If

$$\mathbf{B} = \begin{pmatrix} 7 \\ 3 \\ 6 \\ 2 \end{pmatrix} \quad \text{and} \quad \mathbf{A} = (2 \quad 1 \quad 7 \quad 6)$$

find $\mathbf{B} \cdot \mathbf{A}$.

Solution Matrix **B** is 4×1 and matrix **A** is 1×4. We have

4×1 and 1×4

Do they match?
If yes, the product will
be of dimension

4×4

Thus,

$$\begin{pmatrix} 7 \\ 3 \\ 6 \\ 2 \end{pmatrix} \cdot (2 \quad 1 \quad 7 \quad 6) = \begin{pmatrix} 7(2) & 7(1) & 7(7) & 7(6) \\ 3(2) & 3(1) & 3(7) & 3(6) \\ 6(2) & 6(1) & 6(7) & 6(6) \\ 2(2) & 2(1) & 2(7) & 2(6) \end{pmatrix}$$

$$= \begin{pmatrix} 14 & 7 & 49 & 42 \\ 6 & 3 & 21 & 18 \\ 12 & 6 & 42 & 36 \\ 4 & 2 & 14 & 12 \end{pmatrix}$$

Comment Notice that in Example 17 the answer for $\mathbf{A} \cdot \mathbf{B}$ is not the same as the result for $\mathbf{B} \cdot \mathbf{A}$ in Example 18. As a matter of fact, $\mathbf{A} \cdot \mathbf{B}$ is a 1×1 matrix, whereas $\mathbf{B} \cdot \mathbf{A}$ is a 4×4 matrix. As a rule, if $\mathbf{A}$ and $\mathbf{B}$ are two matrices, then it is not necessarily true that $\mathbf{A} \cdot \mathbf{B} = \mathbf{B} \cdot \mathbf{A}$ even when $\mathbf{A} \cdot \mathbf{B}$ and $\mathbf{B} \cdot \mathbf{A}$ can be calculated.

Often, when working with matrices, we need to multiply a matrix by a number. For example, suppose we want

$$3 \cdot \mathbf{A} \qquad \text{where } \mathbf{A} = \begin{pmatrix} 7 & -2 & 9 & -1 \\ 6 & 5 & -3 & 2 \end{pmatrix}$$

This can be done simply by multiplying each entry of matrix $\mathbf{A}$ by the number 3. Thus, we get

$$3\mathbf{A} = \begin{pmatrix} 3(7) & 3(-2) & 3(9) & 3(-1) \\ 3(6) & 3(5) & 3(-3) & 3(2) \end{pmatrix} = \begin{pmatrix} 21 & -6 & 27 & -3 \\ 18 & 15 & -9 & 6 \end{pmatrix}$$

Definition 3.5 If $\mathbf{A}$ is any matrix and c is any real number (usually called a *scalar*), then $c\mathbf{A}$ means "multiply each entry of matrix $\mathbf{A}$ by the number c."

Example 19 Perform the indicated operations.

a) $-3\begin{pmatrix} 4 & 1 & 6 \\ 2 & 3 & 0 \end{pmatrix}$

b) $3\begin{pmatrix} 5 & 1 \\ 2 & 7 \end{pmatrix} + 4\begin{pmatrix} 6 & -9 \\ -2 & 8 \end{pmatrix}$

Solution a) $$-3\begin{pmatrix} 4 & 1 & 6 \\ 2 & 3 & 0 \end{pmatrix} = \begin{pmatrix} -3(4) & -3(1) & -3(6) \\ -3(2) & -3(3) & -3(0) \end{pmatrix} = \begin{pmatrix} -12 & -3 & -18 \\ -6 & -9 & 0 \end{pmatrix}$$

b) $$\begin{aligned} 3\begin{pmatrix} 5 & 1 \\ 2 & 7 \end{pmatrix} + 4\begin{pmatrix} 6 & -9 \\ -2 & 8 \end{pmatrix} &= \begin{pmatrix} 3(5) & 3(1) \\ 3(2) & 3(7) \end{pmatrix} + \begin{pmatrix} 4(6) & 4(-9) \\ 4(-2) & 4(8) \end{pmatrix} \\ &= \begin{pmatrix} 15 + 24 & 3 - 36 \\ 6 - 8 & 21 + 32 \end{pmatrix} \\ &= \begin{pmatrix} 39 & -33 \\ -2 & 53 \end{pmatrix} \end{aligned}$$

There are certain square matrices that have unusual properties. The 2×2 square matrix

$$\begin{pmatrix} 1 & 0 \\ 0 & 1 \end{pmatrix}$$

is one such matrix. When we multiply any other 2×2 matrix by it, the answer will be exactly the same matrix that we started with.

Example 20 Multiply.

a) $\begin{pmatrix} 7 & 3 \\ -1 & 2 \end{pmatrix} \cdot \begin{pmatrix} 1 & 0 \\ 0 & 1 \end{pmatrix}$ b) $\begin{pmatrix} 1 & 0 \\ 0 & 1 \end{pmatrix} \cdot \begin{pmatrix} 7 & 3 \\ -1 & 2 \end{pmatrix}$

Solution a) $\begin{pmatrix} 7 & 3 \\ -1 & 2 \end{pmatrix} \begin{pmatrix} 1 & 0 \\ 0 & 1 \end{pmatrix} = \begin{pmatrix} 7(1) + 3(0) & 7(0) + 3(1) \\ -1(1) + 2(0) & -1(0) + 2(1) \end{pmatrix} = \begin{pmatrix} 7 & 3 \\ -1 & 2 \end{pmatrix}$

b) $\begin{pmatrix} 1 & 0 \\ 0 & 1 \end{pmatrix} \begin{pmatrix} 7 & 3 \\ -1 & 2 \end{pmatrix} = \begin{pmatrix} 1(7) + 0(-1) & 1(3) + 0(2) \\ 0(7) + 1(-1) & 0(3) + 1(2) \end{pmatrix} = \begin{pmatrix} 7 & 3 \\ -1 & 2 \end{pmatrix}$

In Example 20 you will notice that no matter which way we perform the multiplication, our answer is always

$$\begin{pmatrix} 7 & 3 \\ -1 & 2 \end{pmatrix}.$$

The matrix

$$\begin{pmatrix} 1 & 0 \\ 0 & 1 \end{pmatrix}$$

is called the 2×2 *identity matrix* for multiplication.

Example 21 Multiply.

a) $\begin{pmatrix} 7 & 1 & 3 \\ 6 & 0 & 4 \\ 5 & 1 & 2 \end{pmatrix} \cdot \begin{pmatrix} 1 & 0 & 0 \\ 0 & 1 & 0 \\ 0 & 0 & 1 \end{pmatrix}$

b) $\begin{pmatrix} 1 & 0 & 0 \\ 0 & 1 & 0 \\ 0 & 0 & 1 \end{pmatrix} \cdot \begin{pmatrix} 7 & 1 & 3 \\ 6 & 0 & 4 \\ 5 & 1 & 2 \end{pmatrix}$

Solution a) $\begin{pmatrix} 7 & 1 & 3 \\ 6 & 0 & 4 \\ 5 & 1 & 2 \end{pmatrix} \cdot \begin{pmatrix} 1 & 0 & 0 \\ 0 & 1 & 0 \\ 0 & 0 & 1 \end{pmatrix}$

$$= \begin{pmatrix} 7(1) + 1(0) + 3(0) & 7(0) + 1(1) + 3(0) & 7(0) + 1(0) + 3(1) \\ 6(1) + 0(0) + 4(0) & 6(0) + 0(1) + 4(0) & 6(0) + 0(0) + 4(1) \\ 5(1) + 1(0) + 2(0) & 5(0) + 1(1) + 2(0) & 5(0) + 1(0) + 2(1) \end{pmatrix}$$

$$= \begin{pmatrix} 7 & 1 & 3 \\ 6 & 0 & 4 \\ 5 & 1 & 2 \end{pmatrix}$$

b) $\begin{pmatrix} 1 & 0 & 0 \\ 0 & 1 & 0 \\ 0 & 0 & 1 \end{pmatrix} \cdot \begin{pmatrix} 7 & 1 & 3 \\ 6 & 0 & 4 \\ 5 & 1 & 2 \end{pmatrix}$

$$= \begin{pmatrix} 1(7)+0(6)+0(5) & 1(1)+0(0)+0(1) & 1(3)+0(4)+0(2) \\ 0(7)+1(6)+0(5) & 0(1)+1(0)+0(1) & 0(3)+1(4)+0(2) \\ 0(7)+0(6)+1(5) & 0(1)+0(0)+1(1) & 0(3)+0(4)+1(2) \end{pmatrix}$$

$$= \begin{pmatrix} 7 & 1 & 3 \\ 6 & 0 & 4 \\ 5 & 1 & 2 \end{pmatrix}$$

In each case our answer is exactly the same matrix with which we started.

We can generalize the results of the previous two examples.

Definition 3.6 An *identity matrix* **I** is a square matrix such that when we multiply any other matrix **A** by it on the left side or on the right side, the answer is the same as matrix **A**. The 2×2 identity matrix is

$$\begin{pmatrix} 1 & 0 \\ 0 & 1 \end{pmatrix}$$

The 3×3 identity matrix is

$$\begin{pmatrix} 1 & 0 & 0 \\ 0 & 1 & 0 \\ 0 & 0 & 1 \end{pmatrix}$$

Comment Definition 3.6 specifies that for

$$\mathbf{I} = \begin{pmatrix} 1 & 0 \\ 0 & 1 \end{pmatrix}$$

to be an identity matrix for any matrix **A**, we must be able to multiply matrix **A** both on the left side and on the right side of **I**. This may not always be possible.

Example 22 If

$$\mathbf{A} = \begin{pmatrix} 1 & 7 \\ 6 & -1 \\ 2 & 3 \end{pmatrix} \quad \text{then} \quad \mathbf{A} \cdot \mathbf{I} = \begin{pmatrix} 1 & 7 \\ 6 & -1 \\ 2 & 3 \end{pmatrix} \cdot \begin{pmatrix} 1 & 0 \\ 0 & 1 \end{pmatrix} = \begin{pmatrix} 1 & 7 \\ 6 & -1 \\ 2 & 3 \end{pmatrix}$$

We cannot perform the multiplication **I·A**, however, since the number of columns in matrix **I** is not the same as the number of rows in matrix **A**. In this case **I** is not the identity matrix.

Photo: United Press International, Inc.

Perhaps you are wondering why matrix multiplication is defined in such a complicated fashion, especially since multiplication is not commutative. As we indicated earlier, it is the applications for matrices that lead us to define multiplication in such an unusual way. The following problem illustrates one simple application. In Section 3.8 we will give many other applications.

Suppose Mary is in a supermarket comparing the prices of several cans of vegetables. She obtains the following information on the cost per can of several nationally advertised brand name items: beans 49¢, corn 59¢, beets 55¢. This information can be written in a 1×3 matrix as

$$\begin{pmatrix} 49 & 59 & 55 \end{pmatrix}$$

If Mary decides to purchase 7 cans of beans, 5 cans of corn, and 6 cans of beets, then we can arrange this information as a 3×1 column matrix

$$\begin{pmatrix} 7 \\ 5 \\ 6 \end{pmatrix}$$

The product of these two matrices represents the total cost. Thus,

$$\begin{aligned} \begin{pmatrix} 49 & 59 & 55 \end{pmatrix} \cdot \begin{pmatrix} 7 \\ 5 \\ 6 \end{pmatrix} &= (49 \cdot 7 + 59 \cdot 5 + 55 \cdot 6) \\ &= (968) \end{aligned}$$

Mary would spend a total of 968¢ or \$9.68 for these items.

The supermarket also carries a no-name label (generic label) for these items. The prices for these items along with the brand name ones can be arranged in matrix form as shown.

$$\begin{array}{l} \text{Nationally advertised brand name} \\ \text{No-name generic label} \end{array} \begin{array}{c} \begin{array}{ccc} \text{Beans} & \text{Corn} & \text{Beets} \end{array} \\ \begin{pmatrix} 49 & 59 & 55 \\ 41 & 53 & 47 \end{pmatrix} \end{array} = \mathbf{A}$$

Since Mary is not sure about the quality of the no-name generic-labeled cans, she decides to purchase several cans of each as indicated in matrix **B.**

$$\begin{array}{l} \text{Beans} \\ \text{Corn} \\ \text{Beets} \end{array} \begin{array}{c} \begin{array}{cc} \text{Nationally advertised brand name} & \text{No-name generic label} \end{array} \\ \begin{pmatrix} 4 & 3 \\ 3 & 2 \\ 4 & 2 \end{pmatrix} \end{array} = \mathbf{B}$$

Let us now perform the multiplication $\mathbf{A} \cdot \mathbf{B}$.

$$\mathbf{A} \cdot \mathbf{B} = \begin{pmatrix} 49 & 59 & 55 \\ 41 & 53 & 47 \end{pmatrix} \cdot \begin{pmatrix} 4 & 3 \\ 3 & 2 \\ 4 & 2 \end{pmatrix}$$

$$= \begin{pmatrix} 49(4) + 59(3) + 55(4) & 49(3) + 59(2) + 55(2) \\ 41(4) + 53(3) + 47(4) & 41(3) + 53(2) + 47(2) \end{pmatrix}$$

$$= \begin{pmatrix} 593 & 375 \\ 511 & 323 \end{pmatrix}$$

Can you interpret these results? The 593, for example, means that she spent \$5.93 on the nationally advertised brand name items, and the 323 means that she spent \$3.23 on the no-name generic-labeled items. What interpretation can you find for the 375 and the 511?

We can also perform the multiplication $\mathbf{B} \cdot \mathbf{A}$.

$$\mathbf{B} \cdot \mathbf{A} = \begin{pmatrix} 4 & 3 \\ 3 & 2 \\ 4 & 2 \end{pmatrix} \cdot \begin{pmatrix} 49 & 59 & 55 \\ 41 & 53 & 47 \end{pmatrix}$$

3×2 and 2×3

Do they match?
Yes. The answer is
of dimension

3×3

We have

$$\mathbf{B} \cdot \mathbf{A} = \begin{pmatrix} 4(49) + 3(41) & 4(59) + 3(53) & 4(55) + 3(47) \\ 3(49) + 2(41) & 3(59) + 2(53) & 3(55) + 2(47) \\ 4(49) + 2(41) & 4(59) + 2(53) & 4(55) + 2(47) \end{pmatrix}$$

$$= \begin{pmatrix} 319 & 395 & 361 \\ 229 & 283 & 259 \\ 278 & 342 & 314 \end{pmatrix}$$

In this matrix the 319 means that she spent \$3.19 for beans, the 283 means that she spent \$2.83 for corn, and the 314 means that she spent \$3.14 for beets. Can you interpret the other numbers?

Comment The preceding problem illustrates that even when two matrices are commutative for multiplication, the results and the corresponding interpretations are completely different.

In Section 3.8 we will discuss further applications of matrices.

In Exercises 19–34 perform the indicated operations (if possible).

19. $\begin{pmatrix} 3 & 1 \\ 2 & 6 \end{pmatrix} \cdot \begin{pmatrix} -2 & 3 \\ 5 & 7 \end{pmatrix}$

20. $\begin{pmatrix} 0 & 2 \\ 5 & 3 \end{pmatrix} \cdot \begin{pmatrix} -1 & -8 \\ 7 & 9 \end{pmatrix}$

21. $\begin{pmatrix} 5 & 7 & 6 \\ 3 & 9 & -1 \\ 0 & 1 & 5 \end{pmatrix} \cdot \begin{pmatrix} 3 & 8 \\ 2 & -6 \\ 7 & 11 \end{pmatrix}$

22. $\begin{pmatrix} 5 & 4 & 7 \\ 6 & 2 & 9 \end{pmatrix} \cdot \begin{pmatrix} 8 & 3 \\ -1 & 5 \\ 0 & 2 \end{pmatrix}$

23. $\begin{pmatrix} 5 & 8 \\ 2 & 9 \\ -2 & -5 \end{pmatrix} \cdot \begin{pmatrix} 6 & -3 & 9 \\ 8 & 5 & 1 \end{pmatrix}$

24. $(7 \quad 6 \quad 3 \quad 8) \cdot \begin{pmatrix} 5 \\ 1 \\ 8 \\ 3 \end{pmatrix}$

25. $\begin{pmatrix} 5 \\ 1 \\ 8 \\ 3 \end{pmatrix} \cdot (7 \quad 6 \quad 3 \quad 8)$

26. $(7 \quad 2) \cdot \begin{pmatrix} 5 \\ -3 \end{pmatrix}$

27. $\begin{pmatrix} 5 & -9 & 8 \\ -3 & 4 & 6 \end{pmatrix} \cdot \begin{pmatrix} 2 & -8 & 3 \\ -6 & 7 & 11 \end{pmatrix}$

28. $5\begin{pmatrix} 8 & -3 & 9 \\ -10 & 5 & 2 \end{pmatrix}$

29. $6\begin{pmatrix} 1 & -8 & 2 \\ 10 & -12 & -3 \\ 6 & 5 & 0 \end{pmatrix}$

30. $3\begin{pmatrix} 5 & -4 & 2 \\ 9 & -8 & 6 \\ 2 & 3 & -5 \end{pmatrix} \cdot \begin{pmatrix} 3 & -9 & -8 \\ 6 & -4 & 7 \\ 8 & 2 & 0 \end{pmatrix}$

31. $4\begin{pmatrix} 6 & -8 \\ 3 & 2 \end{pmatrix} + 5\begin{pmatrix} 9 & 1 \\ 0 & 6 \end{pmatrix}$

32. $5\begin{pmatrix} -7 & 3 \\ 9 & 1 \end{pmatrix} - 2\begin{pmatrix} 7 & -1 \\ 2 & -8 \end{pmatrix}$

33. $8\begin{pmatrix} -7 & -6 & 1 \\ 2 & -9 & 0 \\ 1 & 1 & 8 \end{pmatrix} - 6\begin{pmatrix} 4 & -3 & 9 \\ -2 & 5 & 0 \\ 1 & 3 & -8 \end{pmatrix}$

34. $-5\begin{pmatrix} 3 & 9 & -7 \\ 6 & -2 & 8 \\ 4 & 2 & 6 \end{pmatrix} \cdot \begin{pmatrix} -7 & 3 \\ -2 & 8 \\ 0 & 1 \end{pmatrix}$

★ **35.** Find a 2×2 matrix $\mathbf{A}$ such that

$$\mathbf{A}^2 = \mathbf{A} \cdot \mathbf{A} = \begin{pmatrix} 1 & 0 \\ 0 & 1 \end{pmatrix}$$

Assume that $\mathbf{A}$ is not the identity matrix.

★ **36.** Consider any matrix

$$\mathbf{A} = \begin{pmatrix} a & b \\ c & d \end{pmatrix}$$

Under what conditions will the following be true?

$$\mathbf{A}^2 = \mathbf{A} \cdot \mathbf{A} = \begin{pmatrix} 1 & 0 \\ 0 & 1 \end{pmatrix}$$

★ 37. Let

$$\mathbf{A} = \begin{pmatrix} a & b \\ c & d \end{pmatrix} \quad \text{and} \quad \mathbf{B} = \begin{pmatrix} 1 & 1 \\ -1 & 1 \end{pmatrix}$$

where **A** is not the 2 × 2 identity matrix. Under what conditions will $\mathbf{A} \cdot \mathbf{B} = \mathbf{B} \cdot \mathbf{A}$?

★ 38. Can you find two 2 × 2 matrices **A** and **B** such that

$$\mathbf{A} \neq \begin{pmatrix} 0 & 0 \\ 0 & 0 \end{pmatrix} \quad \mathbf{B} \neq \begin{pmatrix} 0 & 0 \\ 0 & 0 \end{pmatrix} \quad \text{and} \quad \mathbf{A} \cdot \mathbf{B} = \begin{pmatrix} 0 & 0 \\ 0 & 0 \end{pmatrix}$$

★ 39. Can you find a 2 × 2 matrix **A** such that $\mathbf{A} \cdot \mathbf{A} = \mathbf{A}$? Assume that

$$\mathbf{A} \neq \begin{pmatrix} 0 & 0 \\ 0 & 0 \end{pmatrix}$$

and that **A** is not the identity matrix.

40. *Quick food.* Heather stops off at a local Frankfurter King store and buys the items shown in the accompanying table.

Item	Quantity	Cost per item
Franks	3	60¢
French fries	2	40¢
Corn on the cob	4	55¢
Drinks	5	35¢

a) Write the quantity (row) vector for the items.
b) Write the cost (column) vector for the items.
c) Using matrix multiplication, find Heather's total bill.

41. *Calories.* Henry goes to a restaurant and orders a steak dinner consisting of a 12-ounce steak, four pieces of garlic bread, $\frac{1}{3}$ cup of vegetables, two scoops of mashed potatoes, three slices of blueberry pie, and two soft drinks. Henry keeps a record of the calories in each item.

Item	Calories
Steak	102 per ounce
Garlic bread	110 per piece
Vegetables	66 per cup
Mashed potatoes	100 per scoop
Blueberry pie	325 per slice
Soft drink	90 per cup

a) Write the quantity ordered of each item as a row matrix.
b) Write the calorie content of each item as a column matrix.
c) Using matrix multiplication, find the total number of calories in the dinner.

42. *Auto parts.* The Gem Parts Company manufactures auto parts that must be processed by each of its two factories. The following matrix indicates the hourly production by each factory of hoses and fan belts.

$$\begin{array}{l} \\ \text{Hoses} \\ \text{Belts} \end{array} \begin{array}{c} \begin{array}{cc} \text{Factory I} & \text{Factory II} \end{array} \\ \begin{pmatrix} 9 & \quad 5 \\ 11 & \quad 6 \end{pmatrix} \end{array} = \mathbf{A}$$

Let

$$\mathbf{B} = \begin{pmatrix} 10 \\ 9 \end{pmatrix}$$

be the number of hours per day that each factory operates. Compute $\mathbf{A} \cdot \mathbf{B}$ and interpret the result.

43. *Water pollution.* The Acme Novelty Company is accused of polluting a river by dumping industrial wastes (mostly chemical) into it. The company manufactures four items, and the amount and kind of pollution generated in the production of each are shown in the accompanying matrix.

$$\begin{array}{l} \\ \text{Item I} \\ \text{Item II} \\ \text{Item III} \\ \text{Item IV} \end{array} \begin{array}{c} \begin{array}{ccc} \text{Pollutant I} & \text{Pollutant II} & \text{Pollutant III} \end{array} \\ \begin{pmatrix} 8 & \quad 2 & \quad 6 \\ 3 & \quad 9 & \quad 11 \\ 5 & \quad 1 & \quad 10 \\ 6 & \quad 7 & \quad 12 \end{pmatrix} \end{array} = \mathbf{A}$$

To comply with regulations, the company has determined that the cost (per unit) to install equipment to remove each pollutant as each item is produced is given by

$$\mathbf{B} = \begin{pmatrix} 7500 \\ 4600 \\ 9300 \end{pmatrix}$$

Compute $\mathbf{A} \cdot \mathbf{B}$ and interpret the results.

44. *Stocks.* On January 10 Bill bought 80 shares of BIM stock, 70 shares of TAT stock, and 150 shares of SOS stock. Mike bought 95, 150, and 250 shares of these stocks, respectively. BIM sold at \$20 a share, TAT at \$61 a share, and SOS at \$48 a share. Use matrix multiplication to find the amount of money spent by Bill and Mike on these stocks.

45. *Cost averaging.* Refer to Exercise 44. In order to average their costs, Bill and Mike buy the same amounts of these stocks on February 5. However, the prices are now \$25 a share for BIM, \$62 a share for TAT, and \$40 a share for SOS.

a) Using matrix multiplication, find the amount of money spent by Bill and Mike on these stocks.

b) How much was spent by Bill and Mike for these stocks over the two-month period? (Use matrix techniques.)

46. *Legal costs.* Valerie is a lawyer. Over a four-month period she reported handling the number of cases shown in the accompanying matrix. Furthermore, Valerie charges \$150 for each uncontested

	Uncontested divorces	Writing wills	House closings
Month 1	2	5	4
Month 2	1	6	2
Month 3	6	3	1
Month 4	3	5	3

divorce, \$75 to write a will, and \$375 for each house closing. Using matrix multiplication, find Valerie's income from each category for each month. What is Valerie's total income for this period?

3.5 Systems of Linear Equations in Two Variables

Often we are given two linear equations in two variables, x and y, and are interested in finding the values of x and y that the equations have in common. For example, suppose we are given the equations

$$\begin{aligned} 4x + 3y &= 29 \\ 2x - 3y &= 1 \end{aligned}$$

and want to find values of x and y that satisfy both equations. In Chapter 2 we discussed graphical techniques for solving the problem. In this section we discuss an algebraic technique for solving the problem. In the next few sections we will use matrix operations to solve such problems.

Example 23 Solve the following system of equations and check your answers.

$$\begin{aligned} 4x + 3y &= 29 \\ 2x - 3y &= 1 \end{aligned}$$

Solution We notice that the number in front of the variable y is the same in both equations. Thus, we add the equations so that the variable y will drop out. We get

$$6x = 30$$

Now we divide both sides of this equation by 6, getting

$$\frac{6x}{6} = \frac{30}{6}$$

or

$$x = 5$$

We now replace this value of x in either of the two original equations involving x and y. We get

$$x = 5$$
$$4x + 3y = 29$$
$$4(5) + 3y = 29$$
$$20 + 3y = 29$$
$$3y = 9$$
$$y = 3$$

Thus, our answer is $x = 5$, $y = 3$.

We can check the answer by substituting 5 for x and 3 for y in each of the original equations. We have

$$4x + 3y = 29 \qquad 2x - 3y = 1$$
$$4(5) + 3(3) \stackrel{?}{=} 29 \qquad 2(5) - 3(3) \stackrel{?}{=} 1$$
$$29 = 29 \qquad 1 = 1$$
$$\text{True} \qquad \text{True}$$

The correct answer is $x = 5$, $y = 3$.

Example 24 Solve the system of equations.

$$4x + 5y = -7$$
$$2x - 3y = 13$$

Solution We notice that if we add both equations together, neither variable vanishes. Thus, let us first multiply the second equation by -2 so that the resulting number in front of x will be -4. We leave the first equation unchanged. We have

$$\left.\begin{aligned} 4x + 5y &= -7 \\ 2x - 3y &= 13 \end{aligned}\right\} \quad \text{Original equations}$$

$$\left.\begin{aligned} 4x + 5y &= -7 \\ -4x + 6y &= -26 \end{aligned}\right\} \quad \text{New equations}$$

When these new equations are added, the x terms drop out to give $11y = -33$. Solving for y gives $y = -3$.

Now we replace this value of y in either of the two original equations involving x and y. We get

$$y = -3$$
$$4x + 5y = -7$$
$$4x + 5(-3) = -7$$
$$4x - 15 = -7$$
$$4x = 8$$
$$x = 2$$

Thus, our answer is $x = 2$, $y = -3$. The reader should check that these are indeed correct solutions.

Example 25 Solve the system of equations.

$$3x + 4y = -5$$
$$4x + 5y = -7$$

Solution In this case we have to multiply each equation by a different number so that the numbers in front of one of the variables in both equations become the same but of opposite sign. To accomplish this, let us multiply the first equation by -5 and the second equation by 4. This gives us the new, but equivalent, set of equations

$$\left.\begin{aligned} -15x - 20y &= 25 \\ 16x + 20y &= -28 \end{aligned}\right\} \quad \text{New equations}$$

When these new equations are added, the variable y drops out and we get $x = -3$. Substituting this value of x into either of the original equations gives

$$\begin{aligned} x &= -3 \\ 3x + 4y &= -5 \\ 3(-3) + 4y &= -5 \\ -9 + 4y &= -5 \\ 4y &= 4 \\ y &= 1 \end{aligned}$$

Our answer is $x = -3$, $y = 1$.

Comment Perhaps you are wondering why we chose to eliminate the y in Example 25. There is no special reason. We could have multiplied the first equation by -4 and the second equation by $+3$, resulting in the equivalent set of equations:

$$-12x - 16y = 20$$
$$12x + 15y = -21$$

When these equations are added, we get

$$-y = -1 \quad \text{or} \quad y = 1$$

Then we proceed as before to find the value of x.

We summarize the procedure for solving a system of linear equations in two variables.

Rule To solve a system of linear equations:

1. Rearrange the equations so that both unknowns appear on one side and the constants appear on the other side.

2. As needed, multiply each equation by a number so that the resulting equivalent equations have numbers in front of one of the variables in both equations that are the same but of opposite sign.
3. Add the two equations obtained in step 2. One of the variables will drop out.
4. Solve the resulting equation, which now contains only one variable, for the value of that variable.
5. Substitute the value of the variable obtained in step 4 in either of the original equations containing both variables. Solve the resulting equation for the other variable.
6. Check your answer in *both* of the original equations.

3.6 Matrices and Linear Systems

In this section we apply some of the ideas of matrices to study systems of linear equations.

To get us started, consider the following situation: The National School Lunch Act was enacted into law by the U.S. Congress in June 1946 to "safeguard the health and well-being of the nation's children and to encourage domestic consumption of nutritional agricultural commodities and other foods." Under this program both public and nonprofit private schools are eligible for federal and state aid, provided they operate the lunch program under the guidelines set up by the program administrators. For the 1978–1979 school year, schools were permitted to charge certain needy students 10¢ for each lunch and 5¢ for each breakfast. Careful and detailed records are required for reimbursements.

For September 15 the records for a particular class read as follows:

Twelve students were late and did not buy breakfast.

The total amount of money collected for breakfast and lunch was \$3.60.

Find how many breakfasts and how many lunches were purchased.

This problem can be solved by setting up a system of two linear equations in two unknowns. Let

x = the number of breakfasts purchased

y = the number of lunches purchased

From the given information (since 12 students were late), there were 12 more lunches purchased than breakfasts. Thus,

$$y - x = 12$$

Each breakfast costs 5¢ and each lunch costs 10¢. Since the total amount of money collected was 360 (in cents),

$$10y + 5x = 360$$

It is clear that we must solve the system of equations

$$y - x = 12$$
$$10y + 5x = 360$$

for a value of x and a value of y that makes both equations true at the same time. While there are many methods of solving a system of linear equations (some of which we will study in the next section), in this section we use matrix methods.

To solve this system using matrix theory, we need to perform what are known as elementary row operations.

Definition 3.7 An *elementary row operation* on any matrix consists of any one of the following three operations:

1. Interchanging any two rows.
2. Multiplying all elements of any row by a given nonzero number.
3. Multiplying each element of a row by a number and then adding the results to the corresponding elements of another row.

Example 26 The matrix

$$\begin{pmatrix} 5 & 6 \\ 3 & 2 \end{pmatrix}$$

can be transformed into the matrix

$$\begin{pmatrix} 3 & 2 \\ 5 & 6 \end{pmatrix}$$

by interchanging the first and second rows.

Example 27 The matrix

$$\begin{pmatrix} 70 & 60 & 30 \\ 20 & 50 & 90 \end{pmatrix}$$

can be transformed into the matrix

$$\begin{pmatrix} 35 & 30 & 15 \\ 20 & 50 & 90 \end{pmatrix}$$

by multiplying each element of row 1 by $\frac{1}{2}$.

Example 28 The matrix

$$\begin{pmatrix} 6 & 9 & 3 \\ -8 & -2 & 0 \\ 7 & 2 & 1 \end{pmatrix}$$

can be transformed into the matrix

$$\begin{pmatrix} 6 & 9 & 3 \\ -20 & -20 & -6 \\ 7 & 2 & 1 \end{pmatrix}$$

by multiplying each element of row 1 by -2 and then adding it to the elements of row 2. Note that row 1 remains unchanged.

Comment The elementary row operations correspond to operations that we can perform with a system of equations. For example, we may interchange any two equations, or we may multiply both sides of an equation by a number and then add it to the other equation.

Let us now return to our system of equations:

$$\begin{aligned} y - x &= 12 \\ 10y + 5x &= 360 \end{aligned}$$

The expression on the left side of the equations can be written in matrix form as

$$\begin{pmatrix} 1 & -1 \\ 10 & 5 \end{pmatrix} \cdot \begin{pmatrix} y \\ x \end{pmatrix}$$

since their product is

$$\begin{pmatrix} y - x \\ 10y + 5x \end{pmatrix}$$

Remember that if no number is indicated in front of a letter, it is understood to be 1. The numbers on the right side of the equation can be written in matrix form as

$$\begin{pmatrix} 12 \\ 360 \end{pmatrix}$$

Thus, we have the matrices

$$\begin{pmatrix} 1 & -1 \\ 10 & 5 \end{pmatrix} \quad \text{and} \quad \begin{pmatrix} 12 \\ 360 \end{pmatrix}$$

We combine these two matrices into an *augmented matrix* using a dashed line to separate them.

$$\left(\begin{array}{cc:c} 1 & -1 & 12 \\ 10 & 5 & 360 \end{array}\right)$$

Now we perform a series of elementary row operations on the augmented matrix to arrive at the answer. We indicate the operations applied to the equations and to the augmented matrix so that the reader can see the correspondence.

Equations	*Matrix*
$y - x = 12$ $10y + 5x = 360$	Our objective is to apply elementary row operations to the augmented matrix so that we get the identity matrix to the left of the dashed line. Thus, we try to get a 1 in the row 1, column 1 position and then a 0 in the row 2, column 1 position. To accomplish this, we multiply each element of row 1 by -10 and then add the results to row 2. This results in
$y - x = 12$ $10y - 10y + 5x + 10x = 360 - 120$ or $y - x = 12$ $0y + 15x = 240$	$\left(\begin{array}{cc\|c} 1 & -1 & 12 \\ 0 & 15 & 240 \end{array}\right)$ Next we try to get a 1 in the row 2, column 2 position. To accomplish this, we multiply each element of row 2 by $\frac{1}{15}$. We get
$y - x = 12$ $0y + x = 16$	$\left(\begin{array}{cc\|c} 1 & -1 & 12 \\ 0 & 1 & 16 \end{array}\right)$ Finally we want to get a 0 in the row 1, column 2 position. To accomplish this, we add row 2 to row 1. This gives
$y + 0x = 28$ $0y + x = 16$	$\left(\begin{array}{cc\|c} 1 & 0 & 28 \\ 0 & 1 & 16 \end{array}\right)$ The final augmented matrix gives us the solution to the problem. We have $1y + 0x = 28$ $0y + 1x = 16$
Thus $y = 28$ $x = 16$	or simply $y = 28 \qquad x = 16$

Since x represents the number of breakfasts purchased and y represents the number of lunches purchased, we conclude that 16 breakfasts and 28 lunches were purchased.

We illustrate the above technique, which is known as the *Gauss-Jordan method* in honor of the two famous mathematicians Carl F. Gauss (1777–1855) and Camille Jordan (1838–1922), with several examples.

Example 29 Using matrix methods, solve the following system of equations for x and y.

$$3x + 7y = 33$$
$$2x + 5y = 23$$

Solution Again we will indicate the operations applied to the equations and to the augmented matrix so that the reader can see the correspondence.

Equations

We arrange the equation in standard form

$$3x + 7y = 33$$
$$2x + 5y = 23$$

Then we perform several operations so that one of the variables drops out. Divide the first equation by 3 to get

$$x + \frac{7}{3}y = 11$$
$$2x + 5y = 23$$

Multiply the first equation by -2 and add the result to the second equation. We get

$$x + \frac{7}{3}y = 11$$
$$0x + \frac{1}{3}y = 1$$

Matrix

We first form the augmented matrix:

$$\left(\begin{array}{cc|c} 3 & 7 & 33 \\ 2 & 5 & 23 \end{array}\right)$$

Then we perform several elementary row operations so that the numbers to the left of the dashed line will make up an identity matrix. To get started, we try to get a 1 in the row 1, column 1 position. This can be accomplished by multiplying each element of row 1 by $\frac{1}{3}$. We get

$$\left(\begin{array}{cc|c} 1 & \frac{7}{3} & 11 \\ 2 & 5 & 23 \end{array}\right)$$

Then we try to get a 0 in the row 2, column 1 position. To accomplish this, we multiply each element of row 1 by -2 and add the results to row 2, getting

$$\left(\begin{array}{cc|c} 1 & \frac{7}{3} & 11 \\ 0 & \frac{1}{3} & 1 \end{array}\right)$$

Equations	*Matrix*
Multiply the second equation by 3: $$x + \frac{7}{3}y = 11$$ $$0x + y = 3$$	Now we try to get a 1 in the row 2, column 2 position. We multiply each element of row 2 by 3, getting $$\left(\begin{array}{cc:c} 1 & \frac{7}{3} & 11 \\ 0 & 1 & 3 \end{array}\right)$$
Multiply the second equation by $-\frac{7}{3}$ and add the results to the first equation. We get $$x + 0y = 4$$ $$0x + y = 3$$	Finally, we try to get a 0 in the row 1, column 2 position. We multiply each element of row 2 by $-\frac{7}{3}$ and add the results to row 1, getting $$\left(\begin{array}{cc:c} 1 & 0 & 4 \\ 0 & 1 & 3 \end{array}\right)$$
Our answer is $$1x + 0y = 4$$ $$0x + 1y = 3$$ or $x = 4 \qquad y = 3$	We notice that we have an identity matrix to the left of the dashed line. We can now read our answer from the extreme right column. $$x = 4 \qquad y = 3$$

Example 30 *Admission charges.* A family consisting of 2 adults and 5 children paid \$34 admission to the Animal Town Amusment Park. Another family consisting of 3 adults and 2 children paid \$29 for admission. (The admission price includes all rides.) Find the admission price for an adult and the price for a child.

Solution To solve this problem, we must first find a pair of simultaneous equations representing the given information. Then we can apply the Gauss-Jordan method by using matrices. Let

$x =$ the admission price for a child

$y =$ the admission price for an adult

The first family consisted of 5 children and 2 adults and paid \$34 admission, so $5x + 2y = 34$. The second family consisted of 2 children and 3 adults and paid \$29 admission, so $2x + 3y = 29$. We must now find a common x and y value for the two equations

$$5x + 2y = 34$$
$$2x + 3y = 29$$

We begin by writing the augmented matrix. We then perform the indicated elementary row operations. (We suggest that the reader write down the equations and indicate how the row operations affect these equations.) The augmented matrix is

$$\left(\begin{array}{cc|c} 5 & 2 & 34 \\ 2 & 3 & 29 \end{array}\right)$$

Multiply row 1 by $\frac{1}{5}$. This will give us a 1 in the first row, first column position:

$$\left(\begin{array}{cc|c} 1 & \frac{2}{5} & \frac{34}{5} \\ 2 & 3 & 29 \end{array}\right)$$

Multiply row 1 by -2 and add the results to row 2. This will give us a 0 in the second row, first column position:

$$\left(\begin{array}{cc|c} 1 & \frac{2}{5} & \frac{34}{5} \\ 0 & \frac{11}{5} & \frac{77}{5} \end{array}\right)$$

Multiply row 2 by $\frac{5}{11}$:

$$\left(\begin{array}{cc|c} 1 & \frac{2}{5} & \frac{34}{5} \\ 0 & 1 & 7 \end{array}\right)$$

Multiply row 2 by $-\frac{2}{5}$ and add the results to row 1. We get

$$\left(\begin{array}{cc|c} 1 & 0 & 4 \\ 0 & 1 & 7 \end{array}\right)$$

We now read our answer from the extreme right column. It is $x = 4$, $y = 7$. Thus, the admission charge for children is \$4 each and the admission charge for adults is \$7 a person.

Comment As we indicated earlier, after we obtain the augmented matrix, our main objective is to change the numbers to the left of the dashed line into the identity matrix, so that the numbers that are to the right of the line will become solutions for x and y.

Example 31 *Check cashing.* Casandra is a bank teller. Yesterday she cashed a \$200 check and gave the customer 25 bills in \$10 and \$5 bills. How many \$5 bills and how many \$10 bills did she give the customer?

Solution Let

F = the number of \$5 bills

T = the number of \$10 bills

Since there was a total of 25 bills, we have $F + T = 25$. Also, each \$5 bill is worth five dollars and each \$10 bill is worth ten dollars, so

$5F + 10T = 200$. Now we must solve the system of equations

$$F + T = 25$$
$$5F + 10T = 200$$

for a value of F and T.

The augmented matrix is

$$\left(\begin{array}{cc|c} 1 & 1 & 25 \\ 5 & 10 & 200 \end{array}\right)$$

We then perform the following elementary row operations. Multiply row 1 by -5 and add the results to row 2. We get

$$\left(\begin{array}{cc|c} 1 & 1 & 25 \\ 0 & 5 & 75 \end{array}\right)$$

Multiply row 2 by $\frac{1}{5}$, getting

$$\left(\begin{array}{cc|c} 1 & 1 & 25 \\ 0 & 1 & 15 \end{array}\right)$$

Multiply row 2 by -1 and add the results to row 1:

$$\left(\begin{array}{cc|c} 1 & 0 & 10 \\ 0 & 1 & 15 \end{array}\right)$$

We now read our answer. It is $F = 10$, $T = 15$. Thus, Casandra gave the customer 10 \$5 bills and 15 \$10 bills.

Example 32 Using matrix methods, solve the following system of equations for x and y.

$$3x + 6y = 15$$
$$-x - 2y = 12$$

Solution We first form the augmented matrix:

$$\left(\begin{array}{cc|c} 3 & 6 & 15 \\ -1 & -2 & 12 \end{array}\right)$$

Then we perform the following elementary row operations. Multiply row 1 by $\frac{1}{3}$ to get

$$\left(\begin{array}{cc|c} 1 & 2 & 5 \\ -1 & -2 & 12 \end{array}\right)$$

Add row 1 to row 2, getting

$$\left(\begin{array}{cc|c} 1 & 2 & 5 \\ 0 & 0 & 17 \end{array}\right)$$

The last row says that $0x + 0y = 17$. This is impossible no matter what values x and y have. Thus, the original set of equations has no common solution. As a matter of fact, no matter what technique is used, we can never find a common solution. If we draw both these equations on the same coordinate axes, we find that the lines are parallel.

The same procedure that we used for solving a set of simultaneous equations involving two unknowns can be used to solve a linear system of three equations in three unknowns.

Example 33 Using matrices, solve the following system of equations for x, y, and z.

$$\begin{aligned} 3x + y + 2z &= 6 \\ x + y + 4z &= 3 \\ 2x + 3y + 2z &= 2 \end{aligned}$$

Solution We first form the augmented matrix:

$$\left(\begin{array}{ccc|c} 3 & 1 & 2 & 6 \\ 1 & 1 & 4 & 3 \\ 2 & 3 & 2 & 2 \end{array}\right)$$

Then we try to obtain the 3×3 identity matrix to the left of the dashed line by performing the indicated elementary row operations. Multiply row 1 by $\frac{1}{3}$, getting

$$\left(\begin{array}{ccc|c} 1 & \frac{1}{3} & \frac{2}{3} & 2 \\ 1 & 1 & 4 & 3 \\ 2 & 3 & 2 & 2 \end{array}\right)$$

Multiply row 1 by -1 and add to row 2:

$$\left(\begin{array}{ccc|c} 1 & \frac{1}{3} & \frac{2}{3} & 2 \\ 0 & \frac{2}{3} & \frac{10}{3} & 1 \\ 2 & 3 & 2 & 2 \end{array}\right)$$

Multiply row 1 by -2 and add to row 3:

$$\left(\begin{array}{ccc|c} 1 & \frac{1}{3} & \frac{2}{3} & 2 \\ 0 & \frac{2}{3} & \frac{10}{3} & 1 \\ 0 & \frac{7}{3} & \frac{2}{3} & -2 \end{array}\right)$$

Multiply row 2 by $\frac{3}{2}$:

$$\left(\begin{array}{ccc|c} 1 & \frac{1}{3} & \frac{2}{3} & 2 \\ 0 & 1 & 5 & \frac{3}{3} \\ 0 & \frac{7}{3} & \frac{2}{3} & -2 \end{array}\right)$$

Multiply row 2 by $-\frac{7}{3}$ and add the results to row 3:

$$\left(\begin{array}{ccc|c} 1 & \frac{1}{3} & \frac{2}{3} & 2 \\ 0 & 1 & 5 & \frac{3}{2} \\ 0 & 0 & -11 & -\frac{11}{2} \end{array}\right)$$

Multiply row 3 by $-\frac{1}{11}$:

$$\left(\begin{array}{ccc|c} 1 & \frac{1}{3} & \frac{2}{3} & 2 \\ 0 & 1 & 5 & \frac{3}{2} \\ 0 & 0 & 1 & \frac{1}{2} \end{array}\right)$$

Multiply row 3 by -5 and add the results to row 2:

$$\left(\begin{array}{ccc|c} 1 & \frac{1}{3} & \frac{2}{3} & 2 \\ 0 & 1 & 0 & -1 \\ 0 & 0 & 1 & \frac{1}{2} \end{array}\right)$$

Multiply each element of row 2 by $-\frac{1}{3}$ and add the results to row 1:

$$\left(\begin{array}{ccc|c} 1 & 0 & \frac{2}{3} & \frac{7}{3} \\ 0 & 1 & 0 & -1 \\ 0 & 0 & 1 & \frac{1}{2} \end{array}\right)$$

Finally (at last!), multiply each element of row 3 by $-\frac{2}{3}$ and add the results to row 1, getting

$$\left(\begin{array}{ccc|c} 1 & 0 & 0 & 2 \\ 0 & 1 & 0 & -1 \\ 0 & 0 & 1 & \frac{1}{2} \end{array}\right)$$

We now read our answer from the extreme right column. It is $x = 2$, $y = -1$, $z = \frac{1}{2}$.

Comment It is possible for a system of equations to have more than one solution or for there to be no solutions at all. We will not discuss the conditions that must be satisfied for this to happen.

Comment In Example 33 you might have decided to first interchange rows 1 and 2 instead of multiplying row 1 by $\frac{1}{3}$. Your ultimate answer would be the same.

Exercises

In Exercises 47–54 determine the appropriate elementary row operation(s) that can be used to transform matrix **A** into matrix **B**.

47. $\mathbf{A} = \begin{pmatrix} 2 & 4 \\ -5 & 3 \end{pmatrix} \qquad \mathbf{B} = \begin{pmatrix} 10 & 20 \\ -5 & 3 \end{pmatrix}$

48. $\mathbf{A} = \begin{pmatrix} 12 & 30 & 4 & 2 \\ -2 & 5 & 6 & 9 \end{pmatrix} \qquad \mathbf{B} = \begin{pmatrix} 18 & 45 & 6 & 3 \\ 0 & 90 & 60 & 84 \end{pmatrix}$

49. $\mathbf{A} = \begin{pmatrix} 10 & 15 \\ 6 & 9 \\ 20 & 24 \end{pmatrix} \quad \mathbf{B} = \begin{pmatrix} 2 & 3 \\ 2 & 3 \\ 5 & 6 \end{pmatrix}$

50. $\mathbf{A} = \begin{pmatrix} -4 & 100 & 20 \\ 5 & 8 & 13 \\ 16 & 2 & 3 \end{pmatrix} \quad \mathbf{B} = \begin{pmatrix} 1 & -25 & -5 \\ 0 & 133 & 38 \\ 0 & 402 & 83 \end{pmatrix}$

51. $\mathbf{A} = \begin{pmatrix} -5 & -6 & 8 \\ 3 & 12 & -15 \\ 8 & 4 & -7 \end{pmatrix} \quad \mathbf{B} = \begin{pmatrix} 1 & 4 & -5 \\ 0 & -28 & 33 \\ 0 & 14 & -17 \end{pmatrix}$

52. $\mathbf{A} = \begin{pmatrix} 6 & -12 & -24 \\ 5 & -2 & -36 \\ 3 & -2 & 8 \end{pmatrix} \quad \mathbf{B} = \begin{pmatrix} 1 & 0 & 0 \\ 0 & 1 & 0 \\ 0 & 0 & 1 \end{pmatrix}$

53. $\mathbf{A} = \begin{pmatrix} 4 & 28 & 20 & 12 \\ 6 & 10 & -7 & 3 \\ 5 & 9 & 13 & 1 \end{pmatrix} \quad \mathbf{B} = \begin{pmatrix} 1 & 7 & 5 & 3 \\ 0 & -13 & -6 & -7 \\ 0 & -32 & -37 & -15 \end{pmatrix}$

54. $\mathbf{A} = \begin{pmatrix} 6 & 18 & 42 & 24 \\ 8 & 28 & 72 & 44 \\ 3 & 9 & 22 & 12 \\ 5 & 15 & 35 & 27 \end{pmatrix} \quad \mathbf{B} = \begin{pmatrix} 1 & 0 & 0 & 0 \\ 0 & 1 & 0 & 0 \\ 0 & 0 & 1 & 0 \\ 0 & 0 & 0 & 1 \end{pmatrix}$

Write the augmented matrix for each set of equations in Exercises 55–64. Do not attempt to solve them.

55. $\begin{aligned} 4x + 3y &= 7 \\ 3x + 5y &= 11 \end{aligned}$

56. $\begin{aligned} x - y &= 2 \\ 2x + 3y &= 5 \end{aligned}$

57. $\begin{aligned} 3x - 2y &= 5 \\ 5x + 3y &= -8 \end{aligned}$

58. $\begin{aligned} 7x - 2y &= 8 \\ 3x - 5y &= -9 \end{aligned}$

59. $\begin{aligned} 7x - 5y + 3z &= 8 \\ 4x + 7y - 9z &= -3 \end{aligned}$

60. $\begin{aligned} 5x + 7y - 2z &= 19 \\ 4x - 3y + 8z &= 17 \\ 3x + 9y - 5z &= 11 \end{aligned}$

61. $\begin{aligned} x + 2y &= 7 \\ 3x + 4y &= 8 \\ 7x - 8y &= 1 \end{aligned}$

62. $\begin{aligned} x &= 3 \\ y + 4z &= 8 \\ x - 5z &= 10 \end{aligned}$

63. $\begin{aligned} x + 2y &= 8 \\ 3y - 4z &= 10 \\ 3x + 5z &= 11 \end{aligned}$

64. $\begin{aligned} 2x + 3y + 4z + 5w &= 3 \\ -3x + 7y - 2z + 8w &= 9 \\ 4x - 9y + 3z + 2w &= 7 \\ 9x + 2z - 3w &= 0 \end{aligned}$

65. Write a linear system to correspond to each of the following augmented matrices.

a) $\left(\begin{array}{cc|c} 7 & 2 & 12 \\ 3 & -4 & 14 \end{array}\right)$

b) $\left(\begin{array}{rr|r} 2 & 5 & 15 \\ -3 & 6 & 1 \end{array}\right)$

c) $\left(\begin{array}{rrrr|r} 7 & -2 & -9 & 6 & 1 \\ -3 & 2 & 5 & 8 & 10 \end{array}\right)$

d) $\left(\begin{array}{rrr|r} 7 & 6 & 2 & 6 \\ 4 & -5 & -9 & 6 \\ 5 & 3 & 2 & 4 \end{array}\right)$

e) $\left(\begin{array}{rrrr|r} 1 & 3 & 2 & 5 & 5 \\ 3 & 4 & 1 & 0 & -2 \\ 3 & -2 & 7 & 6 & 4 \end{array}\right)$

In Exercises 66–79 use matrix operations to find solutions (if they exist) for the set of equations.

66. $\begin{aligned} 3x - y &= 3 \\ x + 3y &= 11 \end{aligned}$

67. $\begin{aligned} 4x + 5y &= 22 \\ -4x + 3y &= -6 \end{aligned}$

68. $\begin{aligned} 3x + 2y &= 41 \\ 4x - y &= 40 \end{aligned}$

69. $\begin{aligned} 4y - 6x &= -26 \\ 6y - 4x &= -24 \end{aligned}$

70. $\begin{aligned} 3x + 2y &= 6 \\ x + y &= 1 \end{aligned}$

71. $\begin{aligned} 6x + 10y &= 7 \\ 15x - 4y &= 3 \end{aligned}$

72. $\begin{aligned} x - y &= 11 \\ 3x + 2y &= 3 \end{aligned}$

73. $\begin{aligned} 2x + 2y - z &= 6 \\ x - 2y + 3z &= 11 \\ x + 3y + 2z &= 17 \end{aligned}$

74. $\begin{aligned} x + 3y + z &= -2 \\ x + 6y - z &= -14 \\ 4x + 9y - 2z &= -20 \end{aligned}$

75. $\begin{aligned} 3x - 2y - 3z &= 16 \\ 9x + 3y + 4z &= 29 \\ 6x + y + 2z &= 23 \end{aligned}$

76. $\begin{aligned} x - 2y + 20z &= 4 \\ 2x + y - 8z &= 3 \\ 3x - 4y - 2z &= 10 \end{aligned}$

★77. $\begin{aligned} 3x + 2z &= 22 \\ 4y + z &= 17 \\ 2x + 3y &= 17 \end{aligned}$

★78. $\begin{aligned} 2x + y &= 0 \\ x + 3z &= 8 \\ y - z &= -1 \end{aligned}$

★79. $\begin{aligned} 7x + 3y - 2z + w &= -9 \\ 3x + y + 2z + 3w &= -5 \\ x + y + 4z - 2w &= 19 \\ 2x + 3y + 2z - 7w &= 30 \end{aligned}$

Use matrix techniques to solve the following exercises.

80. *Stamp collecting.* Phil is a stamp collector. In one group of 75 stamps, he has only 3¢ and 5¢ stamps that have a face value of \$2.55. How many stamps of each kind does he have?

81. *Little League baseball.* On Monday the coach of a Little League baseball team purchased six balls and four bats for the team at a cost of \$31. The following day the coach purchased (at the same prices) eight additional balls and three additional bats at a cost of \$32. Find the cost of a bat and the cost of a ball.

82. *Combating mosquitoes.* The owners of the White Lake Recreation Area wish to spray the shores of the lake with a chemical to kill the mosquitoes, which bother the sunbathers. Local environmental regulations

require that the chemical spray contain only 20% Dursban. The owners have two chemical solutions, one that contains 4% Dursban and one that contains 40% Dursban. How many gallons of each should the owners use to produce 36 gallons of a solution that contains 20% Dursban?

83. *Feeding mice.* A psychologist is performing different experiments with white and black mice. All the white mice are identical and each requires the same amount of food daily. The same is true for the black mice. One day the psychologist used 36 units of food to feed eight white mice and six black mice. The following day the psychologist needed 23 units of food to feed five white mice and four black mice. How many units of food are needed to feed a white mouse and a black mouse daily?

84. *Wage discrimination.* A large farmer in the West is accused of paying female sharecroppers lower wages than male sharecroppers for the identical job. During one week the farmer paid out $550 in wages for a work force consisting of four men and five women. In another week the farmer paid out $1150 for a work force consisting of ten men and eight women. Find the weekly wages paid by the farmer for a male sharecropper and for a female sharecropper.

★ 85. *Anthropology.* In 1953 anthropologists discovered a group of 2200 people living on a small island in the Pacific. Of these 2200 people, 400 were considered functionally illiterate. This represented 20% of the male population and 16% of the female population. How many males and how many females were in the group?

86. *College admissions.* A midwestern college received 6000 applications for admission to the premedical or prelaw program. The number of applications for the premedical program was three times the number of applications for the prelaw program. How many students applied to the premedical program and how many applied to the prelaw program?

87. *Motorboat renting.* The Ace Amusement Company rents out motorboats and rowboats for the afternoon. The rental fee received for seven rowboats and six motorboats is $63. The rental fee for four rowboats and three motorboats is $33. Find the rental fee for a motorboat and for a rowboat.

Photo: Grosser, Monkmeyer

★ 88. *Back-to-school outfits.* Three mothers are comparing their costs for outfitting their fifth-grade children with back-to-school outfits. Tracy purchased three pairs of jeans, four sweaters, and five shirts and spent \$98 for the items. Fran spent \$60 for two, three, and two of these items, respectively, and Lisa spent \$90 for four, three, and three, of these items respectively. All the items were the same size and brand name and were purchased in the same store on the same day. Find the cost of a pair of jeans, a sweater, and a shirt.

3.7 Inverses of Matrices

In this section we discuss another method of solving simultaneous systems of linear equations. It involves computing the inverse of a matrix.

Recall that

$$\mathbf{I} = \begin{pmatrix} 1 & 0 \\ 0 & 1 \end{pmatrix}$$

is the 2×2 identity matrix for multiplication. Thus, if

$$\mathbf{B} = \begin{pmatrix} 5 & 3 \\ 9 & 6 \end{pmatrix}$$

then

$$\begin{array}{ccccccccc} \mathbf{I} & \cdot & \mathbf{B} & = & \mathbf{B} & \cdot & \mathbf{I} & = & \mathbf{B} \\ \begin{pmatrix} 1 & 0 \\ 0 & 1 \end{pmatrix} & \cdot & \begin{pmatrix} 5 & 3 \\ 9 & 6 \end{pmatrix} & = & \begin{pmatrix} 5 & 3 \\ 9 & 6 \end{pmatrix} & \cdot & \begin{pmatrix} 1 & 0 \\ 0 & 1 \end{pmatrix} & = & \begin{pmatrix} 5 & 3 \\ 9 & 6 \end{pmatrix} \end{array}$$

Similarly,

$$\begin{pmatrix} 1 & 0 & 0 \\ 0 & 1 & 0 \\ 0 & 0 & 1 \end{pmatrix}$$

is the 3×3 identity matrix. Now consider the 2×2 matrix

$$\mathbf{A} = \begin{pmatrix} 3 & 1 \\ 0 & 2 \end{pmatrix}$$

If we multiply this matrix by

$$\mathbf{A}^{-1} = \begin{pmatrix} \frac{1}{3} & -\frac{1}{6} \\ 0 & \frac{1}{2} \end{pmatrix}$$

we get

$$\begin{array}{ccccccccc} \mathbf{A} & \cdot & \mathbf{A}^{-1} & = & \mathbf{A}^{-1} & \cdot & \mathbf{A} & = & \mathbf{I} \\ \begin{pmatrix} 3 & 1 \\ 0 & 2 \end{pmatrix} & \cdot & \begin{pmatrix} \frac{1}{3} & -\frac{1}{6} \\ 0 & \frac{1}{2} \end{pmatrix} & = & \begin{pmatrix} \frac{1}{3} & -\frac{1}{6} \\ 0 & \frac{1}{2} \end{pmatrix} & \cdot & \begin{pmatrix} 3 & 1 \\ 0 & 2 \end{pmatrix} & = & \begin{pmatrix} 1 & 0 \\ 0 & 1 \end{pmatrix} \end{array}$$

(The reader should verify this multiplication.) In this case the product of **A** and $\mathbf{A}^{-1}$ is the identity matrix. We call $\mathbf{A}^{-1}$ the inverse of matrix **A**.

Definition 3.8 Let $\mathbf{A}$ be any square matrix of dimension $m \times m$. The square matrix $\mathbf{A}^{-1}$ also of dimension $m \times m$ is called the *inverse* of $\mathbf{A}$ if $\mathbf{A} \cdot \mathbf{A}^{-1} = \mathbf{A}^{-1} \cdot \mathbf{A} = \mathbf{I}$, where $\mathbf{I}$ is the $m \times m$ identity matrix.

Comment Only square matrices have inverses. However, not all square matrices have inverses.

How do we find the inverse of any square matrix? To find the inverse of any matrix, we apply the elementary row operations discussed in the preceding section. We first form an augmented matrix by combining the given matrix with the identity matrix. The original matrix goes to the left of the dashed line and the identity matrix is written to the right.

For example, to find the inverse of the matrix

$$\mathbf{A} = \begin{pmatrix} 3 & 4 \\ 6 & 2 \end{pmatrix}$$

we first form the augmented matrix

$$\left(\begin{array}{cc|cc} 3 & 4 & 1 & 0 \\ 6 & 2 & 0 & 1 \end{array}\right)$$

Then we perform elementary row operations so that the original matrix to the left of the dashed line becomes the identity matrix. The resulting matrix to the right of the line will be the inverse of the original matrix $\mathbf{A}$.

In our case we perform the following elementary row operations: Multiply each element of row 1 by $\frac{1}{3}$. This will give a 1 in the row 1, column 1 position. We get

$$\left(\begin{array}{cc|cc} 1 & \frac{4}{3} & \frac{1}{3} & 0 \\ 6 & 2 & 0 & 1 \end{array}\right)$$

Next we multiply each element of row 1 by -6 and add the results to row 2. This will give a 0 in the row 2, column 1 position:

$$\left(\begin{array}{cc|cc} 1 & \frac{4}{3} & \frac{1}{3} & 0 \\ 0 & -6 & -2 & 1 \end{array}\right)$$

Now we multiply each element of row 2 by $-\frac{1}{6}$. This will give a 1 in the row 2, column 2 position:

$$\left(\begin{array}{cc|cc} 1 & \frac{4}{3} & \frac{1}{3} & 0 \\ 0 & 1 & \frac{1}{3} & -\frac{1}{6} \end{array}\right)$$

Finally we multiply each element of row 2 by $-\frac{4}{3}$ and add the results to row 1. This will give a 0 in the row 1, column 2 position:

$$\left(\begin{array}{cc|cc} 1 & 0 & -\frac{1}{9} & \frac{2}{9} \\ 0 & 1 & \frac{1}{3} & -\frac{1}{6} \end{array}\right)$$

The original matrix has now been transformed into the identity matrix. The matrix to the right of the dashed line is the inverse of **A**. Thus,

$$\mathbf{A}^{-1} = \begin{pmatrix} -\frac{1}{9} & \frac{2}{9} \\ \frac{1}{3} & -\frac{1}{6} \end{pmatrix}$$

The reader should verify that $\mathbf{A}^{-1}$ is indeed the inverse of **A** by computing $\mathbf{A} \cdot \mathbf{A}^{-1}$ and $\mathbf{A}^{-1} \cdot \mathbf{A}$. The results should be the identity matrix **I**.

The procedure just described can be used to find the inverse of any square matrix. Let us illustrate the technique with some additional examples.

Example 34 Find the inverse of the matrix.

$$\mathbf{A} = \begin{pmatrix} 6 & -1 \\ 2 & 3 \end{pmatrix}$$

Solution We first form the augmented matrix, which is

$$\left(\begin{array}{cc|cc} 6 & -1 & 1 & 0 \\ 2 & 3 & 0 & 1 \end{array}\right)$$

Then we apply elementary row operations on the matrix to the left of the dashed line to transform it into the identity matrix. The matrix that results to the right of the line is the inverse of **A**. First, multiply each element of row 1 by $\frac{1}{6}$. This will give a 1 in the row 1, column 1 position:

$$\left(\begin{array}{cc|cc} 1 & -\frac{1}{6} & \frac{1}{6} & 0 \\ 2 & 3 & 0 & 1 \end{array}\right)$$

Next we multiply each element of row 1 by -2 and add the results to row 2, getting

$$\left(\begin{array}{cc|cc} 1 & -\frac{1}{6} & \frac{1}{6} & 0 \\ 0 & \frac{10}{3} & -\frac{1}{3} & 1 \end{array}\right)$$

Then we multiply each element of row 2 by $\frac{3}{10}$:

$$\left(\begin{array}{cc|cc} 1 & -\frac{1}{6} & \frac{1}{6} & 0 \\ 0 & 1 & -\frac{1}{10} & \frac{3}{10} \end{array}\right)$$

Now we multiply each element of row 2 by $\frac{1}{6}$ and add the results to row 1:

$$\left(\begin{array}{cc|cc} 1 & 0 & \frac{3}{20} & \frac{1}{20} \\ 0 & 1 & -\frac{1}{10} & \frac{3}{10} \end{array}\right)$$

The original matrix has now been transformed into the identity matrix. The matrix to the right of the dashed line is the inverse of **A**. Thus,

$$\mathbf{A}^{-1} = \begin{pmatrix} \frac{3}{20} & \frac{1}{20} \\ -\frac{1}{10} & \frac{3}{10} \end{pmatrix}$$

Example 35 Find the inverse of the matrix.

$$\mathbf{A} = \begin{pmatrix} 4 & 12 & 8 \\ 2 & 7 & 7 \\ 9 & 27 & 19 \end{pmatrix}$$

Solution In this case we work with a 3×3 square matrix; however, the procedure is exactly the same as it is for a 2×2 matrix. We first form the augmented matrix:

$$\left(\begin{array}{ccc|ccc} 4 & 12 & 8 & 1 & 0 & 0 \\ 2 & 7 & 7 & 0 & 1 & 0 \\ 9 & 27 & 19 & 0 & 0 & 1 \end{array}\right)$$

Then we multiply each element of row 1 by $\frac{1}{4}$, getting

$$\left(\begin{array}{ccc|ccc} 1 & 3 & 2 & \frac{1}{4} & 0 & 0 \\ 2 & 7 & 7 & 0 & 1 & 0 \\ 9 & 27 & 19 & 0 & 0 & 1 \end{array}\right)$$

Next we multiply each element of row 1 by -2 and add the results to row 2, getting

$$\left(\begin{array}{ccc|ccc} 1 & 3 & 2 & \frac{1}{4} & 0 & 0 \\ 0 & 1 & 3 & -\frac{1}{2} & 1 & 0 \\ 9 & 27 & 19 & 0 & 0 & 1 \end{array}\right)$$

Then we multiply each element of row 1 by -9 and add the results to row 3:

$$\left(\begin{array}{ccc|ccc} 1 & 3 & 2 & \frac{1}{4} & 0 & 0 \\ 0 & 1 & 3 & -\frac{1}{2} & 1 & 0 \\ 0 & 0 & 1 & -\frac{9}{4} & 0 & 1 \end{array}\right)$$

Next we multiply each element of row 3 by -3 and add the results to row 2:

$$\left(\begin{array}{ccc|ccc} 1 & 3 & 2 & \frac{1}{4} & 0 & 0 \\ 0 & 1 & 0 & \frac{25}{4} & 1 & -3 \\ 0 & 0 & 1 & -\frac{9}{4} & 0 & 1 \end{array}\right)$$

Then we multiply each element of row 3 by -2 and add the results to row 1:

$$\left(\begin{array}{ccc|ccc} 1 & 3 & 0 & \frac{19}{4} & 0 & -2 \\ 0 & 1 & 0 & \frac{25}{4} & 1 & -3 \\ 0 & 0 & 1 & -\frac{9}{4} & 0 & 1 \end{array}\right)$$

Finally we multiply each element of row 2 by -3 and add the results to row 1, getting

$$\left(\begin{array}{ccc|ccc} 1 & 0 & 0 & -14 & -3 & 7 \\ 0 & 1 & 0 & \frac{25}{4} & 1 & -3 \\ 0 & 0 & 1 & -\frac{9}{4} & 0 & 1 \end{array}\right)$$

The original matrix has now been transformed into the 3×3 identity matrix. The matrix to the right of the dashed line is the inverse of **A**. Thus,

$$\mathbf{A}^{-1} = \begin{pmatrix} -14 & -3 & 7 \\ \frac{25}{4} & 1 & -3 \\ -\frac{9}{4} & 0 & 1 \end{pmatrix}$$

Readers should verify that $\mathbf{A}^{-1}$ is indeed the inverse of **A**.

Comment We mentioned earlier that not all square matrices have inverses. The matrix

$$\mathbf{A} = \begin{pmatrix} 2 & 2 \\ 3 & 3 \end{pmatrix}$$

has no inverse. Thus, we cannot always tell in advance whether *any* square matrix has an inverse or not.

Suppose we are asked to solve the system of equations

$$2x - 3y = 15$$
$$3x + 2y = 3$$

for values of x and y that satisfy both equations. We can apply the idea of the inverse of a matrix to find the solution. We first write the numbers on the left side of the equation in matrix form as

$$\mathbf{A} = \begin{pmatrix} 2 & -3 \\ 3 & 2 \end{pmatrix}$$

The numbers on the right side of the equation can be written as matrix **B**, where

$$\mathbf{B} = \begin{pmatrix} 15 \\ 3 \end{pmatrix}$$

The unknowns, x and y, can be written as matrix **D**, where

$$\mathbf{D} = \begin{pmatrix} x \\ y \end{pmatrix}$$

If we multiply matrix **A** by matrix **D**, we get

$$\mathbf{A} \cdot \mathbf{D} = \begin{pmatrix} 2 & -3 \\ 3 & 2 \end{pmatrix} \cdot \begin{pmatrix} x \\ y \end{pmatrix} = \begin{pmatrix} 2x - 3y \\ 3x + 2y \end{pmatrix}$$

We can then write the original problem in matrix form as

$$\mathbf{A} \cdot \mathbf{D} = \mathbf{B}$$

Let us multiply both sides of this equation by $\mathbf{A}^{-1}$ (the inverse of matrix $\mathbf{A}$). We get

$$\mathbf{A}^{-1}(\mathbf{A} \cdot \mathbf{D}) = \mathbf{A}^{-1}\mathbf{B}$$

$$(\mathbf{A}^{-1} \cdot \mathbf{A})\mathbf{D} = \mathbf{A}^{-1}\mathbf{B} \qquad \text{(By the associative law)}$$

Since $\mathbf{A}^{-1} \cdot \mathbf{A}$ is the identity matrix, we have have

$$\mathbf{I} \cdot \mathbf{D} = \mathbf{A}^{-1}\mathbf{B}$$

or

$$\mathbf{D} = \mathbf{A}^{-1}\mathbf{B} \qquad \text{(The identity matrix does nothing when you multiply by it.)}$$

Thus, to find

$$\mathbf{D} \quad \text{or} \quad \begin{pmatrix} x \\ y \end{pmatrix}$$

we first find the inverse of **A** and then multiply the results by **B**.

In our case

$$\mathbf{A} = \begin{pmatrix} 2 & -3 \\ 3 & 2 \end{pmatrix}$$

To find the inverse of **A**, we form the augmented matrix and perform the indicated elementary row operations.

$$\left(\begin{array}{cc|cc} 2 & -3 & 1 & 0 \\ 3 & 2 & 0 & 1 \end{array}\right) = \text{augmented matrix}$$

Multiply each element of row 1 by $\frac{1}{2}$. This gives

$$\left(\begin{array}{cc|cc} 1 & -\frac{3}{2} & \frac{1}{2} & 0 \\ 3 & 2 & 0 & 1 \end{array}\right)$$

Multiply each element of row 1 by -3 and add the results to row 2:

$$\left(\begin{array}{cc|cc} 1 & -\frac{3}{2} & \frac{1}{2} & 0 \\ 0 & \frac{13}{2} & -\frac{3}{2} & 1 \end{array}\right)$$

Multiply each element of row 2 by $\frac{2}{13}$:

$$\left(\begin{array}{cc|cc} 1 & -\frac{3}{2} & \frac{1}{2} & 0 \\ 0 & 1 & -\frac{3}{13} & \frac{2}{13} \end{array}\right)$$

Finally multiply each element of row 2 by $\frac{3}{2}$ and add the results to row 1. This gives

$$\left(\begin{array}{cc|cc} 1 & 0 & \frac{2}{13} & \frac{3}{13} \\ 0 & 1 & -\frac{3}{13} & \frac{2}{13} \end{array}\right)$$

Thus,

$$\mathbf{A}^{-1} = \begin{pmatrix} \frac{2}{13} & \frac{3}{13} \\ -\frac{3}{13} & \frac{2}{13} \end{pmatrix}$$

Now we multiply $\mathbf{A}^{-1}$ by $\mathbf{B}$ to get our answer. We have

$$\mathbf{D} = \mathbf{A}^{-1} \cdot \mathbf{B}$$

$$\begin{pmatrix} x \\ y \end{pmatrix} = \begin{pmatrix} \frac{2}{13} & \frac{3}{13} \\ -\frac{3}{13} & \frac{2}{13} \end{pmatrix} \cdot \begin{pmatrix} 15 \\ 3 \end{pmatrix}$$

$$= \begin{pmatrix} \frac{2}{3}(15) + \frac{3}{13}(3) \\ -\frac{3}{13}(15) + \frac{2}{13}(3) \end{pmatrix}$$

$$= \begin{pmatrix} \frac{30}{13} + \frac{9}{13} \\ -\frac{45}{13} + \frac{6}{13} \end{pmatrix} = \begin{pmatrix} 3 \\ -3 \end{pmatrix}$$

Thus, we have $x = 3$, $y = -3$. The reader should verify that these are indeed solutions by substituting $x = 3$, $y = -3$ into the original equations.

Example 36 Solve the following equations for x and y using matrix inverses.

$$4x + 9y = 4$$
$$8x - 15y = 30$$

Solution In this case

$$\mathbf{A} = \begin{pmatrix} 4 & 9 \\ 8 & -15 \end{pmatrix} \qquad \mathbf{D} = \begin{pmatrix} x \\ y \end{pmatrix} \qquad \mathbf{B} = \begin{pmatrix} 4 \\ 30 \end{pmatrix}$$

To find $\mathbf{A}^{-1}$ we form the augmented matrix and then perform the indicated elementary row operations.

$$\left(\begin{array}{cc|cc} 4 & 9 & 1 & 0 \\ 8 & -15 & 0 & 1 \end{array}\right) = \text{augmented matrix}$$

Multiply each element of row 1 by $\frac{1}{4}$. This gives

$$\left(\begin{array}{cc|cc} 1 & \frac{9}{4} & \frac{1}{4} & 0 \\ 8 & -15 & 0 & 1 \end{array}\right)$$

Multiply each element of row 1 by -8 and add the results to row 2:

$$\left(\begin{array}{cc|cc} 1 & \frac{9}{4} & \frac{1}{4} & 0 \\ 0 & -33 & -2 & 1 \end{array}\right)$$

Multiply each element of row 2 by $-\frac{1}{33}$, getting

$$\left(\begin{array}{cc|cc} 1 & \frac{9}{4} & \frac{1}{4} & 0 \\ 0 & 1 & \frac{2}{33} & -\frac{1}{33} \end{array}\right)$$

Now multiply each element of row 2 by $-\frac{9}{4}$ and add results to row 1:

$$\left(\begin{array}{cc|cc} 1 & 0 & \frac{15}{132} & \frac{9}{132} \\ 0 & 1 & \frac{2}{33} & -\frac{1}{33} \end{array}\right)$$

Thus,

$$\mathbf{A}^{-1} = \begin{pmatrix} \frac{15}{132} & \frac{9}{132} \\ \frac{2}{33} & -\frac{1}{33} \end{pmatrix}$$

Now we multiply $\mathbf{A}^{-1}$ by $\mathbf{B}$ to get the answer:

$$\mathbf{D} = \mathbf{A}^{-1} \cdot \mathbf{B}$$

$$\begin{pmatrix} x \\ y \end{pmatrix} = \begin{pmatrix} \frac{15}{132} & \frac{9}{132} \\ \frac{2}{33} & -\frac{1}{33} \end{pmatrix} \cdot \begin{pmatrix} 4 \\ 30 \end{pmatrix}$$

$$= \begin{pmatrix} \frac{15}{132}(4) + \frac{9}{132}(30) \\ \frac{2}{33}(4) + (-\frac{1}{33})(30) \end{pmatrix}$$

$$= \begin{pmatrix} \frac{60}{132} + \frac{270}{132} \\ \frac{8}{33} - \frac{30}{33} \end{pmatrix} = \begin{pmatrix} \frac{5}{2} \\ -\frac{2}{3} \end{pmatrix}$$

Therefore, $x = \frac{5}{2}$, $y = -\frac{2}{3}$.

Example 37 Solve for x, y, and z using matrix inverses.

$$2x + 6y + 4z = 22$$
$$4x + 13y - 5z = 33$$
$$5x + 15y + 11z = 56$$

Solution In this case

$$\mathbf{A} = \begin{pmatrix} 2 & 6 & 4 \\ 4 & 13 & -5 \\ 5 & 15 & 11 \end{pmatrix} \qquad \mathbf{D} = \begin{pmatrix} x \\ y \\ z \end{pmatrix} \qquad \mathbf{B} = \begin{pmatrix} 22 \\ 33 \\ 56 \end{pmatrix}$$

We first find $\mathbf{A}^{-1}$ by forming the augmented matrix and performing the indicated elementary row operations.

$$\left(\begin{array}{ccc|ccc} 2 & 6 & 4 & 1 & 0 & 0 \\ 4 & 13 & -5 & 0 & 1 & 0 \\ 5 & 15 & 11 & 0 & 0 & 1 \end{array}\right) = \text{augmented matrix}$$

Multiply each element of row 1 by $\frac{1}{2}$:

$$\left(\begin{array}{ccc|ccc} 1 & 3 & 2 & \frac{1}{2} & 0 & 0 \\ 4 & 13 & -5 & 0 & 1 & 0 \\ 5 & 15 & 11 & 0 & 0 & 1 \end{array}\right)$$

Multiply each element of row 1 by -4 and add the results to row 2:

$$\left(\begin{array}{ccc|ccc} 1 & 3 & 2 & \frac{1}{2} & 0 & 0 \\ 0 & 1 & -13 & -2 & 1 & 0 \\ 5 & 15 & 11 & 0 & 0 & 1 \end{array}\right)$$

Multiply each element of row 1 by -5 and add the results to row 3:

$$\left(\begin{array}{ccc|ccc} 1 & 3 & 2 & \frac{1}{2} & 0 & 0 \\ 0 & 1 & -13 & -2 & 1 & 0 \\ 0 & 0 & 1 & -\frac{5}{2} & 0 & 1 \end{array}\right)$$

Multiply each element of row 3 by 13 and add the results to row 2:

$$\left(\begin{array}{ccc|ccc} 1 & 3 & 2 & \frac{1}{2} & 0 & 0 \\ 0 & 1 & 0 & -\frac{69}{2} & 1 & 13 \\ 0 & 0 & 1 & -\frac{5}{2} & 0 & 1 \end{array}\right)$$

Multiply each element of row 3 by -2 and add the results to row 1:

$$\left(\begin{array}{ccc|ccc} 1 & 3 & 0 & \frac{11}{2} & 0 & -2 \\ 0 & 1 & 0 & -\frac{69}{2} & 1 & 13 \\ 0 & 0 & 1 & -\frac{5}{2} & 0 & 1 \end{array}\right)$$

Finally multiply each element of row 2 by -3 and add the results to row 1:

$$\left(\begin{array}{ccc|ccc} 1 & 0 & 0 & 109 & -3 & -41 \\ 0 & 1 & 0 & -\frac{69}{2} & 1 & 13 \\ 0 & 0 & 1 & -\frac{5}{2} & 0 & 1 \end{array}\right)$$

Then

$$\mathbf{A}^{-1} = \begin{pmatrix} 109 & -3 & -41 \\ -\frac{69}{2} & 1 & 13 \\ -\frac{5}{2} & 0 & 1 \end{pmatrix}$$

Now we multiply $\mathbf{A}^{-1}$ by $\mathbf{B}$ to get our answer. We have

$$\mathbf{D} = \mathbf{A}^{-1} \cdot \mathbf{B}$$

$$\begin{pmatrix} x \\ y \\ z \end{pmatrix} = \begin{pmatrix} 109 & -3 & -41 \\ -\frac{69}{2} & 1 & 13 \\ -\frac{5}{2} & 0 & 1 \end{pmatrix} \cdot \begin{pmatrix} 22 \\ 33 \\ 56 \end{pmatrix}$$

$$= \begin{pmatrix} 109(22) + (-3)(33) + (-41)(56) \\ -\frac{69}{2}(22) + 1(33) + 13(56) \\ -\frac{5}{2}(22) + 0(33) + 1(56) \end{pmatrix}$$

$$= \begin{pmatrix} 3 \\ 2 \\ 1 \end{pmatrix}$$

Thus, $x = 3$, $y = 2$, $z = 1$. Again, the reader should check these answers by substituting $x = 3$, $y = 2$, $z = 1$ into the original equations.

Exercises

Use matrix inverses to solve the systems of equations in Exercises 89–103.

89. $5x + 3y = 21$
$3x - 4y = 1$

90. $2x + 3y = 5$
$6x - 4y = 28$

91. $8x + 3y = 7$
$6x + y = 9$

92. $6x + 5y = -18$
$3x - 2y = 18$

93. $4x + 9y = -7$
$2x - 3y = 19$

94. $x - 2y = 11$
$3x - y = 13$

95. $x - 2y = 11$
$3x - 5y = 31$

96. $3x - 2y = 11$
$5x + 3y = 69$

97. $8x + 9y = 7$
$2x - 3y = 0$

98. $10x - 2y = 31$
$6x + 3y = 6$

99. $3x - 2y + z = 9$
$x - 2y + 3z = -1$
$6x + 5y - 4z = 21$

100. $5x + 7y - 3z = 48$
$3x - 4y + 2z = 1$
$8x + 5y - 4z = 62$

101. $5x + 3y - 7z = 57$
$2x + 4y - z = 25$
$3x - 5y + 2z = -24$

102. $3x - 2y - 3z = 14$
$6x + y + 2z = 22$
$9x + 3y + 4z = 34$

103. $2x + 3y + 4z = 37$
$8x - 6y + 12z = 102$
$-10x + 9y + 8z = 35$

For Exercises 104–108 write a system of equations and then solve the problem using matrix inverses.

★104. *Sociology class.* A student in a sociology class notices that there are seven-tenths as many men as women in the class. When another man and another woman register for the class, there are only five-sevenths as many men as women in the class. How many people were in the class before the two additional students registered?

★105. *Nutrition.* Alexis is on a special diet. Her doctor told her to limit her daily intake of calcium and vitamin C to 800.08 and 65.395 milligrams, respectively. (These are the minimum daily requirements.) Medical research indicates that each cup of skim milk contains 284 milligrams of calcium and 1.25 milligrams of vitamin C. A cup of orange juice contains 22 milligrams of calcium and 129 milligrams of vitamin C. How much of each of these foods should she consume to maintain her special diet?

★106. *Overstock.* A manufacturer of disinfectants has an oversupply of chemicals X and Y. He wishes to use up his entire stock of 250 gallons of chemical X and 306 gallons of chemical Y by producing his most popular disinfectants Vanish and Fragrance. To produce a vat of Vanish, 8 gallons of chemical X and 9 gallons of chemical Y must be used, and to produce a vat of Fragrance 10 gallons of chemical X and 14 gallons of chemical Y must be used. How many of each disinfectant will be produced in using up the entire overstock?

★**107.** *Student charters.* The Alpha Fraternity at South College has arranged for a skiing vacation during the winter recess. Each fraternity member has agreed to pay an equal share. If there were four more members, however, the cost to each would have been decreased by \$10, and if there were six fewer members, the cost to each would have been increased by \$20. How many members are there in the fraternity and what is the cost to each?

★**108.** *Furniture production.* The number of worker-hours required in each stage of production for each type of furniture produced by the Rickety Furniture Corporation is shown in the accompanying table. During one month, 5400 worker-hours were spent in assembling, 1850 were spent in sanding, and 1250 were spent in painting these items. How many of each item were produced during the month?

	Hours per item		
Item	Assembling	Sanding	Painting
Chair	12	5	3
Table	15	4	2
Bookcase	10	3	3

3.8 Further Applications of Matrices

In addition to the applications already discussed, matrices can be applied to many other situations. The following examples illustrate how this can be done.

Example 38 *Faculty evaluations.* Several students in the Beta Fraternity were asked to indicate which of their professors they would recommend to their friends. The results were then tabulated and put into matrix form. An entry of 1 corresponds to a favorable recommendation, and an entry of 0 corresponds to an unfavorable recommendation. Based on this matrix, answer the following questions:

	Prof. Hoffman	Prof. Lichtenfeld	Prof. Blois	Prof. Rogers	Prof. Kennedy
José	1	1	1	0	1
Richard	0	0	1	1	0
Ralph	1	0	1	1	1
John	1	1	0	0	0
Phil	0	0	0	0	1
Jamie	0	1	1	0	0

Based on this matrix, answer the following questions:

a) Which teacher is most recommended?
b) Which teacher is least recommended?
c) Which teacher(s) does John recommend?
d) Which teacher(s) does Richard not recommend?

Solution a) Professor Blois is the most recommended teacher since there are four 1's in her column.
b) Professor Rogers is least recommended since there are only two 1's in his column.
c) John recommends only Professors Hoffman and Lichtenfeld.
d) Richard does not recommend Professors Hoffman, Lichtenfeld, and Kennedy

Example 39 *Tracing mononucleosis.* Bill has been admitted to the hospital with symptoms of mononucleosis. Doctors are anxious to find out the names of all the people who have had contact with Bill within the past few days. Questioning him, the doctors discover the following information: Bill has seen his wife Marion, his brother Jeff, and his daughter Pam with her boyfriend. Marion has seen her parents. Pam has seen only her boyfriend. Jeff has seen his wife Molly. Bill, Marion, Pam, Molly, Jeff, the boyfriend, and Marion's parents have seen no one else. Based on this information, we can set up the accompanying matrix to indicate who can transmit the disease to whom.

	Bill	Marion	Marion's parents	Jeff	Molly	Pam	Pam's boyfriend
Bill	0	1	0	1	0	1	1
Marion	1	0	1	0	0	0	0
Marion's parents	0	1	0	0	0	0	0
Jeff	1	0	0	0	1	0	0
Molly	0	0	0	1	0	0	0
Pam	1	0	0	0	0	0	1
Pam's boyfriend	1	0	0	0	0	1	0

When information is presented in this form, it can be used much more efficiently by medical authorities in tracking down carriers and/or transmitters of diseases.

Example 40 *Radio communications.* Five soldiers—Jed, Max, Catfish, Augustus, and Frenchie—have parachuted into different parts of an island. Their radio receivers and transmitters are not functioning properly. The following is the condition of each man's equipment:

Jed can receive everybody but can transmit only to Catfish.

Max can receive and transmit to everybody.

Catfish can transmit only to Max and Augustus and cannot receive anybody.

Augustus can receive transmissions only from Catfish and Frenchie but cannot transmit.

Frenchie can transmit only to Max but cannot receive anybody.

We can set up a matrix to indicate the communications that are possible. All entries are either 0 or 1. An entry of 1 means that the soldier in that row can transmit to the soldier in that column. If this is not the case, we enter a 0. (Of course, no one transmits messages to himself.) We have the following *communications matrix*.

	Jed	Max	Catfish	Augustus	Frenchie
Jed	0	0	1	0	0
Max	1	0	1	1	1
Catfish	0	1	0	1	0
Augustus	0	0	0	0	0
Frenchie	0	1	0	0	0

The communications matrix indicates many interesting things. For example, if Jed sends a message to Catfish, then this message can be transmitted to Max and Augustus. Frenchie will not be able to get it. Can Frenchie send a message so that everybody will get it?

Comment Matrices provide a quick and easy way of analyzing communications networks. By analyzing the entry in a particular row or column, we can tell at a glance who can communicate with whom.

Example 41 *Camp records.* A camp director is reviewing the preference records of the campers and the corresponding data file. The accompanying coded information has been obtained from the camp office for six campers. An entry of 1 corresponds to a yes.

	Swimming enthusiast	Boating enthusiast	Baseball enthusiast	Hiking enthusiast
Mark	1	1	1	1
Steve	0	0	0	0
Mary	0	1	1	0
Sue	1	1	1	1
Louis	0	1	0	1
Gail	1	0	0	1

$= \mathbf{A}$

	Submitted complete medical history records	Has Red Cross swimmers certificate	Has backpack equipment	Has baseball equipment
Swimming enthusiast	1	0	0	0
Boating enthusiast	0	1	0	0
Baseball enthusiast	0	0	0	1
Hiking enthusiast	0	0	1	0

$= \mathbf{B}$

The product of these matrices, $\mathbf{A} \cdot \mathbf{B}$, is

$$\begin{pmatrix} 1 & 1 & 1 & 1 \\ 0 & 0 & 0 & 0 \\ 0 & 1 & 1 & 0 \\ 1 & 1 & 1 & 1 \\ 0 & 1 & 0 & 1 \\ 1 & 0 & 0 & 1 \end{pmatrix} \cdot \begin{pmatrix} 1 & 0 & 0 & 0 \\ 0 & 1 & 0 & 0 \\ 0 & 0 & 0 & 1 \\ 0 & 0 & 1 & 0 \end{pmatrix} = \begin{pmatrix} 1 & 1 & 1 & 1 \\ 0 & 0 & 0 & 0 \\ 0 & 1 & 0 & 0 \\ 1 & 1 & 1 & 1 \\ 0 & 1 & 1 & 0 \\ 1 & 0 & 1 & 0 \end{pmatrix}$$

Can you interpret the entries in the product matrix $\mathbf{A} \cdot \mathbf{B}$?

An important application of matrices and their inverses is in *cryptography*, or *code theory*. Secret codes are used by governments when sending messages. There are many specialists who are experts at breaking codes. Yet secret codes are quite difficult to break when they are written in matrix form. They can be decoded by using matrix inverses.

One way of coding messages is to associate each letter of the alphabet with some other letter or with some number. Such codes are rather easy to break since certain letters of the alphabet occur more frequently than others. Nevertheless, this is done quite often. When this system is used, we associate a different coding symbol with each letter to avoid confusion.

Example 42 *Coding.* Use the following coding scheme to code the message DESTROY.

A B C D E F G H I J K L M N O P Q R S T U V W X Y Z
↓ ↓
F H Q S A T J B N W C U I O V D L X E Y R G Z M P K

Solution Using the coding scheme, the message DESTROY would be transmitted as SAEYXVP

Example 43 *Coding-alternate scheme.* Use the following coding scheme to code the message DESTROY.

A	B	C	D	E	F	G	H	I	J	K	L	M	N	O	P	Q	R	S	T	U	V	W	X	Y	Z
↓	↓	↓	↓	↓	↓	↓	↓	↓	↓	↓	↓	↓	↓	↓	↓	↓	↓	↓	↓	↓	↓	↓	↓	↓	↓
8	2	9	13	20	1	10	17	5	18	22	3	15	19	11	26	21	25	4	24	12	23	7	14	16	6

Solution The message DESTROY would be transmitted as 13 20 4 24 25 11 16.

Example 44 Use the coding scheme given in Example 42 to decode the following message: SV-OVY-JNGA-RD.

Solution The message SV-OVY-JNGA-RD is DO NOT GIVE UP.

Messages can also be coded using matrices. We must decide in advance whether the message is to be grouped in pairs of letters, triplets of letters, or some other pattern. This determines the dimension of the matrix that has to be used. For example, suppose we wish to send the message

THE DEAL IS OFF

We would rewrite this message in groups of two letters as

TH ED EA LI SO FF

With the code given in Example 43, this message can be written in matrix form as

$$\begin{pmatrix}24\\17\end{pmatrix}\begin{pmatrix}20\\13\end{pmatrix}\begin{pmatrix}20\\8\end{pmatrix}\begin{pmatrix}3\\5\end{pmatrix}\begin{pmatrix}4\\11\end{pmatrix}\begin{pmatrix}1\\1\end{pmatrix}$$

Now we select any 2×2 matrix that has an inverse and contains no fractions. One such matrix is

$$\mathbf{A} = \begin{pmatrix}2 & 3\\1 & 2\end{pmatrix}.$$

Its inverse is

$$\mathbf{A}^{-1} = \begin{pmatrix}2 & -3\\-1 & 2\end{pmatrix}$$

The product of $\mathbf{A}$ and each of the code matrices is

$$\begin{pmatrix}2 & 3\\1 & 2\end{pmatrix}\cdot\begin{pmatrix}24\\17\end{pmatrix} \quad \begin{pmatrix}2 & 3\\1 & 2\end{pmatrix}\cdot\begin{pmatrix}20\\13\end{pmatrix} \quad \begin{pmatrix}2 & 3\\1 & 2\end{pmatrix}\cdot\begin{pmatrix}20\\8\end{pmatrix} \quad \begin{pmatrix}2 & 3\\1 & 2\end{pmatrix}\cdot\begin{pmatrix}3\\5\end{pmatrix} \quad \begin{pmatrix}2 & 3\\1 & 2\end{pmatrix}\cdot\begin{pmatrix}4\\11\end{pmatrix} \quad \begin{pmatrix}2 & 3\\1 & 2\end{pmatrix}\cdot\begin{pmatrix}1\\1\end{pmatrix}$$

The coded message then is transmitted as

99, 58, 79, 46, 64, 36, . . .

The process of translating a message into a matrix is called *encoding*. To *decode*, we simply regroup the numbers into a matrix and then multiply by the matrix $\mathbf{A}^{-1}$.

If we decide to arrange the letters of the message in groups of three letters, the original message becomes

THE DEA LIS OFF

With the code of Example 43, this message can be written in matrix form as

$$\begin{pmatrix} 24 \\ 17 \\ 20 \end{pmatrix} \begin{pmatrix} 13 \\ 20 \\ 8 \end{pmatrix} \begin{pmatrix} 3 \\ 5 \\ 4 \end{pmatrix} \begin{pmatrix} 11 \\ 1 \\ 1 \end{pmatrix}$$

Now we select any 3×3 matrix that has an inverse and contains no fractions. One such matrix is

$$\mathbf{B} = \begin{pmatrix} 1 & 2 & 1 \\ -1 & 1 & 1 \\ 2 & 3 & 1 \end{pmatrix}$$

Its inverse is

$$\mathbf{B}^{-1} = \begin{pmatrix} 2 & -1 & -1 \\ -3 & 1 & 2 \\ 5 & -1 & -3 \end{pmatrix}$$

The product of **B** and each of the code matrices is

$$\begin{pmatrix} 1 & 2 & 1 \\ -1 & 1 & 1 \\ 2 & 3 & 1 \end{pmatrix} \cdot \begin{pmatrix} 24 \\ 17 \\ 20 \end{pmatrix} \quad \begin{pmatrix} 1 & 2 & 1 \\ -1 & 1 & 1 \\ 2 & 3 & 1 \end{pmatrix} \cdot \begin{pmatrix} 13 \\ 20 \\ 8 \end{pmatrix} \quad \begin{pmatrix} 1 & 2 & 1 \\ -1 & 1 & 1 \\ 2 & 3 & 1 \end{pmatrix} \cdot \begin{pmatrix} 3 \\ 5 \\ 4 \end{pmatrix} \quad \begin{pmatrix} 1 & 2 & 1 \\ -1 & 1 & 1 \\ 2 & 3 & 1 \end{pmatrix} \cdot \begin{pmatrix} 11 \\ 1 \\ 1 \end{pmatrix}$$

The coded message is then transmitted as

78, 13, 119, . . .

To decode this message, simply regroup the numbers into a matrix and then multiply by $\mathbf{B}^{-1}$.

Exercises

109. *Planning college courses.* There are three colleges in the State University, each offering degrees in different programs as indicated in matrix **A.** Furthermore, each different program requires that certain math courses be completed as a prerequisite for graduation as indicated in matrix **B**.

	Liberal arts majors	Engineering majors	Business majors	
College I	1	1	0	
College II	1	0	1	= **A**
College III	0	1	1	

	Only math proficiency courses	Only a college algebra course	Calculus course	
Liberal arts	1	0	0	
Engineering	0	0	1	= **B**
Business	0	1	0	

Find the product $\mathbf{A} \cdot \mathbf{B}$ and explain how this helps the college decide which courses to offer.

110. *Radio communications.* Refer to Example 40. Suppose the radios have been adjusted so that the accompanying matrix represents the communications that are now possible.

a) Interpret these entries by indicating who can communicate with whom.

b) Can a message be transmitted so that everyone can receive it? By whom must it be transmitted?

	Jed	Max	Catfish	Augustus	Frenchie
Jed	0	1	0	1	0
Max	0	0	1	1	1
Catfish	1	1	0	1	0
Augustus	1	0	1	0	1
Frenchie	1	1	1	1	0

111. *Dating.* Calvin is interested in obtaining the phone number of Felice, whom he met at Pete's Disco. His friend Doug does not have her number but calls Andrea and Trudy, who came with Felice to the disco. Trudy calls Mike, Felice's ex-boyfriend, who has her number. In this manner Calvin finally obtains Felice's phone number. Set up a communications matrix to indicate how Calvin can obtain Felice's phone number.

112. *Interlocking directorates.* The attorney general of a state is investigating complaints about several insurance companies that have interlocking directorates. The following is a partial list of the directors of five companies that do business within the state.

Company 1: Jones, Gatsby, Carter, McAllister, Pearson

Company 2: Gatsby, Carter, Pearson, Rogers

Company 3: Gatsby, Jones, McAllister

Company 4: Ruiz, Gatsby, Jones, Rogers

Company 5: Ruiz, Carter, Pearson, McAllister

a) Set up a matrix to indicate which board(s) of directors each person serves on.
b) Which director(s) serve on the most boards?
c) Which director(s) serve with each and every other director?

113. *Sending messages.* Use the code given in Example 42 to encode the message DESTROY-THE-BRIDGE-FIRST-NOW.

★ **114.** *Sending secret messages.* Arrange the message given in Exercise 113 in groups of two letters and then in groups of three letters. Find appropriate 2×2 and 3×3 matrices and indicate how the coded message can be transmitted.

★ **115.** *Enemy spy.* A foreign agent transmits the following message, which has been coded with

$$\begin{pmatrix} 3 & 17 \\ 1 & 6 \end{pmatrix}$$

as the coding matrix. Use matrix inverses to decode the message.

187 65 409 143 361 126 77 26 66 23 113 39

3.9 Summary

In this chapter we discussed some ideas of matrices. After defining what a matrix is, we indicated how to add, subtract, and multiply matrices. Although the matrices must be of the same dimension for us to perform the operations of addition and subtraction, this is not true for multiplication. Our unusual method of multiplying matrices was motivated by the applications we had in mind.

We then indicated how the theory of matrices can be applied to solving sets of simultaneous linear equations. Either we work with the augmented matrices or we compute the inverse of a matrix. We limited our discussion to cases where there is one and only one solution for each set of equations. This need not always be the case. The technique we discussed is called the Gauss-Jordan method for solving systems of linear equations.

Finally, we studied situations that give rise to matrices that have only 0's and 1's as their entries. Such matrices are often used in analyzing communications networks, since they indicate who can communicate with whom.

The interested reader can find further applications for matrices by consulting the references given at the end of the chapter.

Study Guide

You should be able to demonstrate your knowledge of the following ideas presented in this chapter by giving definitions, descriptions, or specific examples. Page references are given for each term so that you can check your answer.

Matrix (p. 93)
Dimension of a matrix (p. 93)
Row matrix (p. 94)
Row vector (p. 94)
Matrix multiplication (p. 102)
Elementary row operation (p. 118)
Inverse of a matrix (p. 131)
Communications matrix (p. 142)
Encoding (p. 145)
Entries or elements of a matrix (p. 93)
Square matrix (p. 93)
Column matrix (p. 94)
Column vector (p. 94)
Identity matrix (p. 108)
Augmented matrix (p. 119)
Gauss-Jordan method (p. 121)
Cryptography or code theory (p. 143)

Formulas to Remember

The following formula was given in this chapter.

If matrix **A** is of dimension $m \times n$ and matrix **B** is of dimension $n \times p$, then $\mathbf{A} \cdot \mathbf{B}$ is of dimension $m \times p$.

$$\mathbf{A} \cdot \mathbf{B}$$

$m \times n$ $n \times p$

Do they match?
If yes, the answer
is of dimension
$m \times p$

Mastery Tests

FORM A

1. True or false: Every square matrix has an inverse.
2. The matrix

$$\begin{pmatrix} 1 & 0 \\ 0 & 1 \end{pmatrix}$$

is the identity matrix for

$$\begin{pmatrix} 3 & 1 \\ 6 & -2 \\ 8 & 0 \end{pmatrix}$$

since

$$\begin{pmatrix} 3 & 1 \\ 6 & -2 \\ 8 & 0 \end{pmatrix} \cdot \begin{pmatrix} 1 & 0 \\ 0 & 1 \end{pmatrix} = \begin{pmatrix} 3 & 1 \\ 6 & -2 \\ 8 & 0 \end{pmatrix}$$

Do you agree?

3. The inverse of the matrix

$$\begin{pmatrix} a & b \\ c & d \end{pmatrix}$$

is the matrix

$$\begin{pmatrix} d & -b \\ -c & a \end{pmatrix} \cdot \frac{1}{ad - bc}$$

provided that $ad - bc \neq 0$. Do you agree?

4. True or false: A matrix that has m rows and n columns is of dimension $n \times m$.

5. If a matrix has the same number of rows and columns, what is it called?

6. Can we add these matrices?

$$\begin{pmatrix} 4 & 1 \\ 6 & 3 \end{pmatrix} \text{ and } \begin{pmatrix} 7 & 6 & 1 \\ 9 & -3 & 0 \end{pmatrix}$$

Explain.

7. True or false: In any communications matrix, the diagonal from upper left to lower right has only 0's.

8. What is the 4×4 identity matrix?

9. The matrix

$$\begin{pmatrix} 7 & 1 \\ 8 & -2 \end{pmatrix}$$

can be transformed into the matrix

$$\begin{pmatrix} 1 & 7 \\ -2 & 8 \end{pmatrix}$$

by performing elementary *row* operations. Do you agree?

10. What is the augmented matrix for this set of equations?

$$5x - 3y = 7$$
$$3x + 8y = 26$$

FORM B

1. *Women's gains.* Consider the following table listing the number of elective jobs held by women.

Job	1974	1978
Members of Congress	16	20
Governors	0	2
Lieutenant governors	0	3
Secretaries of state	7	11
State treasurers	8	8
State legislators	466	702
Mayors	566	735
County commissioners	456	660
Municipal-council members	5,365	6,961

Source: Center for the American Woman and Politics.

a) Rewrite the information in matrix form.
b) Compute the total number of elective jobs held by women in 1974 and in 1978. Compare the figures and comment.

2. Let

$$\mathbf{A} = \begin{pmatrix} 1 \\ 2 \\ 3 \\ 4 \\ 5 \\ 6 \\ 7 \\ 8 \\ 9 \\ 10 \end{pmatrix} \quad \text{and} \quad \mathbf{B} = (1 \quad 2 \quad 3 \quad 4 \quad 5 \quad 6 \quad 7 \quad 8 \quad 9 \quad 10)$$

a) Find $\mathbf{A} \cdot \mathbf{B}$.
b) Does this matrix seem familiar? Comment.

3. As we indicated in the comment on page 106, matrix multiplication is not commutative. You may have concluded that this is true because the matrices may not be of the same dimension. Even when the matrices are of the same dimension, however, it is possible that multiplication is not commutative. For example, if

$$\mathbf{A} = \begin{pmatrix} 4 & 14 \\ -2 & 6 \end{pmatrix} \quad \text{and} \quad \mathbf{B} = \begin{pmatrix} 3 & 1 \\ 6 & 2 \end{pmatrix}$$

then $\mathbf{A} \cdot \mathbf{B} \neq \mathbf{B} \cdot \mathbf{A}$. Can you find two 3×3 matrices **A** and **B** such that $\mathbf{A} \cdot \mathbf{B} \neq \mathbf{B} \cdot \mathbf{A}$?

4. *Associative law for multiplication.* If

$$\mathbf{A} = \begin{pmatrix} 5 & -1 \\ 6 & 2 \end{pmatrix} \quad \mathbf{B} = \begin{pmatrix} 3 & 7 \\ -6 & 8 \end{pmatrix} \quad \mathbf{C} = \begin{pmatrix} 1 & -5 \\ 4 & -8 \end{pmatrix}$$

verify that $\mathbf{A}(\mathbf{B} \cdot \mathbf{C}) = (\mathbf{A} \cdot \mathbf{B})\mathbf{C}$.

5. *Dates.* There are 430 girls in the Gamma Sorority. Some information about them is summarized in the accompanying matrix.

	Blonde	Brunette	Redhead
Freshman	71	97	27
Sophomore	69	67	16
Junior	21	23	14
Senior	33	48	11

a) Mathew is interested only in a redhead. How many girls from the Gamma Sorority can he date?
b) Bruce will date only seniors. How many girls from the Gamma Sorority can he date?

6. *Vitamins.* A dietitian is analyzing the vitamin content of three foods. The accompanying matrix gives this information (in convenient units).

	A	B	C	D
Food I	0.4	0.3	0.2	0.1
Food II	0.4	0	0.1	0.5
Food III	0.7	0.1	0.2	0

a) If a person is served 8 units of food I, 7 units of food II, and 3 units of food III, how much of each type of vitamin will be eaten?
b) Suppose that there is a charge only for the vitamin content of each unit of food that is eaten and that this charge is 18¢, 16¢, 20¢, and 30¢, respectively, for vitamins A, B, C, and D. Find the cost of a unit of each food.

★ 7. *Mixed fruit.* A large supermarket sells three different packages of mixed fruit in 1-pound bags labeled I, II, and III. These are packaged according to the accompanying chart. An order is received for a mixture of 126 ounces of apricots, 120 ounces of apples, and 90 ounces of prunes. How can this order be filled?

	Content (ounces)		
Package	Apricots	Apples	Prunes
I	6	5	5
II	5	8	3
III	8	4	4

8. *Cost of packaging mixed fruit.* Refer to question 7. It costs the dealer 7¢ for an ounce of apricots, 4¢ for an ounce of apples, and 5¢ for an ounce of prunes. Find the total cost to the dealer in producing each different size of package.

9. *Weighted averages.* Matrix **A** shows the grades received by five students in Professor Mean's class. Furthermore, the teacher assigns different

$$\begin{array}{lc} & \text{Test} \\ & \begin{array}{ccccc} 1 & \;\;2 & \;\;3 & \;\;4 & \;\;5 \end{array} \\ \begin{array}{l} \text{Mary} \\ \text{Manya} \\ \text{Pessy} \\ \text{Seymour} \\ \text{Alfred} \end{array} & \begin{pmatrix} 65 & 82 & 45 & 30 & 90 \\ 80 & 70 & 60 & 68 & 95 \\ 100 & 75 & 65 & 72 & 58 \\ 78 & 64 & 82 & 75 & 90 \\ 100 & 75 & 86 & 94 & 88 \end{pmatrix} = \mathbf{A} \end{array}$$

weights to each of these tests as shown in matrix **B**.

$$\begin{pmatrix} 20\% \\ 15\% \\ 30\% \\ 25\% \\ 10\% \end{pmatrix} = \mathbf{B}$$

Find the total weighted average for each student. (*Hint:* Find the product $\mathbf{A} \cdot \mathbf{B}$.)

10. *Health statistics.* The death rates in the United States per 100,000 population for various causes are given in the accompanying table.

	Year													
Cause	1910	1915	1920	1925	1930	1935	1940	1945	1950	1955	1960	1965	1970	1975
Cancer	76	81	83	93	97	108	120	134	139	144	149	157	163	169
Cirrhosis	13	12	7	7	7	8	9	9	9	11	12	13	14	16
Diabetes	15	17	16	17	19	22	26	27	17	16	17	18	18	19
Diphtheria	22	16	15	8	5	3	1	1	0	0	0	0	0	0

Source: Bureau of Vital Statistics, Washington, D.C.

a) Rewrite this information in matrix form.
b) How many people died as a result of cancer (per 100,000) in the years indicated?
c) How many people died as a result of diphtheria (per 100,000) in the years indicated?

11. If

$$\mathbf{A} = \begin{pmatrix} 5 & 6 \\ 2 & 1 \end{pmatrix} \quad \mathbf{B} = \begin{pmatrix} 7 & 9 \\ -1 & 3 \end{pmatrix} \quad \mathbf{C} = \begin{pmatrix} 7 & -6 \\ 1 & 2 \end{pmatrix}$$

verify that the *distributive law*

$$\mathbf{A}(\mathbf{B} + \mathbf{C}) = \mathbf{A} \cdot \mathbf{B} + \mathbf{A} \cdot \mathbf{C}$$

is valid.

★ **12.** Show that if

$$\mathbf{A} = \begin{pmatrix} 5 & -6 \\ 2 & 1 \end{pmatrix}$$

then $(\mathbf{A}^{-1})^{-1} = \mathbf{A}$.

★ **13.** Show that the matrix

$$\mathbf{A} = \begin{pmatrix} a & b & c \\ 0 & 0 & 0 \\ d & e & f \end{pmatrix}$$

has no inverse. Can you generalize?

Suggested Further Reading

Campbell, H., *An Introduction to Matrices, Vectors and Linear Programming*, 2d ed., New York: Appleton-Century-Crofts, 1971.

Festinger, L., "The Analysis of Sociograms Using Matrix Algebra," *Human Relations*, vol. 2, pp. 153–158, 1949.

Finkbeiner, D. T., *Elements of Linear Algebra*, San Francisco: Freeman, 1972.

Forsyth, E., and L. Katz, "A Matrix Approach to the Analysis of Sociometric Data: Preliminary Report," *Sociometry*, vol. 9, pp. 340–347, 1946.

Kemeny, J., L. Snell, and J. Thompson, *Introduction to Finite Mathematics*, 3d ed., Englewood Cliffs, N.J.: Prentice-Hall, 1974.

Mathematics in the Modern World (readings from *Scientific American*), article 7, San Francisco: Freeman, 1968.

Newmark, J., and F. Lake, *Mathematics As a Second Language*, 2d ed., Reading, Mass.: Addison-Wesley, 1977.

Sawyer, W. W., *Prelude to Mathematics*, chap. 8, Baltimore: Penguin Books, 1959.

Searle, S. R., *Matrix Algebra for the Biological Sciences*, New York: Wiley, 1966.

Searle, S. R., and W. H. Hausman, *Matrix Algebra for Business and Economics*, New York: Wiley, 1970.

CHAPTER 4 Linear Programming: A Graphical Approach

Objectives to be Achieved in this Chapter

- *To discuss* what a linear programming problem actually is.
- *To review* the algebra of inequalities, including rules for working with them, how to graph them, and how to solve systems of inequalities.
- *To learn* how to apply graphical techniques to help solve linear programming problems.
- *To apply* the ideas of linear programming to solve problems in many different areas of study.

Farmers Protest Rising Costs and Low Farm Food Prices

DAILY PRESS JANUARY 19, 1979

Washington (Jan. 20)—Many of the nation's farmers staged a demonstration for the second day in a row to protest rising costs. For two hours, traffic on heavily traveled Constitution Avenue was at a standstill as farmers parked their tractors on the roads and stopped the engines.

The farmers are protesting the rising prices that they are being forced to pay for their farm supplies and equipment while their profit margin decreases. Said one protesting farmer from Wisconsin, "The rising costs of labor and materials together with a decrease in the amount received for these farm foods are driving me out of business. Although I am attempting to maximize the yields from my farm, I still cannot make ends meet."

The farmers vowed not to leave Washington until they meet with President Carter to voice their plight.

In recent years linear programming has been applied to analysis in many different areas of study, including agriculture and the energy shortage. Is it possible to determine how to maximize acreage yield when we have certain restrictions?

In this chapter we indicate how we can apply graphical techniques to solve some of the linear programming problems associated with the ideas discussed in the newspaper article.

4.1 Introduction

Linear programming is a relatively new branch of mathematics. It was developed by George B. Dantzig during World War II to help American military strategists meet urgent military needs despite the limited resources available. At that time Dantzig was working for the U.S. government. The simplex method that he developed was of great use in solving such problems.

At first the new theory was used by the U.S. Air Force. Specifically, Marshall Wood and his staff applied the theory to the Air Force's secret SCOOP (Scientific Computation of Optimum Program) project. It was also used during the Berlin air lift.

As a result of the initial work by this Air Force group, new applications were found for the theory of linear programming. Today it is used to solve many wide-ranging problems in the military, business, agricultural, medical, and engineering sciences. Discussion of all these applications is beyond the scope of this book; nevertheless, our main objective in this chapter and the next chapter is to introduce you to the basic ideas of linear programming and to show its usefulness by discussing numerous problems and case studies.

As an example of a rather simple linear programming problem and to familiarize you with the nature of the difficulties involved, let us consider the following example.

Example 1 *Nutritional supplement.* A mountain climber has been advised by his doctor to supplement his diet with two special types of foods, A and B, and to take at least 1000 grams of protein and 500 grams of carbohydrates daily. Each unit of food A contains 40 grams of protein and 30 grams of carbohydrates, while each unit of food B contains 60 grams of protein and 20 grams of carbohydrates. Food A can be purchased for 45¢ a unit and food B can be purchased for 35¢ a unit. What is the least expensive mixture of these two foods that the mountain climber can use and still follow his doctor's orders?

At this point we will merely translate the problem from words into a mathematical expression. The solution will be completed in a later section. We first summarize the information given in the problem in the form of a table as follows:

	Nutritional content (grams)		
	Protein	Carbohydrates	Cost
Food A	40	30	45¢
Food B	60	20	35¢
Minimum needed	1000	500	

Now let

$x =$ the number of units of food A to be used

$y =$ the number of units of food B to be used

The mountain climber can obtain 40 grams of protein from each unit of food A and 60 grams of protein from each unit of food B that he eats. Thus, if he takes x units of food A and y units of food B, his total protein consumption will be

$$\begin{pmatrix}\text{grams of}\\ \text{protein}\\ \text{per unit}\\ \text{of food A}\end{pmatrix} \cdot \begin{pmatrix}\text{units of}\\ \text{food A}\\ \text{eaten}\end{pmatrix} + \begin{pmatrix}\text{grams of}\\ \text{protein}\\ \text{per unit}\\ \text{of food B}\end{pmatrix} \cdot \begin{pmatrix}\text{units of}\\ \text{food B}\\ \text{eaten}\end{pmatrix}$$

or

$$40 \cdot x + 60 \cdot y$$

or simply

$$40x + 60y$$

Since the climber needs a minimum of 1000 grams of protein, we have

$$40x + 60y \geq 1000$$

(We use the symbol $\geq$ to stand for "equal to or greater than.")

In a similar manner, the total amount of carbohydrates obtained from foods A and B is $30x + 20y$. Since the climber needs at least 500 units of carbohydrates, we have

$$30x + 20y \geq 500$$

Obviously, x and y cannot be negative; they must be 0 or greater. (Why?) We express this fact mathematically by writing $x \geq 0$ and $y \geq 0$.

Finally, we consider the cost. Each unit of food A costs 45¢ and each unit of food B costs 35¢, so that x units of food A and y units of food B cost

$$\begin{pmatrix}\text{cost per}\\ \text{unit of}\\ \text{food A}\end{pmatrix} \cdot \begin{pmatrix}\text{units of}\\ \text{food A}\\ \text{used}\end{pmatrix} + \begin{pmatrix}\text{cost per}\\ \text{unit of}\\ \text{food B}\end{pmatrix} \cdot \begin{pmatrix}\text{units of}\\ \text{food B used}\end{pmatrix}$$

or

$$45 \quad \cdot \quad x \quad + \quad 35 \quad \cdot \quad y$$

or simply

$$45x + 35y$$

Obviously, the climber wishes to minimize his costs, so we wish to minimize cost $= 45x + 35y$.

Thus, the problem is to find values of x and y such that $x \geq 0$ and $y \geq 0$ subject to the conditions

$$40x + 60y \geq 1000$$
$$30x + 20y \geq 500$$

and that minimize cost $= 45x + 35y$.

A problem of this type in which we wish to minimize or maximize a function subject to certain conditions is called a *linear programming problem*. It is with these types of problems that we concern ourselves in this chapter. The name linear programming is derived from the fact that the problems to which we apply this theory can be translated mathematically into the form of linear equations. (Of course, when the problems translated mathematically are not linear, then we are dealing with nonlinear mathematical programming.)

Since a knowledge of *inequalities* is essential for solving linear programming problems, we will begin our discussion with a review of the algebra of inequalities and their graphical representation. Then we will indicate how we can solve linear programming problems using either a graphical or an algebraic approach. In this chapter we solve linear programming problems graphically. In Chapter 5 we will solve them algebraically.

Historical Note

George Bernard Dantzig was born on November 8, 1914, in Portland, Oregon. He received his bachelor's and master's degrees from the University of Maryland and his doctorate from the University of California at Berkeley.

At first Dantzig worked as a junior statistician for the U.S. Bureau of Labor Statistics from 1937 to 1939. Then he became the chief mathematician for the Combat Analysis Branch, Statistical Control Group, at the U.S. Air Force Headquarters during the years 1941–1946. Later, from 1946 to 1952, he was math advisor and comptroller at the U.S. Air Force Headquarters.

It was during this period, while working for the U.S. Air Force, that he developed the simplex method for linear programming problems.

Currently George Dantzig is a professor of operations research and computer science at Stanford University.

4.2 Linear Inequalities

When we are introduced to the natural numbers, we immediately notice that they come in a particular order; that is, 1 comes first, then 2, then 3, and so on. When a natural number *a precedes* a natural number b in this order of counting, we say that a is less than b and denote this property with the symbol $<$. Observe that $b - a$ is positive. Thus, we write $a < b$. The same is true for any real numbers a and b.

Definition 4.1 If a and b are any real numbers, then *a is less than b*, denoted $a < b$, if $b - a$ is positive.

Definition 4.2 If a and b are any real numbers, then *a is greater than b*, denoted $a > b$, if $a - b$ is positive. Obviously, if $a > b$, then $b < a$.

Notation The symbol $a \leq b$ is read "a is less than or equal to b." Similarly, the symbol $a \geq b$ is read "a is greater than or equal to b."

We can picture the relationship expressed by the inequality $a < b$ by referring to the *number line* shown in Figure 4.1. In this case $a < b$, since a is to the left of b. On the other hand, $c > a$, since c is to the right of a.

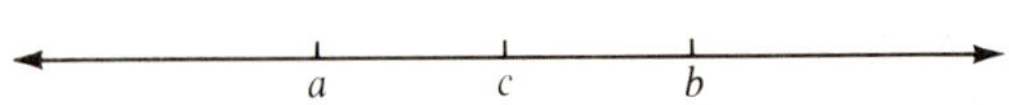

Figure 4.1

The following is a list of certain basic rules of inequalities that we need when working with linear programming problems.

Rules of Inequalities For all real numbers a, b, and c, we have:

1. *Trichotomy rule:* Exactly one of the following is true.

$$a < b \qquad a = b \qquad \text{or} \qquad a > b$$

2. *Transitivity rule:* If $a > b$ and $b > c$, then $a > c$

or

If $a < b$ and $b < c$, then $a < c$

3. *Addition rule:* If $a > b$, then $a + c > b + c$

4. *Multiplication rule:* If $a > b$ and $c > 0$, then $ac > bc$

or

If $a > b$ and $c < 0$, then $ac < bc$

The first three rules seem rather obvious. However, one must be careful when using rule 4. It says that if we multiply both sides of an inequality by a positive number (excluding 0), then the inequality sign remains unchanged. If we multiply both sides by a negative number, then the inequality sign is reversed.

Example 2 Solve the inequality $x + 5 < 8$ for all values of x.

Solution We subtract 5 from both sides using Rule 3. We get

$$\begin{array}{r} x + 5 < \;\;\, 8 \\ -\, 5 = -5 \\ \hline x < \;\;\, 3 \end{array}$$

Thus, any x less than 3 is a solution for the inequality $x + 5 < 8$. Our solution can be represented graphically by the number line shown in Figure 4.2.

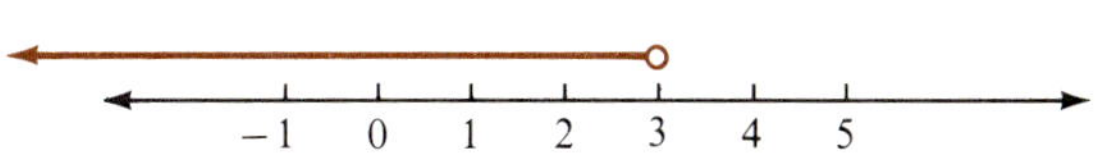

Figure 4.2

Example 3 Solve the inequality $2x + 7x < 3 - 21$.

Solution We first combine terms. This gives

$$9x < -18$$

Then we divide both sides by 9 (a positive number), getting

$$\frac{9x}{9} < \frac{-18}{9}$$

or

$$x < -2$$

Our solution can be represented graphically by the number line shown in Figure 4.3.

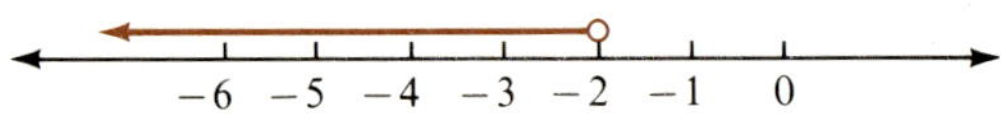

Figure 4.3

Example 4 Solve the inequality $2x - 5x \leq 12 - 27$.

Solution We first combine terms. This gives

$$-3x \leq -15$$

Now we divide both sides by -3. Since we are dividing by a negative number, the inequality sign is reversed. Thus,

$$\frac{-3x}{-3} \geq \frac{-15}{-3}$$

or

$$x \geq 5$$

Our solution can be represented graphically by the line shown in Figure 4.4.

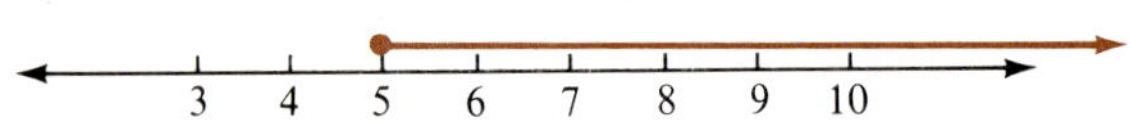

Figure 4.4

In linear programming we are usually concerned with linear inequalities, that is, inequalities in two unknowns of the type $2x + 3y < 6$ or $y - x > 5$ or $2x \geq 6$ or $5y \leq 15$. (The inequalities we discuss in this chapter will always have a positive sign in front of the y term. If this is not the case, we apply Rule 4 for inequalities to change the sign in front of the y term to a positive sign.)

We can draw the graph of such inequalities by the following procedure (we assume that the y term of an inequality containing both x and y terms is on the left side of the inequality):

1. Change the inequality sign to an equals sign.
2. Graph the line for the equation given in step 1.
3. If the inequality is "less than," shade the area below the line.
4. If the inequality is "greater than," shade the area above the line.
5. If the line is vertical and the inequality is "less than," shade the area to the left of the line.
6. If the line is vertical and the inequality is "greater than," shade the area to the right of the line.
7. If the inequality is $\leq$ or $\geq$, include the line. Otherwise, do not include the line. This situation is usually indicated by a dashed line.

Example 5 Draw the graph of $2x + 3y > 6$.

Solution We first change the $>$ sign to an $=$ sign, getting $2x + 3y = 6$. We then draw the graph of this equation by using a table of values and plotting the points. Since the inequality is $>$, we shade the area above the line $2x + 3y = 6$ but do *not* include the line itself. This is shown in Figure 4.5.

x	y
0	2
3	0
6	−2
−3	4
−6	6

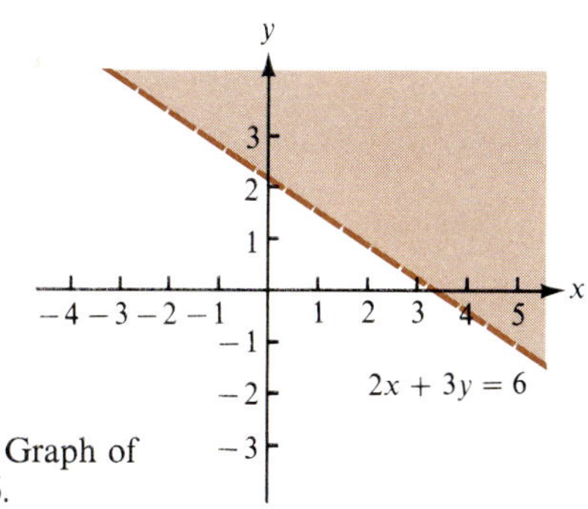

Figure 4.5 Graph of $2x + 3y > 6$.

Example 6 Draw the graph of $3x + 4y \leq 12$.

Solution We first change the $\leq$ sign to an $=$ sign, getting $3x + 4y = 12$. Then we graph this line and shade the area below the line. The line itself *is* included. See Figure 4.6.

x	y
4	0
0	3
8	−3
12	−6
−4	6

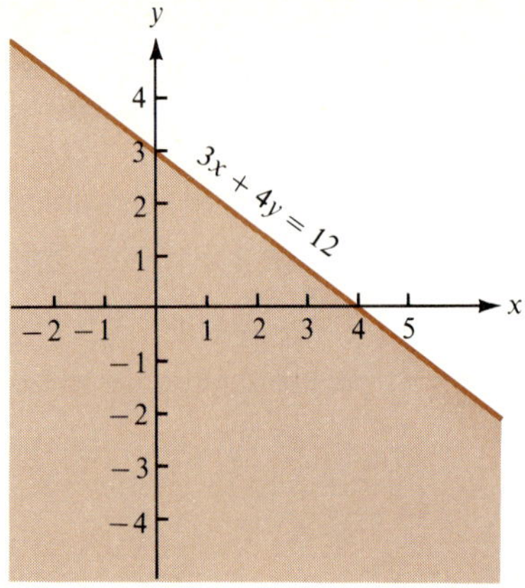

Figure 4.6 Graph of $3x + 4y \leq 12$.

Example 7 Draw the graphs of

a) $x \geq 0$ b) $y \geq 0$

Solution The graphs of $x \geq 0$ and $y \geq 0$ are shown in Figures 4.7 and 4.8, respectively.

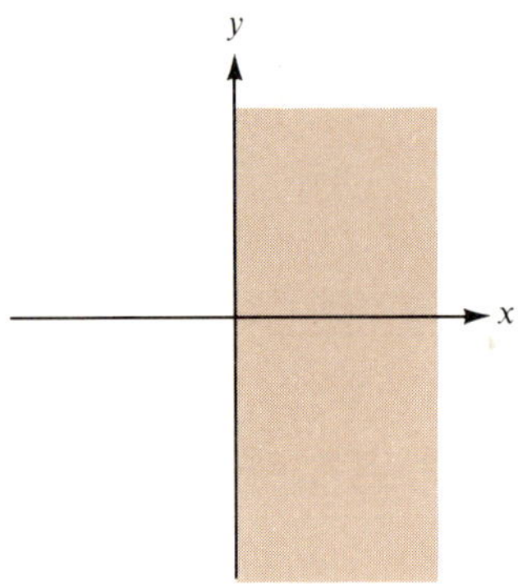

Figure 4.7 Graph of $x \geq 0$.

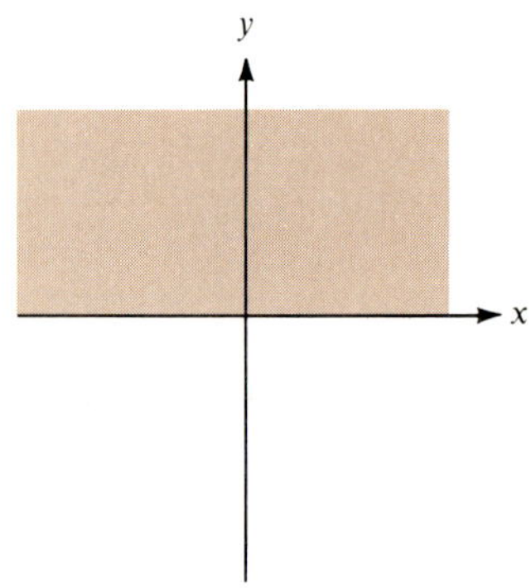

Figure 4.8 Graph of $y \geq 0$.

Systems of Linear Inequalities In Chapters 2 and 3 we discussed numerous methods (both algebraic and graphical) for solving a system of linear equations involving two unknowns. Suppose we are given a system of linear inequalities and wish to find the *solution set*, that is, solve them for values of x and y that satisfy both inequalities. We can solve such a system by using graphical techniques as illustrated in the following examples.

Example 8 Graph the solution set for the set of inequalities.

$$y > x + 2$$
$$y > 6 - x$$

Solution We first draw the graph of the inequality $y > x + 2$. To do this, we change the $>$ sign to an $=$ sign, getting $y = x + 2$. Then we graph the line and shade the area above the line. This is shown in Figure 4.9, where we have shaded horizontally the area above the line.

On the same set of axes we draw the graph of $y > 6 - x$ and shade vertically the area above the line $y = 6 - x$. Notice that one part of the diagram has both kinds of shading. Any point in this region satisfies both inequalities. For example, the point (2, 5) is in this region. The values $x = 2$ and $y = 5$ satisfy both inequalities, since

$$\begin{array}{lcl} y > x + 2 & & y > 6 - x \\ 5 > 2 + 2 & \text{and} & 5 > 6 - 2 \\ 5 > 4 & & 5 > 4 \end{array}$$

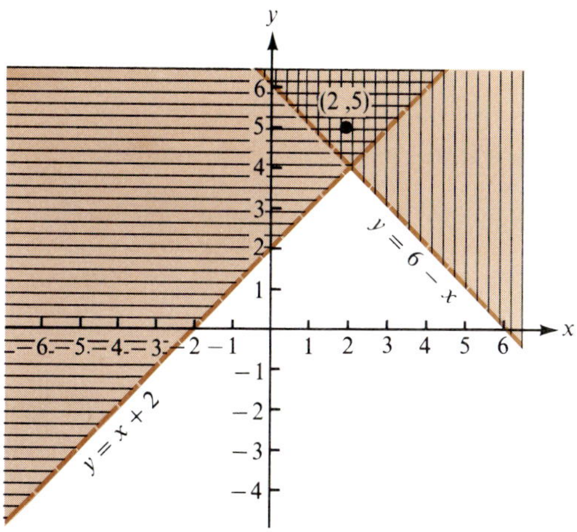

Figure 4.9

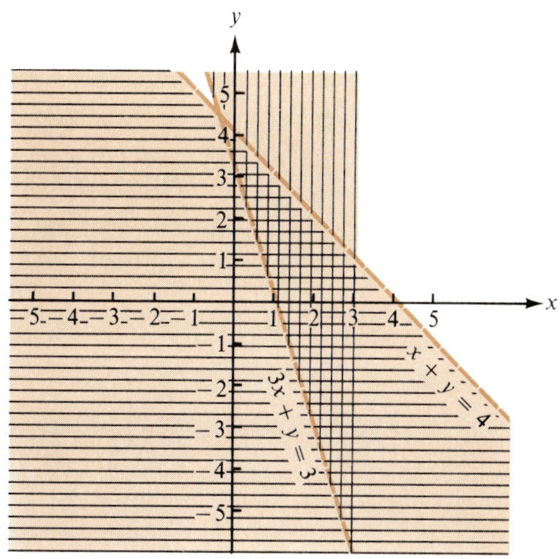

Figure 4.10

Example 9 Graph the solution set for the set of inequalities.

$$x + y < 4$$
$$3x + y > 3$$

Solution The graph of $x + y < 4$ is found by drawing the graph of the line $x + y = 4$ and shading the area below the line. This region is shaded horizontally in Figure 4.10. On the same set of axes we draw the graph of $3x + y > 3$ by shading the area above the line $3x + y = 3$. This region is shaded vertically in Figure 4.10. The region that has both kinds of shading represents the solution to both inequalities.

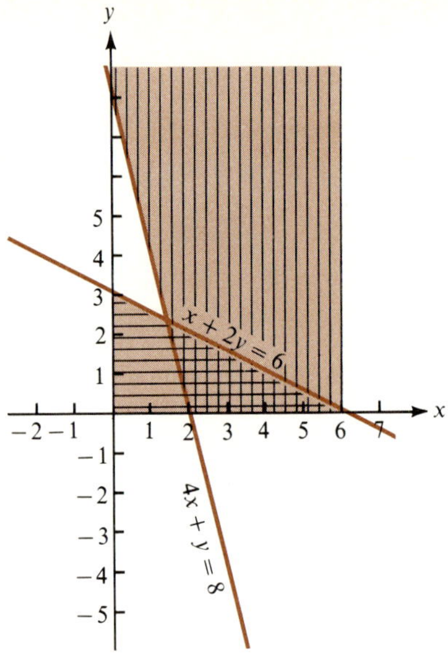

Figure 4.11

Example 10 Graph the solution set for the following system of inequalities.

$$4x + y \geq 8$$
$$x + 2y \leq 6$$
$$x \geq 0, y \geq 0$$

Solution We draw the graph of $4x + y \geq 8$ by shading the area above the line $4x + y = 8$. This region is shaded vertically. The graph of $x + 2y \leq 6$ is found by shading the area below the line $x + 2y = 6$. This region is shaded horizontally. The additional inequalities tell us that both x and y must be positive. Thus, we shade nothing below the x axis and nothing to the left of y axis. The region that has both kinds of shading satisfies all the inequalities. See Figure 4.11.

Exercises

Graph the solution set for the following sets of inequalities.

1. $x + 2y \geq 4$
$3x + 4y < 12$

2. $-x + 2y \leq 4$
$x + 4y \leq 8$

3. $x + y < 6$
$2x + y > 1$

4. $4x + 3y \leq 12$
$x + y > 3$

5. $x + y < 4$
$3x + y > 2$

6. $x + y < 0$
$5x - 3y < 15$

7. $x + y > 2$
$4x + 7 < 3y$

8. $2x + 3y \le 18$
$3x + 2y \le 12$
$x \ge 0, y \ge 0$

9. $3x + 2y \le 6$
$2x + 3y \le 6$
$x \ge 0, y \ge 0$

10. $3x + 5y \le 15$
$2x + y \ge 1$
$x \ge 0, y \ge 0$

11. $3x + 4y \le 15$
$-x + 2y \le 5$
$x \ge 0, y \ge 0$

12. $x + y \le 8$
$3x + y \le 9$
$x \ge 0, y \ge 0$

13. $x + y \le 1$
$2x + 8y \le 6$
$x \ge 0, y \ge 0$

14. $2x + 3y \ge 12$
$4x + 3y \ge 18$
$x \ge 0, y \ge 0$

15. $2x + 3y \ge 12$
$x + 2y \le 14$
$x - y \ge 6$
$x \ge 0, y \ge 0$

4.3 Graphical Solution of Linear Programming Problems

In this section we discuss one technique for finding maximum or minimum values of linear functions subject to certain conditions called *constraints*. This geometric approach to solving linear programming problems can be used primarily when there are only two unknowns (x and y). If there are more than three unknowns (x, y, and z) or if there are many inequalities, however, this method becomes either very complicated or totally impossible. A more general approach for solving linear programming problems that can be used in every case will be given in Chapter 5. This alternate approach is known as the *simplex method*.

The following examples illustrate the graphical technique for solving linear programming problems.

Example 11 *MacDougald's burgers.* A local branch of the MacDougald's national fast food chain sells two kinds of burgers, the superburger and the babyburger. Each babyburger requires 6 ounces of meat and 8 ounces of cheese and produces a profit of 30¢. Each superburger requires 8 ounces of meat and 2 ounces of cheese and produces a profit of 45¢. The manager can use at most $7\frac{1}{2}$ pounds (120 ounces) of meat and 6 pounds (96 ounces) of cheese for the production of these two types of burgers. How many superburgers and babyburgers should be prepared to maximize profit?

Solution We first translate the problem into mathematical symbols. We let

x = the number of babyburgers to be produced

y = the number of superburgers to be produced

We can then make a table that shows all the information given in the problem.

	Amount needed (ounces)		
	Meat	Cheese	Profit
Babyburger	6	8	30¢
Superburger	8	2	45¢
Limitations	120	96	

Each babyburger takes 6 ounces of meat and each superburger takes 8 ounces of meat. Since the manager can use at most 120 ounces of meat in the production of x babyburgers and y superburgers, we have

$$6x + 8y \leq 120$$

Similarly, each babyburger takes 8 ounces of cheese and each superburger takes 2 ounces of cheese. Since the manager can spare at most 96 ounces of cheese in the production of x babyburgers and y superburgers, we have

$$8x + 2y \leq 96$$

Obviously, $x \geq 0$ and $y \geq 0$. (Why?) The manager makes a profit of 30¢ on each babyburger and 45¢ on each superburger sold so that when x babyburgers and y superburgers are sold, the profit is $30x + 45y$.

The mathematical formulation of the problem is to find values of x and y such that

$$6x + 8y \leq 120$$
$$8x + 2y \leq 96$$
$$x \geq 0, y \geq 0$$

and that maximize profit $= 30x + 45y$.

Now we are ready to solve this problem by using the graphical method. We first draw the graph of $6x + 8y \leq 120$ by vertically shading the area below the line $6x + 8y = 120$. Since x and y must both be positive, we shade nothing below the x axis and nothing to the left of the y axis. This is shown in Figure 4.12. On the same set of axes, we then draw the graph of $8x + 2y \leq 96$ by horizontally shading the area below the line $8x + 2y = 96$. This is also shown in Figure 4.12.

One region of the graph has been shaded twice. This region satisfies all the inequalities. We have redrawn this region in Figure 4.13. You will notice that the region has four corners, which we have labeled

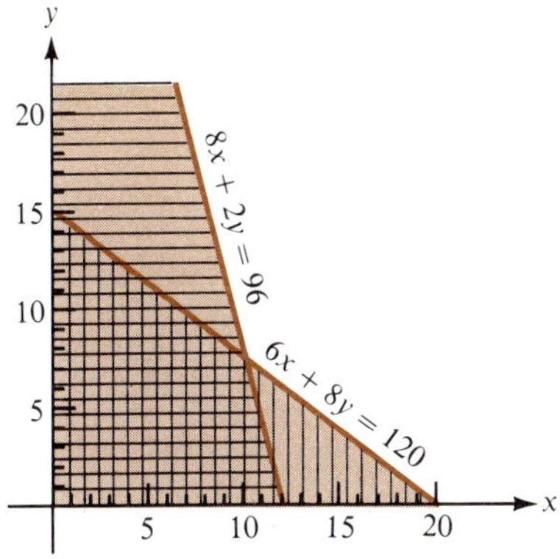

Figure 4.12

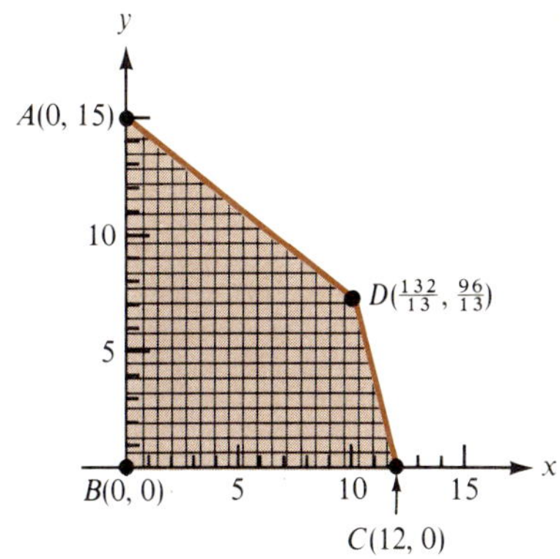

Figure 4.13

A, B, C, and D. It is shown in more advanced texts that a profit expression such as the one we are working with will always have a maximum or minimum value at a *corner* or *extreme point* of the region. Thus, we can obtain the extreme points by reading them off the graph or by considering the inequalities as equations and solving them two at a time. We then set up the following table.

When extreme point or corner is at the point	Then the profit is
(x, y)	$30x + 45y$
$A = (0, 15)$	$30(0) + 45(15) = 675$¢ or \$6.75
$B = (0, 0)$	$30(0) + 45(0) = 0$
$C = (12, 0)$	$30(12) + 45(0) = 360$¢ or \$3.60
$D = (\frac{132}{13}, \frac{96}{13})$	$30(\frac{132}{13}) + 45(\frac{96}{13}) = 637$¢ rounded off, or \$6.37

Thus, the maximum profit occurs when $x = 0, y = 15$. This means that the manager should make no babyburgers and 15 superburgers. His profit will be \$6.75.

Comment If the manager makes any other quantity of superburgers and babyburgers, he will either earn less money or need more than the $7\frac{1}{2}$ pounds of meat and 6 pounds of cheese that are available.

Example 12 *Electric toy trains.* The Electric Train Toy Company manufactures two kinds of electric train sets: the G & O railroad set and the Pensy railroad set. To produce these train sets, each must be processed by two factories, A and B. Factory A assembles the trains and remains open at most 9 hours a day. Factory B paints and packages the trains and remains open at most 8 hours a day. Each G & O railroad train set requires 1 hour in factory A and 2 hours in factory B and yields a profit of \$14. Each Pensy railroad train set requires 3 hours in factory A and only 1 hour in factory B and yields a profit of \$18. How many train sets of each type should be manufactured to maximize profit?

Solution We can make a box that contains all the information given in the problem.

	Factory A	Factory B	Profit
G & O train set	1	2	\$14
Pensy train set	3	1	\$18
Limitations	9	8	

Now let

x = the number of G & O railroad train sets to be produced

y = the number of Pensy railroad train sets to be produced

Each G & O train set requires 1 hour in factory A and each Pensy train set requires 3 hours in factory A. Since factory A remains open at most 9 hours a day in the production of x G & O train sets and y Pensy train sets, we have

$$x + 3y \leq 9$$

Also, each G & O train set requires 2 hours in factory B and each Pensy train set requires 1 hour in factory B. Since factory B can remain open at most 8 hours a day in the production of x G & O train sets and y Pensy train sets, we have

$$2x + y \leq 8$$

Obviously, $x \geq 0$ and $y \geq 0$. When x G & O train sets and y Pensy train sets are produced, the profit is $14x + 18y$.

The mathematical formulation of the problem is to find values of x and y such that

$$\begin{aligned} x + 3y &\leq 9 \\ 2x + y &\leq 8 \\ x \geq 0,\ y &\geq 0 \end{aligned}$$

and that maximize profit $= 14x + 18y$.

We solve this problem by the graphical method. We first draw the graph of $x + 3y \leq 9$ by vertically shading the area below the line $x + 3y = 9$. Since x and y must both be positive, we shade nothing below the x axis and nothing to the left of the y axis. This is shown in Figure 4.14. On the same set of axes we draw the graph of $2x + y \leq 8$ and horizontally shade the area below the line $2x + y = 8$. This is also shown in Figure 4.14.

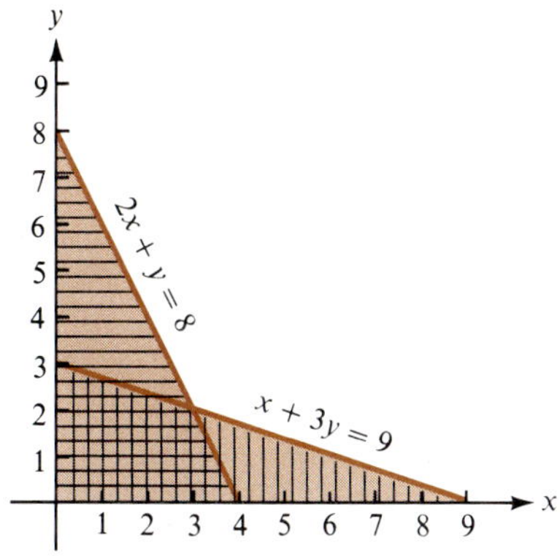

Figure 4.14

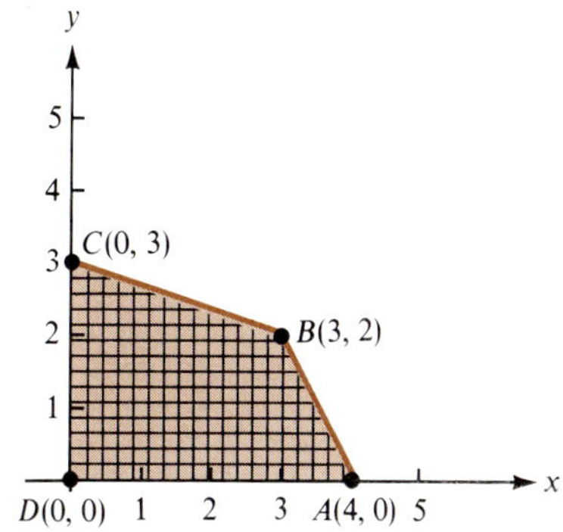

Figure 4.15

Again, we redraw the doubly shaded region as shown in Figure 4.15. There are four extreme points, which we have labeled A, B, C, and D. We then set up the following table.

When extreme point or corner is at the point	Then the profit is
(x, y)	$14x + 18y$
$A = (4, 0)$	$14(4) + 18(0) = \$56$
$B = (3, 2)$	$14(3) + 18(2) = \$78$
$C = (0, 3)$	$14(0) + 18(3) = \$54$
$D = (0, 0)$	$14(0) + 18(0) = \$0$

Thus, the maximum profit occurs when $x = 3, y = 2$. This means that the company should manufacture three G & O railroad train sets and two Pensy railroad train sets. The profit will be $78.

Example 13 Refer to Example 1. The mathematical formulation of the problem was to find values of x and y subject to the constraints (conditions):

$$40x + 60y \geq 1000$$
$$30x + 20y \geq 500$$
$$x \geq 0, y \geq 0$$

and that minimize the cost $= 45x + 35y$. Let us actually find the least expensive mixture of the two foods that the mountain climber can use and still follow his doctor's orders.

Solution Since we already have the mathematical formulation of the problem, we can begin by first drawing the graph of $40x + 60y \geq 1000$ and vertically shading the area above the line $40x + 60y = 1000$. Again we shade nothing below the x axis and nothing to the left of the y axis, as both x and y must be positive. This is shown in Figure 4.16. On the same set of axes we draw the graph of $30x + 20y \geq 500$. We shade

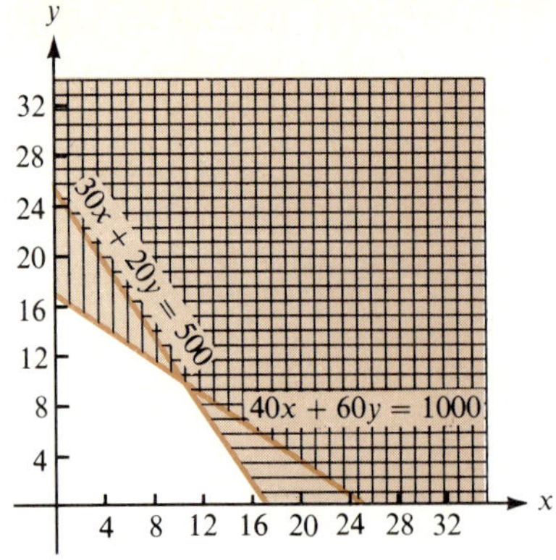

Figure 4.16

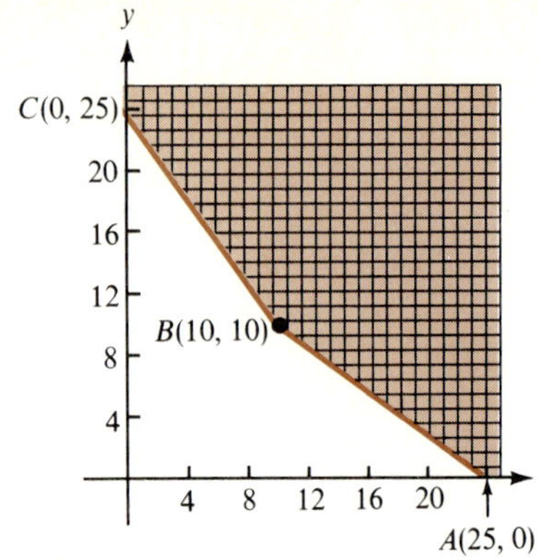

Figure 4.17

horizontally the area above the line $30x + 20y = 500$. This is also shown in Figure 4.16.

Now we redraw the doubly shaded region as shown in Figure 4.17. There are three extreme points, which we have labeled A, B, and C. We then have the following table.

When the extreme point or corner is at the point	Then the cost is
(x, y)	$45x + 35y$
$A = (25, 0)$	$45(25) + 35(0) = 1125$¢, or \$11.25
$B = (10, 10)$	$45(10) + 35(10) = 800$¢, or \$8.00
$C = (0, 25)$	$45(0) + 35(25) = 875$¢, or \$8.75

Thus, the minimum cost occurs when $x = 10$, $y = 10$. This means that the climber should use 10 units of food A and 10 units of food B. His cost will be \$8.00.

Linear programming can also be used in the age-old problem of manpower forecasting, that is, in determining the correct size of a labor force to have available. Forecasting the best size of labor force to have available for meeting production schedules is a difficult task. Furthermore, costs have to be minimized. Future attrition among old employees as well as among new employees is an important factor that cannot be overlooked.

Nevertheless, in recent years mathematicians have been successfully applying linear programming techniques to enable a company to minimize its overall labor costs. The following example is a considerably simplified version of the difficulties involved.

Example 14 *Manpower forecasting with linear programming.* A large department store estimates that it will have to hire at least 150 new sales personnel for the upcoming holiday season. Based on past experience, the company knows that it should hire at least 30 more part-time workers than full-time workers. Furthermore, at least 40 new full-time workers should be hired. The company pays each new full-time employee \$250 a week and each part-time employee \$40 a week. How many employees of each type should the department store hire to minimize its cost while satisfying all requirements?

Solution Let

$$x = \text{the number of full-time workers to be hired}$$
$$y = \text{the number of part-time workers to be hired}$$

Since at least 150 new workers are to be hired, we have $x + y \geq 150$. Also, there must be at least 30 more part-time workers than full-time workers, so $y \geq x + 30$. We are told that at least 40 full-time workers must be hired. This means that $x \geq 40$. The mathematical formulation of the problem, then, is to find values of x and y such that

$$x + y \geq 150$$
$$y \geq x + 30$$
$$x \geq 40$$
$$x \geq 0, y \geq 0$$

and for which the cost $= 250x + 40y$ is minimized.

We first draw the graph of $x + y \geq 150$ by shading horizontally the area above the line $x + y = 150$. This is shown in Figure 4.18. Then we draw the graph of $y \geq x + 30$ by shading vertically the area above the line $y = x + 30$. The graph of $x \geq 40$ is drawn by darkening the area to the right of the vertical line $x = 40$. These graphs are all shown in Figure 4.18.

Notice that one region has been shaded three times. This region satisfies all the inequalities. We redraw the triply shaded region as shown in Figure 4.19. There are two extreme points, which we have

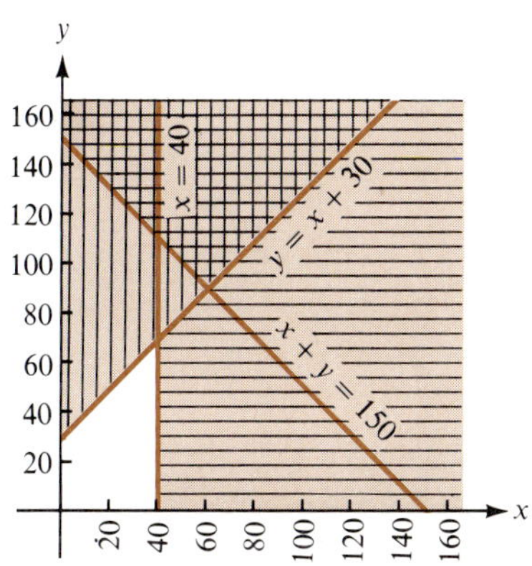

Figure 4.18

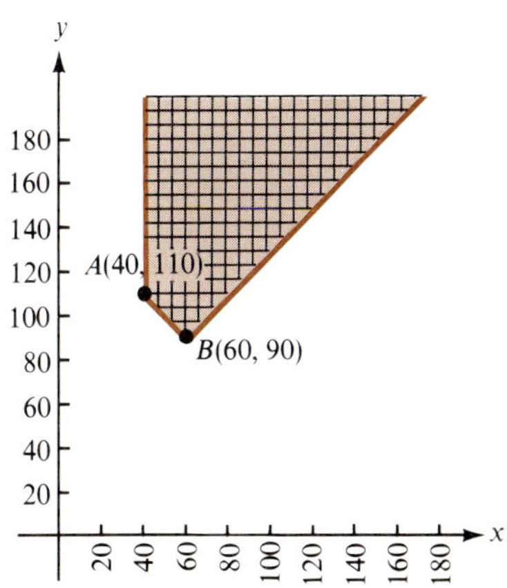

Figure 4.19

labeled A and B. Then we make up the following table:

When the extreme point or corner is at the point	Then the cost is
(x, y)	$250x + 40y$
$A = (40, 110)$	$250(40) + 40(110) = \$14{,}400$
$B = (60, 90)$	$250(60) + 40(90) = \$18{,}600$

Thus, the minimum cost occurs when $x = 40$ and $y = 110$. This means that the department store should hire 40 new full-time workers and 110 new part-time workers. The total cost will be \$14,400.

Comment As we indicated earlier, the graphical technique for solving linear programming problems is very simple and convenient to use when there are few unknowns. If there are more than three unknowns or if there are many inequalities, then this method becomes quite complicated or simply unmanageable. It is for this reason that we will study a more general technique, the simplex method, in the next chapter. This more general method can be used conveniently even when the graphical method is unmanageable.

Exercises

In Exercises 16–40 use the graphical method discussed in this section to maximize or minimize each function as indicated.

16. Maximize $z = -2x + 4y$
subject to
$$\begin{aligned} 4x + 2y &\le 6 \\ 2x + 4y &\le 6 \\ x \ge 0,\ y &\ge 0 \end{aligned}$$

17. Maximize $z = 5x + 6y$
subject to
$$\begin{aligned} -2x + 2y &\le 4 \\ 3x + 4y &\le 12 \\ x \ge 0,\ y &\ge 0 \end{aligned}$$

18. Maximize $z = 3x + 2y$
subject to
$$\begin{aligned} x + y &\le 3 \\ x + 2y &\le 4 \\ x \ge 0,\ y &\ge 0 \end{aligned}$$

19. Maximize $z = 5x + 3y$
subject to
$$\begin{aligned} 8x + 5y &\le 40 \\ 2x + 3y &\le 18 \\ x \ge 0,\ y &\ge 0 \end{aligned}$$

20. Minimize $z = 3x + 2y$
subject to
$$\begin{aligned} 2x + y &\ge 10 \\ x + 4y &\ge 12 \\ x \ge 0,\ y &\ge 0 \end{aligned}$$

21. Minimize $z = 6x + 4y$
subject to
$$\begin{aligned} 2x + 3y &\ge 6 \\ 4x + 24y &\ge 24 \\ x \ge 0,\ y &\ge 0 \end{aligned}$$

22. Minimize $z = 6x + 2y$
subject to
$$\begin{aligned} 2x + 3y &\ge 6 \\ 4x + y &\ge 4 \\ x \ge 0,\ y &\ge 0 \end{aligned}$$

23. Maximize $z = 5x + 7y$
subject to
$$\begin{aligned} x + y &\le 5 \\ 8x + 2y &\le 16 \\ x \ge 0,\ y &\ge 0 \end{aligned}$$

24. Minimize $z = 5x + 3y$
subject to
$$\begin{aligned} x + 4y &\ge 8 \\ 2x + y &\le 4 \\ x \ge 0,\ y &\ge 0 \end{aligned}$$

25. Minimize $z = 2x - 3y$
subject to
$$\begin{aligned} 4x + y &\ge 4 \\ 3x + 5y &\ge 15 \\ x \ge 0,\ y &\ge 0 \end{aligned}$$

26. Maximize $z = 3y + 2x$
subject to
$$\begin{aligned} x + y &\le 8 \\ 2x + y &\le 8 \\ x + 3y &\le 12 \\ x \ge 0,\ y &\ge 0 \end{aligned}$$

27. Minimize $z = 3x - 2y$
subject to
$$\begin{aligned} x + y &\le 4 \\ x + 2y &\ge 2 \\ x \ge 0,\ y &\ge 0 \end{aligned}$$

28. Maximize $z = 5x + 8y$
subject to
$$\begin{aligned} x + y &\ge 2 \\ 2x + 3y &\le 15 \\ 3x + y &\le 12 \\ x \ge 0,\ y &\ge 0 \end{aligned}$$

29. Minimize $z = 9x - 2y$
subject to
$$\begin{aligned} x + 3y &\le 18 \\ 2x + 3y &\ge 12 \\ 2x + y &\le 14 \\ x \ge 0,\ y &\ge 0 \end{aligned}$$

30. Maximize $z = 4x - 3y$
subject to
$$\begin{aligned} 2x + y &\ge 6 \\ x + y &\le 8 \\ x + 2y &\le 6 \\ x \ge 0,\ y &\ge 0 \end{aligned}$$

★ **31.** Minimize $z = 3x + 7y$
subject to
$$\begin{aligned} x + y &\ge 4 \\ 2x + 3y &\le 15 \\ 1 \le x &\le 4 \\ y &\ge 0 \end{aligned}$$

★ **32.** Maximize $z = 3x + 8y$
subject to
$$\begin{aligned} 2x - y - 2 &\ge 0 \\ x + y - 2 &\le 0 \\ y &< 7 \\ x &\ge 0 \end{aligned}$$

33. *Summer camp.* The local chapter of the Fresh Air Fund plans to provide money so that at most 400 elementary and high school children from underprivileged families will be able to go to summer camp. There are to be at least 70 more high school students than elementary school students sent to camp. Furthermore, there must be a minimum of 100 high school students sent to camp. The fund pays \$175 for each high school student and \$150 for each elementary school student sent to camp. How many students of each type should be sent to camp to minimize the cost while satisfying all requirements?

34. *Hot or cold lunches.* The manager of Staten College cafeteria serves at most 600 lunches each day, some hot and some cold. Based on past experience, she knows that the number of cold lunches is at most 75 less than the number of hot lunches served. Furthermore, there must be a minimum of 50 cold lunches available on any day. The cafeteria earns a profit of 15¢ on each hot lunch and 10¢ on each cold lunch served. How many of each must the cafeteria sell to maximize its profit?

35. *Sales tax audits.* Each week the State Sales Tax Bureau conducts at most 100 sales tax audits. Some are office audits and the rest are on-site audits. There are at least twice as many office audits as on-site audits; however, there are also at least 30 on-site audits per week. The state earns approximately \$200 from each office audit and \$350 from each on-site audit. How many audits of each type should be conducted to maximize revenue to the state?

36. *Amusement park rides.* A sign on a thrill-seeking boat ride at an amusement park reads as follows:

> *Maximum capacity:* 50 adults
> 30 children
>
> *Warning:* Each child under 12 years of age must be accompanied by at least one paying adult passenger.

The operator of the boat ride earns a profit of 90¢ on each adult ticket sold and 35¢ on each child ticket sold. How many tickets of each type must be sold to maximize profit?

37. *Advertising.* A large real estate developer in Florida plans to spend at most $200,000 for advertising his new condominiums in newspapers and magazines. The developer wishes to reach as large an audience as possible. It is estimated that each page of newspaper ads is read by 200,000 people, while each page of magazine ads is read by 150,000 people. The developer has decided to buy at least one page of advertisement in the newspapers and at least one page in the magazines. If the cost for the advertisement is $5000 per page in the newspaper and $8000 per page in the magazine, how many pages of advertisement in each should the developer buy to reach as large an audience as possible and still satisfy all the budget restrictions?

38. *Eight-track stereo sets.* The ABC Music Corporation manufactures two types of 8-track stereo sets, a cheaper model and a more expensive model. In manufacturing these sets the cheaper model requires 15 minutes in plant I and 30 minutes in plant II whereas the more expensive model requires 30 minutes in plant I and 24 minutes in plant II. The company earns a profit of $81 from each cheaper stereo set manufactured and $132 from each expensive stereo set manufactured. Each plant can remain open at most 40 hours per week. How many of each should be manufactured to maximize profit?

39. *Protesting farmers.* Refer to the newspaper article given at the beginning of this chapter. One of the protesting farmers owns a large farm on which he plants alfalfa and spinach. The farmer estimates that alfalfa requires three man-days of labor and $50 of capital for each acre planted, whereas spinach requires four man-days of labor and $20 of capital for each acre planted. The farmer can supply at most 600 man-days of labor for the job and the bank refuses to loan the farmer more than $4000. Due to rising costs, the farmer can expect to earn a profit of $90 from each acre of alfalfa planted and $75 from each acre of spinach planted. How many acres of each crop should this particular farmer plant to maximize his profits?

40. *Nutrition.* A hospital nutritionist is preparing two types of foods for the patients in the hospital. Food A contains 3 units of vitamins and 5 units of protein, whereas food B contains 4 units of vitamins and 2 units of protein. Each unit of food A costs 72¢ and each unit of food B costs 96¢. The minimum daily nutritional requirements are 21 units of vitamins and 15 units of protein. How many foods of each type should be produced to minimize cost?

4.4 Summary

In this chapter we introduced the new and expanding branch of mathematics called linear programming. In order to solve linear programming problems by the graphical method, we reviewed numerous algebraic rules for working with inequalities.

We applied the graphical technique to solve a variety of problems, both of the minimization and of the maximization type. Nevertheless, as we indicated, if the number of variables is more than two or if there are many constraints, then this method becomes more complicated. It is for this reason that we will discuss the simplex method for solving linear programming problems in the next chapter. This method is suitable for all types of problems.

Study Guide

You should be able to demonstrate your knowledge of the following ideas presented in this chapter by giving definitions, descriptions, or specific examples. Page references are given for each term so that you can check your answer.

Linear programming problem (p. 157)
Linear inequalities (p. 158)
a less than *b* (p. 158)
a greater than *b* (p. 158)
Number line (p. 158)
Trichotomy rule for inequalities (p. 159)
Transitivity rule for inequalities (p. 159)
Addition rule for inequalities (p. 159)
Multiplication rule for inequalities (p. 159)
Systems of linear inequalities (p. 162)
Solution set (p. 162)
Constraints (p. 165)
Corner or extreme point (p. 167)

Formulas to Remember

This chapter contains two main groups of formulas with which you should be familiar.

Rules for working with inequalities (p. 159)

Procedure for graphing inequalities (p. 161)

FORM A

1. In the graphical solution of linear programming problems, where will the minimum and/or maximum values always occur?
2. Graph the solution set for the set of inequalities.

$$x + 4y \leq 18$$
$$2x + y \leq 12$$
$$x \geq 0, y \geq 0$$

3. Graph the solution set for the set of inequalities.

$$y - 2x - 4 \geq 0$$
$$y - x \leq 1$$

4. If $a > b$ and $c < 0$, then what can be said about ac?
5. If an inequality is of the "greater than or equal to" type, which area should be shaded?
6. When using the "greater than" or "less than" symbols, the tip of the arrow always points toward which number?
7. What are the conditions to which linear functions are subject called?
8. When solving linear programming problems graphically, where will the profit expression always have a maximum or minimum value?
9. True or false: A linear programming problem can have more than one optimal solution. Explain your answer.
10. True or false: The set of solutions to a system of linear inequalities is an infinite set.

FORM B

1. Maximize the function $5x + 7y$ subject to the constraints

$$10x + 12y \leq 1920$$
$$5x + 3y \leq 780$$
$$x \geq 0, y \geq 0$$

2. Minimize the function $30x + 40y$ subject to the constraints

$$2x + 4y \geq 10$$
$$5x + y \geq 16$$
$$x \geq 0, y \geq 0$$

3. Minimize the function $4x + 5y$ subject to the constraints

$$x + 3y \geq 19$$
$$4x + 5y \geq 34$$
$$3x + 2y \geq 22$$
$$x \geq 0, y \geq 0$$

4. Minimize the function $3x + y$ subject to the constraints

$$\begin{aligned} 3x - 2y &\geq 0 \\ 3x + y &\leq 18 \\ y &\geq 3 \\ x \geq 0, y &\geq 0 \end{aligned}$$

5. Consult the readings given at the end of this chapter for an explanation as to why an optimal solution for linear programming problems must occur at a corner or extreme point.

6. A company markets two types of deodorants, Safe and Secure. To produce 1 gallon of Safe requires 5 units of ingredient A and 3 units of ingredient B. It yields a profit of \$300. In producing 1 gallon of Secure, 2 units of ingredient A and 3 units of ingredient B are used. The profit is \$200. If the company has 180 units of ingredient A and 135 units of ingredient B available, how many gallons of each deodorant should it produce to maximize profit?

7. A large electronics company produces two kinds of radios, a cheaper model and a more expensive model. The cheaper model requires 2 hours of work in factory A and 4 hours of work in factory B. Each cheaper radio sold produces a profit of \$3. Manufacturing the expensive radios requires 4 hours in factory A and 2 hours in factory B. Each expensive radio sold produces a profit of \$5. If each factory can perform no more than 12 hours of work a day, how many of each should be produced to maximize profit?

8. A man has just inherited \$10,000. He wishes to invest as much of the the money as possible in two types of bonds. Bond I pays 8% a year and bond II pays 10% per year. His broker advises him not to invest more than 50% of the amount of money invested in type I bonds in purchasing type II bonds. How much should he invest at each rate to maximize his profit?

Suggested Further Reading

Charnes, A., W. W. Cooper, and M. H. Miller, "Application of Linear Programming to Financial Budgeting and the Costing of Funds," *Journal of Business*, vol. 32, pp. 20–46, 1959.

Ijiri, Y., F. K. Levy, and R. C. Lyon, "A Linear Programming Model for Budgeting and Financial Planning," *Journal of Accounting Research*, vol. 1, pp. 198–212, 1963.

Kemeny, J., L. Snell, and J. Thompson, *Introduction to Finite Mathematics*, 3d ed., chap. 7, Englewood Cliffs, N.J.: Prentice-Hall, 1974.

Kemeny, J., et al., *Finite Mathematics with Business Applications*, chap. 7, Englewood Cliffs, N.J.: Prentice-Hall, 1968.

Wheeler, R., *Modern Mathematics for Business Students*, chap. 8, Monterey, Calif.: Brooks/Cole, 1968.

CHAPTER 5 Linear Programming: An Algebraic Approach

Objectives to be Achieved in this Chapter

- *To learn* how to convert inequalities associated with linear programming problems into equalities (by adding slack or surplus variables) so that we can solve the problem by the simplex method.
- *To study* the simplex method. This technique was developed by George Dantzig for solving all kinds of linear programming problems.
- *To apply* the ideas of linear programming to solve problems in many different areas.
- *To discuss* several actual case studies to which linear programming ideas have been applied.

NEW YORK DAILY NEWS MARCH 14, 1979

Red Tape Kills Pipeline Plans

Cleveland (AP)—Standard Oil Co. of Ohio yesterday abandoned its effort to build a pipeline to move Alaska oil east from California, blaming regulatory delays it said it could no longer accept.

"The project-killing delays were endless government permit procedures, pending and threatened litigation, and the prospective unavailability of two natural gas lines which Sohio proposed to convert to the oil pipeline," said Chairman Alton W. Whitehouse in announcing the decision.

The project involves shipping 500,000 barrels of Alaskan oil a day from Long Beach, California, to a national pipeline distribution center in Midland, Texas.

The project would use 670 miles of existing pipeline and a tanker unloading terminal at Long Beach.

Environmental and other citizen groups opposed to the project have challenged it on grounds that the terminal and pipeline would add more air pollution to an area which already has some of the poorest air in the nation.

Is it possible to determine how to allocate our dwindling oil supplies so as to maximize their use and still comply with certain restrictions? Fortunately, the answer is yes!

In this chapter we indicate how the simplex method can be used to solve some linear programming problems associated with the ideas discussed in the newspaper article.

5.1 Introduction

In Chapter 4 we discussed what a linear programming problem actually is and we analyzed the graphical technique for solving such problems. Nevertheless, we indicated that although this method is very simple and convenient to use when there are few unknowns, if there are more than three unknowns or if there are many inequalities, this method becomes quite complicated or simply unmanageable. For this reason we need a more general technique. This is the *simplex method*, which can be used conveniently even when the graphical method is unmanageable. It was developed by George Dantzig during World War II.[1] It is with this technique that we concern ourselves in this chapter.

5.2 Slack and Surplus Variables

Recall that in the mathematical formulation of a linear programming problem, inequalities are almost always involved. Since the simplex method works only when we have equalities, it is necessary to convert each inequality into an equality by adding or subtracting what are known as *slack or surplus variables*. This is done as follows.

[1] Some historical notes about George Dantzig were given at the beginning of Chapter 4.

Consider a linear programming problem whose mathematical formulation is to find values of x and y such that

$$x + y \leq 8 \tag{1}$$

$$2x + y \geq 6 \tag{2}$$

$$x \geq 0, y \geq 0 \tag{3}$$

for which

$$z = 7x + 8y \tag{4}$$

is to be maximized.

Statement (1) says that $x + y$ is less than or equal to 8. We can eliminate the "less than" part by adding on the *slack variable* r. Thus, we get

$$x + y + r = 8$$

The r takes up the slack or difference (if any) between $x + y$ and 8.

Statement (2) says that $2x + y$ is greater than or equal to 6. This time we have no slack to take up. We want to eliminate the "greater than" part. We can accomplish this by subtracting the *surplus variable* s. This gives

$$2x + y - s = 8$$

Statement (3) says that $x \geq 0$ and $y \geq 0$. To this we add the additional requirement that both $r \geq 0$ and $s \geq 0$. Can you see why?

The *objective function*, that is, the expression we wish to minimize or maximize, has to be changed also. Thus, we must change statement (4). We assign a profit of 0 to each slack or surplus variable that we introduce in a problem. This is because we make no profit from r and s. Thus, statement (4) becomes

$$\max z = 7x + 8y + 0r + 0s$$

With the slack and surplus variables added, the mathematical formulation of the problem is

$$\begin{aligned} x + y + r \quad\quad &= 8 \\ 2x + y \quad\quad - s &= 6 \\ x \geq 0, y \geq 0, r \geq 0, s &\geq 0 \\ \max z = 7x + 8y + 0r + 0s& \end{aligned}$$

Comment When rewriting a linear programming problem with slack or surplus variables added, we leave spaces for missing letters as was done in the first two equations above. This was done purposely and the reason for it will be explained in the next section.

We illustrate the procedure further with several examples.

Example 1 Rewrite the following linear programming problem using slack and/or surplus variables. Maximize $z = 3y + 2x$ subject to

$$\begin{aligned} x + y &\leq 6 \\ 2x + y &\leq 8 \\ x + 3y &\leq 12 \\ x \geq 0,\ y &\geq 0 \end{aligned}$$

Solution Since the first three inequalities are of the "less than or equal to" type, we have to add slack variables. With slack variables added, the problem becomes

$$\begin{aligned} &x + y + r = 6 \\ &2x + y + s = 8 \\ &x + 3y + t = 12 \\ &x \geq 0,\ y \geq 0,\ r \geq 0,\ s \geq 0,\ t \geq 0 \\ &\max z = 3y + 2x + 0r + 0s + 0t \end{aligned}$$

Comment Since the three inequalities in Example 1 are different, we must use a different slack variable for each. Thus, we use r, s, and t rather than only one slack variable. If we use the same letter for each, then we are assuming that they are the same.

Example 2 Rewrite the following linear programming problem using slack and surplus variables. Maximize $z = 5x + 8y$ subject to

$$\begin{aligned} x + y &\geq 2 \\ 2x + 3y &\leq 15 \\ 3x + y &\leq 12 \\ x \geq 0,\ y &\geq 0 \end{aligned}$$

Solution In this case we have two types of inequalities. For the inequalities of the "less than or equal to" type, we need slack variables, and for the inequalities of the "greater than or equal to" type, we need surplus variables. With these variables added, the problem becomes

$$\begin{aligned} &x + y - r = 2 \\ &2x + 3y + s = 15 \\ &3x + y + t = 12 \\ &x \geq 0,\ y \geq 0,\ r \geq 0,\ s \geq 0,\ t \geq 0 \\ &\max z = 5x + 8y + 0r + 0s + 0t \end{aligned}$$

Exercises

Rewrite the following linear programming problems using slack and/or surplus variables.

1. Exercise 19 of Chapter 4.
2. Exercise 20 of Chapter 4.
3. Exercise 21 of Chapter 4.
4. Exercise 22 of Chapter 4.
5. Exercise 23 of Chapter 4.
6. Exercise 24 of Chapter 4.
7. Exercise 25 of Chapter 4.
8. Exercise 26 of Chapter 4.
9. Exercise 27 of Chapter 4.
10. Exercise 28 of Chapter 4.
11. Exercise 29 of Chapter 4.
12. Exercise 30 of Chapter 4.

5.3 Forming the Initial Tableau

In this section we begin our discussion of the simplex method for solving linear programming problems. We caution the reader that the explanations are rather lengthy. This is because every step is discussed in detail. Do not get discouraged. The procedure may seem somewhat complicated; however, the method is easy to learn and use.

We restrict our discussion to problems of the type in which all the inequalities are of the form less than or equal to or greater than or equal to *a positive number* and in which all the variables are positive.[2]

The simplex method involves identity matrices and matrix multiplication, so if you have forgotten those ideas we suggest that you review them before reading on.

Suppose we are given the linear programming problem: maximize $z = 8x + 7y$ subject to

$$\begin{aligned} 2x + 3y &\le 15 \\ 3x + y &\le 12 \\ x \ge 0, y &\ge 0 \end{aligned}$$

With slack variables added, this problem becomes

$$\begin{aligned} 2x + 3y + r \qquad &= 15 \\ 3x + y \qquad + s &= 12 \\ x \ge 0, y \ge 0, r \ge 0, s &\ge 0 \\ \max z = 8x + 7y + 0r + 0s& \end{aligned}$$

Now we rewrite this problem in matrix form using only the numbers in front of the letters. There are three matrices associated with this and any linear programming problem. These are denoted by **A**, **b**, and

[2] A complete discussion of the solution of more general linear programming problems can be found in *Linear Programming and Extensions* by George Dantzig (see "Suggested Further Reading" at the end of this chapter).

c. Matrix **A** consists of the numbers in front of the letters. Thus, we have

$$\begin{matrix} & a_1 & a_2 & a_3 & a_4 \\ \mathbf{A} = & \begin{pmatrix} 2 & 3 & 1 & 0 \\ 3 & 1 & 0 & 1 \end{pmatrix} \end{matrix}$$

When forming matrix **A**, we do not use the *objective function* (the z equation) or the statement that says all the letters are greater than or equal to zero. We label the columns of matrix **A** by a_1, a_2, a_3, and a_4 so that we will be able to refer to them later.

Comment If no number is indicated in front of a letter, then the number is understood to be 1. On the other hand, if the letter is missing completely, then the number in front of the letter is understood to be 0.

Comment In the previous section when we added slack and surplus variables, we indicated that a space should be left for any missing letter. This was done so that when we set up the **A** matrix no mistakes are made.

The **b** matrix consists of the numbers to the right of the equal sign. Thus,

$$\mathbf{b} = \begin{pmatrix} 15 \\ 12 \end{pmatrix}$$

Again, in determining the **b** matrix, we exclude the objective function, that is, the z equation. We write the entries of the **b** matrix in *column form* as a *column vector.*

The **c** matrix consists of the numbers obtained from the objective function *after we have multiplied each entry by* -1. We write the entries of the **c** matrix in *row form* as

$$\mathbf{c} = (-8 \quad -7 \quad 0 \quad 0)$$

Comment Do not forget to multiply each number of the objective function by -1 before setting up the **c** matrix. Note that 0 multiplied by -1 is 0.

We first examine the **A** matrix to determine whether it contains an identity matrix or not. It does. It is a 2×2 matrix made up of columns 3 and 4. (If matrix **A** does not contain an identity matrix, then one has to be created artificially. However, we will not discuss that possibility in this book.)

$$\begin{matrix} & a_1 & a_2 & a_3 & a_4 \\ \mathbf{A} = & \begin{pmatrix} 2 & 3 & \boxed{\begin{matrix} 1 & 0 \\ 0 & 1 \end{matrix}} \\ 3 & 1 & \end{pmatrix} \end{matrix}$$

The 2×2 identity matrix

After we have determined that we have an identity matrix, we transfer all the numbers from the three matrices onto our first chart, which is known as *tableau 1*:

Tableau 1

	a_1	a_2	a_3	a_4	b	Vectors in basis[3]
	2	3	1	0	15	a_3
	3	1	0	1	12	a_4
	−8	−7	0	0	0	

The entries for this initial tableau are obtained as follows: The a_1; a_2, a_3, and a_4 columns are exactly the **A** matrix. Similarly, the b column is the original **b** matrix. The last line of the tableau corresponds to the **c** matrix. Since we had an identity matrix consisting of columns a_3 and a_4, we write a_3 and a_4 in the "vectors in basis" column in that order. Finally, in the b column of the third row, you will notice that the entry is 0. This number is obtained by multiplying two matrices. One matrix consists of the profit obtained from the vectors in basis, that is, a_3 and a_4. The other is the **b** matrix. Since a_3 and a_4 correspond to the slack variables r and s, they do not yield any profit. Thus, profit obtained from them is written in matrix form as the row vector $(0 \quad 0)$. When we multiply this matrix by matrix **b**, we get

$$(0 \quad 0)\cdot\begin{pmatrix}15\\12\end{pmatrix} = 0\cdot 15 + 0\cdot 12 = 0$$

(When multiplying a row vector by a column vector, we write the answer without parentheses.) This completes tableau 1.

Our main objective when working with the simplex tableaus is to apply certain algebraic steps that transform the initial tableau into one that has no negative values on the last line of the tableau. If this can be done, then we can read our answer directly from the b column.

We further illustrate the technique of finding the initial tableau with several examples.

Example 3 Find the initial tableau for the following linear programming problem: Maximize $z = 5x + 9y$ subject to

$$\begin{aligned} x + 3y &\le 24 \\ 4x + y &\le 28 \\ x \ge 0,\ y &\ge 0 \end{aligned}$$

Solution We first add the slack variables r and s. The mathematical formulation of the problem is then

$$\begin{aligned} x + 3y + r \quad\quad &= 24 \\ 4x + y \quad\quad + s &= 28 \\ x \ge 0,\ y \ge 0,\ r \ge 0,\ s &\ge 0 \\ \max z = 5x + 9y + 0r + 0s & \end{aligned}$$

[3] The term *vectors in basis* is commonly used for this column.

We form the **A** matrix, which is

$$\mathbf{A} = \begin{array}{c} \begin{matrix} a_1 & a_2 & a_3 & a_4 \end{matrix} \\ \left(\begin{array}{cc:cc} 1 & 3 & 1 & 0 \\ 4 & 1 & 0 & 1 \end{array}\right) \end{array} \longrightarrow \text{The } 2 \times 2 \text{ identity matrix}$$

The **A** matrix contains a 2×2 identity matrix formed by columns a_3 and a_4. Matrix **b** consists of the numbers to the right of the equal sign (excluding the z equation), so that

$$\mathbf{b} = \begin{pmatrix} 24 \\ 28 \end{pmatrix}$$

The **c** matrix is

$$\mathbf{c} = (-5 \quad -9 \quad 0 \quad 0)$$

Now we are ready for tableau 1. It is

$$\text{Tableau 1} \quad \begin{array}{c} \begin{array}{ccccc} a_1 & a_2 & a_3 & a_4 & b \end{array} \\ \left(\begin{array}{cccc|c} 1 & 3 & 1 & 0 & 24 \\ 4 & 1 & 0 & 1 & 28 \\ \hdashline -5 & -9 & 0 & 0 & 0 \end{array}\right) \end{array} \quad \begin{array}{c} \text{Vectors} \\ \text{in basis} \\ a_3 \\ a_4 \\ \\ \end{array}$$

The profit obtained from the vectors in basis, that is, from a_3 and a_4, is 0 and 0, respectively. In matrix form we write this as the row vector $(0 \quad 0)$. This is multiplied by the **b** matrix to get:

$$(0 \quad 0) \cdot \begin{pmatrix} 24 \\ 28 \end{pmatrix} = 0 \cdot 24 + 0 \cdot 28 = 0$$

This becomes the entry in the third row of the b column. This completes tableau 1.

Example 4 Find the initial tableau for the following linear programming problem: Maximize $z = 3x + 5y$ subject to

$$\begin{aligned} x + y &\leq 6 \\ 2x + y &\leq 8 \\ x + 3y &\leq 12 \\ x \geq 0,\ y &\geq 0 \end{aligned}$$

Solution Since we have three inequalities, excluding the $x \geq 0$ and $y \geq 0$ inequalities, we must add three slack variables, r, s, and t. The mathematical formulation of this problem with the slack variables added is

$$\begin{aligned} x + y + r \qquad\qquad &= 6 \\ 2x + y \qquad + s \qquad &= 8 \\ x + 3y \qquad\qquad + t &= 12 \\ x \geq 0,\ y \geq 0,\ r \geq 0,\ s \geq 0,\ t \geq 0 \\ \max z = 3x + 5y + 0r + 0s + 0t \end{aligned}$$

The **A** matrix is

$$\mathbf{A} = \begin{array}{c} \begin{array}{ccccc} a_1 & a_2 & a_3 & a_4 & a_5 \end{array} \\ \left(\begin{array}{cc|ccc|} \cline{3-5} 1 & 1 & 1 & 0 & 0 \\ 2 & 1 & 0 & 1 & 0 \\ 1 & 3 & 0 & 0 & 1 \\ \cline{3-5} \end{array}\right) \end{array} \longleftarrow \text{The } 3 \times 3 \text{ identity matrix}$$

Since there are three equations, we examine the **A** matrix to find a 3×3 identity matrix. There is a 3×3 identity matrix consisting of columns a_3, a_4, and a_5. The **b** matrix is

$$\mathbf{b} = \begin{pmatrix} 6 \\ 8 \\ 12 \end{pmatrix}$$

The **c** matrix is

$$\mathbf{c} = (-3 \quad -5 \quad 0 \quad 0 \quad 0)$$

The profit obtained from the vectors in basis, that is, a_3, a_4, and a_5, is 0, 0, and 0, respectively. In matrix form, we write this as the row vector $(0 \quad 0 \quad 0)$. This is multiplied by the **b** matrix to get:

$$(0 \quad 0 \quad 0) \cdot \begin{pmatrix} 6 \\ 8 \\ 12 \end{pmatrix} = 0 \cdot 6 + 0 \cdot 8 + 0 \cdot 12 = 0$$

This becomes the entry in the third row of the **b** column.

We now have the complete tableau 1:

$$\begin{array}{lcc} & \begin{array}{rrrrr:r} a_1 & a_2 & a_3 & a_4 & a_5 & b \end{array} & \begin{array}{c} \text{Vectors} \\ \text{in basis} \end{array} \\ \text{Tableau 1} & \left(\begin{array}{rrrrr:r} 1 & 1 & 1 & 0 & 0 & 6 \\ 2 & 1 & 0 & 1 & 0 & 8 \\ 1 & 3 & 0 & 0 & 1 & 12 \\ \hdashline -3 & -5 & 0 & 0 & 0 & 0 \end{array}\right) & \begin{array}{c} a_3 \\ a_4 \\ a_5 \\ \\ \end{array} \end{array}$$

Example 5 Find the initial tableau for the following linear programming problem: Maximize $z = 7x + 3y + 8p + 2w$ subject to

$$\begin{aligned} 3x + 3y + 4p + 5w &\leq 12 \\ 2x + 2y + 3p - 2w &\leq 15 \\ 4x - 3y - 3p + 7w &\leq 20 \\ x + y - p \qquad &\leq 21 \\ x \geq 0, y \geq 0, p \geq 0, w \geq 0 & \end{aligned}$$

Solution In this case we have four inequalities, excluding the ≥ 0 inequalities, so we must add four slack variables. The mathematical formulation of the problem with the slack variables added is

$$\begin{aligned}
3x + 3y + 4p + 5w + r \quad\quad\quad &= 12\\
2x + 2y + 3p - 2w \quad + s \quad\quad &= 15\\
4x - 3y - 3p + 7w \quad\quad + t \quad &= 20\\
x + y - p \quad\quad\quad\quad + u &= 21\\
x \geq 0, y \geq 0, p \geq 0, w \geq 0, r \geq 0, s \geq 0, t \geq 0, u \geq 0 &\\
\max z = 7x + 3y + 8p + 2w + 0r + 0s + 0t + 0u &
\end{aligned}$$

The **A** matrix is

$$\mathbf{A} = \begin{pmatrix} 3 & 3 & 4 & 5 & 1 & 0 & 0 & 0 \\ 2 & 2 & 3 & -2 & 0 & 1 & 0 & 0 \\ 4 & -3 & -3 & 7 & 0 & 0 & 1 & 0 \\ 1 & 1 & -1 & 0 & 0 & 0 & 0 & 1 \end{pmatrix}$$

(columns $a_1, a_2, a_3, a_4, a_5, a_6, a_7, a_8$; columns a_5 through a_8 are boxed: The 4×4 identity matrix)

In this case we try to find a 4×4 identity matrix. Why? There is a 4×4 identity matrix consisting of columns a_5, a_6, a_7, and a_8. The **b** matrix is

$$\mathbf{b} = \begin{pmatrix} 12 \\ 15 \\ 20 \\ 21 \end{pmatrix}$$ and the **c** matrix is

$$\mathbf{c} = (-7 \quad -3 \quad -8 \quad -2 \quad 0 \quad 0 \quad 0 \quad 0)$$

We now have the complete tableau:

a_1	a_2	a_3	a_4	a_5	a_6	a_7	a_8	b	Vectors in basis
3	3	4	5	1	0	0	0	12	a_5
2	2	3	−2	0	1	0	0	15	a_6
4	−3	−3	7	0	0	1	0	20	a_7
1	1	−1	0	0	0	0	1	21	a_8
−7	−3	−8	−2	0	0	0	0	0	

Exercises

Make up the initial tableau for the following exercises.

13. Exercise 16 of Chapter 4.
14. Exercise 17 of Chapter 4.
15. Exercise 18 of Chapter 4.
16. Exercise 19 of Chapter 4.
17. Exercise 23 of Chapter 4.
18. Exercise 26 of Chapter 4.

19. Maximize $z = 5x + 3y$ subject to

$$x + y \leq 4$$
$$x - y \leq 2$$
$$x \geq 0, \ y \geq 0$$

20. Maximize $z = 2x + 3y$ subject to

$$x + 2y \leq 3$$
$$2x - \ \ y \leq 6$$
$$x \geq 0, \ y \geq 0$$

21. Maximize $z = 3x + 4y + 5p$ subject to

$$x + 2y + p \leq 1$$
$$3x + 4y - p \leq 8$$
$$2x + 3y + p \leq 6$$
$$x \geq 0, \ y \geq 0, \ z \geq 0$$

22. Maximize $5x + 2y + 3p + 9w$ subject to

$$8x + 2y - 3p + 2w \leq 12$$
$$x + \ \ y - \ \ p + \ \ w \leq \ \ 6$$
$$2x + 5y + 2p - \ \ w \leq 10$$
$$3x + 4y - \ \ p + 4w \leq 20$$
$$x \geq 0, \ y \geq 0, \ p \geq 0, \ w \geq 0$$

5.4 The Simplex Method Applied to Maximization Problems

In the preceding section we indicated how to form the initial tableau. Once we have completed the initial tableau, we examine the bottom row to see whether any of the numbers are negative. If there are no negative numbers, then we will discuss later how to proceed.

If there are negative numbers in the last row, we have to go on to a second tableau. To best illustrate the procedure, let us again look at tableau 1 given on page 184, which is repeated here. In this case there are two negative numbers in the bottom row. The simplex method tells us that we select the smallest negative number.

	a_1	a_2	a_3	a_4	b	Vectors in basis
Tableau 1	2	3	1	0	15	a_3
Pivotal row	3	1	0	1	12	a_4
	−8	−7	0	0	0	

Pivotal column

The column that contains this entry is called the *pivotal column*. (If there is a tie, you can select either one.) We have circled this vertically in the tableau. Now we divide each *positive* number of the pivotal column (the one circled vertically) into the number in the b column that is on the same row as the pivotal column. We get

$$\frac{15}{2} \quad \text{and} \quad \frac{12}{3}$$

We choose as the *pivotal row* the row that results in the *smallest* quotient. (If two or more quotients are the same, we can select any one of them as the pivotal row.) In this case $\frac{12}{3} = 4$ is smaller than $\frac{15}{2} = 7.5$, so that the second row (the one containing a_4 in the basis) is the pivotal row. We circle this row horizontally. Notice that one number is in both the pivotal row and the pivotal column. It is the number 3, that has been circled twice. This number, which is called the *pivotal element*, plays an important role in the formation of tableau 2. The pivotal element is never in the bottom row.

In the second tableau, a_1 will replace a_4 in the "vectors in basis" column since it represents the pivotal column. Thus, our second tableau starts off as:

	a_1	a_2	a_3	a_4	b	Vectors in basis
Tableau 2						a_3
						a_1

Now we must find the entries for this tableau. First, each element of the pivotal row of the first tableau is divided by the pivotal number. In our case we get

$$\frac{3}{3} = 1 \qquad \frac{1}{3} = \frac{1}{3} \qquad \frac{0}{3} = 0 \qquad \frac{1}{3} = \frac{1}{3} \qquad \frac{12}{3} = 4$$

We enter these in the second row of tableau 2, getting

a_1	a_2	a_3	a_4	b	Vectors in basis
					a_3
1	$\frac{1}{3}$	0	$\frac{1}{3}$	4	a_1

For the entries in the first and third rows (the nonpivotal rows), we use the following rule:

Pivotal Rule

1. Pick a number from a nonpivotal row.
2. Divide the number that is on the same row and in the pivotal column by the pivotal number.
3. Multiply the result of step 2 by the entry that is in the pivotal row and the same column with which we started.

4. Substract the results of step 3 from the number in step 1.
5. Enter the answer in the new tableau in the same position as the number in step 1.
6. Repeat the same procedure to find all the numbers in the non-pivotal rows.

This rule is rather simple to use. For the entry in the b column of row 1:

Number in same row and in pivotal column

$$15 \quad - \quad \frac{2}{3} \cdot 12 \rightarrow$$

Number in same column as 15 and also in pivotal row

Original number — Pivotal number

Thus, we get $15 - \frac{2}{3} \cdot 12 = 15 - 8 = 7$. For all the other entries in the first row, we follow the same procedure.

For the a_1 column: $2 - \dfrac{2}{3} \cdot 3 = 2 - 2 = 0$

For the a_2 column: $3 - \dfrac{2}{3} \cdot 1 = \dfrac{7}{3}$

For the a_3 column: $1 - \dfrac{2}{3} \cdot 0 = 1$

For the a_4 column: $0 - \dfrac{2}{3} \cdot 1 = -\dfrac{2}{3}$

These will be the entries for row 1 of the second tableau.

To find the entries for the third row, we repeat the same procedure.

For the b column: $0 - \left(-\dfrac{8}{3}\right) \cdot 12 = 0 + \dfrac{8}{3} \cdot 12 = 32$

For the a_1 column: $-8 - \left(-\dfrac{8}{3}\right) \cdot 3 = -8 + 8 = 0$

For the a_2 column: $-7 - \left(-\dfrac{8}{3}\right) \cdot 1 = -7 + \dfrac{8}{3} = -\dfrac{13}{3}$

For the a_3 column: $0 - \left(-\dfrac{8}{3}\right) \cdot 0 = 0 + 0 = 0$

For the a_4 column: $0 - \left(-\dfrac{8}{3}\right) \cdot 1 = 0 + \dfrac{8}{3} = \dfrac{8}{3}$

These entries are now entered on the chart. This completes tableau 2.

$$\text{Tableau 2} \quad \begin{array}{ccccc} a_1 & a_2 & a_3 & a_4 & b \end{array} \quad \begin{array}{c}\text{Vectors}\\ \text{in basis}\end{array}$$

$$\left(\begin{array}{cccc|c} 0 & \frac{7}{3} & 1 & -\frac{2}{3} & 7 \\ 1 & \frac{1}{3} & 0 & \frac{1}{3} & 4 \\ \hline 0 & -\frac{13}{3} & 0 & \frac{8}{3} & 32 \end{array}\right) \quad \begin{array}{c} a_3 \\ a_1 \\ \\ \end{array}$$

Since the bottom row has a negative number, the simplex method tells us that we need another tableau. The next tableau is found in exactly the same manner that we found this one.

First we select the minimum number of the bottow row. This is $-\frac{13}{3}$. Thus, the a_2 column will be the pivotal column and we circle it vertically. To find the pivotal row, we must find the minimum of

$$7 \div \frac{7}{3} \quad \text{and} \quad 4 \div \frac{1}{3}$$

We have

$$7 \div \frac{7}{3} = \frac{21}{7} = 3 \quad \text{and} \quad 4 \div \frac{1}{3} = 4 \cdot \frac{3}{1} = 12$$

So $7 \div \frac{7}{3}$ is smaller. Thus, the a_3 row will be the pivotal row. We circle it horizontally. Also, $\frac{7}{3}$ will be the pivotal number.

In the third tableau a_2 will replace a_3 in the "vectors in basis" column as shown here.

$$\text{Tableau 3} \quad \begin{array}{ccccc} a_1 & a_2 & a_3 & a_4 & b \end{array} \quad \begin{array}{c}\text{Vectors}\\ \text{in basis}\end{array}$$

$$\left(\begin{array}{cccc|c} & & & & \\ & & & & \\ \hline & & & & \end{array}\right) \quad \begin{array}{c} a_2 \\ a_1 \\ \\ \end{array}$$

To find the entries for this tableau, we first divide each element of the pivotal row by the pivotal number, getting

$$0 \div \frac{7}{3} = 0 \cdot \frac{3}{7} = 0$$

$$\frac{7}{3} \div \frac{7}{3} = \frac{7}{3} \cdot \frac{3}{7} = 1$$

$$1 \div \frac{7}{3} = 1 \cdot \frac{3}{7} = \frac{3}{7}$$

$$\left(-\frac{2}{3}\right) \div \left(\frac{7}{3}\right) = \left(-\frac{2}{3}\right) \cdot \left(\frac{3}{7}\right) = -\frac{2}{7}$$

$$7 \div \frac{7}{3} = 7 \cdot \frac{3}{7} = 3$$

To find the entries for the nonpivotal rows (that is, rows 2 and 3), we use the procedure outlined in the pivotal rule. Thus, in the second row:

For the b column: $4 - \frac{1/3}{7/3} \cdot 7 = 4 - \left(\frac{1}{3}\right) \cdot \left(\frac{3}{7}\right) \cdot (7) = 3$

For the a_1 column: $1 - \frac{1/3}{7/3} \cdot (0) = 1 - \left(\frac{1}{3}\right) \cdot \left(\frac{3}{7}\right) \cdot (0) = 1$

For the a_2 column: $\frac{1}{3} - \frac{1/3}{7/3} \cdot \left(\frac{7}{3}\right) = \frac{1}{3} - \left(\frac{1}{3}\right) \cdot \left(\frac{3}{7}\right) \cdot \left(\frac{7}{3}\right) = 0$

For the a_3 column: $0 - \frac{1/3}{7/3} \cdot (1) = 0 - \left(\frac{1}{3}\right) \cdot \left(\frac{3}{7}\right) \cdot (1) = -\frac{1}{7}$

For the a_4 column: $\frac{1}{3} - \frac{1/3}{7/3} \cdot \left(-\frac{2}{3}\right) = \frac{1}{3} - \left(\frac{1}{3}\right) \cdot \left(\frac{3}{7}\right) \cdot \left(-\frac{2}{3}\right)$

$$= \frac{3}{7}$$

In the third row, we have

For the b column: $32 - \left(\frac{-13/3}{7/3}\right) \cdot (7) = 32 + \left(\frac{13}{3}\right) \cdot \left(\frac{3}{7}\right) \cdot (7)$

$$= 45$$

For the a_1 column: $0 - \left(\frac{-13/3}{7/3}\right) \cdot (0) = 0 + \left(\frac{13}{3}\right) \cdot \left(\frac{3}{7}\right) \cdot (0)$

$$= 0$$

For the a_2 column:

$$-\frac{13}{3} - \left(\frac{-13/3}{7/3}\right) \cdot \left(\frac{7}{3}\right) = \left(-\frac{13}{3}\right) + \left(\frac{13}{3}\right) \cdot \left(\frac{3}{7}\right) \cdot \left(\frac{7}{3}\right) = 0$$

For the a_3 column: $0 - \left(\frac{-13/3}{7/3}\right) \cdot (1) = 0 + \left(\frac{13}{3}\right) \cdot \left(\frac{3}{7}\right) \cdot 1$

$$= \frac{13}{7}$$

For the a_4 column:

$$\frac{8}{3} - \left(\frac{-13/3}{7/3}\right) \cdot \left(-\frac{2}{3}\right) = \frac{8}{3} + \left(\frac{13}{3}\right) \cdot \left(\frac{3}{7}\right) \cdot \left(-\frac{2}{3}\right) = \frac{10}{7}$$

These entries are now entered on the chart. This completes tableau 3.

	a_1	a_2	a_3	a_4	b	Vectors in basis
Tableau 3	0	1	$\frac{3}{7}$	$-\frac{2}{7}$	3	a_2
	1	0	$-\frac{1}{7}$	$\frac{3}{7}$	3	a_1
	0	0	$\frac{13}{7}$	$\frac{10}{7}$	45	

Since the bottom row has no negative numbers, we are done. Now we read the answer from the b column of the chart and the "vectors in basis" column. Note that a_2 corresponds to y, and a_1 corresponds to x. Thus, the maximum value of the original problem is obtained when $x = 3, y = 3$. Furthermore, the bottom row of the tableau also gives the maximum value of z; it is 45.

Comment Although the procedure for obtaining the entries for the new tableau as described in the pivotal rule may seem a bit unusual, it is actually similar to the row operations for matrices described in Chapter 3. Dividing each entry of the pivotal row by the pivotal number a is nothing more than multiplying each entry in the row of a matrix by $1/a$. Similarly, for the nonpivotal rows, we are actually adding suitable multiples of the pivotal row to the other rows so each of the remaining entries in the pivotal column becomes a zero. You should review the row operations given in Chapter 3 to convince yourself that our procedure is nothing more than a series of row operations designed to reduce the pivotal column to a column that has a one in the pivot entry and zeros elsewhere. The technique is similar to the elimination procedure for solving systems of simultaneous equations using matrices discussed in Chapter 3. You may wish to work with the equations alongside the matrices to see the similarity.

We will illustrate the procedure further with several examples.

Example 6 Maximize $z = 10x + 12y$ subject to

$$\begin{aligned} x + y &\leq 8 \\ 3x + 4y &\leq 24 \\ x \geq 0, \ y &\geq 0 \end{aligned}$$

Solution We first rewrite the problem by adding the slack variables r and s. We get

$$\begin{aligned} x + y + r &= 8 \\ 3x + 4y + s &= 24 \\ x \geq 0, \ y \geq 0, \ r \geq 0, \ s &\geq 0 \\ \max z = 10x + 12y + 0r + 0s & \end{aligned}$$

To form the initial tableau, we set up the three matrices **A**, **b**, and **c**. We have

$$\mathbf{A} = \begin{pmatrix} \overset{a_1}{1} & \overset{a_2}{1} & \overset{a_3}{1} & \overset{a_4}{0} \\ 3 & 4 & 0 & 1 \end{pmatrix} \quad \mathbf{b} = \begin{pmatrix} 8 \\ 24 \end{pmatrix} \quad \mathbf{c} = (-10 \quad -12 \quad 0 \quad 0)$$

Since the **A** matrix contains a 2×2 matrix consisting of columns a_3 and a_4, we can complete tableau 1:

Tableau 1

	a_1	a_2	a_3	a_4	b	Vectors in basis
	1	1	1	0	8	a_3
Pivotal row →	3	4	0	1	24	a_4
	−10	−12	0	0	0	

Pivotal number (4); Pivotal column (a_2)

The 0 in the third row of the b column is obtained by multiplying the profit matrix associated with a_3 and a_4, that is, $(0 \quad 0)$, by the **b** matrix. We get

$$(0 \quad 0) \cdot \begin{pmatrix} 8 \\ 24 \end{pmatrix} = 0 \cdot 8 + 0 \cdot 24 = 0$$

Since the bottom row of the tableau contains negative numbers, we will have to go on to another tableau. We select the smallest value. It is -12, and it is in the a_2 column. Thus, the a_2 column will be the pivotal column. We circle it vertically. To find the pivotal row, we divide the numbers in the **b** column by the positive entries in the same row that are in the pivotal column. We get $\frac{8}{1} = 8$ and $\frac{24}{4} = 6$. The smallest is 6 and corresponds to the a_4 row. This is circled horizontally. Thus, the second row will be the pivotal row and 4 will be the pivotal number. In the second tableau, a_2 will replace a_4 in the "vectors in basis" column.

To find the entries for the second tableau, we first divide each element of the pivotal row by the pivotal number, getting

$$\frac{3}{4} = \frac{3}{4} \qquad \frac{4}{4} = 1 \qquad \frac{0}{4} = 0 \qquad \frac{1}{4} = \frac{1}{4} \qquad \frac{24}{4} = 6$$

To find the entries for the nonpivotal rows (that is, rows 1 and 3), we use the pivotal rule outlined earlier. In the first row, we have

For the b column: $8 - \frac{1}{4} \cdot 24 = 8 - 6 = 2$

For the a_1 column: $1 - \frac{1}{4} \cdot 3 = 1 - \frac{3}{4} = \frac{1}{4}$

For the a_2 column: $1 - \frac{1}{4} \cdot 4 = 1 - 1 = 0$

For the a_3 column: $1 - \frac{1}{4} \cdot 0 = 1$

For the a_4 column: $0 - \frac{1}{4} \cdot 1 = -\frac{1}{4}$

In the third row, we have

$$\text{For the } b \text{ column:} \quad 0 - \left(\frac{-12}{4}\right) \cdot 24 = 0 + 72 = 72$$

$$\text{For the } a_1 \text{ column:} \quad -10 - \left(\frac{-12}{4}\right) \cdot 3 = -10 + 9 = -1$$

$$\text{For the } a_2 \text{ column:} \quad -12 - \left(\frac{-12}{4}\right) \cdot 4 = -12 + 12 = 0$$

$$\text{For the } a_3 \text{ column:} \quad 0 - \left(\frac{-12}{4}\right) \cdot 0 = 0 + 0 = 0$$

$$\text{For the } a_4 \text{ column:} \quad 0 - \left(\frac{-12}{4}\right) \cdot 1 = 0 + 3 = 3$$

These entries are now entered on the tableau. This completes tableau 2:

Tableau 2	a_1	a_2	a_3	a_4	b	Vectors in basis
	$\frac{1}{4}$	0	1	$-\frac{1}{4}$	2	a_3
	$\frac{3}{4}$	1	0	$\frac{1}{4}$	6	a_2
	-1	0	0	3	72	

Pivotal row

Pivotal column

Since the bottom row of this tableau contains a negative number, we have to go on to another tableau. The smallest number in the bottom row is -1 and is in the a_1 column. Thus, a_1 will be the pivotal column. We circle this vertically. To find the pivotal row, we find the minimum of

$$2 \div \frac{1}{4} \quad \text{and} \quad 6 \div \frac{3}{4}$$

We have

$$2 \div \frac{1}{4} = 2 \cdot \frac{4}{1} = 8 \quad \text{and} \quad 6 \div \frac{3}{4} = 6 \cdot \frac{4}{3} = 8$$

Since there is a tie, we can select either a_3 or a_2 as the pivotal row. We will select a_3 as the pivotal row. We circle this horizontally. Also, $\frac{1}{4}$ will be the pivotal number.

In the third tableau, the entries for the pivotal row are

$$\frac{1}{4} \div \frac{1}{4} = 1 \qquad 0 \div \frac{1}{4} = 0 \qquad 1 \div \frac{1}{4} = 4 \qquad -\frac{1}{4} \div \frac{1}{4} = -1 \qquad 2 \div \frac{1}{4} = 8$$

The entries for the second row are:

$$\text{For the } b \text{ column:} \quad 6 - \frac{3/4}{1/4} \cdot 2 = 6 - \frac{3}{4} \cdot \frac{4}{1} \cdot 2 = 0$$

$$\text{For the } a_1 \text{ column:} \quad \frac{3}{4} - \frac{3/4}{1/4} \cdot \frac{1}{4} = 0$$

For the a_2 column: $1 - \frac{3/4}{1/4} \cdot 0 = 1$

For the a_3 column: $0 - \frac{3/4}{1/4} \cdot 1 = -3$

For the a_4 column: $\frac{1}{4} - \frac{3/4}{1/4} \cdot \left(-\frac{1}{4}\right) = \frac{1}{4} - \frac{3}{4} \cdot \frac{4}{1} \cdot \left(-\frac{1}{4}\right) = 1$

The entries for the third row are:

For the b column: $72 - \left(\frac{-1}{1/4}\right) \cdot 2 = 72 + 1 \cdot \frac{4}{1} \cdot 2 = 80$

For the a_1 column: $-1 - \left(\frac{-1}{1/4}\right) \cdot \frac{1}{4} = -1 + 1 = 0$

For the a_2 column: $0 - \left(\frac{-1}{1/4}\right) \cdot 0 = 0 - (-1) \cdot \frac{4}{1} \cdot 0 = 0$

For the a_3 column: $0 - \left(\frac{-1}{1/4}\right) \cdot 1 = 0 - (-1) \cdot \frac{4}{1} \cdot 1 = 4$

For the a_4 column: $3 - \left(\frac{-1}{1/4}\right) \cdot \left(-\frac{1}{4}\right) = 3 - (-1) \cdot \frac{4}{1} \cdot \left(-\frac{1}{4}\right)$
$= 2$

This completes tableau 3:

$$\text{Tableau 3} \quad \begin{array}{c} \\ \\ \end{array} \begin{matrix} a_1 & a_2 & a_3 & a_4 & & b \end{matrix} \qquad \begin{matrix}\text{Vectors} \\ \text{in basis}\end{matrix}$$

$$\left(\begin{array}{cccc|c} 1 & 0 & 4 & -1 & 8 \\ 0 & 1 & -3 & 1 & 0 \\ \hline 0 & 0 & 4 & 2 & 80 \end{array}\right) \quad \begin{matrix} a_1 \\ a_2 \\ \\ \end{matrix}$$

Since there are no negative numbers in the bottom row, we are done. As a_1 corresponds to x and a_2 corresponds to y, our answer is $x = 8$, $y = 0$. The maximum value of z is 80.

Example 7 *CB radios.* The Acme Manufacturing Company produces two kinds of CB radio sets, an economy model and a deluxe model. The economy model requires 5 hours of work on the assembly line, 2 hours of time for testing, and 1 hour for painting and packaging and yields a profit of \$15. The deluxe model requires 6 hours of work on the assembly line, 4 hours of time for testing, and 1 hour for painting and packaging and yields a profit of \$25. For the week of February 1–5, the Acme Company has available 4560 work hours on the assembly line, 3000 hours for testing, and 1000 hours for painting and packaging. How many sets of each type should be produced to maximize profit? Find the maximum profit.

Solution We first arrange the information given in the problem in the form of a table.

Model	Hours on assembly line	Hours for testing	Hours for painting and packaging	Profit
Economy	5	2	1	\$15
Deluxe	6	4	1	\$25
Limitations	4560	3000	1000	

Now we let

$x =$ the number of economy models to be produced

$y =$ the number of deluxe models to be produced

The problem then is to find values of x and y that maximize $z = 15x + 25y$ subject to

$$5x + 6y \leq 4560$$
$$2x + 4y \leq 3000$$
$$x + y \leq 1000$$
$$x \geq 0, y \geq 0$$

With slack variables added, the problem becomes

$$5x + 6y + r = 4560$$
$$2x + 4y + s = 3000$$
$$x + y + t = 1000$$
$$x \geq 0, y \geq 0, r \geq 0, s \geq 0, t \geq 0$$
$$\max z = 15x + 25y + 0r + 0s + 0t$$

The three matrices are

$$\mathbf{A} = \begin{pmatrix} 5 & 6 & 1 & 0 & 0 \\ 2 & 4 & 0 & 1 & 0 \\ 1 & 1 & 0 & 0 & 1 \end{pmatrix} \quad \text{(columns } a_1, a_2, a_3, a_4, a_5\text{)}$$

$$\mathbf{b} = \begin{pmatrix} 4560 \\ 3000 \\ 1000 \end{pmatrix} \qquad \mathbf{c} = (-15 \quad -25 \quad 0 \quad 0 \quad 0)$$

Thus, the initial tableau is

Tableau 1

	a_1	a_2	a_3	a_4	a_5	b	Vectors in basis
	5	6	1	0	0	4560	a_3
Pivotal row →	2	4	0	1	0	3000	a_4
	1	1	0	0	1	1000	a_5
	−15	−25	0	0	0	0	

Pivotal number (4); Pivotal column (a_2)

The smallest number in the bottom row is -25. Thus, a_2 will be the pivotal column. To find the pivotal row, we select the minimum of $\frac{4560}{6}$, $\frac{3000}{4}$, and $\frac{1000}{1}$. Since $\frac{4560}{6} = 760$, $\frac{3000}{4} = 750$, and $\frac{1000}{1} = 1000$, the second row, which corresponds to a_4, will be the pivotal row and will be replaced by a_2 in the second tableau. The pivotal number will be 4. The entries for the pivotal row of the second tableau are

$$\frac{2}{4} = \frac{1}{2} \qquad \frac{4}{4} = 1 \qquad \frac{0}{4} = 0 \qquad \frac{1}{4} = \frac{1}{4} \qquad \frac{0}{4} = 0 \qquad \frac{3000}{4} = 750$$

Using the pivotal rule, the entries for the other rows are:

Column	First row	Third row	Fourth row
b	$4560 - \frac{6}{4}(3000) = 60$	$1000 - \frac{1}{4}(3000) = 250$	$0 - \left(\frac{-25}{4}\right)(3000) = 18{,}750$
a_1	$5 - \frac{6}{4}(2) = 2$	$1 - \frac{1}{4}(2) = \frac{1}{2}$	$-15 - \left(\frac{-25}{4}\right)(2) = -\frac{5}{2}$
a_2	$6 - \frac{6}{4}(4) = 0$	$1 - \frac{1}{4}(4) = 0$	$-25 - \left(\frac{-25}{4}\right)(4) = 0$
a_3	$1 - \frac{6}{4}(0) = 1$	$0 - \frac{1}{4}(0) = 0$	$0 - \left(\frac{-25}{4}\right)(0) = 0$
a_4	$0 - \frac{6}{4}(1) = \frac{-3}{2}$	$0 - \frac{1}{4}(1) = -\frac{1}{4}$	$0 - \left(\frac{-25}{4}\right)(1) = \frac{25}{4}$
a_5	$0 - \frac{6}{4}(0) = 0$	$1 - \frac{1}{4}(0) = 1$	$0 - \left(\frac{-25}{4}\right)(0) = 0$

This completes tableau 2:

Tableau 2

	a_1	a_2	a_3	a_4	a_5	b	Vectors in basis
Pivotal row →	2	0	1	$-\frac{3}{2}$	0	60	a_3
	$\frac{1}{2}$	1	0	$\frac{1}{4}$	0	750	a_2
	$\frac{1}{2}$	0	0	$-\frac{1}{4}$	1	250	a_5
	$-\frac{5}{2}$	0	0	$\frac{25}{4}$	0	18,750	

Pivotal column: a_1

There is still one negative number in the bottom row. Thus, a_1 will now be the pivotal column for the next tableau. To find the pivotal row, we find the minimum of

$$\frac{60}{2} \qquad \frac{750}{1/2} \qquad \frac{250}{1/2}$$

Thus, $\frac{60}{2} = 30$ is the minimum, and a_3 will be the pivotal row, as we have indicated in the tableau. The pivotal number will be 2.

For the third tableau, the numbers in the pivotal row are

$$\frac{2}{2}=1 \qquad \frac{0}{2}=0 \qquad \frac{1}{2}=\frac{1}{2} \qquad \frac{-3/2}{2}=\frac{-3}{4} \qquad \frac{0}{2}=0 \qquad \frac{60}{2}=30$$

Using the pivotal rule, the entries for the remaining rows of the third tableau are:

Column	Second row	Third row	Fourth row
b	$750-\frac{1/2}{2}(60)=735$	$250-\frac{1/2}{2}(60)=235$	$18{,}750-\left(\frac{-5/2}{2}\right)(60)=18{,}825$
a_1	$\frac{1}{2}-\frac{1/2}{2}(2)=0$	$\frac{1}{2}-\frac{1/2}{2}(2)=0$	$\left(\frac{-5}{2}\right)-\left(\frac{-5/2}{2}\right)(2)=0$
a_2	$1-\frac{1/2}{2}(0)=1$	$0-\frac{1/2}{2}(0)=0$	$0-\left(\frac{-5/2}{2}\right)(0)=0$
a_3	$0-\frac{1/2}{2}(1)=-\frac{1}{4}$	$0-\frac{1/2}{2}(1)=-\frac{1}{4}$	$0-\left(\frac{-5/2}{2}\right)(1)=1.25$
a_4	$\frac{1}{4}-\frac{1/2}{2}\left(\frac{-3}{2}\right)=\frac{5}{8}$	$\left(-\frac{1}{4}\right)-\frac{1/2}{2}\left(\frac{-3}{2}\right)=\frac{1}{8}$	$\left(\frac{25}{4}\right)-\left(\frac{-5/2}{2}\right)\left(\frac{-3}{2}\right)=4.375$
a_5	$0-\frac{1/2}{2}(0)=0$	$1-\frac{1/2}{2}(0)=1$	$0-\left(\frac{-5/2}{2}\right)(0)=0$

This completes tableau 3:

$$\text{Tableau 3}\qquad \begin{array}{c} \\ \\ \\ \\ \\ \end{array}\begin{array}{ccccc|c} a_1 & a_2 & a_3 & a_4 & a_5 & b \\ 1 & 0 & \frac{1}{2} & -\frac{3}{4} & 0 & 30 \\ 0 & 1 & -\frac{1}{4} & \frac{5}{8} & 0 & 735 \\ 0 & 0 & -\frac{1}{4} & \frac{1}{8} & 1 & 235 \\ \hline 0 & 0 & 1.25 & 4.375 & 0 & 18{,}825 \end{array} \qquad \begin{array}{c} \text{Vectors in basis} \\ a_1 \\ a_2 \\ a_5 \\ \\ \end{array}$$

Since there are no negative numbers in the bottom row, we are done. We know that a_1 corresponds to x and a_2 corresponds to y, so our answer is $x = 30$, $y = 735$, and maximum $z = 18{,}825$. Thus, the manufacturer should produce 30 economy model CB sets and 735 deluxe model CB sets for a maximum profit of \$18,825.

Comment Although a_5 appears as part of the solution, we are not interested in it as it represents a slack variable. We neglect it.

Example 8 *Health foods.* A certain health foods store sells four different kinds of foods, each fortified with different quantities of zinc, lecithin, and brewer's yeast. The accompanying table indicates the number of units of vitamins or minerals added to each food as well as the profit earned from each package of food that is sold.

Food	Amount of zinc added	Amount of lecithin added	Amount of brewer's yeast added	Profit
X	50	40	30	60¢
Y	40	20	50	35¢
P	10	30	50	50¢
W	50	20	20	45¢
Total amount available	6,000	8,000	12,000	

If the store has 6,000 units of zinc, 8,000 units of lecithin, and 12,000 units of brewer's yeast available, how many packages of each food should be produced to maximize profit?

Solution Let

$x =$ the number of packages of food X to be produced
$y =$ the number of packages of food Y to be produced
$p =$ the number of packages of food P to be produced
$w =$ the number of packages of food W to be produced

The mathematical formulation of the problem is to find values of x, y, p, and w such that

$$\begin{aligned} 50x + 40y + 10p + 50w &\le 6{,}000 \\ 40x + 20y + 30p + 20w &\le 8{,}000 \\ 30x + 50y + 50p + 20w &\le 12{,}000 \\ x \ge 0,\ y \ge 0,\ p \ge 0 \text{ and } w &\ge 0 \end{aligned}$$

and for which $z = 60x + 35y + 50p + 45w$ is maximized. With slack variables added, we get

$$\begin{aligned} 50x + 40y + 10p + 50w + r \qquad\qquad &= 6{,}000 \\ 40x + 20y + 30p + 20w \qquad + s \qquad &= 8{,}000 \\ 30x + 50y + 50p + 20w \qquad\qquad + t &= 12{,}000 \\ x \ge 0,\ y \ge 0,\ p \ge 0,\ w \ge 0,\ r \ge 0,\ s \ge 0,\ t &\ge 0 \\ \max z = 60x + 35y + 50p + 45w + 0r + 0s + 0t & \end{aligned}$$

Thus, we have the following matrices:

$$\mathbf{A} = \begin{pmatrix} \overset{a_1}{50} & \overset{a_2}{40} & \overset{a_3}{10} & \overset{a_4}{50} & \overset{a_5}{1} & \overset{a_6}{0} & \overset{a_7}{0} \\ 40 & 20 & 30 & 20 & 0 & 1 & 0 \\ 30 & 50 & 50 & 20 & 0 & 0 & 1 \end{pmatrix} \qquad \mathbf{b} = \begin{pmatrix} 6{,}000 \\ 8{,}000 \\ 12{,}000 \end{pmatrix}$$

$$\mathbf{c} = (-60 \quad -35 \quad -50 \quad -45 \quad 0 \quad 0 \quad 0)$$

The initial tableau with columns a_5, a_6, and a_7 of matrix **A** forming an identity matrix is

$$\text{Tableau 1} \quad \begin{array}{ccccccc|c} a_1 & a_2 & a_3 & a_4 & a_5 & a_6 & a_7 & b \\ \hline 50 & 40 & 10 & 50 & 1 & 0 & 0 & 6{,}000 \\ 40 & 20 & 30 & 20 & 0 & 1 & 0 & 8{,}000 \\ 30 & 50 & 50 & 20 & 0 & 0 & 1 & 12{,}000 \\ \hline -60 & -35 & -50 & -45 & 0 & 0 & 0 & 0 \end{array} \quad \begin{array}{c} \text{Vectors in basis} \\ a_5 \\ a_6 \\ a_7 \\ \\ \end{array}$$

Readers should verify that a_1 is the pivotal column, a_5 is the pivotal row, and that 50 is the pivotal number. In the second tableau, with a_1 replacing a_5 in the "vectors in basis" column, the entries for the pivotal row are

$$\frac{50}{50} = 1 \qquad \frac{40}{50} = 0.8 \qquad \frac{10}{50} = 0.2 \qquad \frac{50}{50} = 1 \qquad \frac{1}{50} = 0.02$$

$$\frac{0}{50} = 0 \qquad \frac{0}{50} = 0 \qquad \frac{6{,}000}{50} = 120$$

Using the pivotal rule, the entries for the remaining rows are

Column	Second row	Third row	Fourth row
b	$8{,}000 - \frac{40}{50}(6{,}000) = 3{,}200$	$12{,}000 - \frac{30}{50}(6{,}000) = 8{,}400$	$0 - \left(\frac{-60}{50}\right)(6{,}000) = 7{,}200$
a_1	$40 - \frac{40}{50}(50) = 0$	$30 - \frac{30}{50}(50) = 0$	$-60 - \left(\frac{-60}{50}\right)(50) = 0$
a_2	$20 - \frac{40}{50}(40) = -12$	$50 - \frac{30}{50}(40) = 26$	$-35 - \left(\frac{-60}{50}\right)(40) = 13$
a_3	$30 - \frac{40}{50}(10) = 22$	$50 - \frac{30}{50}(10) = 44$	$-50 - \left(\frac{-60}{50}\right)(10) = -38$
a_4	$20 - \frac{40}{50}(50) = -20$	$20 - \frac{30}{50}(50) = -10$	$-45 - \left(\frac{-60}{50}\right)(50) = 15$
a_5	$0 - \frac{40}{50}(1) = -0.8$	$0 - \frac{30}{50}(1) = -0.6$	$0 - \left(\frac{-60}{50}\right)(1) = 1.2$
a_6	$1 - \frac{40}{50}(0) = 1$	$0 - \frac{30}{50}(0) = 0$	$0 - \left(\frac{-60}{50}\right)(0) = 0$
a_7	$0 - \frac{40}{50}(0) = 0$	$1 - \frac{30}{50}(0) = 1$	$0 - \left(\frac{-60}{50}\right)(0) = 0$

This completes tableau 2:

Tableau 2

a_1	a_2	a_3	a_4	a_5	a_6	a_7	b	Vectors in basis
1	0.8	0.2	1	0.02	0	0	120	a_1
0	−12	22	−20	−0.8	1	0	3,200	a_6
0	26	44	−10	−0.6	0	1	8,400	a_7
0	13	−38	15	1.2	0	0	7,200	

Since there is a negative number in the bottom row, we have to go on to a third tableau. (In applied cases, we may have to calculate many tableaus before we arrive at a solution.) In this case a_3 is the pivotal column, a_6 is the pivotal row, and 22 is the pivotal number. Verify this.

The third tableau with a_3 replacing a_6 is

Tableau 3

a_1	a_2	a_3	a_4	a_5	a_6	a_7	b	Vectors in basis
1	0.91	0	1.18	0.03	−0.009	0	90.91	a_1
0	−0.55	1	−0.91	−0.04	0.05	0	145.45	a_3
0	50	0	30	1	−2	1	2,000	a_7
0	−7.73	0	−19.55	−0.18	1.73	0	12,727.27	

The reader should verify that all the entries in this tableau are accurate. Again in this case there are negative numbers in the bottom row, so we have to go on to a fourth tableau.

Comment In determining the pivotal row, we must find the minimum of *only* $\frac{90.91}{1.18}$ and $\frac{2,000}{30}$. We do not consider $\frac{145.45}{-0.91}$, since -0.91 is a negative number. Similarly, if there had been a zero in the a_4 column, the row that contained it would not be eligible to be the pivotal row.

The fourth tableau, with a_4 replacing a_7, is

Tableau 4

a_1	a_2	a_3	a_4	a_5	a_6	a_7	b	Vectors in basis
1	−1.06	0	0	−0.009	0.07	−0.04	12.24	a_1
0	0.97	1	0	−0.0097	−0.01	0.03	206.12	a_3
0	1.67	0	1	0.03	−0.07	0.03	66.67	a_4
0	24.85	0	0	0.47	0.43	0.65	14,030.60	

Since there are no negative numbers in the bottom row, we are done. As a_2 does not appear in the "vectors in basis" column of the final tableau, its value is zero. The same is true for any letter that is missing from the "vectors in basis" column of the final tableau. Thus,

$x = 12.24$, $y = 0$, $p = 206.12$, and $w = 66.67$. The maximum z is 14,030.60. This means that the health foods store should produce

12.24 packages of food X

0 packages of food Y

206.12 packages of food P

66.67 packages of food W

The profit is $14,030.60

Comment In Example 8, all the numbers were rounded to two decimal places. This is standard procedure in most applied problems.

Exercises

Use the simplex method to solve the following exercises.

23. Maximize $z = 7x + 3y$ subject to
$$\begin{aligned} x + 3y &\le 24 \\ 3x + y &\le 36 \\ x \ge 0,\ y &\ge 0 \end{aligned}$$

24. Maximize $z = 4x + 6y$ subject to
$$\begin{aligned} 2x + y &\le 18 \\ 5x + 6y &\le 30 \\ x \ge 0,\ y &\ge 0 \end{aligned}$$

25. Maximize $z = 9x + 3y$ subject to
$$\begin{aligned} 3x + 5y &\le 15 \\ x + 2y &\le 8 \\ x \ge 0,\ y &\ge 0 \end{aligned}$$

26. Maximize $z = 6x + 2y$ subject to
$$\begin{aligned} 4x + 5y &\le 20 \\ 2x + 3y &\le 12 \\ x \ge 0,\ y &\ge 0 \end{aligned}$$

27. Maximize $z = 4x + 5y$ subject to
$$\begin{aligned} x + 3y &\le 12 \\ 2x + y &\le 8 \\ x \ge 0,\ y &\ge 0 \end{aligned}$$

28. Maximize $z = 2x + 3y$ subject to
$$\begin{aligned} x + 4y &\le 8 \\ 5x + 3y &\le 30 \\ x + y &\le 7 \\ x \ge 0,\ y &\ge 0 \end{aligned}$$

29. Maximize $z = 2x - 3y + 4w$ subject to
$$\begin{aligned} x + 2y + 3w &\le 24 \\ x - y + w &\le 8 \\ 2x + 3y + 4w &\le 36 \\ x \ge 0,\ y \ge 0,\ w &\ge 0 \end{aligned}$$

30. Maximize $z = 3x + 4y + 7w$ subject to
$$\begin{aligned} x + 2y - 3w &\le 12 \\ 5y + 6w &\le 60 \\ 2x - 3y + 4w &\le 48 \\ x \ge 0,\ y \ge 0,\ w &\ge 0 \end{aligned}$$

31. Maximize $z = 3x + 4y + w$ subject to
$$\begin{aligned} 2x + 3y + w &\le 18 \\ 7x - 2y - w &\le 14 \\ x \ge 0,\ y \ge 0,\ w &\ge 0 \end{aligned}$$

32. Maximize $z = 9x + 2y$
subject to
$$\begin{aligned} 2x + 3y &\leq 10 \\ 3x + 5y &\leq 15 \\ x - y &\leq 9 \\ x \geq 0, y &\geq 0 \end{aligned}$$

★33. Maximize $z = 25x + 30y$
subject to
$$\begin{aligned} 2x + 3y &\leq 96 \\ 5x + 7y &\leq 140 \\ x &\leq 15 \\ y &\leq 20 \\ x \geq 0, y &\geq 0 \end{aligned}$$

34. *Jogging outfits.* The Mark Novelty Corporation produces two kinds of jogging outfits. Each must be processed by two assembly lines, A and B. To complete one dozen of the cheaper jogging outfits, line A must work 3 hours and line B must work 2 hours. To complete one dozen of the expensive outfits, line A must work 3 hours and line B must work 4 hours. Assembly line A cannot be operated for more than 12 hours, whereas assembly line B cannot be operated for more than 10 hours. The company earns a profit of $16 from each dozen of the cheaper outfit and $30 from each dozen of the expensive outfit that it sells. How many of each should be manufactured to maximize profit?

35. *Stereo sets.* The Stereo Corporation manufactures cassette tape recorders and 8-track stereo sets. Each tape recorder that is manufactured requires 2 hours on assembly line A and 3 hours on assembly line B and yields a profit of $18. Each stereo set that is manufactured requires 6 hours on line A and 4 hours on line B and yields a profit of $48. Furthermore, each set requires 2 hours in the testing shop. Assembly line A can be operated for at most 15 hours a day, line B for at most 12 hours a day, and the testing shop for at most 8 hours a day. How many of each should be manufactured to maximize profit?

36. *Blank tapes.* The Stereo Corporation in Exercise 35 also manufactures blank cassette tapes and blank 8-track stereo tapes in another division of the company. The time (in minutes) required on each assembly line and the profit from each are shown in the table. Assembly line C can be operated for at most 18 hours (1080 minutes) per day, and line D can be operated for at most 15 hours (900 minutes) per day. How many of each should be manufactured to maximize profit?

Type of tape	Assembly line C	Assembly line D	Profit
60-minute cassette tape	20	10	50¢
90-minute cassette tape	30	20	75¢
84-minute 8-track tape	36	24	120¢
90-minute 8-track tape	48	30	150¢

37. *Transportation.* The S & R Company sells three different products that can be used for transportation: a tricycle, a ten-speed bicycle, and a moped. The number of hours that each product must be processed through the various shops is shown in the table. The molding shop operates at most 90 hours per week, the assembly shop at most 100

hours per week, and the paint shop at most 70 hours per week. How many of each should be produced per week to maximize profit?

Item	Molding shop	Assembly shop	Paint shop	Profit
Tricycle	2	1	2	\$12
Ten-speed bicycle	4	3	3	\$20
Moped	3	6	2	\$36

38. *Printing.* A printing company prints advertising circulars and posters. A truckload of circulars takes 20 worker-hours to make and costs \$2000, whereas a truckload of posters takes 10 worker-hours to make and costs \$3000. Each truckload of circulars printed yields a profit of \$3500, and each truckload of posters yields a profit of \$5000. If there are 400 worker-hours available in a week and the weekly budget is \$50,000, how many of each type should be printed to maximize profit?

39. *Farming.* A farmer has a large farm on which he wishes to plant sunflower seeds and soybeans. For each acre of sunflower seeds planted, the farmer must invest \$20 in capital and \$50 in labor. The profit is \$40. For each acre of soybeans planted, the farmer must invest \$30 in capital and \$45 in labor. The profit is \$50. The farmer can afford to invest at most \$500 for labor and \$600 in capital. How many acres of each crop should he plant to maximize profit?

Photo: Monkmeyer

40. *Calculator games.* The Precision Instrument Company produces two kinds of calculator games, the baseball model and the hockey model. Each calculator must be processed by two factories, A and B. Factory A assembles the calculators and remains open at most 8 hours a day. Factory B, which tests, paints, and packages the calculators, can stay open at most 8 hours a day. Each baseball model requires 1 hour in factory A and 2 hours in factory B and yields a profit of \$7, whereas each hockey game model requires 3 hours in factory A and only 1 hour in factory B and yields a profit of \$6. How many calculators of each kind should be manufactured to maximize profit?

41. *Mixed fruit.* A wholesale grocer has just received a shipment of 6000 ounces of dried apricots, 4000 ounces of dried apples, and 3000 ounces of dried prunes. The grocer plans to package the dried fruit in one of three possible combinations of mixed fruit as shown in the table. How many of each should be produced to maximize profit?

Product	Ounces of dried apricots needed	Ounces of dried apples needed	Ounces of dried prunes needed	Profit
Package A	2	3	4	38¢
Package B	3	4	2	72¢
Package C	5	4	3	80¢

42. *Electrical installations.* The All-Boro Maintenance Company consists of three electricians, José, Pedro, and Tony, who work on their jobs as a team. They install electrical wiring in apartment houses, one-family houses, and two-family houses. They do not do any industrial or commercial installations. José needs 5, 6, and 3 hours, respectively, to do his part of the job on each apartment house, one-family house, and two-family house that is being rewired. Pedro needs 7, 8, and 5 hours, respectively, to do his part of the job, and Tony needs 3, 4, and 2 hours, respectively. The profits from each job are approximately \$150, \$80, and \$110, respectively. Neither man works more than 40 hours per week. How many electrical installations of each type should be performed to maximize profit?

43. *Mining.* The Yellow Creek Mine Company operates two mines, each of which produces three different qualities of ore. Mine A produces 7 tons of high-grade, 6 tons of medium-grade, and 10 tons of low-grade ore per day of operations and results in a profit of \$900. Mine B produces 8, 6, and 14 tons, respectively, of these ores per day and results in a profit of \$700. The company is renovating one of its refining plants. As a result, it can refine at most 150 tons of high-grade ore, 100 tons of medium-grade ore, and 200 tons of low-grade ore per week. How many days per month should each mine be operated to maximize profit?

5.5 The Simplex Method Applied to Minimization Problems: Duality[4]

Suppose we are given the following linear programming problem: Find the values of x and y that minimize $w = 24x + 32y$ subject to

$$3x + 2y \geq 12$$
$$4x + 8y \geq 16$$
$$x \geq 0, \ y \geq 0$$

[4] This section is more difficult and assumes a background knowledge of algebra. It may be omitted without affecting the continuity of the text.

At the moment we are unable to solve this problem since the linear function $w = 24x + 32y$ has to be minimized rather than maximized. As a matter of fact, all the problems we have solved so far by the simplex method have the following characteristics:

1. The variables are nonnegative.
2. The inequalities involving the variables are less than or equal to a positive number.
3. The objective function has to be maximized.

Fortunately, the simplex method can be used to minimize as well as to maximize a linear function. The only thing we have to do is to find the *dual* of the problem. Associated with any linear programming problem to be maximized is a minimization problem called its *dual*. Similarly, each minimization problem has a corresponding maximization problem called its dual.

Let us rewrite our problem using the variables x_1 and x_2 instead of x and y. The problem, then, is to minimize $w = 24x_1 + 32x_2$ subject to

$$3x_1 + 2x_2 \geq 12$$
$$4x_1 + 8x_2 \geq 16$$
$$x_1 \geq 0, x_2 \geq 0$$

In matrix form this problem can be written as

$$\min \mathbf{W} = \mathbf{CX}$$

subject to

$$\mathbf{AX} \geq \mathbf{B} \quad \text{and} \quad \mathbf{X} \geq 0$$

where

$$\mathbf{A} = \begin{pmatrix} 3 & 2 \\ 4 & 8 \end{pmatrix} \qquad \mathbf{B} = \begin{pmatrix} 12 \\ 16 \end{pmatrix} \qquad \mathbf{C} = (24 \quad 32) \qquad \mathbf{X} = \begin{pmatrix} x_1 \\ x_2 \end{pmatrix}$$

The dual of this problem is determined by forming *transposes* of these matrices (rows become columns and vice versa) so that the resulting problem will be of the form

$$\max \mathbf{Z} = \mathbf{B}^t\mathbf{Y}$$

subject to

$$\mathbf{A}^t\mathbf{Y} \leq \mathbf{C}^t$$
$$\mathbf{Y} \geq 0$$

where $\mathbf{A}^t$, $\mathbf{B}^t$, and $\mathbf{C}^t$ are transposes of the original matrices. In general, the matrix of a dual problem is formed from the matrix of the original by writing the rows as columns and the columns as rows. Thus, in our case, since

$$\mathbf{A} = \begin{pmatrix} 3 & 2 \\ 4 & 8 \end{pmatrix} \quad \text{then} \quad \mathbf{A}^t = \begin{pmatrix} 3 & 4 \\ 2 & 8 \end{pmatrix}$$

and since

$$\mathbf{B} = \begin{pmatrix} 12 \\ 16 \end{pmatrix} \quad \text{then} \quad \mathbf{B}^t = (12 \quad 16)$$

and since

$$\mathbf{C} = (24 \quad 32) \quad \text{then} \quad \mathbf{C}^t = \begin{pmatrix} 24 \\ 32 \end{pmatrix}$$

The matrix **Y** is a column matrix consisting of new unknowns y_1, y_2, and so on, which have to be maximized by the standard simplex method. The number of unknowns is exactly the same as the number of inequalities in the original problem. The new problem, without matrix notation, becomes: Maximize $z = 12y_1 + 16y_2$ subject to

$$\begin{aligned} 3y_1 + 4y &\le 24 \\ 2y_1 + 8y_2 &\le 32 \\ y_1 \ge 0, y_2 &\ge 0 \end{aligned}$$

We can now use the standard simplex method to solve the problem. We first add slack variables r and s to get

$$\begin{aligned} 3y_1 + 4y_2 + r \quad &= 24 \\ 2y_1 + 8y_2 \quad + s &= 32 \\ y_1 \ge 0, y_2 \ge 0, r &\ge 0, s \ge 0 \\ \max z = 12y_1 + 16y_2 &+ 0r + 0s \end{aligned}$$

The initial tableau is

$$\text{Tableau 1} \quad \begin{array}{c} \\ \\ \\ \\ \end{array} \begin{array}{ccccc|c} & a_1 & a_2 & a_3 & a_4 & b \\ \hline & 3 & 4 & 1 & 0 & 24 \\ & 2 & 8 & 0 & 1 & 32 \\ \hline & -12 & -16 & 0 & 0 & 0 \end{array} \quad \begin{array}{c} \text{Vectors in basis} \\ a_3 \\ a_4 \\ \\ \end{array}$$

The second tableau, with a_2 replacing a_4 in the "vectors in basis" column, is

$$\text{Tableau 2} \quad \begin{array}{ccccc|c} & a_1 & a_2 & a_3 & a_4 & b \\ \hline & 2 & 0 & 1 & -\frac{1}{2} & 8 \\ & \frac{1}{4} & 1 & 0 & \frac{1}{8} & 4 \\ \hline & -8 & 0 & 0 & 2 & 64 \end{array} \quad \begin{array}{c} \text{Vectors in basis} \\ a_3 \\ a_2 \\ \\ \end{array}$$

The third tableau, with a_1 replacing a_3 in the "vectors in basis" column, is

$$\text{Tableau 3}\quad \begin{array}{c} \\ \\ \\ \end{array}\begin{array}{cccc|c} a_1 & a_2 & a_3 & a_4 & b \\ \left(\begin{array}{c}1\\0\\\hline 0\end{array}\right. & \begin{array}{c}0\\1\\\hline 0\end{array} & \begin{array}{c}\frac{1}{2}\\-\frac{1}{8}\\\hline 4\end{array} & \begin{array}{c}-\frac{1}{4}\\\frac{3}{16}\\\hline 0\end{array} & \left.\begin{array}{c}4\\3\\\hline 96\end{array}\right) \end{array} \quad \begin{array}{c}\text{Vectors}\\\text{in basis}\\ a_1 \\ a_2 \\ \\ \end{array}$$

Since there are no negative numbers in the bottom row, we are done. The maximum z value is 96. It is shown in more advanced texts that the minimum value of the objective function w is the same as the maximum value of the objective function z and that the solutions are found as the last entries in the *columns* corresponding to the slack variables. Thus, in our case, the minimum value of w is 96, which occurs when $x_1 = 4$ and $x_2 = 0$.

Example 9 Minimize $w = 18x_1 + 20x_2$ subject to

$$-2x_1 - 5x_2 \leq -24$$
$$3x + 4x_2 \geq 28$$
$$x_1 \geq 0, x_2 \geq 0$$

Solution Since this is a minimization problem, we must form its dual to convert it to a maximization problem. Before doing this, let us multiply the first constraint inequality by -1 so that the inequality will change from $\leq$ to $\geq$. (See the multiplication rule for inequalities on page 159 of Chapter 4.) When this is done, we have the following linear programming problem:

$$\min w = 18x_1 + 20x_2$$

subject to

$$2x_1 + 5x_2 \geq 24$$
$$3x_1 + 4x_2 \geq 28$$
$$x_1 \geq 0, x_2 \geq 0$$

We then have the following matrices

$$\mathbf{A} = \begin{pmatrix} 2 & 5 \\ 3 & 4 \end{pmatrix} \qquad \mathbf{B} = \begin{pmatrix} 24 \\ 28 \end{pmatrix} \qquad \mathbf{C} = (18 \quad 20) \qquad \mathbf{X} = \begin{pmatrix} x_1 \\ x_2 \end{pmatrix}$$

The transposes of these matrices are

$$\mathbf{A}^t = \begin{pmatrix} 2 & 3 \\ 5 & 4 \end{pmatrix} \qquad \mathbf{B}^t = (24 \quad 28) \qquad \mathbf{C}^t = \begin{pmatrix} 18 \\ 20 \end{pmatrix} \qquad \mathbf{Y} = \begin{pmatrix} y_1 \\ y_2 \end{pmatrix}$$

Comment To avoid getting confused, we introduce the variables y_1 and y_2 instead of working with the variables x_1 and x_2. The same is true for the variables w and z.

The dual of the original problem, then is

$$\max z = 24y_1 + 28y_2$$

subject to

$$2y_1 + 3y_2 \leq 18$$
$$5y_1 + 4y_2 \leq 20$$
$$y_1 \geq 0,\ y_2 \geq 0$$

With slack variables r and s added, we get

$$2y_1 + 3y_2 + r \qquad = 18$$
$$5y_1 + 4y_2 \qquad + s = 20$$
$$y_1 \geq 0,\ y_2 \geq 0,\ r \geq 0,\ s \geq 0$$
$$\max z = 24y_1 + 28y_2 + 0r + 0s$$

The initial tableau is

Tableau 1

	a_1	a_2	a_3	a_4	b	Vectors in basis
	2	3	1	0	18	a_3
	5	4	0	1	20	a_4
	−24	−28	0	0	0	

The second tableau, with a_2 replacing a_4, is

Tableau 2

	a_1	a_2	a_3	a_4	b	Vectors in basis
	−1.75	0	1	−0.75	3	a_3
	1.25	1	0	0.25	5	a_2
	11	0	0	7	140	

Since the bottom row has no negative numbers, we are done. The maximum z value is 140. This is also the minimum value of w. Now we look in the bottom row and in the columns corresponding to slack variables. Note that a_3 corresponds to x_1 and a_4 corresponds to x_2. Thus, the answer to the original problem is $x_1 = 0$, $x_2 = 7$. The minimum value of w is 140.

Example 10 *Linear programming applied to good nutrition.*[5] A convalescing patient has been ordered by her doctor to eat certain foods so that she gets a minimum of 6 units of protein, 4 units of carbohydrates, and 5 units of vitamins. She plans to purchase fish, vegetables, and crackers to satisfy these requirements. The nutritional content of these foods is

[5] This application is a variation of the article "The Cost of Subsistence" by George J. Stigler, which appeared in *The Journal of Farm Economics*, vol. 27, 1972. We suggest that you read the article for a more thorough discussion.

shown in the accompanying table. How much should be purchased to minimize costs and still satisfy the doctor's orders?

Food	Units of protein	Units of carbohydrates	Units of vitamins	Cost
Fish	5	4	3	\$2.00
Vegetables	1	2	2	\$1.00
Crackers	2	8	1	\$1.50

Solution Let

x_1 = the number of pounds of fish to be purchased

x_2 = the number of pounds of vegetables to be purchased

x_3 = the number of pounds of crackers to be purchased

The mathematical formulation of the problem then becomes

$$\min w = 2x_1 + x_2 + 1.5x_3$$

subject to

$$\begin{aligned} 5x_1 + x_2 + 2x_3 &\geq 6 \\ 4x_1 + 2x_2 + 8x_3 &\geq 4 \\ 3x_1 + 2x_2 + x_3 &\geq 5 \\ x_1 \geq 0, x_2 \geq 0, x_3 &\geq 0 \end{aligned}$$

We have the following matrices

$$\mathbf{A} = \begin{pmatrix} 5 & 1 & 2 \\ 4 & 2 & 8 \\ 3 & 2 & 1 \end{pmatrix} \qquad \mathbf{B} = \begin{pmatrix} 6 \\ 4 \\ 5 \end{pmatrix} \qquad \mathbf{C} = (2 \quad 1 \quad 1.5) \qquad \mathbf{X} = \begin{pmatrix} x_1 \\ x_2 \\ x_3 \end{pmatrix}$$

The transposes of these matrices are

$$\mathbf{A}^t = \begin{pmatrix} 5 & 4 & 3 \\ 1 & 2 & 2 \\ 2 & 8 & 1 \end{pmatrix} \qquad \mathbf{B}^t = (6 \quad 4 \quad 5) \qquad \mathbf{C}^t = \begin{pmatrix} 2 \\ 1 \\ 1.5 \end{pmatrix} \qquad \mathbf{Y} = \begin{pmatrix} y_1 \\ y_2 \\ y_3 \end{pmatrix}$$

The dual of the original problem then is

$$\max z = 6y_1 + 4y_2 + 5y_3$$

subject to

$$\begin{aligned} 5y_1 + 4y_2 + 3y_3 &\leq 2 \\ y_1 + 2y_2 + 2y_3 &\leq 1 \\ 2y_1 + 8y_2 + y_3 &\leq 1.5 \\ y_1 \geq 0, y_2 \geq 0, y_3 &\geq 0 \end{aligned}$$

Three tableaus are needed to solve this problem as shown. The reader should verify that the entries given are accurate.

Tableau 1

a_1	a_2	a_3	a_4	a_5	a_6	b	Vectors in basis
5	4	3	1	0	0	2	a_4
1	2	2	0	1	0	1	a_5
2	8	1	0	0	1	1.5	a_6
−6	−4	−5	0	0	0	0	

Tableau 2

a_1	a_2	a_3	a_4	a_5	a_6	b	Vectors in basis
1	0.8	0.6	0.2	0	0	0.4	a_1
0	1.2	1.4	−0.2	1	0	0.6	a_5
0	6.4	−0.2	−0.4	0	1	0.7	a_6
0	0.8	−1.4	1.2	0	0	2.4	

Tableau 3

a_1	a_2	a_3	a_4	a_5	a_6	b	Vectors in basis
1	0.29	0	0.29	−0.43	0	0.14	a_1
0	0.86	1	−0.14	0.71	0	0.43	a_3
0	6.57	0	−0.43	0.14	1	0.79	a_6
0	2	0	1	1	0	3	

There are no negative values in the bottom row, so we are done. The answer is $x_1 = 1$, $x_2 = 1$, $x_3 = 0$. The minimum value of w is 3. Thus, the patient should buy 1 pound of fish, 1 pound of vegetables, and 0 pounds of crackers. The minimum cost is \$3.

Exercises

Use the simplex method to solve the following exercises.

44. Minimize $w = 9x_1 + 6x_2$
subject to
$x_1 + 4x_2 \geq 8$
$2x_1 + 5x_2 \geq 10$
$x_1 \geq 0, x_2 \geq 0$

45. Minimize $w = 5x_1 + 6x_2$
subject to
$x_1 + x_2 \geq 6$
$3x_1 + 4x_2 \geq 24$
$x_1 \geq 0, x_2 \geq 0$

46. Minimize $w = 3x_1 + x_2$
subject to
$x_1 + x_2 \geq 4$
$2x_1 + 3x_2 \geq 6$
$x_1 \geq 0, x_2 \geq 0$

47. Minimize $w = 8x_1 + 2x_2 + 3x_3$
subject to
$x_1 + x_2 + 2x_3 \geq 16$
$3x_1 - 2x_2 + 6x_3 \geq 12$
$3x_1 + x_2 - 3x_3 \geq 10$
$x_1 \geq 0, x_2 \geq 0, x_3 \geq 0$

48. Minimize $w = 15x_1 + 21x_2$
subject to
$$\begin{aligned} x_1 + 3x_2 &\geq 2 \\ x_1 + x_2 &\geq 4 \\ -x_1 + x_2 &\geq 1 \\ x_1 \geq 0, x_2 &\geq 0 \end{aligned}$$

49. Minimize $w = 4x_1 + 5x_2$
subject to
$$\begin{aligned} 3x_1 + x_2 &\geq 12 \\ 3x_1 - 4x_2 &\geq 8 \\ x_1 + x_2 &\geq 15 \\ x_1 \geq 0, x_2 &\geq 0 \end{aligned}$$

50. Minimize $w = 3x_1 + x_2 + 8x_3$
subject to
$$\begin{aligned} 4x_1 + x_2 + 3x_3 &\geq 48 \\ x_1 - 2x_2 + 8x_3 &\geq 36 \\ x_1 \geq 0, x_2 \geq 0, x_3 &\geq 0 \end{aligned}$$

51. Minimize $w = 3x_1 - 2x_2 + 4x_3$
subject to
$$\begin{aligned} x_1 + x_2 \qquad &\leq 6 \\ 3x_1 + x_2 - x_3 &\leq 10 \\ x_2 + 3x_3 &\leq 4 \\ x_1 \geq 0, x_2 \geq 0, x_3 &\geq 0 \end{aligned}$$

52. Minimize $w = 2x_1 + 2x_2$
subject to
$$\begin{aligned} -x_1 + x_2 &\geq -2 \\ x_1 + x_2 &\leq -3 \\ x_1 \geq 0, x_2 &\geq 0 \end{aligned}$$

★ 53. Minimize $w = 5x_1 + 2x_2 + 9x_3$
subject to
$$\begin{aligned} 2x_1 + 3x_2 + 4x_3 &= 5 \\ x_1 + 3x_2 + 2x_3 &\geq 4 \\ 3x_1 + 2x_2 + x_3 &\geq 2 \\ x_1 \geq 0, x_2 \geq 0, x_3 &\geq 0 \end{aligned}$$

54. *Oil refinery.* An oil company operates two oil refineries, each of which processes and produces low-grade, medium-grade, and high-grade oil. Refinery A can produce 400, 500, and 300 barrels of oil per day of these oils, respectively, whereas refinery B can produce 200, 300, and 100 barrels per day. Furthermore, the operating expenses for these refineries are \$200 per day for refinery A and \$300 per day for refinery B. Due to the energy crisis, the oil company must supply at least 1500, 2000, and 1800 barrels per day of low-grade, medium-grade, and high-grade oil per day, respectively. How many days should each refinery operate to minimize costs and still satisfy the above requirements?

55. *Baby meals.* A pediatrician has advised a mother to feed her infant meals that contain at least 2400 calories, at least 20 grams of carbohydrates, and at least 36 grams of protein per day. The mother has two foods available, which she wishes to mix together in the most economical manner to satisfy the baby's nutritional requirements. The accompanying table gives the nutritional information and costs of these foods. How many units of each food should the mother use to minimize costs?

	Calories	Carbohydrates	Protein	Cost
Food A	300	8	18	70¢
Food B	400	10	12	60¢

56. *Weekend hiking.* A camp is scheduling a weekend hike in the woods and is planning to take along three different kinds of prepacked meals. The accompanying table gives the nutritional information and the costs of these meals. The minimum daily requirement for protein is 90 units and for calories, 1000 units. How many meals of each type should be served to minimize costs?

	Protein	Calories	Cost
Meal A	40	300	80¢
Meal B	30	400	60¢
Meal C	36	250	70¢

57. *Operating gas stations.* Mr. Smith operates two gas stations on State Highway I, an Exxon station and a Mobil station. For each station he estimates that he must sell at least the indicated number of gallons of gasoline per week so as to make a decent living. If it costs Mr. Smith \$200 per week to operate the Exxon station and \$250 per week to operate the Mobil station, find the number of days each station should be open to minimize costs.

	Regular leaded	Regular unleaded	Premium
Exxon	2000	4000	3000
Mobil	4000	3000	5000

58. *Mining.* The Yellow Creek Mining Company operates two mines from which it obtains gold, silver, and copper. Mine A costs \$2,000 a day to operate and yields 70 ounces of gold, 7,000 ounces of copper, and 10,000 ounces of silver per day. Mine B costs \$1,500 a day to operate and yields 60 ounces of gold, 6,600 ounces of copper, and 8,000 ounces of silver per day. To satisfy production schedules, the company must supply at least 200 ounces of gold, at least 50,000 ounces of copper, and at least 48,000 ounces of silver. How many days should each mine be operated to meet the production schedules at a minimum cost?

5.6 Further Applications: The Transportation Problems[6]

As we indicated in previous sections, the simplex method was invented by Dantzig to solve problems in business and industry. Numerous applications have already been given. Recently linear programming has been applied to other disciplines also. Several such case studies are presented in the following examples.

[6] This section is more difficult and assumes a background knowledge of algebra. It may be omitted without affecting the continuity of the text.

Example 11 *Linear programming applied to collecting garbage.* The sanitation department of Woodsburgh consists of 15 trucks and 36 employees. Due to budgetary limitations, no new employees can be hired and no new trucks can be purchased. In an effort to maximize the amount of garbage collected, sanitation department officials recently began experimenting with half crews (one truck, two workers) as opposed to full crews (one truck, three workers). On the average, each full crew collects 1 ton of garbage, whereas each half crew can collect only $\frac{1}{2}$ ton of garbage. How should the sanitation department allocate its resources so as to maximize the amount of garbage collected?

Solution We will merely present the mathematical formulation of the problem. In Exercise 59 the reader will be asked to complete the solution. We first summarize the information given in this problem in the form of a table.

Photo: De Wys Inc.

	Workers needed	Trucks needed	Tons of garbage collected
Half crew	2	1	1/2
Full crew	3	1	1
Limitations	36	15	

Let

x = the number of half crews assigned

y = the number of full crews assigned

The problem is to find values of x and y for which $z = \frac{1}{2}x + y$ is maximized subject to

$$x + y \leq 15$$
$$2x + 3y \leq 36$$
$$x \geq 0, y \geq 0$$

Example 12 *Linear programming and water pollution.*[7] In October 1975 the New York State Bureau of Environmental Protection (Division of Fish and Wildlife) launched an all-out campaign to clean up the Hudson River. Two companies that had been polluting the river at different points installed treatment plants to help remove some of the PCBs from the chemical wastes dumped into the river. Linear programming was used to minimize the cost of this cleanup. (Our discussion is merely intended to indicate many of the difficulties or constraints involved.)

[7] This simplified example is based on the publication "Monitoring of PCB's in Fish Taken from the Hudson River" by the New York State Bureau of Environmental Protection, Division of Fish and Wildlife, October 1975.

Since both polluting companies wanted to minimize the cost of removing the PCBs from the chemical wastes, the objective function to be minimized is the cost of removing the PCB:

$$C = \begin{pmatrix}\text{cost to}\\ \text{company A}\\ \text{of removing}\\ \text{the PCBs}\end{pmatrix} + \begin{pmatrix}\text{cost to}\\ \text{company B}\\ \text{of removing}\\ \text{the PCBs}\end{pmatrix}$$

Assume that company A generates w_1 pounds of PCB wastes a day and that p_1 percent will be removed by its treatment plant. Also assume that company B generates w_2 pounds of PCB wastes a day and that p_2 percent will be removed by its treatment plant. It is a known fact that nature will attempt to remove some of the chemical pollutants by the natural processes of the river as the water flows downstream. Let n_1 be the percentage of the pollutants removed by nature from company A's wastes and let n_2 be the percentage of the pollutants removed by nature from company B's wastes. If R represents the total amount of PCBs arriving downstream, then

$$R = (1 - p_1)w_1 + (1 - n_1)w_1 + (1 - p_2)w_2 + (1 - n_2)w_2$$

The legal limit on the value of R is usually set by environmental protection agencies.

Now we must find constraints that represent the cost of operating the treatment plants. In most cases, simply increasing the amount of money spent on pollution control will not necessarily result in a proportionate increase in the percentage of PCBs removed from the water. Furthermore, each company has its own cost function. Obviously, as the number of polluting companies increases, the number of constraints increases and the problem becomes quite different. Thus, determining the cost of removing p percent of a company's PCB pollution depends on a number of factors, some of which have to be determined experimentally.[8]

Linear programming can be used to determine the minimum cost of future additional labor when production schedules for the next few months indicate that more labor will be needed than is available and when either new employees can be hired or overtime wages can be paid.

Example 13 *Linear programming and manpower forecasting.*[9] The ARZT Corporation will require 100, 120, and 130 employees for the summer months of June, July, and August, respectively. The personnel director estimates

[8] One detailed discussion of the problem and its solution can be found in "An Economic Approach to Water Quality Control" by Kenneth D. Kerri in *Journal of the Water Pollution Control Federation* (vol. 38, no. 12, pp. 1883–1897, 1966).

[9] This example is adapted from "Manpower Forecasting with Linear Programming" by Myles A. Vogel in *Industrial Engineering* (vol. 8, pp. 43–45, 1976).

that there will be only 90, 80, and 70 employees currently on the payroll who are available during these three months. Moreover, new employees have a high attrition rate. For these three months it is estimated that this rate will be 0.9, 0.8, and 0.6, respectively, and production from these new workers is estimated to be 0.2, 0.4, and 0.6. No more than 15 new employees can be hired per month and each is paid a salary of \$150 a month. Furthermore, there is \$1120 available in the budget for overtime per month, although the limit on overtime is 50 man-month-hours. How many new workers should be hired and how many hours of overtime should be assigned per month? Also find the minimum cost.

Solution If we let

x_1 = the number of new workers to be hired in June

x_2 = the number of new workers to be hired in July

x_3 = the number of new workers to be hired in August

y_1 = the number of man-month-hours of overtime to be assigned in June

y_2 = the number of man-month-hours of overtime to be assigned in July

y_3 = the number of man-month-hours of overtime to be assigned in August

then the objective function to be minimized is

$$z = 150x_1 + 150x_2 + 150x_3 + 1120y_1 + 1120y_2 + 1120y_3$$

subject to

$$(0.9)(0.2)x_1 + y_1 = 100 - 90 = 10$$
$$(0.9)(0.2)x_1 + (0.8)(0.4)x_2 + y_2 = 120 - 80 = 40$$
$$(0.9)(0.2)x_1 + (0.8)(0.4)x_2 + (0.6)(0.6)x_3 + y_3 = 130 - 70 = 60$$
$$x_1 \le 15 \qquad x_2 \le 15 \qquad x_3 \le 15$$
$$y_1 \le 50 \qquad y_2 \le 50 \qquad y_3 \le 50$$

Using a computer equipped with a modified cyphernetics program, it can be shown that the following solutions are obtained:

Month	Number of new employees to be hired	Number of overtime hours	Cost
June	15	7.30	\$105,108.25
July	15	32.5	\$104,852.00
August	15	47.1	\$104,078.00

Thus, the total additional labor cost for the three months is \$105,108.25 + \$104,852.00 + \$104,078.00, or \$314,038.25.

Example 14 *Linear programming and oil allocation in California.* In recent years America has been experiencing a severe energy crisis. There are some environmentalists who oppose (m)any new energy developments. Donald W. Barnett indicates that the environmentalists' blanket opposition to all new energy developments could be counterproductive because "they neglect the differing impacts of various energy sources."[10] Specifically, he indicates how linear programming can be used to rank the various oils that could supply the California market. In this case the objective function to be minimized is the cost of supplying California with its oil needs subject to resource, sulfur, and oil spill constraints.

Another interesting application of linear programming arises in the "transportation problem," where products must be transported from different "warehouses" to various "stores." Linear programming can be used to determine a routing that will minimize the total transportation cost, that is, specify which warehouse should ship a specified number of items to a particular store at minimum cost.

Example 15 *Linear programming applied to the transportation problem.*[11] The Miller Car Rental Agency rents cars in six cities on the East Coast: New York City, Hartford, Boston, Manchester, Albany, and Quebec, Canada. Since many of these rentals are one way, the company often finds that some of its facilities have a surplus of cars, whereas others may need cars. On one particular day the facilities in Albany, Manchester, and Quebec have a surplus of 35, 25, and 40 cars, respectively. On the other hand, Hartford needs 28 cars, Boston needs 46 cars, and New York City needs 26 cars. The approximate mileage between these cities is shown in the accompanying table. How many cars should each facility with a surplus of cars send to Hartford, Boston, and New York City so as to minimize the total distance driven by these cars?

	To		
From	New York City	Hartford	Boston
Albany	140	95	170
Manchester	240	135	55
Quebec	510	420	385

Solution If we interpret the cities of Albany, Manchester, and Quebec as "warehouses" and the cities of New York, Hartford, and Boston as

[10] From "The Ranking of Energy Resources Subject to Environmental Constraints: The Case of Crude Oil Supplies for California" by Donald W. Barnett in *Energy Systems and Policy* (vol. 2, no. 1, pp. 31–58, 1977).

[11] My thanks to Sam Hoffman of the Miller Car Rental Agency for this problem.

"stores," then this problem can be viewed as a transportation problem where we want to transport the rental cars from warehouses to stores with a minimum of driving.

Let us use A for Albany, M for Manchester, Q for Quebec, N for New York City, B for Boston, and H for Hartford. Similarly, let

AH be the number of cars transferred from Albany to Hartford

AB be the number of cars transferred from Albany to Boston

AN be the number of cars transferred from Albany to New York City

MH be the number of cars transferred from Manchester to Hartford

MB be the number of cars transferred from Manchester to Boston

MN be the number of cars transferred from Manchester to New York City

QH be the number of cars transferred from Quebec to Hartford

QB be the number of cars transferred from Quebec to Boston

QN be the number of cars transferred from Quebec to New York City

We then have the following three equations (constraints) representing the number of cars to be sent from Albany, Manchester, and Quebec to Hartford, Boston, and New York City:

$$\begin{aligned} AH + AB + AN &= 35 \\ MH + MB + MN &= 25 \\ QH + QB + QN &= 40 \end{aligned}$$

We also have three additional constraints (equations) representing the number of cars received by Hartford, Boston, and New York City from Albany, Manchester, and Quebec:

$$\begin{aligned} AH + MH + QH &= 28 \\ AB + MB + QB &= 46 \\ AN + MN + QN &= 26 \end{aligned}$$

The objective function we wish to minimize represents the number of miles driven from the "warehouse" cities to the "store" cities. Thus, the objective function to be minimized is

$$\begin{aligned} &95AH + 170AB + 140AN + 135MH + 55MB + 240MN \\ &\quad + 420QH + 385QB + 510QN \end{aligned}$$

We can now use the simplex method to solve this problem. Since problems of this type come up repeatedly, however, mathematicians have developed a special algorithm (technique) for solving transportation problems. The interested reader can consult the references given at the end of this chapter.

Example 16 *Linear programming and bank portfolio management.*[12] The Midtown National Bank has investment assets totaling \$100 million, some in the form of outstanding business loans and mortgages and some in the form of U.S. treasury obligations. Government regulations and bank rules limit the amount of money that a bank can lend out. Specifically, Midtown can lend out at most \$30 million for business loans and mortgages at a 10% interest rate, and can invest at least \$50 million in U.S. treasury obligations paying 8% interest. Furthermore, the board of directors of the bank requires that at least \$15 million be used on business loans and mortgages. How much should the bank invest in each category (business loans and U.S. treasury obligations) to maximize its investment income?

Solution Let

x = the amount of money invested in business loans and mortgages

y = the amount of money invested in U.S. treasury obligations

Since \$100 million is available for investment purposes, we have $x + y \leq 100$. Since the bank will lend out at least \$15 million and at most \$30 million for business loans, we have $15 \leq x \leq 30$. Also, the bank's investment in U.S. treasury obligations is at least \$50 million, so we have $y \geq 50$. Finally, the bank receives 10% interest on business loans and 8% interest on U.S. treasury obligations, so its income is $z = 0.10x + 0.08y$. Thus, the mathematical formulation of the problem is to find values of x and y for which $z = 0.10x + 0.08y$ is maximized subject to

$$
\begin{aligned}
x + y &\leq 100 \\
15 \leq x &\leq 30 \\
y &\geq 50 \\
x \geq 0,\ y &\geq 0
\end{aligned}
$$

The simplex method can now be applied to solve this linear programming problem.

Additional applications of the simplex method of linear programming will be given when we discuss game theory in Chapter 7.

[12] This example is based on the articles "Linear Programming: A New Approach to Bank Portfolio Management" by A. Broaddus in *Federal Reserve Bank of Richmond: Monthly Review* (vol. 58, no. 11, pp. 3–11, 1972); and "Linear Programming and Optimal Bank Asset Management Decisions" by K. J. Cohen and F. S. Hammer in *Journal of Finance* (vol. 22, pp. 147–165, 1967).

59. Complete the solution of Example 11 in this chapter.

60. Read the article "A Linear Programming Model for Budgeting and Financial Planning" by U. Ijiri, F. K. Levy, and R. C. Lyon in *Journal of Accounting Research I* (pp. 198–212, 1963) for a discussion on how linear programming can be used in budgeting and financial planning.

61. Read the article "Maintaining a Profitable Ecological Balance" by G. M. Van Dyne and Kenneth R. Rebman in *Some Mathematical Models in Biology*, edited by Robert M. Thrall (Ann Arbor: University of Michigan Press, 1967) for an interesting application of linear programming to a bioeconomic situation.

62. Read the article "A Linear Programming Approach to Cultural Intensity" by Hans Hoffman in *Game Theory in the Behavioral Sciences* (Pittsburgh: University of Pittsburgh Press, 1969) for an interesting discussion of how linear programming can be used to measure cultural intensity.

63. Read the article "Improving the Performance of an Intelligent Data Management System" by J. G. Kollias and others in *Computer Journal* (vol. 20, pp. 302–307, 1977) for a discussion on how linear programming can be applied to data management.

64. Read the article "Different Approaches to Advertising Media Selection" by Dennis Gensch in *Operational Research Quarterly* (vol. 21, no. 2, pp. 193–219, 1970) for a discussion on how linear programming can be used in media selection.

65. Read the article "Some Applications of Linear Programming to the Inverse Gravity Problem" by C. Safon and others in *Geophysics* (vol. 42, pp. 1215–1220, 1977).

5.7 Summary

In this chapter and Chapter 4 we have attempted to introduce the important and ever-growing field of mathematics dealing with linear programming. Although we indicated how to solve such problems by both the geometric and the algebraic methods, we did not consider all the possible difficulties we might encounter when solving such problems. For example, if no identity matrix can be found within the **A** matrix, then one has to be created artificially. Since the problem becomes complicated when this has to be done, we have purposely avoided it.

Sometimes, when solving a linear programming problem, we find that a variable that has been eliminated from the "vectors in basis" column in a previous tableau returns in a later tableau. This occurrence, which is known as *cycling*, has also been eliminated.

Of course, it is possible that a linear programming problem has no solution. This occurs when there is a minimum z value in the bottom row of tableau 1 and yet all the numbers in the same column are negative. Again, we have avoided such problems in our discussion.

Our objective was to introduce you to the techniques used in solving most linear programming problems. The interested reader can consult the references given for a detailed discussion of linear programming in general and for specific ways of dealing with the difficulties mentioned.

Study Guide

You should be able to demonstrate your knowledge of the following ideas presented in this chapter by giving definitions, descriptions, or specific examples. Page references are given for each term so that you can check your answer.

Simplex method (p. 179)	Slack or surplus variables (p. 179)
Objective function (p. 180)	Matrix **A** (p. 183)
Matrix **b** (p. 183)	Matrix **c** (p. 183)
Column vector (p. 183)	Tableau (p. 184)
Vectors in basis (p. 184)	Pivotal column (p. 188)
Pivotal row (p. 189)	Pivotal element (p. 189)
Pivotal rule (p. 189)	Minimization problem (p. 207)
Dual (p. 207)	Transpose of a matrix (p. 207)
Transportation problem (p. 218)	Cycling (p. 221)

Formulas to Remember

This chapter contains two main groups of formulas with which you should be familiar.

Pivotal rule (p. 189)

Duality and the transpose of a matrix (p. 207)

Mastery Tests

FORM A

1. To convert inequalities to equalities we add or subtract ____________ ____________.

2. True or false: The simplex method is a graphical method that can be used to solve linear programming problems.

3. Consider the problem: Maximize $7x + 2y + 3$ subject to

$$x + 4y \leq 10$$
$$x \geq 0, \; y \geq 0$$

a) Find the constraints.
b) Find the objective function.
c) Can the simplex method be used to solve it? Explain your answer.

4. A student attempted to convert the inequalities

$$3x + 4y \le 7$$
$$2x + 9y \le 12$$
$$7x + 4y \le 20$$

into equalities by adding the slack variables r and s, obtaining

$$\begin{aligned} 3x + 4y + r \qquad &= 7 \\ 2x + 9y + r \qquad &= 12 \\ 7x + 4y \qquad + s &= 20 \end{aligned}$$

Is this a valid procedure? Explain your answer.

5. The linear expression that we wish to maximize or minimize is called the ____________________ function.

6. The column that contains the smallest negative number on the bottom row of a tableau is known as the ____________________.

7. What happens if the bottom row of tableau 1 has no negative numbers?

8. Find the dual of the following linear programming problem:

$$\begin{aligned} \text{Minimize} \quad & w = 2x + 3h + 4p + 2m \\ \text{Subject to} \quad & x + h + p + m \ge 25 \\ & 2x + h + 3p + m \ge 175 \\ & x \ge 0, h \ge 0, p \ge 0, m \ge 0 \end{aligned}$$

9. What is the transpose of this matrix?

$$\begin{pmatrix} 2 & 1 & 7 & 9 \\ 3 & 2 & 8 & 6 \\ 5 & -2 & 0 & 1 \end{pmatrix}$$

10. Can you find one condition under which a linear programming problem will have no solution?

FORM B

1. Maximize $z = 4x + 3y + 7w$ subject to

$$\begin{aligned} x + 2y + w &\le 24 \\ 3x + y + 2w &\le 36 \\ 2x + 2y + 3w &\le 48 \\ x \ge 0, y \ge 0, w &\ge 0 \end{aligned}$$

2. Maximize $z = 2x + 3y + 4w$ subject to

$$\begin{aligned} 4x + 7y + 8w &\le 24 \\ 8x + 7y + 4w &\le 24 \\ x \ge 0, y \ge 0, w &\ge 0 \end{aligned}$$

★3. Minimize $z = x + y + w$ subject to

$$\begin{aligned} 3x + 4y + 3w &\ge 36 \\ 3x + y + 2w &\ge 24 \\ x \ge 0, y \ge 0, w &\ge 0 \end{aligned}$$

4. Complete the solution of Example 15 in this chapter.

5. *Mixed nuts packages.* A nut company has 960 ounces of almonds and 1600 ounces of walnuts, which it wishes to mix and sell in 1-pound packages. Package I is to contain 6 ounces of almonds and 10 ounces of walnuts and will yield a profit of 48¢, whereas package II is to contain 5 ounces of almonds and 11 ounces of walnuts and will yield a profit of 60¢. How many packages of each type should be made to maximize profit?

★ 6. *Transportation problem.* A publishing company has two printing shops and two distribution points for its books. Printing shop I has 80 cartons of a book and printing shop II has 110 cartons of the book. Distribution point I needs 70 cartons of the book and distribution point II needs 65 cartons of the book. It costs the publishing company $4 to transport a carton of books from shop I to distribution point I and $6 to distribution point II. Also, it costs the publishing company $5 to transport a carton of books from shop II to distribution point I and $8 to distribution point II. How many cartons should be transported from each printing shop to minimize cost?

★ 7. *Cat food.* The Pussy Cat Food Company plans to market a new type of cat food that is made by mixing two types of foods. Each can of food is to provide a cat with at least 80 grams of protein, 8 grams of fat, and 150 grams of carbohydrates. The nutritional breakdown of the foods used in making the cat food is shown in the table. How many units of each food should be used to minimize cost?

	Protein	Fat	Carbohydrate	Cost
Food A	30	2	30	75¢
Food B	10	3	25	60¢

8. *Extra credit assignment.* Mary is working on some extra credit problems for her professor. Some are theory problems, some are application problems, and some involve a combination of both theory and application. For each theory problem correctly solved Mary will receive 8 credits, for each application problem she will receive 4 credits, and for each combination problem she will receive 6 credits. Her professor will accept up to 150 problems of these three types. Mary knows that each theory problem takes her 20 minutes to solve, each application problem 10 minutes, and each combination problem 12 minutes. She can devote at most 12 hours (720 minutes) for all the extra credit problems. Since theory problems have always been difficult for Mary, she has decided not to spend more than 500 minutes on them. How many problems of each type should Mary do to maximize her credit?

★ 9. *Linear programming and investment strategies.* Mr. Sagio has just inherited $100,000, which he wishes to invest in either medium-grade bonds with a 10% yield or in speculative stocks with a 18% yield. He insists that at least 75% of the money be invested in bonds and at least

$10,000 be invested in speculative stocks. How should he invest his money so as to maximize his yield and still satisfy the specified conditions?

10. Using the simplex method, complete the solution of the bank portfolio management problem in Example 16 of this chapter.

Suggested Further Reading

Dantzig, George B., *Linear Programming and Extensions*, Princeton, N.J.: Princeton University Press, 1963.

Gass, S. I., *Linear Programming*, 3d ed., New York: McGraw-Hill, 1969.

Lanchester, F. W., "Mathematics in Warfare," in James R. Newman, ed., *The World of Mathematics*, vol. IV, pt. 20, New York: Simon & Schuster, 1956.

Mathematics in the Modern World (readings from *Scientific American*), articles 37, 38, 42, and 43, San Francisco: Freeman, 1968.

Morse, P., and G. E. Kimball, "How to Hunt a Submarine," in James R. Newman, ed., *The World of Mathematics*, vol. IV, pt. 20, New York: Simon & Schuster, 1956.

Painter, R. J., and R. P. Yantis, *Elementary Matrix Algebra with Linear Programming*, Boston: Prindle, Weber and Schmidt, 1971.

Vazsonyi, A., *Scientific Programming in Business and Industry*, New York: Wiley 1958.

CHAPTER 6 Probability

Objectives to be Achieved in this Chapter

- *To discuss* several ways of defining what is meant by the word *probability* and how to calculate probabilities.
- *To study* two ways of determining how many possible outcomes an experiment may have; one is with a formula and the other is with tree diagrams.
- *To distinguish* between two different ways of arranging objects, one in which order counts (permutations) and the other in which order does not count (combinations).
- *To indicate* how Pascal's triangle can be used in calculating the number of combinations possible.
- *To present* rules for calculating probabilities involving mutually exclusive events and events that are not mutually exclusive.
- *To learn* how to calculate probabilities when one event depends on whether or not another happens (conditional probability). Events also may be independent. Each of these cases gives rise to different multiplication rules.
- *To understand* how to determine the probability that a particular event was caused by some other event (Bayes' rule).
- *To discuss* and apply an important probability distribution (the binomial distribution), which is used when an experiment has only two possible outcomes.
- *To apply* the ideas of probability to many different situations including mathematical expectation, odds, genetics, law, and the effectiveness of new educational techniques, to mention but a few.

Statistical Probability of Divorce for Persons Born 1945–1949 Correlated with Years of Education

Years of school completed	Percent whose first marriage had ended in divorce by 1975		Percent whose first marriage may eventually end in divorce	
	Men	Women	Men	Women
0 to 11	15	24	34	44
12	15	17	36	37
13 to 15	15	19	42	49
16	8	8	29	29
17 or more	8	9	30	33
All ever-married persons born 1945–1949	13	17	34	38

Source: U.S. Bureau of the Census, 1975.

Estimated 1972 Employment, Projected 1985 Requirements, and Average Annual Openings by Occupation, 1972–1985

Occupation	Estimated employment (1972)	Projected requirements (1985)	Percentage of change (1972–1985)	Annual average openings (1972–1985)		
				Total	Employment change	Replacement needs
Computer and related occupations:						
Operating personnel	480,000	531,000	10.6	27,000	4,000	23,000
Programmers	186,000	290,000	55.9	13,000	8,000	5,000
System analysts	103,000	185,000	79.6	8,300	6,300	2,000
Insurance occupations:						
Actuaries	5,500	9,000	62.0	500	300	200
Claim adjusters	128,000	152,000	19.0	5,800	1,900	3,900
Claim examiners	31,000	28,500	−7.8	600	−200	800
Underwriters	61,000	71,000	16.6	2,500	800	1,700
Administrative and related occupations:						
Accountants	714,000	935,000	31.0	41,900	17,000	24,900
City managers	2,500	3,700	54.0	150	100	50
Credit officers	114,000	160,000	40.9	7,500	3,600	3,900
Lawyers	303,000	380,000	25.8	16,500	6,000	10,500
Personnel workers	240,000	400,000	69.2	20,800	12,200	8,600
Public relations workers	87,000	110,000	28.6	5,000	1,900	3,100
Hotel manager and assistants	110,000	160,000	41.2	7,500	3,600	3,900

Source: U.S. Department of Labor, Bureau of Labor Statistics, *Occupational Manpower and Training Needs*, Bulletin 1824, rev. ed., 1974.

If you were born between the years of 1945 and 1949 and are married, what is the statistical probability that your marriage will end in divorce? As incredible as it may sound, the first table indicates that the probability of divorce for both men and women is related to the number of years of education attained by the people involved. Furthermore, the table indicates the percentage of the marriages that *have already ended* in divorce as well as the percentage of first marriages that *will eventually end* in divorce. How are such probabilities calculated? Are they accurate predictions of future divorce rates?

Now consider the second table, which indicates the projected employment requirements in several occupations for the year 1985. How are such projections made? Are they accurate?

In this chapter we first define what probability is and how we calculate it. After we study statistics in Chapter 9, we will see how probability and statistics can be used together to analyze such problems. Moreover, we will indicate how to determine whether our probability predictions are reasonable, that is, how they compare with actual observed events.

6.1 Introduction

Although the word *probability* may sound strange to you, it is not so unfamiliar as you may think. As a matter of fact, it is used frequently in everyday situations. We often make decisions and take actions as a result of the probability of certain events. For example, consider the following situation: You are getting ready to go to school and hear the weather forecast for the day: "Rain with a probability of 85%." Undoubtedly, you prepare yourself accordingly.

Let us, however, analyze this weather forecast. What is really meant is that based on past records, 85% of the time when the weather conditions have been as they are today, rain has followed. This forecast is based on the assumption that since in the *past* rain has occurred a certain percentage of the time, it will occur the same percentage of the time in the *future*. This is but one use of probability; it is based on the *relative frequency* interpretation of probability. Since weather forecasts are sometimes (or often?) inaccurate, we immediately realize that probability theory cannot be used to predict future events; it can only indicate the chances of the different possible outcomes.

Probability can also be considered in terms of expectation. For example, if a fair coin is tossed many times,[1] we expect that the coin will land heads up half the time. We would all agree that the probability is $\frac{1}{2}$ that heads comes up. This is because there are only two possible outcomes when we flip a coin, heads or tails.

Probability is also contained in many statements that express a personal opinion. This is illustrated in the following conversation

[1] A fair coin is one that has the same chance of landing on heads as on tails. Throughout this book we will assume that we have fair coins unless we are told otherwise.

between two motorists:

Mary Ann: I am buying gas today because the gas station will *probably* be closed on the weekend.

Heather: I will give you 2-to-1 odds that gas has gone up again another 2¢ a gallon.

Mary Ann: You're *probably* right, but what can I do.

Heather: You know, Mary Ann, I'll bet that the oil companies are purposely concealing some of their supplies until prices rise. After that, you'll *probably* get all the gas you want.

Mary Ann: *Chances* are that you're right.

In the preceding dialogue each motorist is making a statement that is actually a personal belief based on probability.

Basically, the theory of probability deals with the study of uncertainties. Thus, it has been found to have wide applications in the following situations:

1. It is used by insurance companies when they calculate insurance premiums and the *probable* life expectancies of their policyholders. See the discussion in Chapter 8. As a matter of fact, the entire Chapter 8 ("Decision Analysis") is based on a knowledge of the principles of probability.
2. It is used (formally or informally) by a gambler who decides to bet 10 to 1 on a particular horse.
3. It is used by industry officials in determining the reliability of certain equipment.
4. It is used by medical researchers who claim that smoking increases your *chance* of getting lung cancer.
5. It is used by biologists in their study of genetics.
6. It is used by pollsters in such polls as the Harris poll, the Gallup poll, and the Nielsen ratings to determine the reliability of their polls.
7. It is used by an investor who decides that a particular stock has a a greater chance for future growth than any other stock.
8. It is used by business managers in determining which products to manufacture, which products to advertise, and through which media: television, radio, magazines, newspapers, subway and bus advertisements, etc.
9. It is used by psychologists in predicting reactions or behavioral patterns under certain stimuli.
10. It is used by government economists in predicting that the inflation rate will increase or decrease in the future.

These are but a few of the applications of probability ideas. Some of these are discussed in this book. Since probability has so many possible meanings and uses, in this chapter we analyze the nature of probability and discuss various rules that allow us to calculate probabilities for many different situations. We explain what is meant by the words *probability*, *odds*, and *chance*. In the process we indicate how these ideas have found wide application in many different areas other than in the gambling rooms.

Historical Note

It is interesting to note that although probability and statistics are two important branches of mathematics that are usually studied together, this was not always the case, as evidenced by the fact that statistics was often studied from a mathematical point of view independent of a mathematical study of probability.

Historically, probability has its origin in the gambling room. The Chevalier de Méré, a professional French gambler, had asked his friend Blaise Pascal (1623–1662) to solve the following problem: In what proportion should two players of equal skill divide the stakes remaining on the gambling table if they are forced to stop before finishing the game? Pascal wrote to and began an active correspondence with Pierre Fermat (1601—1665) concerning the problem. Although Pascal and Fermat agreed on the answer, they gave different proofs. It was in this series of correspondences during the year 1652 that they developed the modern theory of probability.

A century earlier the Italian mathematician and gambler Girolomo Cardan (1501–1576) wrote *The Book on Games of Chance*. This is really a complete textbook for gamblers, since it contains many tips on how to cheat successfully. The origins of the study of probability are found in this book. Cardan was also an astrologer. According to legend, he predicted his own death astrologically and to protect his reputation as the country's best astrologer, he committed suicide on that day. All of Cardan's sons were gamblers and in constant trouble with the law. When one of his sons murdered his wife, Cardan cut off the son's ears as a punishment for the crime.

Another famous mathematician who contributed to the theory of probability was Abraham De Moivre (1667–1754). Like de Méré, De Moivre spent many hours with London gamblers and wrote a manual for gamblers entitled *Doctrine of Chances*. Like Cardan, De Moivre predicted the day of his death. A rather interesting story is told of De Moivre's death. De Moivre was ill and each day he noticed that he was sleeping 15 minutes longer than he did on the preceding day. Using progressions, he computed that he would die in his sleep on the day after he slept 23 hours and 45 minutes. On the day following a sleep of 24 hours, De Moivre died.

The French mathematician Pierre Simon de Laplace (1749–1827) also contributed much to the historical development of probability. In 1812 he published *Théorie Analytique des Probabilités*, in which he referred to probability as a science that began with games but which had wide-ranging applications. In particular, he applied probability theory not to gambling situations but as an aid in astronomy.

Over the course of many years probability theory has left the gambling rooms and has grown to be an important and ever-expanding branch of mathematics.

6.2 What Is Probability?

Since many ideas of probability are derived from gambling situations, let us consider the following experiment: An honest die (see Figure 6.1) is rolled once and the number of dots (or number) that appears face up is recorded. When such an experiment was performed 6000 times by a group of students, the following results were recorded:

Outcome	Number of rolls
a 1 comes up	998
a 2 comes up	1001
a 3 comes up	999
a 4 comes up	1002
a 5 comes up	1003
a 6 comes up	997
Total	6000

Notice that each of the possible outcomes occurred approximately $\frac{1}{6}$ of the time. It would then seem reasonable to conclude that the probability of a 1 coming up is $\frac{1}{6}$, the probability of a 2 coming up is $\frac{1}{6}$, and so on.

Figure 6.1 Possible outcomes when a die is rolled once.

Let us analyze this more carefully. When a die is rolled, there are six possible outcomes as shown in Figure 6.1. If we are interested in the number of threes that appear, then each time that a three appears, we call it a *favorable outcome*. Since there are six possible outcomes, the probability of the die showing three dots would then be the number of favorable outcomes divided by the number of possible outcomes, which is 1 divided by 6, or $\frac{1}{6}$. By referring to the previous table, we notice that the number of times three came up was $\frac{999}{6000}$, or approximately $\frac{1}{6}$. We are saying that if we have an experiment that is repeated n times and if during these repetitions some event occurs f times, then the probability of that event happening again is f/n.

Now suppose we flip a coin. We know that there are two possible outcomes, heads or tails. It would seem reasonable to say that the probability of the coin coming up heads is $\frac{1}{2}$, since there are 2 possible outcomes, only 1 of which is favorable.

More generally, when an experiment is performed, all the possible outcomes that occur are referred to as the *sample space* of the experiment. The outcome that is of interest to us is referred to as an *event*. Thus, if a die is thrown once, then the sample space consists of the six possible outcomes shown in Figure 6.1. When a coin is tossed once, the sample space consists of the two possible outcomes, heads and tails.

Once we determine the sample space, we can define what we mean by probability.

Definition 6.1 If an event can occur in n equally likely ways and if f of these ways are considered favorable, then the *probability* of getting a favorable outcome is

$$\frac{\text{Number of favorable outcomes}}{\text{Total number of outcomes}} = \frac{f}{n}$$

Thus, the probability of an event equals the number of favorable outcomes divided by the total number of possible outcomes. We use the symbol Prob(A) to stand for "probability of event A."

Comment Definition 6.1 is only one of several definitions of probability that exist. It is based on the assumption that if an event has probability f/n, then in the long run there will be f favorable outcomes of the event out of n trials. For this reason probability is referred to as the *theoretical frequency* of the event, since it represents the percentage of times that the event will happen in repeated experiments.

Now suppose we were working for an insurance company and were interested in determining the probability that a 30-year-old person will live to be age 75. (This will be done in Chapter 8.) By researching birth and death records of individuals and determining the number of people who actually lived to age 75, we can estimate this probability. Such information can be obtained from mortality tables. In this case the probability of event A occurring can be approximated by

$$\text{Prob}(A) = \frac{\text{number of times event } A \text{ actually happened}}{\text{total number of observations}}$$

Obviously, as the number of observations increases, so does the accuracy of this estimate of the probability.

Let us illustrate the concept of probability with several examples.

Example 1 A die is rolled once. Find the probability that the number shown is an even number.

Solution When a die is rolled, there are 6 possible outcomes as shown in Figure 6.1. Of these 6 outcomes, 3 are favorable. These occur when the die shows a two, a four, or a six. Thus,

$$\text{Prob(die shows an even number)} = \frac{3}{6} = \frac{1}{2}$$

Example 2 *Altered die.* A die is altered by painting an additional dot on the side that originally had only one dot. (See Figure 6.2.) Find the probability that the number that shows is an even number.

Figure 6.2 An altered die.

Solution There are 6 possible outcomes, although only 5 different ones are noticeable. Now there are 4 favorable outcomes, since a two can occur twice, a four can occur once, and a six can occur once. Thus,

$$\text{Prob(die shows an even number)} = \frac{4}{6} = \frac{2}{3}$$

Example 3 Mary and her friend Gwendolyn are visiting the Sear's Tower in Chicago. They each enter a different elevator that can let them off on any floor from 1 to 6. What is the probability that they both get off on the same floor?

Solution Since both Mary and Gwendolyn can get off at any floor between 1 and 6, there are 36 possible outcomes as in the table on page 234. A favorable outcome occurs if both get off on floor 1, floor 2, and so on as shown. Thus, there are 6 favorable outcomes out of 36 possible outcomes, so that

$$\text{Prob(both get off on same floor)} = \frac{6}{36} = \frac{1}{6}$$

Mary gets off on floor	Gwendolyn gets off on floor	Mary gets off on floor	Gwendolyn gets off on floor
1	1 ← Favorable outcome	4	1
1	2	4	2
1	3	4	3
1	4	4	4 ← Favorable outcome
1	5	4	5
1	6	4	6
2	1	5	1
2	2 ← Favorable outcome	5	2
2	3	5	3
2	4	5	4
2	5	5	5 ← Favorable outcome
2	6	5	6
3	1	6	1
3	2	6	2
3	3 ← Favorable outcome	6	3
3	4	6	4
3	5	6	5
3	6	6	6 ← Favorable outcome

Example 4 *Escaped animal.* A traveling circus consists of 52 performing animals as follows:

8 elephants	11 monkeys
4 lions	15 horses
2 tigers	12 dogs

A report is received on the police teletype stating that an animal has escaped. Find the following probabilities.

a) It is an elephant.
b) It is not a monkey.

Solution a) There are 52 animals in the circus, 8 of which are elephants. Thus, Prob (escaped animal is an elephant) $= \frac{8}{52} = \frac{2}{13}$

b) There are 52 animals in the circus, 41 of which are not monkeys. Thus, Prob (escaped animal is not a monkey) $= \frac{41}{52}$

Example 5 Mr. and Mrs. Avery have four children. Find the probability that they have two boys and two girls.

Solution We first find the sample space; that is, we list the total number of ways in which a family can consist of four children, as on page 235. Of the 16 possible ways, 6 represent outcomes that have two boys and two girls as indicated. Thus,

$$\text{Prob(two boys and two girls)} = \frac{6}{16} = \frac{3}{8}$$

	Child 1	Child 2	Child 3	Child 4
	Boy	Boy	Boy	Boy
	Boy	Boy	Boy	Girl
	Boy	Boy	Girl	Boy
	Boy	Girl	Boy	Boy
Favorable outcome →	Boy	Boy	Girl	Girl
Favorable outcome →	Boy	Girl	Girl	Boy
Favorable outcome →	Boy	Girl	Boy	Girl
	Boy	Girl	Girl	Girl
	Girl	Boy	Boy	Boy
Favorable outcome →	Girl	Boy	Boy	Girl
Favorable outcome →	Girl	Boy	Girl	Boy
Favorable outcome →	Girl	Girl	Boy	Boy
	Girl	Boy	Girl	Girl
	Girl	Girl	Boy	Girl
	Girl	Girl	Girl	Boy
	Girl	Girl	Girl	Girl

Example 6 A card is selected from an ordinary 52-card deck. Find the probability that the card is each of the following.

a) A deuce (a card with a 2 on it)
b) A club
c) A picture card
d) A red card
e) An ace of diamonds

Solution a) There are 4 deuces out of a possible 52 cards. Thus,

$$\text{Prob}(\text{deuce}) = \frac{4}{52} = \frac{1}{13}$$

b) There are 13 clubs out of a possible 52 cards. Thus,

$$\text{Prob}(\text{club}) = \frac{13}{52} = \frac{1}{4}$$

c) There are 12 picture cards out of a possible 52 cards. Thus,

$$\text{Prob}(\text{picture card}) = \frac{12}{52} = \frac{3}{13}$$

d) There are 26 red cards out of a possible 52 cards. Thus,

$$\text{Prob}(\text{red card}) = \frac{26}{52} = \frac{1}{2}$$

e) There is only 1 ace of diamonds out of a possible 52 cards. Thus,

$$\text{Prob}(\text{ace of diamonds}) = \frac{1}{52}$$

Example 7 A pair of dice is rolled and the number of dots appearing on both dice is counted. Find the probability that the sum of the numbers on both dice is each of the following:

a) 7 b) 12 c) More than 12

d) Between 2 and 12, including these two numbers

Solution When a pair of dice is rolled, there are 36 possible outcomes as pictured in Figure 6.3.

a) The events *sum of 7 on both dice together* can happen in 6 ways. Thus,

$$\text{Prob (sum is 7)} = \frac{6}{36} = \frac{1}{6}$$

b) The event *sum of 12 on both dice together* can occur only when both dice show six dots. This can happen in only 1 way. Thus,

$$\text{Prob (sum is 12)} = \frac{1}{36}$$

c) When two dice are rolled, the maximum sum is 12. We cannot obtain a sum that is more than 12. Thus,

$$\text{Prob (sum is more than 12)} = 0$$

Generally, an event that can *never* happen is called a *null event*, and its probability is zero.

d) When two dice are rolled, we *must* obtain a sum that is between 2 and 12, including the numbers 2 and 12. There are 36 possible

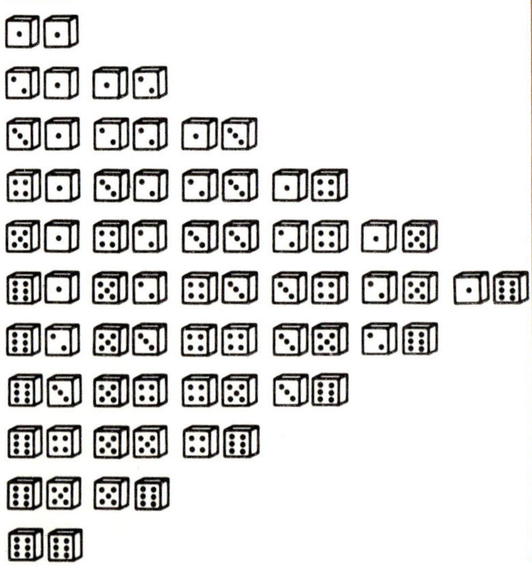

Outcome	Sum	Number of ways in which this can happen
	2	1
	3	2
	4	3
	5	4
	6	5
	7	6
	8	5
	9	4
	10	3
	11	2
	12	1

Figure 6.3

outcomes, all of which are favorable. Thus,

$$\text{Prob(sum is between 2 and 12, including 2 and 12)} = \frac{36}{36} = 1$$

Therefore, in this case, a favorable outcome must occur. Any event that is certain to occur is called a *certain event* or *definite event*, and its probability is one.

Exercises

1. A little boy has ten pennies in his pocket, two of which are shiny. He reaches into his pocket and removes one of the pennies. Find the probability that he removes a shiny penny.

2. A clever counterfeiter has randomly inserted 5 bogus $20 bills among a pile of 95 other genuine $20 bills. The counterfeiter is attempting to deposit the 100 bills in a bank. The teller is very lax in her responsibilities and decides to select only one bill and thoroughly check it for counterfeiting. Find the probability that the counterfeiter gets away with his scheme and is not caught.

3. In Marcia's aquarium there are 48 fish as follows:

6 guppies	7 algae eaters
9 anglefish	5 catfish
8 kissers	13 neons

   One morning Marcia's little sister takes a net and randomly removes one fish. Find the probability that the fish removed is a guppy.

4. *Affirmative action.* The following is an ethnic and minority group breakdown of the 300 new employees hired by the Edison Corporation to reverse its past discriminatory hiring practices. Find the probability that a recently hired employee is:
   a) Black
   b) Oriental or Mexican
   c) Black, Oriental, Puerto Rican, American Indian, Mexican, or white woman

Classification	Number
Black	70
Oriental	37
Puerto Rican	61
American Indian	47
Mexican	22
White women	63
Total	300

5. *Affirmative action* (*continued*). The following is a further breakdown of the 300 new employees hired by the Edison Corporation discussed in Exercise 4. Find the probability that an employee selected at random from this group will be:

a) A male American Indian
b) A man
c) A woman
d) Puerto Rican

Classification	Number of	
	Men	Women
Black	41	29
Oriental	20	17
Puerto Rican	37	24
American Indian	29	18
Mexican	21	1
White women	0	63

6. At the corner of Broadway and Main Street there is a traffic light that remains green for 40 seconds, yellow for 8 seconds, and red for 19 seconds, after which time the same cycle is repeated. John drives up to the corner with his girlfriend. Find the probability that the light is:

a) Red
b) Green

7. A recent survey by Alman and Kobb indicates that many married men do not remember such things as the date of their marriage or their wife's birthday. A case in point, as described in this survey, is Joe who knows that his wife was born in February but does not remember the exact date. If he guesses Valentine's Day (February 14), find the probability that it is indeed her birthday.

8. The Math Club at Staten College consists of 18 men and 4 women. If a member is selected at random, find the probability that it is a woman.

9. Helen Hoffman is completing her 1040 U.S. federal income tax return and remembers the first eight digits of her social security number but has completely forgotten the last digit, so she decides to guess. Find the probability that she guesses the last digit correctly.

10. The table on page 239 indicates how the number of people on government payrolls has increased since 1929.

a) If an employee is selected at random during 1960, what is the probability that the individual is on a government payroll?
b) If an employee is selected at random during 1974, what is the probability that the individual is on a government payroll?
c) Compare the answer obtained in part a with the answer obtained in part b. Comment.

Year	Number of people on government payrolls	Number of people on all payrolls
1929	3,065,000	31,339,000
1940	4,202,000	32,376,000
1950	6,206,000	45,222,000
1960	8,353,000	54,232,000
1970	12,561,000	70,920,000
1974	14,284,000	78,334,000

Source: U.S. Department of Labor, 1977.

11. Blackjack Phil has obtained an altered pair of dice. The dice have an extra dot on each face that originally had only one dot as shown in Figure 6.2. These altered dice are now rolled once.

a) List all the possible sums that can be obtained with these dice.
b) Find the probability of rolling a sum of 2.
c) Find the probability of rolling a sum between 4 and 12, including 4 and 12.
d) Find the probability that the sum is an even number.

12. Which of the following numbers cannot be the probability of some event?

a) $\frac{3}{2}$ b) $\frac{1}{7}$ c) $-\frac{4}{5}$ d) 0 e) $\frac{7}{12}$

13. At a recent New Year's party, four of the guests became drunk and were asked to leave. As each of the drunk guests left the party, he or she picked up a coat that definitely belonged to one of these four people. Since these guests were drunk, however, they did not know whose coat they were taking. Find the probability that each person gets his or her own coat.

14. A slot machine has three wheels and each wheel has a 0, 1, and 2 painted on it. When a button is pushed, each wheel displays a number. Each wheel operates independently. The complete number shown is the amount of money won. Thus, if the wheels together show 120, \$120 is won.

a) List all the possible outcomes for this machine.
b) Find the probability of winning \$222.
c) Find the probability of winning nothing.
d) Find the probability of winning at least \$1.

15. To make sure that his students do their homework assignments on time, Professor Chamides gives weekly quizzes consisting of three true-false questions. Manya is a student in Professor Chamides's class who has not done this week's assignment. Since she is not prepared for the quiz, she decides to guess the answers. Find the probability that at least one of her answers is correct.

16. There were 864 deaths reported by Metropolitan Hospital last year. The cause of death and the age of the person at the time of death are shown in the accompanying table. If one patient is selected at random, find the following probabilities.
  a) The person died as a result of a heart attack.
  b) The person was over 65 years old at the time of death.
  c) The person was between 51 and 65 years old and died of cancer.

	Cause of death				
Age	Heart attack	Cancer	Respiratory and lung infections	Kidney troubles	Other
30–50	112	31	6	5	81
51–65	87	79	53	18	119
Over 65	24	56	88	29	76

17. A wheel has the numbers 1 through 50 painted on it as shown in Figure 6.4. Tickets numbered 1 through 50 have been sold. The wheel will be rotated and the number on which the pointer lands will be the winning number. The pointer cannot stop on the line between the numbers. A prize of $500 will be awarded to the person whose ticket number matches the winning number. Find the following probabilities.
  a) Ticket number 13 wins.
  b) Ticket number 51 wins.
  c) The winning ticket is between the numbers 11 and 31, inclusive.

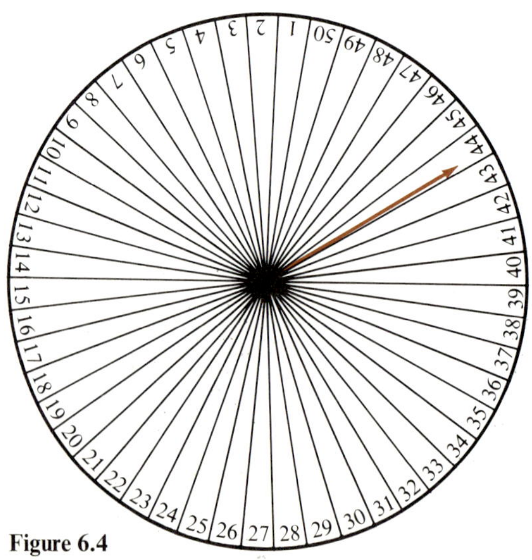

Figure 6.4

18. In a survey of the attitudes of college students toward the use of drugs on campus, the data indicated in the table were obtained. If a student is randomly selected, find the following probabilities.
   a) The student is a man.
   b) The student is a woman who approves of drug use on campus.
   c) The student does not care.

	Approve	Don't approve	Don't care
Man	91	38	56
Woman	82	47	64

19. Professor Morales has told her class that she intends to grade on a curve according to the following scheme:

   6 A's

   8 B's

   14 C's

   8 D's

   6 F's

   a) Sophia is a student in Professor Morales' class. Find the probability that she passes the course.
   b) Alfred is also a student in the class and needs a grade of at least C to avoid being placed on probation. Find the probability that he gets a grade of C or better.

20. *Botany.* A biologist is experimenting with certain plants that produce flowers that are colored dark red, light red, pink, and white in the ratio 9:4:2:1. If a plant is randomly selected, find the probability that it will produce each of the following.
   a) Dark red flowers
   b) White flowers
   c) Yellow flowers

6.3 Counting Techniques

It should be apparent from our discussion and the exercises in the previous section that the main difficulty in assigning probabilities in a problem is to determine the total number of possible outcomes and the number of those outcomes that are considered favorable. In some cases it may not be that difficult to actually list all the possible outcomes, whereas in others that may be too time-consuming. Fortunately, mathematicians have devised a formula that often enables us to determine the number of possible outcomes without actually listing them. We start our discussion with an example.

Example 8 A veterinarian, who limits her practice to only cats and dogs, has four appointments scheduled on a particular morning. In how many different ways can these appointments be arranged?

Solution In order to determine the total number of possible appointments, we can use a convenient counting device known as a *tree diagram*. This is shown in Figure 6.5, where C stands for cat and D for dog. You will notice that there are a total of 16 possible appointment schedules. Let us make the assumption that each of these appointment schedules is equally likely. Then we can determine the probability that the veterinarian will see, for example, three cats and one dog. There are 4 such favorable outcomes (CCCD, CDCC, CCDC, and DCCC) out of a possible 16, so the probability is $\frac{4}{16}$, or $\frac{1}{4}$.

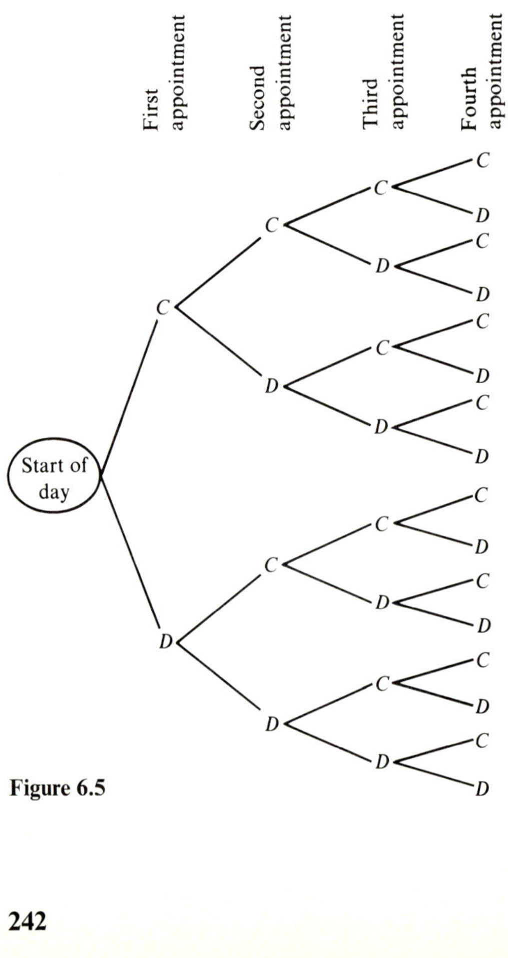

Figure 6.5

Comment Generally speaking, the use of tree diagrams to determine the number of possible outcomes is very impractical when the number of possibilities gets large.

In Example 8 we can also reason in the following manner. The number of different types of animals to be seen by a veterinarian is shown in the table. Multiplying these numbers gives $2 \cdot 2 \cdot 2 \cdot 2$ or 16,

Appointment 1	Appointment 2	Appointment 3	Appointment 4
2	2	2	2

which is the same answer we obtained by actually listing all the possible outcomes in the tree diagram.

Formula 6.1 *Counting rule.* If one thing can be done in m different ways, and if after this is done a second thing can be done in n different ways, then both things can be done in a total of $m \cdot n$ different ways in the specified order. (The same rule extends to more than two things done in sequence.)

Example 9 Heather has been advised by her counselor that in order to graduate in June she must complete one course from each of the three categories shown in the table. All these courses will be offered in the spring

A	B	C
Math 231 Bio 100 Chem 041	Educ 212 Eng 117	Soc 114 Eco 412

semester. Neglecting any other considerations, how many *different* programs can Heather select?

Solution Heather can select any one of the courses from the 3 in category A, any one of the 2 in category B, and any one of the 2 in category C. Thus, using Formula 6.1, the total number of different programs for which she can register is $3 \cdot 2 \cdot 2 = 12$.

We can also determine the 12 possible programs by using the tree diagram shown in Figure 6.6.

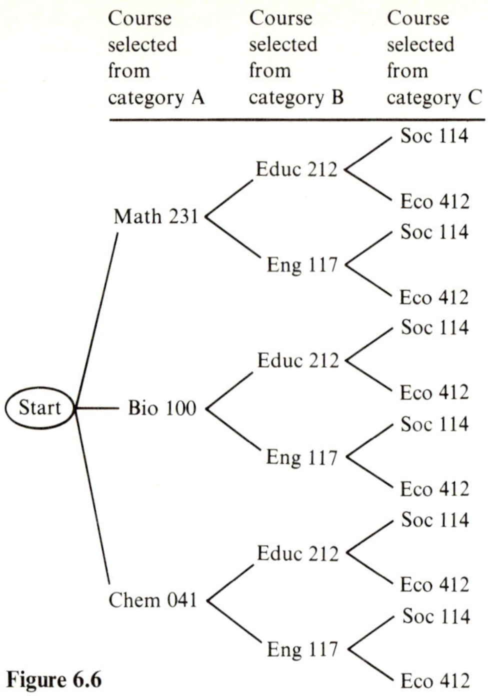

Figure 6.6

Example 10 *Jailbreak.* A prisoner is planning his jailbreak. He plans to go from town A to town B by either walking, taking a bus, hitchhiking, or stealing a bike and bicycling. From town B he plans to go on to town C by taking either a train, a ferryboat, or a bus. In how many different ways can the prisoner escape from town A to town C (by way of town B)?

Solution Since the prisoner can go from town A to town B by 4 different ways and from town B to town C by 3 different ways, Formula 6.1 tells us that he can escape from town A to town C (by way of town B) in $4 \cdot 3$ or 12 different ways.

Example 11 *Labeling experimental mice.* Scientists at the Wagner Laboratories are experimenting with thousands of mice that have been injected with infectious bacteria. Each has an identification tag consisting of two letters followed by three numbers. How many different identification tags are possible under the following conditions.

a) Repetition of numbers or letters is allowed.
b) Repetition of numbers or letters is not allowed.

Solution a) If repetition is allowed, then the same letter or number can be used again, so that the total number of different identification tags is

$$26 \cdot 26 \cdot 10 \cdot 10 \cdot 10 = 676{,}000$$

b) If repetition of letters is not allowed, then there are 26 possibilities for the first letter, but only 25 possibilities for the second letter since once a letter is used it may not be used again. The same is true for the numbers. Thus, the total number of different identification tags is

$$26 \cdot 25 \cdot 10 \cdot 9 \cdot 8 = 468{,}000$$

Exercises

21. There are seven pitchers and four catchers on a baseball team. In how many different ways can the manager select a starting battery (pitcher and catcher) for a game?

22. There are five different roads that lead to a sports stadium. In how many different ways can a person drive to the stadium by one road and leave by a different road?

■23. How many different license plates can be formed if each is to consist of three letters followed by two numbers and if:
   a) Repetition of letters or numbers is allowed.
   b) Repetition of letters or numbers is *not* allowed.

■24. How many different seven-digit telephone numbers are possible if the first digit cannot be a zero? (Assume no other restrictions and that repetition is allowed.)

■25. Almost every individual in the United States has a nine-digit social security number.
   a) How many such numbers are possible? (*Note:* Repetitions are allowed.)
   b) If the government decides to add a letter to the end of each social security number, how many possible social security numbers will there be?

26. *Smoke signals.* The Felquois Indians sent messages from one tribe to another by using the following different types of smoke signals:

   Long, black smoke cloud

   Short, black smoke cloud

   Long, white smoke cloud

   Short, white smoke cloud

   Each of these clouds was followed by a two-minute pause, after which a second smoke cloud was created. Again a two-minute pause followed, after which a third smoke cloud was created. Each message consisted of three smoke clouds.
   a) Using this scheme, how many different messages could the Felquois Indians transmit from one tribe to another?
   b) What is the probability that a transmitted message will consist of three long, black smoke clouds one after another?

27. A computer dating service has 481 men and 523 women on file. How many possible man-woman matches can be arranged?

28. At a recent teacher's convention six teachers, as shown in the table, were honored for their distinguished service to Island College. The

Teacher	Department
Prof. Chi	Mathematics
Prof. Dreyfus	Psychology
Prof. Emeritus	Library
Prof. Hausley	Mathematics
Prof. Nelson	Biology
Prof. Pietrachke	Foreign languages

teachers were seated at a (straight) head table.

a) In how many different ways can the teachers be seated at the table?

★b) If Professors Emeritus and Pietrachke are bitter enemies and cannot be seated together, how many different seating arrangements are possible?

★c) Find the probability that the two math teachers are seated together.

29. During a recent ice storm, a falling tree severed three electric power lines as shown in the accompanying diagram. A repair crew is sent to the scene and must reconnect the wires exactly as they were originally. In how many different ways can the three wires be connected?

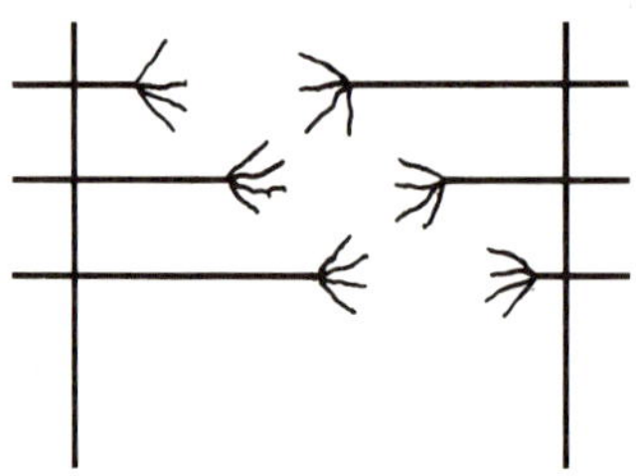

30. In how many different ways can nine men in the army line up for an inspection call?

31. One part of a certain civil service exam consists of ten true-false questions. How many different answer sheets can be turned in?

32. Roland Smith has an important business meeting this morning. He can select any of seven suits, four shirts, five ties, and six pairs of shoes to wear for the meeting. How many different outfits are possible?

★33. Three boys and three girls enter a subway car and sit down on a bench. Find the probability that the children are seated in a row alternately, in the order boy, girl, boy, girl, boy, girl?

34. A man encounters a beggar who is soliciting a handout. The man reaches into his pocket, randomly selects two coins, and gives them to the beggar. How many possible sums of money are possible if the man has a penny, a nickel, a dime, a quarter, a half-dollar, and a dollar piece in his pocket?

★35. Construct a tree diagram showing all the possibilities, male and female, for a family that has five children.

36. *Consumer protection.* In an effort to protect the consumer, the agriculture department of a certain state requires that all canned foods manufactured within the state be stamped with numbers or letters as follows:

a) The company number as assigned by the department; there are nine companies within the state.
b) The plant location within the state; the state is divided into eight geographic areas labeled A, B, C, D, E, F, G, and H.
c) The data of manufacture: month and last two digits of year.

For example, a can with the code 2 B 0379 means that it was manufactured by company 2 located in area B during March 1979. Under this scheme, how many different codes are possible for cans manufactured during the years 1977–1981?

37. The *New York Daily News* has, as a daily feature, a five- or six-letter word that the reader must unscramble to make a meaningful word. In how many different ways can the letters of the word SOCIETY be arranged? (*Note:* Each arrangement does not necessarily have to form a word.)

6.4 Permutations and Combinations

Consider the following situation: Three vacationing students, Mel, Carl, and Rhoda, have purchased standby tickets on a skytrain transatlantic flight that will take them from London back to New York. There are three aisle seats remaining vacant in the plane; one in the forward section, one in the mid section, and one in the tail section of the airplane. In how many different ways can the three travelers line up to board the plane and pick the best of the remaining seats?

There are six ways as shown in the accompanying table. In this case you will notice that in each arrangement, order is important.

First person to board plane	Second person to board plane	Last person to board plane
Mel	Carl	Rhoda
Mel	Rhoda	Carl
Carl	Mel	Rhoda
Carl	Rhoda	Mel
Rhoda	Carl	Mel
Rhoda	Mel	Carl

Thus, the arrangement Mel, Carl, Rhoda means that Mel boards first, then Carl, and finally Rhoda. If the number of seats is limited, then the person who boards first has first choice of selecting the best of the remaining seats as compared with the person who boards last.

This important idea in mathematics, in which a number of objects can be arranged in a particular order, is known as a permutation.

Definition 6.2 A *permutation* is any arrangement of distinct objects in a particular order.

Let us now examine another such problem:

Example 12 Four young ladies, Stephanie Marie Gallagher, Liz Armstrong, Ann Sullivan, and Patricia Beth O'Connell, are finalists in a state beauty contest. The judges must select a winner and an alternate from among these four contestants. In how many different ways can this be done?

Solution The winner and the alternate can be selected in 12 different ways. We list these possibilities in a table.

Winner	Alternate	Winner	Alternate
Stephanie	Liz	Ann	Stephanie
Stephanie	Ann	Ann	Liz
Stephanie	Patricia	Ann	Patricia
Liz	Stephanie	Patricia	Stephanie
Liz	Ann	Patricia	Liz
Liz	Patricia	Patricia	Ann

Comment Perhaps you are wondering whether there are any formulas that can be used to determine the number of possible permutations. The answer is yes and this will be done shortly.

Example 13 A doctor has 5 examination rooms. There are 5 patients in the waiting room. In how many different ways can the patients be assigned to the examination rooms?

Solution We can solve this problem by numbering the waiting rooms 1 through 5 as shown in Figure 6.7, and then considering them in sequence.

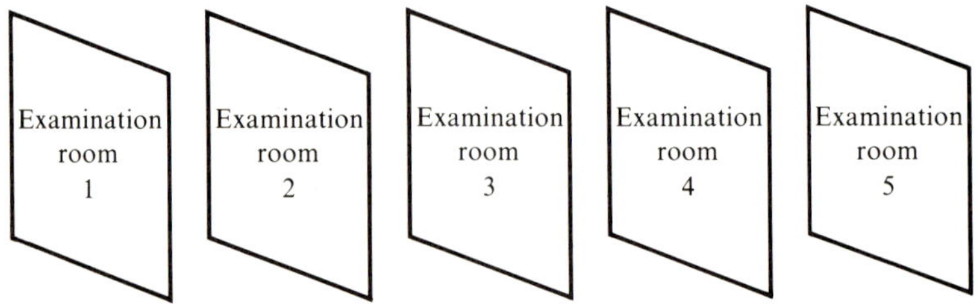

Figure 6.7

Examination room 1 can be used by any one of the 5 patients. Once a patient has been assigned to examination room 1, there are 4 patients who can use examination room 2, so there are (using Formula 6.1)

$$5 \cdot 4 = 20 \text{ ways}$$

of using examination room 1 and examination room 2. Similarly, once examination rooms 1 and 2 are occupied, there are 3 patients remaining who can use examination room 3, so there are

$$5 \cdot 4 \cdot 3 = 60 \text{ ways}$$

of using examination rooms 1, 2, and 3. Continuing in this manner we find that there are

$$5 \cdot 4 \cdot 3 \cdot 2 \cdot 1 = 120 \text{ ways}$$

of using examination rooms 1, 2, 3, 4, and 5. Thus, there are 120 different ways in which the 5 patients can be assigned to the 5 examination rooms.

Notice in Example 13 that we had to multiply $5 \cdot 4 \cdot 3 \cdot 2 \cdot 1$. Often in mathematics we have to multiply a series of numbers together starting from a given number and multiplying by the number that is 1 less, and so on until we get to the number 1, which is where we stop. For this type of multiplication, we introduce *factorial notation* and use the symbol $n!$ read as "n factorial." In this case we would write the product as $5!$. Thus,

$$5! = 5 \cdot 4 \cdot 3 \cdot 2 \cdot 1 = 120$$

Accordingly,

$$1! = 1$$
$$2! = 2 \cdot 1 = 2$$
$$3! = 3 \cdot 2 \cdot 1 = 6$$
$$4! = 4 \cdot 3 \cdot 2 \cdot 1 = 24$$
$$5! = 5 \cdot 4 \cdot 3 \cdot 2 \cdot 1 = 120$$
$$6! = 6 \cdot 5 \cdot 4 \cdot 3 \cdot 2 \cdot 1 = 720$$
$$7! = 7 \cdot 6 \cdot 5 \cdot 4 \cdot 3 \cdot 2 \cdot 1 = 5{,}040$$
$$8! = 8 \cdot 7 \cdot 6 \cdot 5 \cdot 4 \cdot 3 \cdot 2 \cdot 1 = 40{,}320$$
$$9! = 9 \cdot 8 \cdot 7 \cdot 6 \cdot 5 \cdot 4 \cdot 3 \cdot 2 \cdot 1 = 362{,}880$$
$$10! = 10 \cdot 9 \cdot 8 \cdot 7 \cdot 6 \cdot 5 \cdot 4 \cdot 3 \cdot 2 \cdot 1 = 3{,}628{,}800$$

Also, we define $0!$ to be 1. This makes formulas involving $0!$ meaningful.

Let us now return to the beauty contest example (Example 12) discussed earlier. In that example the judges were interested in selecting 2 winners from among the 4 finalists. This actually represents the number of permutations of 4 things taken 2 at a time, or the number of possible permutations of 2 things that can be formed out of a possible 4 things. We abbreviate this by writing ${}_4P_2$. This is read as the number of permutations of 4 things taken 2 at a time.

Notation $_nP_r$ represents the number of permutations of n things taken r at a time, and $_nP_n$ represents the number of permutations of n things taken n at a time.

To determine the number of possible permutations, we use the following formulas.

Formula 6.2 The number of possible permutations of n things taken r at a time is given by

$$_nP_r = \frac{n!}{(n-r)!}$$

Formula 6.3 The number of possible permutations of n things taken n at a time is given by

$$_nP_n = n!$$

Let us illustrate the use of these formulas with several examples.

Example 14 Evaluate.

a) $_5P_3$ b) $_8P_6$ c) $_7P_4$
d) $_6P_1$ e) $_6P_0$ f) $_7P_7$

Solution Using Formulas 6.2 and 6.3, we have:

a) $$_5P_3 = \frac{5!}{(5-3)!} = \frac{5!}{2!} = \frac{5 \cdot 4 \cdot 3 \cdot 2 \cdot 1}{2 \cdot 1} = 60$$

b) $$_8P_6 = \frac{8!}{(8-6)!} = \frac{8!}{2!} = \frac{8 \cdot 7 \cdot 6 \cdot 5 \cdot 4 \cdot 3 \cdot 2 \cdot 1}{2 \cdot 1} = 20{,}160$$

c) $$_7P_4 = \frac{7!}{(7-4)!} = \frac{7!}{3!} = \frac{7 \cdot 6 \cdot 5 \cdot 4 \cdot 3 \cdot 2 \cdot 1}{3 \cdot 2 \cdot 1} = 840$$

d) $$_6P_1 = \frac{6!}{(6-1)!} = \frac{6!}{5!} = \frac{6 \cdot 5 \cdot 4 \cdot 3 \cdot 2 \cdot 1}{5 \cdot 4 \cdot 3 \cdot 2 \cdot 1} = 6$$

e) $$_6P_0 = \frac{6!}{(6-0)!} = \frac{6!}{6!} = \frac{6 \cdot 5 \cdot 4 \cdot 3 \cdot 2 \cdot 1}{6 \cdot 5 \cdot 4 \cdot 3 \cdot 2 \cdot 1} = 1$$

f) $$_7P_7 = 7! = 7 \cdot 6 \cdot 5 \cdot 4 \cdot 3 \cdot 2 \cdot 1 = 5040$$

Example 15 Medical researchers are experimenting with a new drug (in pill form) to be given to patients to cure the common cold. Ten people have volunteered to take the pill. However, the researchers have decided that they will use only 3 of these volunteers. The first will be given the new pill, the second will be given a placebo, and the third will be given an aspirin. In how many different ways can these 3 subjects be selected for the experiment?

Solution In this case, there are 10 people, so $n = 10$. Since 3 volunteers are to be selected, $r = 3$. Also $n - r = 10 - 3 = 7$. Then, using Formula 6.2, we have

$$
\begin{aligned}
{}_nP_r = \frac{n!}{(n-r)!} &= \frac{10!}{(10-3)!} \\
&= \frac{10!}{7!} = \frac{10 \cdot 9 \cdot 8 \cdot 7 \cdot 6 \cdot 5 \cdot 4 \cdot 3 \cdot 2 \cdot 1}{7 \cdot 6 \cdot 5 \cdot 4 \cdot 3 \cdot 2 \cdot 1} \\
&= 720
\end{aligned}
$$

Thus, the 3 subjects can be selected in 720 different ways.

Example 16 Stephanie is taking her 6 children to school in the family station wagon. As the children enter the car, however, trouble develops as each child wants to sit as close as possible to the window on the right side of the car. In an effort to make peace, Stephanie numbers the seats 1 through 6 and tells the children that each day a different seating arrangement will be used. The children agree and everybody is happy. For how many days will this seating scheme work; that is, how many different seating arrangements are possible?

Solution Here, $n = 6$ as there are 6 children. Since there are also 6 seats, we use Formula 6.3. We have

$$
\begin{aligned}
{}_nP_n = {}_6P_6 &= 6! \\
&= 6 \cdot 5 \cdot 4 \cdot 3 \cdot 2 \cdot 1 = 720 \text{ ways}
\end{aligned}
$$

Thus, there are 720 different seating arrangements. (That should afford Stephanie several years of enjoyment with the seating problem!)

Example 17 A baseball scout has received a list of 15 promising prospects for the team's consideration. He is asked to list, in order of preference, the 5 most outstanding of these prospects. In how many different ways can he select the 5 best players?

Solution Since there are 15 promising prospects, $n = 15$. Also only the top 5 are to be considered, so that $r = 5$ and $n - r = 15 - 5 = 10$. Using Formula 6.2, we have

$$
\begin{aligned}
{}_nP_r &= \frac{n!}{(n-r)!} \\
&= \frac{15!}{10!} \\
&= \frac{15 \cdot 14 \cdot 13 \cdot 12 \cdot 11 \cdot \cancel{10} \cdot \cancel{9} \cdot \cancel{8} \cdot \cancel{7} \cdot \cancel{6} \cdot \cancel{5} \cdot \cancel{4} \cdot \cancel{3} \cdot \cancel{2} \cdot \cancel{1}}{\cancel{10} \cdot \cancel{9} \cdot \cancel{8} \cdot \cancel{7} \cdot \cancel{6} \cdot \cancel{5} \cdot \cancel{4} \cdot \cancel{3} \cdot \cancel{2} \cdot \cancel{1}} \\
&= 15 \cdot 14 \cdot 13 \cdot 12 \cdot 11 = 360{,}360
\end{aligned}
$$

Thus, the scout can select the top 5 players in 360,360 different ways.

Now consider the following situation: A team of 5 skin divers is on an underwater expedition when they discover a badly deteriorated shipwreck. It is agreed that only 3 of the divers will enter the wreck and the remaining 2 will remain outside but nearby the ship in case of emergency. How many different 3-person search groups can be formed to explore the shipwreck?

It should be apparent in this situation that we are interested in selecting 3 out of 5 people. However, this time we are not interested in the order in which the selection is made. Such a selection is called a *combination.*

Let us determine how many possible combinations there actually are. If the names of the divers are John, Jacques, Lloyd, Sam, and Kirk, then we are interested in selecting 3 out of a possible 5 divers. Using Formula 6.2, we have

$$
\begin{aligned}
{}_5P_3 &= \frac{5!}{(5-3)!} \\
&= \frac{5!}{2!} = 60
\end{aligned}
$$

Thus, there is a total of 60 possible ways of selecting 3 out of 5 possible people where *order counts.* Since we are *not* interested in order, there cannot be 60 possible combinations. Thus, if the 3 divers selected are John, Jacques, and Lloyd, there would be the following 6 ways of arranging these divers:

John, Jacques, Lloyd	Jacques, Lloyd, John
Jacques, John, Lloyd	Lloyd, John, Jacques
John, Lloyd, Jacques	Lloyd, Jacques, John

Since these permutations consist of the same 3 people, John, Jacques, and Lloyd, they are considered as only 1 combination. The same is true for any other combination of 3 divers. Thus, the 60 that we obtained must be divided by 6, giving $\frac{60}{6}$ or 10 possible different combinations. These are listed here.

1. John, Jacques, Lloyd
2. John, Jacques, Sam
3. John, Jacques, Kirk
4. Jacques, Lloyd, Sam
5. Jacques, Lloyd, Kirk
6. John, Lloyd, Sam
7. John, Jacques, Kirk
8. Jacques, Sam, Kirk
9. Lloyd, Sam, Kirk
10. John, Sam, Kirk

Notice that $6 = 3!$, so that the number of combinations of 5 things taken 3 at a time is

$$\frac{{}_5P_3}{3!} = \frac{5!}{3!2!} = 10$$

More generally, we have the following definition and formula.

Definition 6.3 A *combination* is any selection from a collection of distinct objects where order is not important.

Notation We use the notation ${}_nC_r$ to stand for the number of different ways of selecting r objects from a possible n objects where order is not important. Some books use the symbol

$$\binom{n}{r}$$

instead of ${}_nC_r$.

With this definition and notation in mind, we have the following formula.

Formula 6.4 The number of combinations of n things taken r at a time is

$${}_nC_r = \frac{n!}{r!(n-r)!}$$

Example 18 *Anymore room?* Five people are waiting for an elevator. Suddenly one arrives; however, there is room for only 2 more people to enter. In how many different ways can the 2 people who are to enter the elevator be selected?

Solution Since 5 people are waiting for the elevator, $n = 5$. Also, there is room for only 2 people, so $r = 2$. In this case order is not important, so we are interested in ${}_5C_2$. Using Formula 6.4, we have

$$\begin{aligned} {}_5C_2 &= \frac{5!}{2!(5-2)!} = \frac{5!}{2!3!} \\ &= \frac{5 \cdot 4 \cdot 3 \cdot 2 \cdot 1}{2 \cdot 1 \cdot 3 \cdot 2 \cdot 1} = 10 \end{aligned}$$

Thus, the 2 people who are to enter the elevator can be selected in 10 ways.

Comment In Example 18 we found that ${}_5C_2 = 10$. In the discussion preceding the example, we found that ${}_5C_3 = 10$. In other words, ${}_5C_2 = {}_5C_3$. Can you explain why?

Example 19 *Nuclear accident.* On Wednesday, March 27, 1979, the nuclear generating facility on Three Mile Island near Middletown, Pennsylvania, malfunctioned and began discharging radiation into the air. In an attempt to prevent a nuclear "meltdown," 10 nuclear physicists were contacted to shed some light on the problem. (Later, additional physicists were

The four cooling towers at the Three Mile Island Nuclear Power Plant (*Photo:* United Press International)

called in.) As the 10 scientists were busily engaged on another project, 7 of them agreed to come to the crippled nuclear plant. (Any 3 of them could remain to oversee their other existing project.) In how many different ways could the 7 scientists have been selected?

Solution Since order was not important, 7 scientists had to be selected out of a possible 10. This could be done in ${}_{10}C_7$ ways. Using Formula 6.4, we have

$${}_{10}C_7 = \frac{10!}{7!(10-7)!} = \frac{10!}{7!3!} = 120$$

Thus, the 7 scientists could have been selected in 120 different ways.

Example 20 The Acme Manufacturing Company Board of Directors consists of 15 members, 10 men and 5 women. A committee consisting of 3 men and 3 women is to be set up to investigate charges by affirmative action groups on the discrimination practices of the company.

a) In how many different ways can this committee be formed?
b) What is the probability that Bob and his brother Ray, who are members of the board, will both be on this committee?
c) What is the probability that the committee will contain both Bob and Juanita, who are bitter enemies and do not talk to each other?

Solution a) The 3 male members can be selected from the 10 men on the board in ${}_{10}C_3$ ways. Similarly, the 3 female members can be selected from the 5 women board members in ${}_5C_3$ ways. Applying Formula 6.1, the total number of different ways that this committee can be selected is

$$_{10}C_3 \cdot {}_5C_3 = \frac{10!}{3!7!} \cdot \frac{5!}{3!2!}$$

$$= 120 \cdot 10 = 1200$$

Thus, the committee can be selected in 1200 different ways.

b) Since Bob and his brother Ray must be on the committee, the third male member must be selected from the remaining 8. This can be done in $_8C_1$ ways. The 3 female members can be selected from the 5 female board members in $_5C_3$ ways. Thus, the probability that both Bob and his brother Ray are on the committee is

$$\frac{_8C_1 \cdot {}_5C_3}{1200} = \frac{8 \cdot 10}{1200} = \frac{1}{15}$$

c) Since Bob is to be on the committee, the 2 other male members can be selected from the 9 remaining male board members in $_9C_2$ ways. Also, since Juanita is to be on the committee, the 2 other female members can be selected from the 4 remaining female board members in $_4C_2$ ways. Thus, the probability that both Bob and Juanita are on the committee is

$$\frac{_9C_2 \cdot {}_4C_2}{1200} = \frac{36 \cdot 6}{1200} = \frac{9}{50}$$

Comment In Example 20 there is a greater chance that Bob will be on the committee with Juanita, to whom he does not talk, than with his brother Ray. Can you explain this unusual fact?

Example 21 *Poker anybody?* Felicia is playing 5-card stud poker. This a game in which each of 2 or more players is dealt 5 cards from an ordinary 52-card deck. What is the probability that she is dealt a straight flush? (A straight flush is a hand in which the cards are of the same suit and in order. For example, the 5, 6, 7, 8, and 9 of spades; the ace, 2, 3, 4, and 5 of clubs; etc. Also, the last card cannot be a 2, 3, or 4.)

Solution Since 5 cards are to be selected from a possible 52 cards, the total number of hands that can be dealt to any player is $_{52}C_5$. Now we must determine how many of these are favorable. According to the rules of the game, the last card cannot be a 2, 3, or 4. Thus, there are 10 choices for the lowest card of any straight flush. These are ace, 2, 3, 4, 5, 6, 7, 8, 9, and 10. Since these can be hearts, diamonds, spades, or clubs, there are $4 \cdot 10$ or 40 favorable outcomes. Thus

$$\text{Prob(straight flush)} = \frac{40}{_{52}C_5} = \frac{40}{2{,}598{,}960} = 0.0000153907$$

Thus, Felicia's chances of being dealt a straight flush are very small.

Comment For those familiar with poker, we present the probabilities in Table 6.1.

Table 6.1 Different Poker Hands and Their Probabilities

Type of hand	Probability of it being dealt to you
Royal flush (ace, king, queen jack, 10 in the same suit)	0.0000015
Four of a kind (four of a kind, like four 4's or four queens)	0.00024
Flush (five cards in a single suit but not straight)	0.0020
Two pairs	0.0475
Nothing of interest	0.5012

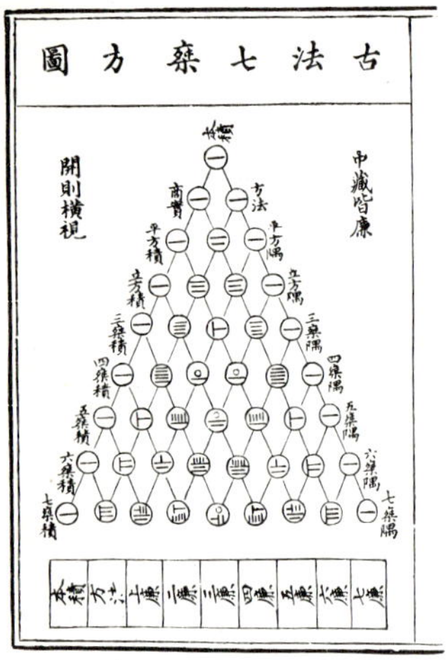

Figure 6.8 [*Source:* Joseph Needham, *Science and Civilization*, Volume III (New York: Cambridge University Press)]

Instead of using Formula 6.4 for evaluating ${}_nC_r$, we can use the following alternate technique, which avoids factorial notation. It is known as *Pascal's triangle* and is based on the fourteenth-century Chinese manuscript entitled by its author *The Old Method Chart of the Seven Multiplying Squares.* See Figure 6.8. An example of Pascal's triangle is shown here.

```
                  1
                1   1
              1   2   1
            1   3   3   1
          1   4   6   4   1
        1   5  10  10   5   1
      1   6  15  20  15   6   1
```

It is easy to see how such a triangle is constructed. Each row has a 1 on either end. All the in-between entries are obtained by adding the numbers immediately above and directly to the right and left of them,

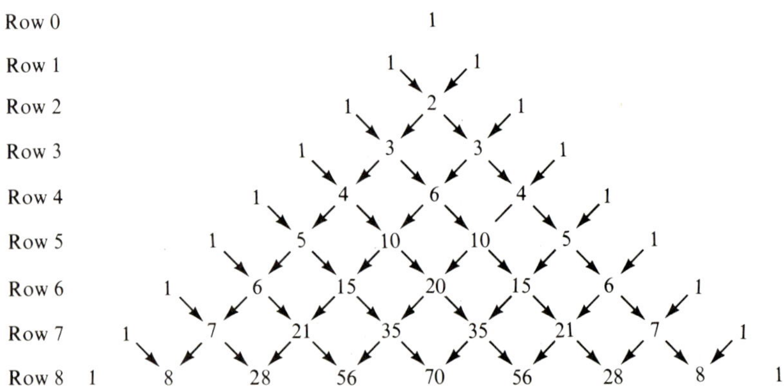

Figure 6.9 Construction of Pascal's triangle.

as shown by the arrows in Figure 6.9. For example, to obtain the entries for the eighth row, we first put a 1 on each end (moving over slightly). Then we add the 1 and the 7 from row 7 to get 8 as shown. We then add 7 and 21 to get 28, then we add 21 to 35 to get 56, and so on. Remember to place the 1's on each end. The numbers must be lined up as shown in the diagram. Such a triangle of numbers is known as *Pascal's triangle.*

Let us use Pascal's triangle to solve problems involving combinations.

Example 22 The Metro Insurance Company receives 8 claims for fire damage during one week. Management decides to select 3 of these claims randomly and to investigate them thoroughly for the possibility of fraud (or arson). In how many different ways can this be done?

Solution Here we must evaluate ${}_8C_3$. Let us use Pascal's triangle. We first write down the 8 rows of the triangle as shown in Figure 6.9. We label them as indicated. Now we look at row 8. Since we are interested in selecting 3 things out of a possible 8 things, we skip the end 1 and move across to the third entry. It is 56. Thus ${}_8C_3 = 56$.

Similarly, if we are interested in finding ${}_8C_6$, we move to the sixth entry after the end 1 on row 8. The entry is 28. Thus, ${}_8C_6 = 28$. If we wanted ${}_8C_8$, we would move to the eighth entry after the 1. The entry is 1. Thus ${}_8C_8 = 1$. Also, to find ${}_8C_0$, we go to row 8 and move to the right 0 columns. This means that we do not go anywhere. We just stay at the 1. Thus, ${}_8C_0 = 1$.

Comment In Example 22 we could also have evaluated ${}_8C_3$ by using Formula 6.4. We would get

$${}_8C_3 = \frac{8!}{3!5!} = 56$$

which is the same answer we obtained using Pascal's triangle.

Exercises

38. A new car dealer has five showroom windows and five different cars. In each showroom window, the dealer wishes to put a different car. In how many different ways can this be done? (Order counts.)

39. There are eight babies in one wing of the maternity ward of a hospital. In how many different ways can these babies be arranged in front of the viewing windows?

40. Eight police officers arrive at Metro Hospital to give blood for a fellow wounded cop. However, the hospital needs only five volunteers. In how many different ways can these volunteers be selected?

41. A chemist mixed six chemicals together in a solution and created a new synthetic drug. Unfortunately, he does not recall the order in which

the drugs were introduced into the solution. He decides to repeat the experiment. How many possibilities are there?

42. There are eight rooms available for use by the math department. This summer the math department is offering eight courses during a given hour. In how many different ways can these courses be assigned to a room?

43. The Calco Food Processing Company randomly selects three loads of every day's production and thoroughly checks them for food contamination. On January 4, the company produced 17 loads of food. In how many different ways can the three loads to be inspected be selected?

■44. A baseball manager has selected his nine starting players.
   a) How many different batting orders are possible?
   b) What is the probability that the pitcher bats last?

■45. There are 12 chapters in this book. The editor decides to rearrange the order in which the chapters will appear in the book. In how many different ways can this be done? (Assume that each chapter is independent of the others and that the knowledge contained in one chapter is not needed to understand another chapter.)

■46. The Marvo Insurance Company is expanding its West Coast operations and expects to hire 5 male and 5 female employees. After interviewing numerous candidates, it is determined that 17 men and 12 women qualify for the job. In how many different ways can the jobs be filled from among these candidates?

47. A television drama critic has been asked to rank the top three of this year's ten best movies. In how many ways can this be done?

■48. A child is playing with a telephone and dials a long-distance (10-digit) number.
   a) How many possible long-distance numbers are there? (Assume that there are no restrictions.)
   b) What is the probability that the child dials his own number?

49. Stephanie Wilson is visiting her aunt and uncle who live in New York City. There are ten different movies playing in town that Stephanie wishes to see. However, Stephanie's schedule will permit her the time to see only four of them. In how many different ways can she select the four movies to see? (Assume that order does not count.)

50. Marcia is taking a math final that consists of seven questions. She must answer only five of them and they can be answered in any order. In how many different ways (orders) can Marcia answer the five questions?

51. Priscilla asks her kid brother (who is too young to read) to place her four-volume dictionary on a shelf.
   a) In how many different ways can the books be placed on the shelf?
   b) Find the probability that the little boy places the books on the shelf in sequential order, first Volume 1, then Volume 2, etc.

52. There are five vacancies on the community school board. In how many ways can these vacancies be filled from among the 12 names placed in nomination?

■ 53. The math club at State University has 20 members. At its next meeting a president, a vice-president, a treasurer, and a secretary are to be elected. The club charter provides that no person can occupy more than one position. In how many different ways can these offices be filled from among the 20 members?

54. A gas station is running short of gas and has only enough to fill up at most a car, a truck, and a bus. In how many different ways can the three vehicles line up for gas?

55. There are 20 faculty members in the School of Technology at State University. According to student evaluations, 10 are rated excellent, 6 are rated good, and 4 are rated poor. Two teachers are selected at random. Find the probability of getting the following:
   a) Two excellent teachers
   b) One good and one poor teacher
   c) Two poor teachers

■ 56. Verify the probabilities given in Table 6.1 for the different poker hands.

6.5 Basic Rules of Probability

Let us look in on Angelo who is rolling dice with a friend. Angelo stands to win \$500 if he rolls a sum of 7 or 11 with a pair of dice. What is the probability that he wins the \$500?

To answer this question, we first notice that Angelo *cannot* roll a pair of dice and get both a sum of 7 and a sum of 11 at the same time. We express this fact mathematically by saying that the events *obtaining a sum of 7* and *obtaining a sum of 11* are *mutually exclusive*.

When a pair of dice is rolled, there are 36 possible outcomes as shown in Figure 6.10. The probability of obtaining a sum of 7 is $\frac{6}{36}$, and the probability of obtaining a sum of 11 is $\frac{2}{36}$. Using the definition of

Figure 6.10 Thirty-six possible outcomes when two dice are rolled at the same time.

probability, we know there are 8 favorable outcomes out of a possible 36, so that

$$\text{Prob}(\text{sum of 7 or 11}) = \frac{8}{36} = \frac{2}{9}$$

Suppose we add the probability of obtaining a sum of 7 to the probability of obtaining a sum of 11. We get

$$\text{Prob}(\text{sum of 7}) + \text{Prob}(\text{sum of 11}) = \frac{6}{36} + \frac{2}{36} = \frac{8}{36} = \frac{2}{9}$$

In other words,

$$\text{Prob}(\text{sum of 7 or 11}) = \text{Prob}(\text{sum of 7}) + \text{Prob}(\text{sum of 11})$$

Can we generalize this result to any two mutually exclusive events? The answer is yes.

Definition 6.4 Consider any two events A and B. If both events *cannot* occur at the same time, we say that events A and B are *mutually exclusive.*

Formula 6.5 If events A and B are mutually exclusive, then

$$\text{Prob}(A \text{ or } B) = \text{Prob}(A) + \text{Prob}(B)$$

Example 23 Mr. and Mrs. MacMillan have 3 children. What is the probability that the family consists of 2 boys and 1 girl or 2 girls and 1 boy as opposed to 3 children of the same sex?

Solution The events *having 2 boys and 1 girl* and *having 2 girls and 1 boy* are mutually exclusive, since they cannot both occur at the same time. Thus, we can apply Formula 6.5. We have

$$\begin{aligned}\text{Prob}(\text{2 boys, 1 girl or 2 girls, 1 boy}) &= \text{Prob}(\text{2 boys, 1 girl}) \\ &\quad + \text{Prob}(\text{2 girls, 1 boy})\end{aligned}$$

To calculate Prob (2 boys, 1 girl), we consider the sample space shown in the table on page 261. Thus,

$$\text{Prob}(\text{2 boys, 1 girl}) = \frac{3}{8} \quad \text{and} \quad \text{Prob}(\text{2 girls, 1 boy}) = \frac{3}{8}$$

so that

$$\text{Prob}(\text{2 boys, 1 girl or 2 girls, 1 boy}) = \frac{3}{8} + \frac{3}{8} = \frac{6}{8} = \frac{3}{4}$$

The probability that all 3 children are of the same sex is $\frac{2}{8} = \frac{1}{4}$.

Child 1	Child 2	Child 3	Family consists of
Boy	Boy	Boy	3 boys
Boy	Boy	Girl	2 boys, 1 girl
Boy	Girl	Boy	2 boys, 1 girl
Boy	Girl	Girl	1 boy, 2 girls
Girl	Boy	Boy	2 boys, 1 girl
Girl	Boy	Girl	1 boy, 2 girls
Girl	Girl	Boy	1 boy, 2 girls
Girl	Girl	Girl	3 girls

Example 24 *Car sales.* The accompanying table summarizes the auto sales of the Jack Chevrolet Corp. for the week of January 2–8. If a car is selected from this group at random, what is the probability that it has 4 or 6 cylinders?

Type of car	Number sold
4 cylinder	34
6 cylinder	20
8 cylinder	15
Total	69

Solution The events *car has 4 cylinders* and *car has 6 cylinders* are mutually exclusive, so we can apply Formula 6.5. We have

$$\text{Prob}(4 \text{ cylinders}) = \frac{34}{69}$$

$$\text{Prob}(6 \text{ cylinders}) = \frac{20}{69}$$

Thus, using Formula 6.5,

$$\text{Prob}(4 \text{ or } 6 \text{ cylinders}) = \text{Prob}(4 \text{ cylinders}) + \text{Prob}(6 \text{ cylinders})$$

$$= \frac{34}{69} + \frac{20}{69} = \frac{54}{69}$$

Example 25 In Example 24 what is the probability that the car does not have 4 or 6 cylinders?

Solution If the car does not have 4 or 6 cylinders, then it must have 8 cylinders. We can calculate this probability directly from the information given in the chart. Thus,

$$\text{Prob}(8 \text{ cylinders}) = \frac{15}{69}$$

We can also use a different approach. Since the car has 8 cylinders or it does not have 8 cylinders, these two events are mutually exclusive. Obviously, one of these two events must occur. Thus,

$$\text{Prob(car has 8 cylinders)} + \text{Prob (car does not have 8 cylinders)} = 1$$

(Can you explain why this sum is 1?) In Example 24 we already calculated the probability that the car does not have 8 cylinders or, equivalently, that it has 4 or 6 cylinders. It was $\frac{54}{69}$. Therefore,

$$\text{Prob (car has 8 cylinders)} + \text{Prob(car does not have 8 cylinders)} = 1$$

$$\text{Prob(car has 8 cylinders)} + \frac{54}{69} = 1$$

Substracting $\frac{54}{69}$ from both sides gives

$$\text{Prob(car has 8 cylinders)} = 1 - \frac{54}{69} = \frac{69}{69} - \frac{54}{69} = \frac{15}{69}$$

More generally, let Prob(A) be the probability that event A will happen and let Prob(A'), read as the probability of A prime, be the probability that event A will not happen. Since A either happens or does not happen, we have

$$\begin{aligned}&\text{Prob}(A \text{ happens or does not happen})\\&\quad = \text{Prob}(A \text{ happens}) + \text{Prob}(A \text{ does not happen})\\&\quad 1 = \text{Prob}(A) + \text{Prob}(A')\end{aligned}$$

so that

$$1 - \text{Prob}(A) = \text{Prob}(A')$$

Some books refer to A' as the *complement* of event A.

Refer to the car sales problem in Example 24. A further breakdown reveals the following information about the cars.

Type of car	Number sold
4 cylinders with air conditioning	0
4 cylinders without air conditioning	34
6 cylinders with air conditioning	14
6 cylinders without air conditioning	6
8 cylinders with air conditioning	12
8 cylinders without air conditioning	3
Total	69

Suppose we are interested in determining the probability of randomly selecting a car from this group that has either 6 cylinders or air conditioning. In this case we notice that the events *obtaining a car with*

6 cylinders and *obtaining a car with air conditioning* are not mutually exclusively. Let us apply Formula 6.5 and see what happens. We have

$$\begin{aligned}&\text{Prob (6 cylinders or air conditioning)}\\&\quad= \text{Prob (6 cylinders)} + \text{Prob (air conditioning)}\\&\quad= \frac{20}{69} + \frac{26}{69} = \frac{46}{69}\end{aligned}$$

Notice, however, that there are 14 cars that have been counted twice because they have 6 cylinders and also have air conditioning. Thus, we have to subtract the 14. When this is done we get

$$\begin{aligned}&\text{Prob (6 cylinders or air conditioning)}\\&\quad= \text{Prob (6 cylinders)} + \text{Prob (air conditioning)}\\&\qquad - \text{Prob (car has both)}\\&\quad= \frac{20}{69} + \frac{26}{69} - \frac{14}{69} = \frac{32}{69}\end{aligned}$$

This leads us to the following more general addition rule.

Formula 6.6 *Addition rule.* If A and B are any events, then the probability of obtaining either of them is equal to the probability of A plus the probability of B minus the probability of both occurring at the same time. Symbolically.

$$\text{Prob}(A \text{ or } B) = \text{Prob}(A) + \text{Prob}(B) - \text{Prob}(A \text{ and } B)$$

Example 26 *Voting.* Consider the accompanying newspaper article. What is the probability that either a married man or his wife will vote in the coming mayoral election?

> Greensburgh (Oct. 20)—According to a poll released yesterday, voter turnout for the coming mayoral election is expected to be at an all-time low. One out of eleven married men said that they would vote. For women, the figure was one out of nine. Only one out of 28 couples said that they would both vote. Both candidates indicated that they were mobilizing their forces to get out more votes.

Solution The events man votes and wife votes are not mutually exclusive events, since both events can occur. Thus, we use Formula 6.6. We have:

$$\begin{aligned}&\text{Prob(man or wife votes)} =\\&\quad \text{Prob(man votes)} + \text{Prob(woman votes)} - \text{Prob(both vote)}\end{aligned}$$

Based on the newspaper article

$$\text{Prob(man votes)} = \frac{1}{11} \qquad \text{Prob(woman votes)} = \frac{1}{9}$$

$$\text{Prob(both vote)} = \frac{1}{28}$$

Therefore,

$$\text{Prob(man or wife votes)} = \frac{1}{11} + \frac{1}{9} - \frac{1}{28}$$

$$= \frac{252}{2772} + \frac{308}{2772} - \frac{99}{2772} = \frac{461}{2772}$$

Thus, Prob(man or wife votes) $= \frac{461}{2772}$.

Example 27 In order for Boris to graduate from college, he must demonstrate English proficiency by passing a two-part exam. Boris estimates that he has a $\frac{3}{5}$ chance of passing part A of the exam and a $\frac{2}{5}$ chance of passing part B of the exam. He figures that he has a probability of $\frac{4}{5}$ of passing either part of the exam. What is the probability that he passes both parts of the exam?

Solution Applying Formula 6.6, we have

Prob(passes part A or B) =

Prob(passes part A) + Prob(passes part B) − Prob(passes both parts)

$$\frac{4}{5} = \frac{3}{5} + \frac{2}{5} - \text{Prob(passes both parts)}$$

$$\frac{4}{5} = 1 - \text{Prob(passes both parts)}$$

Thus, Prob(passes both parts) $= 1 - \frac{4}{5} = \frac{1}{5}$.

Exercises

---

57. A card is selected from an ordinary deck of cards. Find the following probabilities.

a) It is an ace or a club.

b) It is an ace or a picture card.

58. Forty percent of the members of a certain club are Catholic, 55 percent are Democrats, and 19 percent are both Catholic and Democrats. If a club member is selected at random, what is the probability that the member is a Catholic or a Democrat?

59. *Medical research.* A medical researcher obtained the following information about 100 patients with heart problems: $\frac{2}{3}$ have high blood pressure, $\frac{1}{4}$ are overweight, and $\frac{1}{8}$ have high blood pressure and are overweight. If one of these 100 patients is selected at random, find the probability that the patient is overweight or has high blood pressure.

60. Determine whether the following events are mutually exclusive or not.

a) Studying for your math exam and liking music
b) Smoking and having lung cancer
c) Graduating from college and getting a good job
d) Coming to school by car and coming to school by bicycle
e) Getting an A in this course and getting a B in this course

61. The probability that Jeremiah is accepted by a law school is $\frac{1}{8}$. The probability that Gervelyne is accepted by a law school is $\frac{1}{9}$. The probability that both are accepted is $\frac{1}{15}$. Find the probability that either is accepted by a law school.

62. The probability that Juanita remembers to buy beer is $\frac{4}{7}$. The probability that she remembers to buy cigarettes is $\frac{5}{8}$. The probability that she remembers to buy both is $\frac{3}{4}$. Find the probability that she remembers to buy beer or cigarettes.

63. The following information is available about the members of the Rocky Beach Health Club: $\frac{3}{8}$ play tennis, $\frac{4}{7}$ jog, and $\frac{1}{2}$ play tennis and jog. If a club member is selected at random, find the probability that the member plays tennis or jogs.

64. According to state tax officials, $\frac{1}{5}$ of the mistakes made on tax returns are in the arithmetic calculations, $\frac{1}{8}$ of the mistakes are because people use the wrong forms, and $\frac{1}{50}$ of the mistakes involve arithmetic calculations and wrong tax forms. If a rejected tax form is selected at random, find the probability that it is because the wrong tax form was used or because it contains an arithmetic mistake.

For any three events A, B, and C, the probability of A or B or C is given by

$$\begin{aligned}\text{Prob}(A \text{ or } B \text{ or } C) &= \text{Prob}(A) + \text{Prob}(B) + \text{Prob}(C)\\ &\quad - \text{Prob}(A \text{ and } B) - \text{Prob}(A \text{ and } C) - \text{Prob}(B \text{ and } C)\\ &\quad + \text{Prob}(A \text{ and } B \text{ and } C)\end{aligned}$$

Use this formula to solve the following exercises.

65. Sherry is planning to vacation in Europe this summer. The probability that she visits different countries is given in the accompanying table. Find the probability that she visits one of these three countries.

Country	Probability
France	$\frac{18}{39}$
London	$\frac{22}{39}$
Paris	$\frac{2}{3}$
France and Paris	$\frac{11}{39}$
London and Paris	$\frac{1}{3}$
France and London	$\frac{3}{13}$
France, London, and Paris	$\frac{2}{13}$

66. At Jackmore College all business students must complete the following three math-related course: Math 51—Introduction to College Algebra, Math 82—Introduction to Computing, and Math 93—Introduction to Finite Math. The probabilities that a student registers for one or more of these three courses in a freshman semester are given in the accompanying table. Find the probability that a student registers for all three courses in his or her freshman semester.

Course	Probability
Math 51	$\frac{2}{3}$
Math 82	$\frac{37}{60}$
Math 93	$\frac{26}{60}$
Math 51 and 82	$\frac{19}{60}$
Math 51 and 93	$\frac{3}{10}$
Math 82 and 93	$\frac{1}{3}$
Either Math 51, 83, or 93	1

6.6 Conditional Probability

Example 28 Recently the board of trustees of Deermark College established a new policy requiring all future graduates to pass a mathematics proficiency test prior to graduation. As expected, the new policy generated student reaction. To get firsthand information, Gary Peoples, a reporter for the student newspaper, conducted a survey of 500 randomly selected students. The students were asked to indicate their class status and their opinion of the new proposal. The results of the survey are shown in the accompanying table.

Class status	In favor	Opposed	No opinion	Total
Freshman	63	83	21	167
Sophomore	27	61	11	99
Junior	37	43	27	107
Senior	41	29	25	95
Nonmatriculated	23	5	4	32
Total	191	221	88	500

Based on this survey, we can answer the following questions.

a) What is the probability that a randomly selected student from this group voted against the new policy?
b) What is the probability that a freshman selected at random voted against the new policy?
c) What is the probability that a randomly selected nonmatriculated student has no opinion about the new policy?

Solution a) Since there were 500 students involved in the survey, 221 of whom oppose the new policy, we have

$$\text{Prob (student voted against new policy)} = \frac{221}{500}$$

b) To answer this part of the question, we must restrict our attention to freshmen only. In the survey there were 167 freshmen questioned, 83 of whom opposed the new policy. Thus,

$$\text{Prob (student is freshman who voted against new policy)} = \frac{83}{167}$$

c) To answer this part of the question, we restrict our attention to nonmatriculated students only. In the survey there were 32 nonmatriculated students questioned, 4 of whom had no opinion. Thus,

$$\text{Prob (nonmatriculated student has no opinion)} = \frac{4}{32} = \frac{1}{8}$$

The situation described in part b of Example 28 is called a *conditional probability* because we are interested in the probability of a student voting against the new policy given that (or conditional upon the fact that) the student is a freshman. This condition can be expressed mathematically by using the vertical line | to stand for the words "given that" or "if we know that." Using this notation, we write

$$\text{Prob (student voted against policy} \mid \text{student is a freshman)} = \frac{83}{167}$$

Similarly for part c, we write

$$\text{Prob (student has no opinion} \mid \text{student is nonmatriculated)} = \frac{4}{32} = \frac{1}{8}$$

For situations such as the one described in Example 28, mathematicians have developed a conditional probability formula. Before studying this formula, let us summarize some of the information already known.

$$\text{Prob (student is a freshman)} = \frac{167}{500}$$

$$\text{Prob (student is a freshman } \textit{and} \text{ opposes new policy)} = \frac{83}{500}$$

$$\text{Prob (student is nonmatriculated)} = \frac{32}{500} = \frac{8}{125}$$

$$\text{Prob (student is nonmatriculated } \textit{and} \text{ has no opinion)} = \frac{4}{500} = \frac{1}{125}$$

Let us now divide Prob(student is freshman *and* opposes new policy) by Prob(student is freshman). We get

$$\frac{\text{Prob(student is freshman } \textit{and} \text{ opposes new policy)}}{\text{Prob(student is freshman)}}$$

$$= \frac{\frac{83}{500}}{\frac{167}{500}} = \left(\frac{83}{500}\right) \div \left(\frac{167}{500}\right)$$

$$= \left(\frac{83}{500}\right) \cdot \left(\frac{500}{167}\right) = \frac{83}{167}$$

This is the same answer we obtained for Prob(student voted against policy | student is freshman). In both cases the answer is $\frac{83}{167}$.

Similarly, if we divide Prob(student is nonmatriculated *and* has no opinion) by Prob(student is nonmatriculated), we get

$$\frac{\text{Prob(student is nonmatriculated } \textit{and} \text{ has no opinion)}}{\text{Prob(student is nonmatriculated)}}$$

$$= \frac{\frac{1}{125}}{\frac{8}{125}} = \left(\frac{1}{125}\right) \div \left(\frac{8}{125}\right)$$

$$= \left(\frac{1}{125}\right) \cdot \left(\frac{125}{8}\right) = \frac{1}{8}$$

Again this is the same result we obtained for Prob (student has no opinion | student is nonmatriculated). In both cases the answer is $\frac{1}{8}$.

This leads us to the following conditional probability formula.

Formula 6.7 *Conditional probability formula.* If A and B are any events, then

$$\text{Prob}(A \mid B) = \frac{\text{Prob}(A \text{ and } B)}{\text{Prob}(B)}$$

Example 29 In an effort to conserve energy, Amelia often travels to school by means of public transportation. Since such transportation is unpredictable, she is sometimes late to her math class. The probability that she will come to school by public transportation is .68, and the probability that she will come to school by public transportation and be late to class is .17. If Amelia is observed coming to school by public transportation, what is the probability that she will be late for her math class?

Solution In this case we are given

$$\text{Prob(comes by public transportation)} = .68$$

$$\text{Prob(comes by public transportation } \textit{and} \text{ is late to class)} = .17$$

We now apply Formula 6.7. We have

$$\text{Prob(late} \mid \text{public transportation)} = \frac{\text{Prob(public transportation and late)}}{\text{Prob(public transportation)}} = \frac{.17}{.68} = \frac{17}{68} = \frac{1}{4}$$

Thus, the probability that Amelia will be late to class if we know that she comes to school by public transportation is $\frac{1}{4}$, or .25.

Example 30 *Tuition increase.* The board of trustees discussed at the beginning of this section also is considering increasing tuition for the coming school year by 9%. The probability that they vote to raise tuition is .82. Past experience has shown that whenever tuition is increased by 9%, student enrollment drops by approximately 5% with a probability of .95. Find the probability that the board votes to increase tuition by 9% and that the enrollment drops by 5%.

Solution In this case we are given that

$$\text{Prob(raise tuition)} = .82$$

$$\text{Prob(enrollment drops} \mid \text{raise tuition)} = .95$$

We wish to find Prob(raise tuition *and* enrollment drops). We apply Formula 6.7, getting

$$\text{Prob(enrollment drops} \mid \text{raise tuition)} = \frac{\text{Prob(raise tuition } \textit{and} \text{ enrollment drops)}}{\text{Prob(raise tuition)}}$$

To simplify the computations, let us multiply both sides by Prob(raise tuition). We get

$$\text{Prob(enrollment drops} \mid \text{raise tuition)} \cdot \text{Prob(raise tuition)} = \text{Prob(raise tuition and enrollment drops)}$$

Now we can easily solve for Prob(raise tuition and enrollment drops). We have

$$\text{Prob(raise tuition and enrollment drops)} = (.95)(.82) = .779$$

Thus Prob(raise tuition and enrollment drops) = .779.

The results of Example 30 suggest the following rule.

Formula 6.8 *Multiplication rule.* If A and B are any events, then

$$\text{Prob}(A \text{ and } B) = \text{Prob}(A|B) \cdot \text{Prob}(B)$$

Example 31 *Joyriding.* Jake is considering taking his old jalopy out for a drive into the countryside this weekend. The probability that he takes the car out is .72. The probability that the car breaks down given that he takes it out for a drive is .12. Find the probability that Jake takes the car out and that it breaks down.

Solution In this case we use Formula 6.8. We are told that

$$\text{Prob(takes car out)} = .72$$
$$\text{Prob(car breaks down}|\text{took car out)} = .12$$

so that

$$\begin{aligned}\text{Prob}&\text{(takes car out and breaks down)}\\ &= \text{Prob(car breaks down}|\text{took car out)} \cdot \text{Prob(takes car out)}\\ &= (.12)(.72)\\ &= .0864\end{aligned}$$

Thus, Prob(takes car out and it breaks down) = .0864.

Example 32 *Strike.* The workers of the Babet Dairy Corporation are on strike, demanding a large pay raise and a cost of living adjustment clause in their contract (see the accompanying newspaper clipping). There is a .98 probability that the cost of milk will go up at least 2¢ a quart if the workers' demands are met. Furthermore, there is a .19 probability that their demands will be met. Find the probability that their demands are met *and* that the price of milk goes up at least 2¢ a quart.

Milk Truck Drivers Still on Strike

New York (April 11)—The strike by the milk drivers enters its third week today with no end in sight. State mediators have been unable to get both sides together to negotiate on a new contract. The union is demanding a 30% raise spread out over two years and a cost of living adjustment clause. Management considers such demands irresponsible and refuses to negotiate with the union.

Solution We are given that

$$\text{Prob}(\text{price goes up} \mid \text{demands are met}) = .98$$
$$\text{Prob}(\text{demands are met}) = .19$$

Applying Formula 6.8 gives

$$\begin{aligned}&\text{Prob}(\text{demands are met and price goes up})\\ &\quad = \text{Prob}(\text{price goes up} \mid \text{demands are met}) \cdot \text{Prob}(\text{demands are met})\\ &\quad = (.98)(.19)\\ &\quad = .1862\end{aligned}$$

Thus, the probability that their demands are met and that the price of milk goes up at least 2¢ a quart is .1862.

In many cases it turns out that whether or not one event happens does not affect whether another will happen. For example, when two dice are rolled, the outcome for one die has nothing to do with the outcome for the second die. Such events are called *independent events.*

Definition 6.5 Two events A and B are said to be *independent* if the occurrence of event B is in no way affected by the occurrence or non-occurrence of event A. In other words, if A and B are independent events, then

$$\text{Prob}(A \mid B) = \text{Prob}(A)$$

When dealing with independent events, Formula 6.8 can be simplified, as the following example illustrates.

Example 33 A shipment of 100 CB radio sets is received by an electronics store. It is known that 5 of the sets are defective. Two sets are randomly selected and inspected. Find the probability that both sets are defective under the following conditions.

a) The first set is not replaced before the second is selected.
b) The first set is replaced before the second set is selected.

Solution a) Since the first set is not replaced before the second set is selected, we use Formula 6.8. We have

$$\begin{aligned}&\text{Prob}(\text{both sets are defective})\\ &\quad = \text{Prob}(\text{set 2 is defective} \mid \text{set 1 is defective})\\ &\quad\quad \cdot \text{Prob}(\text{set 1 is defective})\end{aligned}$$

Note that since the first set is not replaced, there are only 4 defective sets remaining out of a possible 99 sets. This is because the first set selected was defective. Thus,

$$\text{Prob}(\text{both sets are defective}) = \frac{4}{99} \cdot \frac{5}{100} = \frac{20}{9900} = \frac{1}{495}$$

Therefore, Prob(both sets are defective) $= \frac{1}{495}$.

b) Since the first set is replaced into the shipment before the second selection is made, then whether or not the first set is defective in no way affects what happens on the second selection. (Of course, the sets are thoroughly mixed up before the second set is selected.) The events *defective on second selection* and *defective on first selection* are independent. Thus,

$$\begin{aligned}&\text{Prob(defective set on selection 2}\,|\,\text{defective set on selection 1)}\\&\qquad = \text{Prob(defective set on selection 2)}\end{aligned}$$

so that

$$\begin{aligned}&\text{Prob (both sets defective)}\\&\qquad = \text{Prob(defective set on selection 2}\,|\,\text{defective set on}\\&\qquad\quad \text{selection 1)} \cdot \text{Prob(defective set on selection 1)}\\&\qquad = \text{Prob(defective set on selection 2)}\\&\qquad\quad \cdot \text{Prob(defective set on selection 1)}\\&\qquad = \frac{5}{100} \cdot \frac{5}{100} = \frac{25}{10{,}000} = \frac{1}{400}\end{aligned}$$

Therefore, the probability that both sets are defective is $\frac{1}{400}$.

Example 33 suggests that when two events A and B are independent, then $\text{Prob}(A\,|\,B)$ is the same as $\text{Prob}(A)$, since B in no way affects what happens to A. Formula 6.8 then becomes as follows:

Formula 6.9 If A and B are independent events, then

$$\text{Prob}(A \text{ and } B) = \text{Prob}(A) \cdot \text{Prob}(B)$$

Example 34 A recent survey in Alexville showed that 86% of the population have some form of health insurance. Furthermore, the survey showed that 21% of these people also have some form of life insurance. If an adult resident of Alexville is randomly selected, find the probability that the individual has both health and life insurance. (Assume independence.)

Solution We apply Formula 6.9:

$$\begin{aligned}&\text{Prob(has health and life insurance)}\\&\qquad = \text{Prob(has health insurance)} \cdot \text{Prob(has life insurance)}\\&\qquad = (.86)(.21)\\&\qquad = .1806\end{aligned}$$

Therefore, the probability that the individual has both types of insurance is .1806.

Comment The multiplication rule for independent events (Formula 6.9) can be generalized for more than two independent events. We simply multiply all the respective probabilities. For example, if the three independent events A, B, and C have .8, .7, and .6 probabilities of occurring, respectively, then the probability that all three will occur is

$$\begin{aligned}\text{Prob}(A \text{ and } B \text{ and } C) &= \text{Prob}(A)\cdot\text{Prob}(B)\cdot\text{Prob}(C)\\ &= (.8)(.7)(.6)\\ &= .336\end{aligned}$$

Exercises

67. Twenty-six percent of all students who are business majors at Polytech University take a finite math course. Additionally, 11% of all business majors take both a calculus course and a finite math course. If a student who is a business major and who has taken a finite math course is randomly selected, find the probability that the student will take a calculus course.

68. Refer to the first table given at the beginning of this chapter. Sally Jensen was born in 1947 and holds a master's degree from Harvard University. She is married to Maurice Jensen. Find the probability that her marriage will not end in divorce.

69. Sally Jensen discussed in Exercise 68 already has five children and is planning on having a sixth child; she hopes to have a boy. Her first five children are girls. Find the probability that her sixth child will finally be a boy.

70. *Nuclear safeguards.* A particular nuclear reactor has two safety valves that are designed to prevent a nuclear accident similar to the one that occurred at the Three Mile Island facility in Pennsylvania. The probability that the first safety valve will function satisfactorily in an emergency is .97, and the probability that the second backup valve will function properly if the first fails in an emergency is .95. Find the probability that both valves fail and that a nuclear mishap occurs.

71. *Smuggling.* According to one U.S. customs agent, there is a .13 chance that a smuggler can transport drugs across the border between the United States and Mexico without being apprehended, and a .06 chance that it can be smuggled into Florida by sea coming from South American countries without being apprehended. A drug dealer is planning to smuggle two shipments of drugs into the United States, one by way of Mexico and the other by sea into Florida. Find the probability that both shipments will be confiscated by authorities.

72. *Death penalty.* A sociologist interviewed 1000 individuals from minority groups in New York City to determine their opinion on the death penalty. The results of the survey are shown in the accompanying table.

Group	In favor of death penalty	Opposed to death penalty
Black	176	221
Puerto Rican	118	147
Oriental	94	62
Other minorities	151	31

Find the probability that an individual in the group is the following:

a) Black and opposes the death penalty
b) Opposed to the death penalty given that he is black
c) Black given that he is opposed to the death penalty

73. *Religion and politics.* The sociologist described in Exercise 72 then interviewed 1000 randomly selected individuals to determine whether there is a relationship between a person's religion and the person's political beliefs. The results of the survey are shown in the accompanying table.

	Political preference		
Religion	Democrat	Republican	Conservative
Protestant	310	360	60
Catholic	120	69	11
Jewish	29	13	8
Other	10	6	4

Find the probability that a randomly selected individual in the group is the following:

a) A Republican
b) A Republican given that he is a Catholic
c) A Catholic given that he is a Republican

74. In a certain community the probability that a man over 40 years old is overweight is .37. The probability that his blood pressure is high given that he is overweight is .69. If a man over 40 years of age is selected at random, what is the probability that he is overweight and that he has high blood pressure?

75. Bernard is standing in front of a phone booth. There is a piece of paper on the floor that says "Phone Out of Order." Is this an old sign? Should he deposit his last dime? The probability that the phone is out of order is $\frac{5}{7}$. The probability that the phone is out of order *and* that he loses his dime when he deposits it is $\frac{4}{9}$. If the phone is actually out of order, what is the probability that Bernard loses his dime?

76. In Calvington the probability that a man drinks is .71. If the probability that a man drinks and his wife smokes is .56, what is the probability that his wife smokes given that the man drinks?

77. Insurance company records indicate the following about the driving records of the people of Midville. The probability that both the husband and wife drive a car is .43, and the probability that only the husband drives is .79. If it is known that Mr. Federow drives, find the probability that Mrs. Federow drives also.

78. *Fertilizer.* The Salmo Fertilizer Corporation manufactures eight different kinds of fertilizer, each containing different concentrations of nitrogen as follows:

4 – 4 – 2
4 – 4 – 1
4 – 6 – 2
4 – 6 – 3
5 – 10 – 5
4 – 8 – 3
4 – 2 – 2
15 – 30 – 15

Pete McDonald discovers a bag of fertilizer produced by the Salmo Corp. remaining from last summer in his workshed. The label indicating the nitrogen content concentration is missing. If Pete remembers that the first digit was a 4, find the probability that the bag contains a 4-4-2 mixture.

79. If the probability of an adult having untreated hypertension (high blood pressure) is .15, find the probability that three totally unrelated people *do not* have untreated hypertension.

80. *Weather emergency.* A hospital administration is arranging emergency plans. The weather bureau has forecast a heavy snowfall with a .90 probability. The administration knows that if the snowfall is heavy, then there is an 80% probability that all hospital personnel will not be able to get to the hospital. Find the probability that the snowfall is heavy *and* that all hospital personnel cannot get to the hospital.

81. Leon has volunteered to roast a turkey for the family dinner. The probability that he forgets to use spice and tenderizer is $\frac{1}{3}$. The probability that he forgets to use only spice is $\frac{4}{9}$. If we know that Leon forgot to use spice, what is the probability that he also forgot to use tenderizer?

82. *Drugs on campus.* Twenty percent of all dormitory students at Whipple University have tried drugs at one time or another. Thirty-seven percent of the students are dormitory students. If a student is selected at random, what is the probability that the student lives in the dormitory and has tried drugs?

6.7 Bayes' Rule[2]

Let us look in on Dr. Carey, who has two bottles of sample pills on his desk for the treatment of arthritic pain. One day he gives Madeline a few pills from one of the bottles on his desk for the treatment of her arthritic pain. (All other treatments have failed.) However, he does not remember from which bottle he took the pills. The pills in bottle B_1 claim to be 70% effective with no known side effects. The pills in bottle B_2 claim to be 90% effective with some possible side effects. Bottle B_1 is closer to Dr. Carey on his desk and the probability is $\frac{2}{3}$ that he selected the pills from this bottle. On the other hand, bottle B_2 is farther away from Dr. Carey and the probability is therefore $\frac{1}{3}$ that he selected the pills from this bottle. The problem is to determine the bottle from which the pills were taken.

In many problems we are given situations such as this one, where we know the outcome of the experiment and are interested in concluding that the outcome happened because of a particular event. For situations such as these, we need a formula. This will be done in this section.

Example 35 *Medicine.* In the situation we are discussing, find the probability that the pills are effective in relieving Madeline's pain.

Solution Madeline can be relieved of her pain by taking the pills from either bottle B_1 or bottle B_2. Let A represent the event *Madeline's pain is relieved*, let B_1 represent the event that the pills were taken from bottle B_1, and let B_2 represent the event that the pills were taken from bottle B_2. The accompanying tree diagram illustrates how Madeline's pain can be relieved. From the given information

$$\text{Prob}(A|B_1) = 0.70 \quad \text{and} \quad \text{Prob}(B_1) = \frac{2}{3}$$

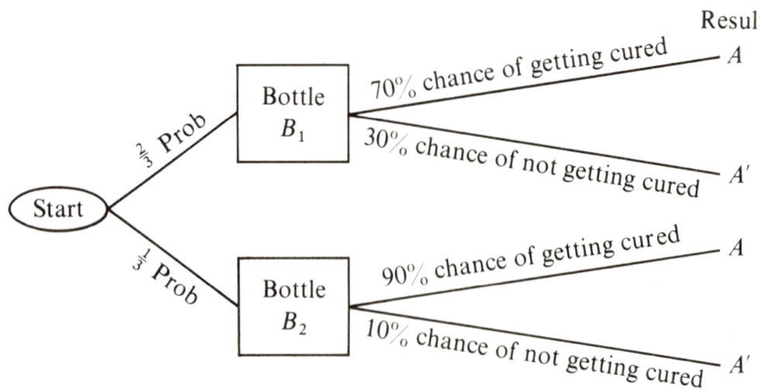

[2] This section assumes a background knowledge of algebra and may be omitted without affecting the continuity of the text.

Applying the multiplication rule (Formula 6.8) gives

$$\begin{aligned}
&\text{Prob(selects } B_1 \text{ and pill from it is effective)}\\
&\quad = \text{Prob(Madeline is cured} \mid \text{pill from } B_1 \text{ was selected)}\\
&\qquad \cdot \text{Prob(selecting } B_1)\\
&\quad = (.70)\frac{2}{3} = \frac{7}{10}\cdot\frac{2}{3}\\
&\quad = \frac{14}{30} = \frac{7}{15}
\end{aligned}$$

Symbolically,

$$\text{Prob}(A \text{ and } B_1) = \text{Prob}(A \mid B_1)\cdot\text{Prob}(B_1) = \frac{7}{15}$$

Also, from the given information

$$\text{Prob}(A \mid B_2) = \frac{9}{10} \quad \text{and} \quad \text{Prob}(B_2) = \frac{1}{3}$$

so that

$$\begin{aligned}
\text{Prob}(A \text{ and } B_2) &= \text{Prob}(A \mid B_2)\cdot\text{Prob}(B_2)\\
&= \frac{9}{10}\cdot\frac{1}{3} = \frac{3}{10}
\end{aligned}$$

Madeline can be cured in one of two mutually exclusive ways:

1. Bottle B_1 is selected and a pill from it is effective.
2. Bottle B_2 is selected and a pill from it is effective.

Thus, using the addition rule for mutually exclusive events (Formula 6.5), we have

$$\begin{aligned}
&\text{Prob(pills are effective)}\\
&\quad = \text{Prob(selects } B_1 \text{ and pills are effective)}\\
&\qquad + \text{Prob(selects } B_2 \text{ and pills are effective)}
\end{aligned}$$

Symbolically,

$$\begin{aligned}
\text{Prob(pills are effective)} &= \text{Prob}(A \text{ and } B_1) + \text{Prob}(A \text{ and } B_2)\\
\text{Prob}(A) &= \text{Prob}(A \mid B_1)\cdot\text{Prob}(B_1) + \text{Prob}(A \mid B_2)\cdot\text{Prob}(B_2)\\
&= \frac{7}{15} + \frac{3}{10}\\
&= \frac{14}{30} + \frac{9}{30} = \frac{23}{30}
\end{aligned}$$

Therefore, the probability that the pills are effective in relieving Madeline of her pain is $\frac{23}{30}$.

Example 36 Two weeks later Madeline returns to Dr. Carey and reports that the pills were extremely effective. Dr. Carey would now like to recommend the same medicine for his other patients who suffer from the same pain. What is the probability that the pills came from:

a) Bottle B_1
b) Bottle B_2

Solution a) Using the conditional probability formula (Formula 6.7), we have

$$\text{Prob}(\text{pills came from } B_1 \mid \text{pills are effective}) = \frac{\text{Prob}(\text{selects } B_1 \text{ and pills are effective})}{\text{Prob}(\text{pills are effective})}$$

Symbolically

$$\text{Prob}(B_1 \mid A) = \frac{\text{Prob}(A \text{ and } B_1)}{\text{Prob}(A)}$$

Substituting the values obtained in Example 35 gives

$$\text{Prob}(B_1 \mid A) = \frac{7/15}{23/30} = \frac{7}{15} \div \frac{23}{30} = \frac{7}{15} \cdot \frac{30}{23} = \frac{14}{23}$$

Thus, the probability that the pills came from bottle B_1 is $\frac{14}{23}$.

b) To find the probability that the pills came from bottle B_2, we use a procedure similar to the one used in part a. We have

$$\text{Prob}(\text{pills came from } B_2 \mid \text{pills are effective}) = \frac{\text{Prob}(\text{selects } B_2 \text{ and pills are effective})}{\text{Prob}(\text{pills are effective})}$$

In symbols,

$$\text{Prob}(B_2 \mid A) = \frac{\text{Prob}(A \text{ and } B_2)}{\text{Prob}(A)} = \frac{3/10}{23/30} = \frac{3}{10} \div \frac{23}{30} = \frac{3}{10} \cdot \frac{30}{23} = \frac{9}{23}$$

Thus, the probability that the pills came from bottle B_2 is $\frac{9}{23}$.

We can combine the results of the preceding two examples as follows:

$$\text{Prob}(B_1 | A) = \frac{\text{Prob}(A \text{ and } B_1)}{\text{Prob}(A)}$$

$$= \frac{\text{Prob}(A \text{ and } B_1)}{\text{Prob}(A \text{ and } B_1) + \text{Prob}(A \text{ and } B_2)}$$

$$= \frac{\text{Prob}(A | B_1) \cdot \text{Prob}(B_1)}{\text{Prob}(A | B_1) \cdot \text{Prob}(B_1) + \text{Prob}(A | B_2) \cdot \text{Prob}(B_2)}$$

$$\text{Prob}(B_2 | A) = \frac{\text{Prob}(A \text{ and } B_2)}{\text{Prob}(A)}$$

$$= \frac{\text{Prob}(A \text{ and } B_2)}{\text{Prob}(A \text{ and } B_1) + \text{Prob}(A \text{ and } B_2)}$$

$$= \frac{\text{Prob}(A | B_2) \cdot \text{Prob}(B_2)}{\text{Prob}(A | B_1) \cdot \text{Prob}(B_1) + \text{Prob}(A | B_2) \cdot \text{Prob}(B_2)}$$

When these results are generalized we have Bayes' rule.

Formula 6.10 *Bayes' rule.* Consider a sample space that is composed of the mutually exclusive events A_1, A_2, $A_3, \ldots, A_n$. Suppose each event has a nonzero probability of occurring and that one must definitely occur. If B is any event in the sample space, then

$$\text{Prob}(A_1|B) = \frac{\text{Prob}(B|A_1)\cdot\text{Prob}(A_1)}{\text{Prob}(B|A_1)\cdot\text{Prob}(A_1) + \text{Prob}(B|A_2)\cdot\text{Prob}(A_2) + \cdots + \text{Prob}(B|A_n)\cdot\text{Prob}(A_n)}$$

$$\text{Prob}(A_2|B) = \frac{\text{Prob}(B|A_2)\cdot\text{Prob}(A_2)}{\text{Prob}(B|A_1)\cdot\text{Prob}(A_1) + \text{Prob}(B|A_2)\cdot\text{Prob}(A_2) + \cdots + \text{Prob}(B|A_n)\cdot\text{Prob}(A_n)}$$

$$\vdots \qquad\qquad \vdots$$

$$\text{Prob}(A_n|B) = \frac{\text{Prob}(B|A_n)\cdot\text{Prob}(A_n)}{\text{Prob}(B|A_1)\cdot\text{Prob}(A_1) + \text{Prob}(B|A_2)\cdot\text{Prob}(A_2) + \cdots + \text{Prob}(B|A_n)\cdot\text{Prob}(A_n)}$$

Formula 6.10 is simple to use, as the following examples illustrate.

Example 37 *Smoke detectors.* A large real estate manager purchased 50,000 smoke detectors to comply with new city ordinances. She purchased 25,000 from company A, 15,000 from company B, and 10,000 from company C. It is known that some of the smoke detectors malfunction and go off spontaneously. It is also known that 4% of the detectors produced by company A are defective, 5% of the detectors produced by company B are defective, and 6% of the detectors produced by company C are defective. A call is received by the management office that one of the detectors is malfunctioning. Find the probability that it was produced by manufacturer A.

Solution From the given information,

$$\text{Prob(produced by company A)} = \frac{25{,}000}{50{,}000} = .5$$

$$\text{Prob(produced by company B)} = \frac{15{,}000}{50{,}000} = .3$$

$$\text{Prob(produced by company C)} = \frac{10{,}000}{50{,}000} = .2$$

$$\text{Prob(defective} \mid \text{produced by company A)} = .04$$
$$\text{Prob(defective} \mid \text{produced by company B)} = .05$$
$$\text{Prob(defective} \mid \text{produced by company C)} = .06$$

Here we must determine Prob(produced by company $A \mid$ defective). We apply Bayes' rule:

$$\text{Prob (produced by A} \mid \text{defective)}$$

$$= \frac{\text{Prob (defective} \mid \text{prod. by A)} \cdot \text{Prob (prod. by A)}}{\text{Prob (def.} \mid \text{prod. by A)} \cdot \text{Prob (prod. by A)} + \text{Prob (def.} \mid \text{prod. by B)} \cdot \text{Prob (prod. by B)} + \text{Prob (def.} \mid \text{prod. by C)} \cdot \text{Prob (prod. by C)}}$$

$$= \frac{(.40)(.5)}{(.40)(.5) + (.05)(.3) + (.06)(.2)}$$

$$= \frac{.02}{.02 + .015 + .012} = \frac{.02}{.047}$$

$$= \frac{20}{47}$$

Thus, the probability that the defective smoke detector was produced by company A is $\frac{20}{47}$.

Example 38 *Measuring the effectiveness of new educational techniques.* All students at the College of Staten Island must pass a Board of Higher Education (BHE) math proficiency exam to graduate. The records of the math department indicate that only 60% of the students pass the BHE proficiency exam on their first try. In an effort to increase the percentage of students who pass the exam, the math department has set up special proficiency courses that students can take. After completing the course, a student must pass a qualifying exam. It is known that 95% of the students who can pass the BHE exam can also pass the qualifying exam, whereas only 20% of the students who fail the BHE exam can pass the qualifying exam. The administration of the college is interested in knowing the effectiveness of this new approach. If a student who passes the qualifying exam is randomly selected, what is the probability that the student will also pass the BHE exam? That is, find the probability that the program is a success in increasing the percentage of students who pass the BHE exam.

Solution We use Bayes' rule. We are interested in finding Prob(pass BHE exam | pass qualifying exam). From the given information:

$$\text{Prob(pass qualifying exam} \mid \text{pass BHE exam)} = .95$$
$$\text{Prob(pass BHE exam)} = .60$$
$$\text{Prob(pass qualifying exam} \mid \text{fail BHE exam)} = .20$$
$$\text{Prob(fail BHE exam)} = .40$$

Then

$$\text{Prob (pass BHE exam} \mid \text{pass qualifying exam)}$$
$$= \frac{\text{Prob (pass qualifying exam} \mid \text{pass BHE exam)} \cdot \text{Prob (pass BHE exam)}}{\text{Prob (pass qualifying exam} \mid \text{pass BHE exam)} \cdot \text{Prob (pass BHE exam)} + \text{Prob (pass qualifying exam} \mid \text{fail BHE exam)} \cdot \text{Prob (fail BHE exam)}}$$
$$= \frac{(.95)(.60)}{(.95)(.60) + (.20)(.40)} = \frac{.57}{.57 + .08} = \frac{.57}{.65}$$
$$\approx .8769$$

Thus, the probability is .8769 that the student will pass the BHE exam. Equivalently, this means that approximately 87.69% of the students who pass the qualifying exam can be expected to pass the BHE exam, as opposed to 60% of the students who passed the BHE exam prior to this new program. (In practice, other factors should also be considered before one arrives at such a conclusion. There may simply be a high correlation between the two tests.)

Exercises

83. *Tracking an escaped prisoner.* A prisoner has just escaped from jail. There are three roads leading away from the jail. If the prisoner selects road A to make good his escape, the probability that he succeeds is $\frac{1}{8}$. If he selects road B, the probability is $\frac{1}{9}$. If he selects road C, the probability that he succeeds is $\frac{1}{10}$. Furthermore, the probability that he selects each of these roads is the same; it is $\frac{1}{3}$. If the prisoner succeeds in his escape, what is the probability that he made good his escape by using road C?

84. *Medicine.* The American Cancer Society as well as the medical profession recommends that people have themselves checked annually for any cancerous growths. If a person has cancer, then the probability is .99 that it will be detected by a test. Furthermore, the probability that the test results will be positive (meaning that cancer is possible) when no cancer actually exists is .10. Government records indicate that 8% of the population in the vicinity of a chemical corporation that produces asbestos have some form of cancer. Donald Williams takes the test and the results are positive. What is the probability that he does *not* have cancer?

85. A large supermarket received two milk deliveries yesterday. In the morning shipment 30 of the 1000 containers delivered were spoiled. In the afternoon shipment 40 of the 600 containers delivered were spoiled. All the milk that is delivered is immediately placed on the shelves in a random manner. Before any other customer purchases any milk from either delivery, Ginette purchases a container of milk from the supermarket and discovers that the milk is sour. What is the probability that Ginette's sour milk came from the morning shipment?

86. A doctor receives 10,000 hypodermic needles a year, 3,000 from company A, 4,000 from company B, 2,000 from company C, and 1,000 from company D. Based on past experience, the doctor knows that 7% of company A's needles are defective, 6% of company B's needles are defective, 8% of company C's needles are defective, and 9% of company D's needles are defective. If a defective needle is found, what is the probability that it was produced by company A?

87. A speech clinic is experimenting with three different approaches, A, B, and C, to cure a particular speech defect. Thus far, method A has been successful with 10 out of 15 patients, method B has been effective with 12 out of 18 patients, and method C has been successful with 8 out of 15 patients. If a little boy who had a speech defect was cured, what is the probability that treatment A was used? (Assume that each method is equally likely to be selected to treat a patient.)

88. A gambler has two dice in his hand. One of them is an ordinary die and the other die has three dots painted on each of its six faces. One of the dice is selected at random and rolled twice. Both times the face with three dots on it comes up. What is the probability that the altered die was used?

89. An author who lives on the East Coast must send a manuscript to a publisher on the West Coast. The manuscript can be sent by any one of four delivery services, A, B, C, or D. The probabilities that the manuscript will arrive at its destination safely by 9:00 A.M. on the next business day when shipped by these four delivery services are .8, .9, .7, and .85, respectively. The author randomly selects a delivery service. If the manuscript *does not* arrive safely at its destination by 9:00 A.M on the next business day, what is the probability that delivery service B was used?

90. In Exercise 89 if the manuscript does arrive safely, find the probability that delivery service D was used.

91. A psychologist is conducting experiments with rats. A rat can enter maze A, maze B, or maze C with probabilities .5, .4, and .1, respectively. Furthermore, the probability that the rat finds its way through each maze is .7, .2, and .1, respectively. If a rat actually finds its way through a maze, what is the probability that it went through maze B?

★92. *Ecology*. Environmentalists have developed a test for determining when the mercury level in fish is above the permissible levels. If the fish actually contain an excessive amount of mercury, then the test is 98% effective in determining this, and only 2% will escape undetected. On the other hand, if the mercury content is within permissible limits, then the test will correctly indicate this 95% of the time. Only 5% of the time will the test incorrectly indicate that the mercury content is not within permissible limits. The test is to be used on fish from a river into which a chemical company has been dumping its wastes. It is estimated that 20% of the fish in the river contain excessive amounts of mercury. A fish is caught and tested by this procedure. The test indicates that the mercury level is within permissible limits. Find the probability that the mercury content is actually greater than the permissible levels.

6.8 The Binomial Probability Distribution and Applications[3]

Consider Felicia who is a quality control engineer for a large electronics company. Her job is to test thoroughly each stereo set manufactured by the company and to classify it as acceptable or unacceptable. Due to the company's rigorous quality standards, each set has an equally likely chance of being acceptable or unacceptable. On a given morning Felicia inspects five sets. Suppose we are interested in determining the probability that she (1) accepts all the sets, (2) rejects all the sets, or (3) accepts three of the five sets. How should we proceed?

There are many probability problems with outcomes that can be grouped into one of two categories, acceptable or unacceptable. An experiment that has only two possible outcomes is referred to as a *binomial probability experiment* or a *Bernoulli experiment* in honor of the mathematician Jacob Bernoulli (1654–1705), who studied it in detail. His contributions to probability are contained in his book *Ars Conjectandi* published after his death in 1713.

Definition 6.6 A *binomial probability experiment* is an experiment that satisfies the following properties:

1. There are n independent repeated trials of the experiment.
2. Each trial results in one of two possible outcomes. We call one outcome a success and denote it by the letter S and the other outcome a failure, denoted by F.
3. The probability of success on a single trial is p and remains the same from trial to trial. The probability of failure is $1 - p = q$. Symbolically, Prob(success) $= p$ and Prob(failure) $= q = 1 - p$.
4. We are interested in the number of successes obtained in n trials of the experiment.

[3] This section can best be understood and studied if you use a hand-held calculator.

Let us return to the problem posed at the beginning of this section. We have a binomial probability experiment and are interested in determining different probabilities. At the moment the only way to calculate these probabilities is to apply the basic definition of probability. We must list all the possible outcomes and then determine how many of them are favorable. In this case there are two possible outcomes for stereo set 1, either it is accepted, a, or it is not accepted and is rejected, r. Similarly, there are two possible outcomes for set 2 regardless of whether set 1 was accepted or rejected. The same is true for sets 3, 4, and 5. Thus, there are $2 \cdot 2 \cdot 2 \cdot 2 \cdot 2$ or 32 possible outcomes, which we list as follows:

aaaaa	araaa	raaaa	rraaa
aaaar	araar	raaar	rraar
aaara	arara	raara	rrara
aaarr	ararr	raarr	rrarr
aaraa	arraa	raraa	rrraa
aarra	arrar	rarar	rrrar
aarar	arrra	rarra	rrrra
aarrr	arrrr	rarrr	rrrrr

We can summarize these outcomes by using a table similar to Table 6.2.

Table 6.2 The Possible Outcomes Determine the Number of Sets Accepted by Felicia

No sets accepted	One set accepted	Two sets accepted	Three sets accepted	Four sets accepted	Five sets accepted
rrrrr	arrrr	aarrr	rraaa	raaaa	aaaaa
	rarrr	raarr	arraa	araaa	
	rrarr	rraar	aarra	aaraa	
	rrrar	rrraa	aaarr	aaara	
	rrrra	arrra	raaar	aaaar	
		ararr	raraa		
		arrar	raara		
		rarar	arara		
		rarra	araar		
		rrara	aarar		

Now we can determine the desired probabilities. We have

$$\text{Prob(Felicia accepts all the sets)} = \frac{1}{32}$$

$$\text{Prob(Felicia rejects all the sets)} = \frac{1}{32}$$

$$\text{Prob(Felicia accepts three of the five sets)} = \frac{10}{32} = \frac{5}{16}$$

Must we go through such calculations in order to determine probabilities when we have a binomial probability experiment? Fortunately, the answer is no. Since all binomial probability experiments are similar in nature and result in either success or failure for each trial of the experiment, we need a formula to give us the probability of obtaining x successes out of n trials of the experiment where the probability of success on any one trial of the experiment is p and the probability of failure is q.

To arrive at such a formula, let us return to Felicia who inspects the stereo sets. Suppose she inspects five sets and we wish to determine the probability that she accepts two of them. Immediately we realize that each set can be accepted or rejected with equal probability as specified in the problem. Thus, Prob(set is accepted) $= \frac{1}{2}$ and Prob (set is rejected) $= \frac{1}{2}$. Since we are interested in only the sets that are accepted, we classify this outcome as a success and write

$$\text{Prob(accepted)} = \text{Prob(success)} = \frac{1}{2}$$

$$\text{Prob(rejected)} = \text{Prob(failure)} = \frac{1}{2}$$

Therefore, in symbols,

$$p = \frac{1}{2} \qquad q = \frac{1}{2} \qquad p + q = \frac{1}{2} + \frac{1}{2} = 1$$

Let us consider one particular way in which we can obtain the two out of five accepted sets. It is given in the table along with the associated probabilities.

Outcome	Accepted	Accepted	Rejected	Rejected	Rejected
Success or failure	Success	Success	Failure	Failure	Failure
Probability	$\frac{1}{2}$	$\frac{1}{2}$	$\frac{1}{2}$	$\frac{1}{2}$	$\frac{1}{2}$
Symbol	p	p	q	q	q

Thus, the probability that this particular event will happen is $p \cdot p \cdot q \cdot q \cdot q$, or p^2q^3. Equivalently, it is

$$\left(\frac{1}{2}\right)^2\left(\frac{1}{2}\right)^3 = \left(\frac{1}{2}\right)\left(\frac{1}{2}\right)\left(\frac{1}{2}\right)\left(\frac{1}{2}\right)\left(\frac{1}{2}\right) = \frac{1}{32}$$

So far we have neglected to consider the possibility that the two accepted stereo sets do not necessarily have to be the first two that are inspected. Table 6.2 indicates that there are ten different ways in which we can obtain two acceptable sets. Thus, we must multiply $\frac{1}{32}$ by 10, getting $\frac{10}{32}$ or $\frac{5}{16}$ for the probability of obtaining two acceptable sets out of a possible five sets. This is exactly the same answer we obtained by actually listing all the possible outcomes.

Notice that in determining how many possible ways there are to obtain two out of the five acceptable sets, we are actually computing the number of combinations of 5 things taken 2 at a time or ${}_5C_2$. Using Formula 6.4, we get

$${}_5C_2 = \frac{5!}{2!(5-2)!} = \frac{5!}{2!3!} = 10$$

Thus, the probability of obtaining 2 successes out of 5 trials of an experiment is given by ${}_5C_2 \cdot p^2q^3$.

More generally, if we are interested in the probability of getting x successes out of n trials of an experiment, then we consider one way in which this can happen as shown:

$$\begin{array}{lll} \text{Outcome} & \text{S S S S S} & \text{F F F F} \\ \text{Probability:} & \underbrace{p \cdot p \cdot p \cdot p \cdot p}_{x \text{ of them}} \cdot \ldots & \cdot \underbrace{q \cdot q \cdot q \cdot q}_{n-x \text{ of them}} \end{array}$$

In this particular case we have assumed (as in the previous example) that all x successes occur first, after which $n - x$ failures occur. This gives p^xq^{n-x}.

We have to multiply this answer by the number of ways in which we can obtain x successes out of n trials of an experiment. Of course, this is ${}_nC_x$, which equals

$${}_nC_x = \frac{n!}{x!(n-x)!}$$

This leads us to the following binomial distribution formula.

Formula 6.11 *Binomial distribution formula.* Consider a binomial experiment that has two possible outcomes, success or failure. Let Prob (success) = p and Prob (failure) = q. If this experiment is repeated n times, then the probability of obtaining x successes out of n trials is

$$\text{Prob } (x \text{ successes}) = \frac{n!}{x!(n-x)!} \cdot p^xq^{n-x}$$

Formula 6.11 is easy to use, as the following examples illustrate.

Example 39 *Admission to medical school.* Ninety percent of the graduates of State University are admitted to a particular medical school. This year 6 graduates from State University have applied for admission to the medical school. Find the probability that only 4 of them will be accepted.

Solution Since 6 students have applied for admission, $n = 6$. We are interested in the probability that 4 are accepted, so $x = 4$. Also, 90% of the graduates are admitted, so $p = .90$, and 10% of the graduates are not admitted, so $q = .10$. Now we apply Formula 6.11. We have

$$\begin{aligned} \text{Prob}(4 \text{ are accepted}) &= \text{Prob}(x = 4) \\ &= \frac{6!}{4!(6-4)!}(.90)^4(.10)^{6-4} \\ &= \frac{6!}{4!2!}(.9)^4(.1)^2 \\ &= \frac{6 \cdot 5 \cdot 4 \cdot 3 \cdot 2 \cdot 1}{4 \cdot 3 \cdot 2 \cdot 1 \cdot 2 \cdot 1}(.9)(.9)(.9)(.9)(.1)(.1) \\ &= .0984 \end{aligned}$$

Thus, the probability that only 4 of them will be accepted is .0984.

Example 40 *On-time arrivals.* An airline company claims that 98% of its scheduled plane flights arrive on time. This Sunday the airline has 10 flights scheduled. Find the probability that 8 of these flights arrive on time.

Solution Since there are 10 flights scheduled for Sunday, $n = 10$. We are interested in the probability that 8 of these flights arrive on time, so that $x = 8$. Also 98% of the planes arrive on time, so that $p = .98$ and $q = .02$. Applying Formula 6.11 gives

$$\begin{aligned} \text{Prob}(8 \text{ flights arrive on time}) &= \text{Prob}(x = 8) \\ &= \frac{10!}{8!(10-8)!}(.98)^8(.02)^{10-8} \\ &= \frac{10!}{8!2!}(.98)^8(.02)^2 \\ &= .0153 \end{aligned}$$

Thus, the probability that 8 of the 10 flights arrive on time is .0153.

Example 41 Consider the accompanying newspaper article. Bill and his friends own 5 cars, which they park in front of his house. Find the probability that none of them will be stolen this year.

Car Thefts on the Rise Again

Washington (April 7)—Look out your window. Is your car still in your driveway or in front of your house? If it is there, then you're lucky. According to the latest FBI study released yesterday, there are an average of 2300 vehicles stolen per day in the United States. This puts the chances of your car being stolen at about 1 in 120. According to the survey, about 60% of the vehicles are stolen from private residences, apartments, or streets in residential areas between the hours of 6.00 P.M. and 6.00 A.M.

Most of the cars are stolen to be stripped for their parts or to be used for joyriding.

Solution Since Bill and his friends own 5 cars, $n = 5$. We are interested in the probability that none of the cars will be stolen, so that $x = 0$. According to the newspaper article, 1 out of every 120 cars is stolen, so that $p = \frac{1}{120} = .008$ and $q = \frac{119}{120} = .992$. Applying Formula 6.11 gives

$$\text{Prob}(0 \text{ stolen cars}) = \text{Prob}(x = 0) = \frac{5!}{0!5!}(.008)^0(.992)^5$$

Remember that $(.008)^0 = 1$, so that

$$\text{Prob}(0 \text{ stolen cars}) = 1(.992)^5 = .9606$$

Thus, the probability that none of these cars will be stolen this year is .9606.

Example 42 Consider the accompanying newspaper article. If 10 people are randomly selected, find the probability that *at least* 8 of them believe nuclear energy should *not* be abandoned.

According to Our Poll . . . Sixty-five Percent Say Nuclear Energy Should Not Be Abandoned

Washington (April 12)—In the wake of the recent nuclear mishap at Three Mile Island in Pennsylvania, the *Daily Press* conducted a nationwide survey to determine the people's grass-roots sentiment about nuclear energy. The survey reported that 65% of those responding said that nuclear energy should not be abandoned but rather that more safeguards should be found to protect the public.

Solution Since 10 people are selected, $n = 10$. Here we are interested in the probability that *at least* 8 of them believe that nuclear energy should not be abandoned. Thus, we must calculate Prob$(x = 8)$, then Prob$(x = 9)$, and finally Prob$(x = 10)$. In each case, $p = .65$ and $q = .35$ We have

$$\text{Prob}(x = 8) = \frac{10!}{8!2!}(.65)^8(.35)^2 = .1757$$

$$\text{Prob}(x = 9) = \frac{10!}{9!1!}(.65)^9(.35)^1 = .0725$$

$$\text{Prob}(x = 10) = \frac{10!}{10!0!}(.65)^{10}(.35)^0 = .0135$$

Now we add these probabilities (why?), getting

$$\text{Prob}\begin{pmatrix}\text{at least 8 of them believe nuclear}\\ \text{energy should not be abandoned}\end{pmatrix}$$
$$= .1757 + .0725 + .0135 = .2617$$

Comment Calculating binomial probabilities can sometimes be quite time-consuming even when a calculator is available. To make the job a little easier, we can use Table 2 in the appendix, which gives the probabilities for different values of n, x, and p. No computations are needed. We need to know only the values of n, x, and p.

Comment Whenever there is a blank in the binomial probability chart, it means that the probability is approximately zero.

Example 43 We are given a binomial distribution with $n = 12$ and $p = .6$. Using the binomial distribution chart (Table 2 in the appendix), find the probability of getting exactly 7 successes.

Solution Using the binomial distribution chart, we first locate $n = 12$ and then move across the top of the chart until we reach the $p = .6$ column. To find the probability of exactly 7 successes, we look for the value given in the table for $x = 7$. It is .227. Thus, when $n = 12$ and $p = .6$, the probability of exactly 7 successes is .227.

Exercises

■ **93.** Forty percent of the blood donated to a hospital on January 12 is type A positive. If 7 pints of blood are randomly selected, find the probability that exactly 2 pints are type A positive.

■ 94. The math department is planning to offer eight sections of Math 100 next semester. However, the teachers assigned to the courses are not posted in advance. Professor Ehlichson always teaches two of these sections. Joanna Snell plans to take Math 100 but does not want to have Professor Ehlichson, who gives very difficult tests. If she randomly registers for one of the sections, find the probability that she does not get Professor Ehlichson.

■ 95. *Illegal parking.* According to personnel records, 60 of the 100 cars parked in the company parking lot are owned by women employees. One day eight cars are towed away because they were parked illegally. Find the probability that five of these cars are owned by women.

■ 96. *Planning the family.* Mr. and Mrs. Washington are planning to have six children. They would like to have four boys and two girls. Find the probability that they get their wish. (Assume that the probability of a child being a boy is $\frac{1}{2}$ and the probability of a girl is $\frac{1}{2}$.)

■ 97. *Trapping.* Harlow Moore is a game warden. He has set eight beaver traps in different sections of a national park. There is an 80% chance that each trap will function properly and catch a beaver. Find the probability that Harlow traps at least six beavers.

■ 98. *Traffic survey.* All cars exiting at Exit 12 of a highway must turn either right or left at the end of the exit ramp. The traffic department estimates that 75% of the cars turn right. To test this claim, eight cars are observed exiting at Exit 12. Find the probability that at most one car turns left.

■ 99. *Botany.* A botanist has developed a new formula that is 90% effective in restoring withered plants to their original healthy state. If nine plants are treated with this new formula, find the probability that at least six of them will be restored to their original health.

■ 100. *Medicine.* A pediatrician finds that approximately 20% of the babies injected with an antipolio vaccine develop some form of reaction. If ten babies are injected with the vaccine, find the probability that at most four of them will develop the reaction.

■ 101. *Tires.* A tire manufacturer claims that 95% of a particular model of tire will last for at least 20,000 miles. Eddie has just purchased a set of six such tires for his pickup truck. Find the probability that all six tires last for at least 20,000 miles.

■ 102. *Winning the teddy bear.* Phil is at an amusement park. At one booth there is a hanging basket. A player is given five balls, which must be tossed into the basket. To win a huge teddy bear as a prize, the player must get at least four balls into the basket. It is estimated that there is a $\frac{2}{3}$ probability that any ball can be tossed into the basket. Find the probability that Phil wins the teddy bear.

■ 103. *Check my oil, please.* Dave Yackel owns a gas station on Interstate 81. He estimates that 12% of the drivers ask to have their oil checked when filling up with gas. If 15 cars are observed filling up at Dave's gas station, find the probability that at most 4 of them ask to have their oil checked.

■ **104.** *Sorry, closed.* The Automobile Club of New York estimated that 90% of the gas stations in the New York City area would be closed on Easter Sunday. There are 12 gas stations in the vicinity of Mabel's house. Find the probability that she finds a gas station that is open.

■ **105.** *Bull's-eye.* A marksman is shooting at a target. The probability that he hits the target is $\frac{7}{8}$. If he shoots at the target eight times, find the probability that he hits the target at least six times.

■ **106.** A baseball player has a .300 batting average. In tomorrow's game the player will be at bat five times. Find the probability that the player gets two hits.

■ **107.** Market research indicates that 75% of all U.S. homes have at least one live plant. Ten homes are randomly selected. Find the probability that there will be *no* live plants in any of these homes.

6.9 Applications of Probability: Odds and Mathematical Expectation

Often when playing various games of chance, we are interested in knowing how much money can be won in the long run from a particular game. This is referred to as the *expectation* or *expected value.* If no money can be won or lost in the long run from a particular game, then we say that it is a *fair game.* Obviously, gambling casinos do not operate fair games or else they would go broke. Such games are in favor of the house and against the player; that is, the player has a negative expected value and the house has a positive expected value. We refer to the amount of money to be won (or lost) as the mathematical expectation of the game.

Definition 6.7 The *mathematical expectation* of a game is the amount of money to be won or lost in the long run on the average.

For any events we have the following formula.

Formula 6.12 Consider any event that has probability p_1 of occurring and has a payoff, that is, an amount won, of m_1. Consider also a second event with probability p_2 and payoff m_2, a third event with probability p_3 and payoff m_3, etc. Then the mathematical expectation of these events is

$$m_1p_1 + m_2p_2 + m_3p_3 + \cdots$$

(The subscripts have no special significance.)

Comment The expected value is not the amount of money to be expected each time the game is played. Rather it is the average expectation over a long period of playing the game.

Example 44 A contractor has submitted bids on two different jobs. If the first bid is accepted, the contractor stands to make a profit of $50,000 with a probability of $\frac{3}{5}$. If the second bid is accepted, the contractor can make a profit of $40,000 with a probability of $\frac{5}{8}$. Find the contractor's mathematical expectation.

Solution Using Formula 6.12, we have that the mathematical expectation is

$$(50{,}000)\left(\frac{3}{5}\right) + (40{,}000)\left(\frac{5}{8}\right) = 30{,}000 + 25{,}000 = 55{,}000$$

Thus, the contractor's mathematical expectation is $55,000.

Example 45 A business executive has invested $50,000 in a new pizza store. There is a $\frac{5}{7}$ chance that the business will be a success, in which case the business can make a profit of $200,000. On the other hand, there is a $\frac{2}{7}$ chance that the business will not succeed, in which case the executive will be forced to sell the equipment and stands to lose $30,000. Find the mathematical expectation.

Solution Using Formula 6.12, with a + sign denoting a gain and a − sign a loss, we have

$$(+200{,}000)\left(\frac{5}{7}\right) + (-30{,}000)\left(\frac{2}{7}\right) = \$142{,}857.14 - \$8{,}571.43$$

$$= \$134{,}285.71$$

Thus, the mathematical expectation is $134,285.71.

Comment If a gambler in Atlantic City were to calculate the expected value from the various slot machines, he would immediately notice that money can only be lost to the house (in the long run) from these machines.

Another idea associated with probability concerns betting. Gamblers frequently speak of the odds against a game. Thus, we often see newspaper articles like the one on the next page. If the probability is $\frac{7}{11}$ that the Yankees win the series, then gamblers would say that the odds are 7:4 in favor of the Yankees. The 7 represents the chances of winning and the 4 represents the chances of not winning.

Yankees Are Favored to Win Series

New York (Sept. 20)—The New York Yankees are favored to win the upcoming baseball world series. Local oddsmakers were betting 7-to-4 odds that the Yankees would win the series in six games.

Definition 6.8 The *odds in favor* of an event occurring are p to q, where p is the number of favorable outcomes and q is the number of unfavorable outcomes. The *odds against* the event occurring are q to p.

Example 46 A card is randomly selected from a deck of cards. What are the odds against getting an ace?

Solution There are 52 cards in the deck; 4 of them are aces and 48 of them are not aces. Thus, the odds against getting an ace are 48 to 4.

Example 47 *Will you be audited by the IRS?* Due to budget cutbacks, the IRS announced that it would be able to audit about 21 of each 1000 1978 federal tax returns filed. What are the odds that your 1978 tax return will be audited?

Solution Out of each 1000 tax returns filed, 21 will be audited and 979 will not be audited. Thus, the odds in favor that you will be audited are 21 to 979.

In Chapter 8 ("Decision Analysis") we will discuss expected value further. Specifically, we will indicate how expected value can be used in the decision-making process.

Exercises

108. A gambler tosses a fair die. If the number of dots that shows is one, four, or five, she wins \$5. If the number of dots showing is three or six, then she loses \$8. If the number of dots showing is two then she wins \$2. What is her mathematical expectation?

109. *Stock investment.* A man invests \$50,000 on the stock market for one week. The particular stock that he bought can go up with a probability of $\frac{5}{9}$, in which case the man will make a profit of \$7,000, or the stock can go down with a probability of $\frac{4}{9}$, in which case the man will lose \$15,000. What is his mathematical expectation?

110. *Perishable fruit.* A fruit dealer receives a shipment of strawberries that must be sold today, otherwise they will have to be thrown away. There is a $\frac{4}{11}$ chance that they will be sold today, in which case the profit will be \$50. There is also a $\frac{7}{11}$ chance that they will not be sold today, in which case the loss will be \$25. Find the dealer's expectation.

111. Bill rolls a die. If an even number comes up, then he will pay his friend \$1 for each dot that comes up. If an odd number comes up, then his friend will pay him \$1 for each dot that comes up. Find Bill's mathematical expectation.

112. A secretary types three letters and addresses three envelopes. Before inserting each letter into the appropriate envelope, she drops them all on the floor. If she picks up all the letters and envelopes and inserts a letter at random (without looking) into an envelope, what are the odds in favor of each letter being inserted in its correct envelope?

113. Louise parks by a meter and does not put in a dime. The probability that she gets a parking ticket is $\frac{4}{7}$. What are the odds against her getting a ticket?

114. The weather forecast for tomorrow is heavy snow with a .90 probability. What are the odds that it does not snow?

115. *Oil exploration.* An oil company is exploring for oil in a particular tract of the Baltimore Canyon on the East Coast. Geologists estimate that there is a probability of only .05 that oil will be found. What are the odds that oil will be found in the area?

116. *Wrong hat.* Along with seven other guests, Isaac places his hat on the table. All the hats look alike. The hostess puts all the hats in the closet. After the party Isaac returns and selects a hat at random. What are the odds against him getting his own hat?

117. A salesman finds that he has to cover five more stores. However, he has time to cover only two of them. His estimate of each store's sales and the probabilities that these sales will be realized is shown in the accompanying table. Using mathematical expectation, help the salesman decide which stores to visit.

Store	Possible sale	Estimated probability that sale will be realized
A	\$3000	.42
B	\$5000	.29
C	\$2000	.60
D	\$6000	.52
E	\$4000	.38

118. In a special lottery 50,000 \$1 tickets are sold. The first prize is \$1000 and the five second-prize winners will share \$1000 equally. Mary buys one ticket for the lottery.

a) What is her mathematical expectation?

b) What are the odds in favor of her being a winner?

c) What are the odds against her being a winner?

119. A medical laboratory receives a varying number of daily requests for urine analysis as shown in the accompanying table. What is the expected number of requests received per day?

Number of requests	Probability
1	.05
2	.43
3	.25
4	.17
5	.06
6	.03
7	.01

120. The traffic department is considering installing a traffic signal at the intersection of Main Street and Broadway because of the numerous accidents reported daily as shown in the accompanying table. What is the expected number of accidents at this intersection?

Number of accidents	Probability
0	$\frac{1}{32}$
1	$\frac{1}{32}$
2	$\frac{9}{32}$
3	$\frac{10}{32}$
4	$\frac{8}{32}$
5	$\frac{3}{32}$

6.10 Further Applications of Probability: Genetics

An important application of probability is in the field of *genetics*. This branch of science is concerned with which traits or characteristics can be inherited. The pioneer in this field was the Austrian botanist and monk Gregor Mendel (1822–1884). He performed many experiments with garden peas and plants in his monastery garden in an attempt to determine which characteristics are passed on from one generation to the next. Much of the modern scientific theory of heredity is based on Mendel's law of heredity.

Specifically, Mendel crossbred plants from wrinkled seeds with plants from smooth seeds. All the resulting plants had smooth seeds. When these new plants were crossbred with one another, however, $\frac{3}{4}$ of the resulting plants had smooth seeds. The remaining $\frac{1}{4}$ of the seeds were wrinkled. He concluded that all traits are determined by *genes*, some of which are *dominant* and others *recessive*. When a gene is dominant, then the trait it represents will always appear no matter what type of gene it is paired with. On the other hand, if a gene is recessive, then the trait it represents will appear only when it is paired with a similar recessive trait. This helped to explain why some peas were smooth and some were wrinkled. The smooth-seed gene is a dominant trait and the wrinkled-seed gene is a recessive trait. When a plant that had both smooth-seed and wrinkled-seed genes was crossbred with a similar plant, then $\frac{3}{4}$ of the resulting plants were pure smooth seeded as shown in the accompanying table. When wrinkled-seed plants were crossbred with smooth-seed plants, all the resulting plants were smooth, since wrinkled seeds is a recessive trait that is dominated by the smooth-seeded plant.

	Smooth seeds	Wrinkled seeds
Smooth seeds	Offspring have smooth seeds	Offspring have smooth seeds
Wrinkled seeds	Offspring have smooth seeds	Offspring have wrinkled seeds

The same analysis was used by Mendel to explain why different flowers of the same plant have different colors. For example, when a red-flowered four o'clock plant carrying a gene R for red was crossbred with another four o'clock plant carrying R, the offspring was RR and always had red flowers. Similarly, when a white-flowered four o'clock plant carrying a gene W for white was crossbred with another four o'clock plant carrying W, the offspring was WW and always had white flowers. On the other hand, when a plant carrying R was crossbred with a plant carrying W, a strange thing happened. The offspring contained the genes RW and was therefore pink, P. Mendel continued with these experiments and recorded the numerical percentages with which characteristics were inherited as shown in the table.

		Genes of one plant		
		Red (RR)	White (WW)	Pink (RW)
			Offspring will be	
	Red (RR)	RR offspring—all red	RW offspring—all pink	RR—50% red RW—50% pink
Genes of other plant	White (WW)	RW offspring—all pink	WW offspring—all white	RW—50% pink WW—50% white
	Pink (RW)	RR—50% red RW—50% pink	RW—50% pink WW—50% white	RR—25% red RW—50% pink WW—25% white

These results further reinforced Mendel's views that certain traits are dominant while others are recessive. The color red in the four o'clock plant is a dominant *trait* and the color white is a recessive *trait*. Using the table, we can establish the following probabilities.

		Probability of offspring's color		
Plant 1	Plant 2	Red	Pink	White
RR	RR	1	0	0
WW	WW	0	0	1
RR	WW	0	1	0
RW	RW	$\frac{1}{4}$	$\frac{1}{2}$	$\frac{1}{4}$
RW	RR	$\frac{1}{2}$	$\frac{1}{2}$	0
RW	WW	0	$\frac{1}{2}$	$\frac{1}{2}$

The same analysis can be applied to human beings. Certain traits are known to be caused by a single gene. For example, eye color and albinism in humans are known to be hereditary and are determined by certain genes, one from each parent.

Example 48 *Albinism.* It is known that in human beings, albinism (no skin color, eye color, or hair color) is an inherited trait that is recessive, and normal color is dominant. Thus, if we let the letter a represent a gene for albinism and N represent a gene for normal color,[4] then a person who has NN genes will have normal color since the person has received normal genes, one from each parent. Similarly, if a person has Na genes, then the person receives a normal gene from one parent and an albino gene from another parent. Since albinism is a recessive trait, the person will be normal. However, if a person has aa genes, then the person will be an albino since the person has received an albino gene from each parent. Martha, who is married to Jack, is pregnant with her first child. Both prospective parents are known to have Na genes. What is the probability that the child will be an albino?

Solution The child's color will be determined by a gene from the father combined with a gene from the mother as shown in the table. The child has four possible gene makeups for albinism.

		Jack: N gene	Jack: a gene
Martha	N gene	Child will be NN—normal	Child will be Na—normal
	a gene	Child will be Na—normal	Child will be aa—albino

[4] It is customary to denote dominant traits by capital letters and recessive traits by lowercase letters.

Only in one of these does albinism appear. Thus, the probability that the child will be an albino is $\frac{1}{4}$.

Example 49 *Genetics and paternity suits.* The inheritance of blood groups has been investigated extensively by biologists. There are four major blood groups in human beings, A, B, AB, and O. A knowledge of the inheritance of blood groups has helped resolve many cases in which a woman falsely accuses a man of fathering her child. The accompanying table indicates how this is done.

Mother's blood type	Child's blood type	Father's blood type cannot be
O, A, or B	O	AB
B	AB	B or O
A	AB	A or O
AB	AB	O
B or O	A	B or O
A or O	B	A or O

As we indicated earlier, certain human characteristics such as red hair are controlled by recessive genes. Red-green color blindness and myopia are controlled by sex-linked genes similar to the gene that causes hemophilia. On the other hand, migraine, Tay-Sachs disease, certain types of hereditary deafness, sickle-cell anemia, eye color, and certain types of feeblemindedness and anemia are caused by recessive genes. By consulting with medical authorities, parents can greatly diminish the probability of these events occurring in their offspring.

Exercises

121. A biologist is experimenting with plants that have the trait T or t. If one plant that has the genes Tt is crossbred with another plant that has the genes Tt, find the probability that the offspring will be the following:
a) Tt b) TT c) tt

122. Suppose the biologist in Exercise 121 crossbred a plant that has the genes Tt with one that has the genes tt. Find the probability that the offspring will be:
a) Tt b) TT c) tt

123. Suppose the biologist in Exercise 121 now crossbreeds a plant that has the genes TT with another that has the genes Tt. Find the probability that the offspring will be:
a) Tt b) TT c) tt

For Exercises 124–126 use the following information: An important human characteristic known to be transmitted by genes is eye color.

It is known that the eye color brown, B, is a dominant trait and the eye color blue, b, is a recessive trait. Thus, a person with BB genes will have brown eyes.

124. Mr. Cardan who has brown eyes, Bb, and Mrs. Cardan who has blue eyes, bb, are planning to have a child. Find the probability that the child will have:
   a) Blue eyes
   b) Brown eyes

125. How are the probabilities in Exercise 124 affected if Mrs. Cardan also has brown, Bb, eyes?

126. Ms. Jones, who is known to have bb genes, gave birth to a child who has Bb genes. What type of eye color genes can Mr. Jones have for him to be the father?

6.11 Summary

In this chapter we introduced some of the basic ideas of probability. Specifically, we gave two possible interpretations of the idea of probability and indicated how we actually calculate the probability of certain events. We defined probability in terms of relative frequency. Thus, we have

$$\text{Prob} = \frac{\text{number of outcomes favoring an event}}{\text{total number of possible outcomes}}$$

We discussed various rules for counting possible outcomes, including tree diagrams, permutations, combinations, Pascal's triangle, and other counting techniques. Permutations are arrangements of objects where order is important, whereas combinations are selections of objects where order is not important.

We presented rules that enabled us to calculate probabilities for mutually exclusive events, independent events, and conditional events.

We also discussed Bayes' rule, which we use when we know the outcome of some experiment and are interested in determining the probability that it was caused by or is the result of one particular event.

Throughout this chapter we applied the ideas of probability to many different situations in addition to the usual gambling problems. Additionally, we applied probability to determine the amount of money to be won or lost and the odds of this happening or not happening. This is called mathematical expectation and odds. Finally we indicated how probability can be used when studying genetics.

Further applications will be given in Chapter 8 when we discuss decision analysis. Specifically, we will indicate how probability is used by the life insurance industry.

Study Guide

You should be able to demonstrate your knowledge of the following ideas presented in this chapter by giving definitions, descriptions, or specific examples. Page references are given for each term so that you can check your answer.

Relative frequency (p. 228)
Sample space (p. 232)
Probability (p. 232)
Null event (p. 236)
Tree diagram (p. 242)
Permutation (p. 248)
Combination (p. 253)
Mutually exclusive events (p. 260)
Conditional probability (p. 267)
Conditional probability formula (p. 268)
Bernoulli experiment (p. 283)
Binomial probability experiment (p. 283)
Expectation (p. 291)
Mathematical expectation (p. 291)
Genes (p. 295)
Recessive gene (p. 295)
Laws of heredity (p. 295)
Favorable outcome (p. 232)
Event (p. 232)
Theoretical frequency (p. 232)
Certain or definite event (p. 237)
Counting rule (p. 243)
Factorial notation (p. 249)
Pascal's triangle (p. 256)
Complement of an event (p. 262)
Addition rule (p. 263)
Multiplication rule (p. 270)
Independent events (p. 271)
Bayes' rule (p. 279)
Binomial distribution formula (p. 286)
Expected value (p. 291)
Fair game (p. 291)
Odds in favor (p. 293)
Odds against (p. 293)
Genetics (p. 295)
Dominant gene (p. 295)
Trait (p. 296)

Formulas to Remember

The following list gives all the formulas studied in this chapter.

$$\text{Probability} = \frac{\text{number of favorable outcomes}}{\text{total number of possible outcomes}} = \frac{f}{n}$$

The number of permutations of n things taken r at a time: ${}_nP_r = \dfrac{n!}{(n-r)!}$

The number of permutations of n things taken n at a time: ${}_nP_n = n!$

The number of combinations of n things taken r at a time:

$${}_nC_r = \frac{n!}{r!(n-r)!}$$

Mathematical expectation of an event: $m_1p_1 + m_2p_2 + m_3p_3 \cdots$

Odds in favor of an event are p to q

Odds against an event are q to p

p = the number of favorable outcomes

q = the number of unfavorable outcomes

Addition rule: Prob(A or B) = Prob(A) + Prob(B) for mutually exclusive events

Addition rule: Prob(A) + Prob(B) − Prob(A and B) for the general case

Conditional probability formula: $\text{Prob}(A|B) = \dfrac{\text{Prob}(A \text{ and } B)}{\text{Prob}(B)}$

Multiplication rule: $\text{Prob}(A \text{ and } B) = \text{Prob}(A|B) \cdot \text{Prob}(B)$

Multiplication rule: $\text{Prob}(A \text{ and } B) = \text{Prob}(A) \cdot \text{Prob}(B)$ for independent events

Bayes' rule:

$$\text{Prob}(A_n|B) = \frac{\text{Prob}(B|A_n) \cdot \text{Prob}(A_n)}{\text{Prob}(B|A_1) \cdot \text{Prob}(A_1) + \text{Prob}(B|A_2) \cdot \text{Prob}(A_2) + \cdots + \text{Prob}(B|A_n) \cdot \text{Prob}(A_n)}$$

Binomial formula: $\text{Prob}(x \text{ successes}) = \dfrac{n!}{x(n-x)!} \cdot p^x q^{n-x}$

Mastery Tests

FORM A

1. What is the probability of drawing a picture card from a deck of cards?
   a) $\frac{12}{52}$ b) $\frac{4}{52}$ c) $\frac{1}{52}$ d) 0
2. A coin is tossed three times. Find the probability of getting three heads?
   a) $\frac{1}{8}$ b) $\frac{1}{6}$ c) $\frac{3}{8}$
   d) $\frac{1}{3}$ e) None of these
3. Which of the following *cannot* be the probability of an event?
   a) .0023 b) $-\frac{2}{3}$ c) 1.00
   d) 0 e) None of these
4. In a binomial probability distribution, the sum of p and q is always
   a) 0 b) 1
   c) .5 d) It depends on the problem.
5. The probability that Mr. Smith has a cavity on any tooth is .4. If he has only 20 teeth, what is the probability that he has exactly 12 cavities?
   a) $\dfrac{20!}{12!}(.4)^{12}(.6)^8$ b) $\dfrac{20!}{8!}(.4)^8(.6)^{12}$ c) $\dfrac{20!}{12!8!}(.4)^{12}(.6)^8$
   d) $\dfrac{20!}{12!8!}(.4)^8(.6)^{12}$ e) None of these
6. The number of permutations of six distinct things taken two at a time is
   a) 120 b) 15 c) 720
   d) 30 e) None of these

7. Evaluate ${}_6P_3$.

   a) 360 b) 720 c) 6
   d) 120 e) None of these

8. The probability that Mary likes her math teacher is .64. What is the probability that she does not like him?

   a) .64 b) .36 c) 1
   d) 0 e) None of these

9. How many different four-person committees can be formed from a possible six people?

   a) 720 b) 360 c) 30
   d) 15 e) None of these

10. How many different outfits can Sharmon put together if she has five blouses, three skirts, and two pairs of shoes?

    a) 30 b) 10 c) 17
    d) 5 e) None of these

11. Five raffles numbered 1 through 5 are placed in a hat. The probability of drawing a 7 is

    a) 1 b) 7 c) 0
    d) $\frac{5}{7}$ e) None of these

12. What is the probability of getting a sum of 6 in tossing two fair dice?

    a) $\frac{1}{6}$ b) $\frac{5}{36}$ c) $\frac{7}{36}$
    d) $\frac{12}{36}$ e) None of these

13. How many numbers larger than 4000 can be formed from the digits 3, 4, 5, and 6 (no repetitions allowed)?

    a) 24 b) 12 c) 18
    d) 6 e) None of these

14. What is the probability that a family that has three children will have *at least* two girls?

    a) $\frac{2}{3}$ b) $\frac{2}{6}$ c) $\frac{2}{8}$
    d) $\frac{3}{8}$ e) None of these

15. A car can be bought with or without air conditioning, with or without automatic transmission, and with one of four different engine sizes. How many different combinations of features can a car have?

    a) 8 b) 24 c) 6
    d) 16 e) None of these

FORM B

1. Read the article "Gambling Doesn't Pay" by Andrew Sterdett in *The Math Teacher* (vol. LX, no. 3, 1967). In this article the author calculates the expected value of various popular gambling games and concludes that gambling does not pay.

2. Is a combination lock really a combination problem? Explain your answer.

3. Scientists have determined that the number of possible genetic makeups of any one child is 2^{48}. If the Joneses have two children, find the probability that both children are exactly alike.

4. Mr. Johnson recently suffered a massive heart attack. He believes that he has a 50% chance of living since there are two possible outcomes, live or die. Do you agree with this reasoning? Explain your answer.

5. If the probability of an event happening is one in a million, is the event unlikely or impossible to happen? Explain.

6. In the baseball world series, the series ends when any one of the teams wins four games. Assume that the New York Yankees are playing in the series. Let the probability that the Yankees win any one game be $\frac{1}{2}$, and assume that this probability remains the same from game to game. Find the probability that the world series will end in:
   a) Four games
   b) Five games
   c) Six games
   d) Seven games

7. Many people believe that an unlucky event occurs when the thirteenth of the month is on a Friday. Furthermore, they believe that this occurs very rarely. To verify this, consider the accompanying table, which indicates the days of the week on which the thirteenth occurred during the past 400 years. Compute the probability that the thirteenth of the month will occur on each day of the week. Which day has the greatest probability of the thirteenth occurring?

Day of the week	Number of times the thirteenth occurred on this day
Sunday	687
Monday	685
Tuesday	685
Wednesday	687
Thursday	684
Friday	688
Saturday	684

8. The following is a list of the birthdays of the presidents of the United States. You will notice that no president was born during June. If a president has an equally likely chance of being born in any of the 12 months, find the probability that a president was born in June. (Base your answer on the data given in the table.)

President	Birthday	President	Birthday
George Washington	Feb. 22, 1732	James Garfield	Nov. 19, 1831
John Adams	Oct. 30, 1735	Chester Arthur	Oct. 5, 1830
Thomas Jefferson	Apr. 13, 1743	Grover Cleveland	Mar. 18, 1837
James Madison	Mar. 16, 1751	Benjamin Harrison	Aug. 20, 1833
James Monroe	Apr. 28, 1758	William McKinley	Jan. 29, 1843
John Quincy Adams	July 11, 1767	Theodore Roosevelt	Oct. 27, 1858
Andrew Jackson	Mar. 15, 1767	William Taft	Sept. 15, 1857
Martin Van Buren	Dec. 5, 1782	Woodrow Wilson	Dec. 28, 1857
William H. Harrison	Feb. 9, 1773	Warren Harding	Nov. 2, 1865
John Tyler	Mar. 29, 1790	Calvin Coolidge	July 4, 1872
James Knox Polk	Nov. 2, 1795	Herbert Hoover	Aug. 10, 1874
Zachary Taylor	Nov. 24, 1784	Franklin D. Roosevelt	Jan. 30, 1882
Millard Fillmore	Jan. 7, 1800	Harry S. Truman	May 8, 1884
Franklin Pierce	Nov. 23, 1804	Dwight D. Eisenhower	Oct. 14, 1890
James Buchanan	Apr. 23, 1791	John F. Kennedy	May 29, 1917
Abraham Lincoln	Feb. 12, 1809	Lyndon B. Johnson	Aug. 27, 1908
Andrew Johnson	Dec. 29, 1808	Richard M. Nixon	Jan. 9, 1913
Ulysses S. Grant	Apr. 27, 1822	Gerald Ford	July 14, 1913
Rutherford Hayes	Oct. 4, 1822	Jimmy Carter	Oct. 1, 1924

9. According to police statistics, there were only 9 accidents reported last year during ice storms on a particular road. Furthermore, there were 87 accidents reported on this road on clear, sunny days. Does this mean that it is safer to drive on this road during an ice storm? Explain your answer.

10. *Law.* In June 1964 an elderly woman was mugged in San Pedro, California. In the vicinity of the crime a bearded, black man sat waiting in a yellow car. Shortly after the crime was committed, a young white woman, wearing her blonde hair in a ponytail, was seen running from the scene of the crime and getting into the car, which sped off. The police broadcast a description of the suspected muggers. Soon afterward a couple fitting the description was arrested and convicted of the crime. Although the evidence in the case was largely circumstantial, the prosecutor based his case on probability and the unlikelihood of another couple having such characteristics. He assumed the probabilities shown in the accompanying table.

Characteristic	Assumed probability
Drives yellow car	$\frac{1}{10}$
Black-white couple	$\frac{1}{1000}$
Black man	$\frac{1}{3}$
Man with bear	$\frac{1}{10}$
Blonde woman	$\frac{1}{4}$
Woman with ponytail	$\frac{1}{10}$

He then multiplied the individual probabilities:

$$\left(\frac{1}{10}\right)\left(\frac{1}{1000}\right)\left(\frac{1}{3}\right)\left(\frac{1}{10}\right)\left(\frac{1}{4}\right)\left(\frac{1}{10}\right)=\frac{1}{12{,}000{,}000}$$

He claimed that the probability is 1/12,000,000 that another couple has such characteristics. The jury agreed and convicted the couple. The conviction was overturned by the California Supreme Court in 1968. The defense attorneys got some professional advice on probability. Serious errors were found in the prosecutor's probability calculations. Some of these involved assumptions about independent events. As a matter of fact, it was demonstrated that the probability is .41 that another couple with the same characteristics existed in the area once it was known that there was at least one such couple.[5]

Suggested Further Reading

Adler, I., *Probability and Statistics for Everyman*, New York: New American Library, 1966.

Bellman, R., and D. Blackwell, "Red Dog Blackjack Poker," *Scientific American*, vol. 184, pp. 44–47, 1951.

Bergamini, S., et al., *Life*, *Mathematics* (*Life* Science Library), New York: Time-Life Books, 1970.

David, F. N., *Games, Gods, and Gambling*, New York: Macmillan (Hafner Press), 1962.

Derman, Cyrus, Glaser Leslie, and Olkin Ingram, *A Guide to Probability and Its Applications*, New York: Holt, Rinehart and Winston, 1973.

Epstein, R. A., *Theory of Gambling and Statistical Logic*, New York: Academic, 1967.

Feller, W., *An Introduction to Probability Theory and Its Applications*, New York: Wiley, 1950.

Gallup, G., *The Sophisticated Poll Watcher's Guide*, Princeton, N.J.: Princeton Opinion Press, 1972.

Gardner, M., *New Mathematical Diversions from the Scientific American*, New York: Simon & Schuster, 1966.

Haverman, E., "Wonderful Wizard of Odds," *Life*, vol. 51, no. 14, p. 30ff., 1961.

Huff, Darrell, *How to Take a Chance*, New York: Norton, 1959.

Kac, M., "Probability," *Scientific American*, September 1964.

Kalb, M., and B. Kalb, "Twenty Days in October," *New York Times Magazine*, June 23, 1974.

Mathematics in the Modern World (readings from *Scientific American*), San Francisco: Freeman, 1968.

Newmark, Joseph, *Statistics and Probability in Modern Life*, 2d ed., New York: Holt, Rinehart and Winston, 1977.

Niven, Ivan, *Mathematics of Choices*, New York: Random House, 1965.

Ore, Øystern, *Cardano, The Gambling Scholar*, New York: Dover, 1965.

Parzen, Emanuel, *Modern Probability Theory and Its Applications*, New York: Wiley, 1960.

[5] For a complete discussion of this probability case, read "Trial by Mathematics" in *Time* magazine (January 8, 1965, p. 42; and April 26, 1968, p. 41).

Polya, G., *Mathematics and Plausible Reasoning*, Princeton, N.J.: Princeton University Press, 1957.

Shaply, L. S., and M. Shubik, "A Method for Evaluating the Distribution of Power in a Committee System," *American Political Science Review*, vol. 48, pp. 787–792, 1954.

Weaver, W., *Lady Luck*, New York: Anchor Books, 1963.

The following articles, which all discuss some aspects of probability and its applications, appear in James R. Newman, ed., *The World of Mathematics*, New York: Simon & Schuster, 1956.

Haldane, J. B. S., "The Mathematics of Natural Selection," vol. 2, pt. 5, article 14.

Haldane, J. B. S., "On Being the Right Size," vol. 2, pt. 5, article 15.

Keynes, J. M., "The Application of Probability to Conduct," vol. 2, pt. 5, article 4.

Laplace, P. S., "Concerning Probability," vol. 2, pt. 7, article 1.

CHAPTER 7

Game Theory

Objectives to be Achieved in this Chapter

- *To study* games involving two players in which there is no kitty or bank. What one player wins the other player loses. These are two-person zero-sum games.
- *To present* a tabular way of indicating what each player can win or lose as a result of any move. This is the payoff matrix.
- *To analyze* the best moves or optimal strategies for each player. A best move is a move that results in the most money won or the least money lost.
- *To compute* a number that enables us to determine whether a game is fair or not.
- *To indicate* how to find optimal strategies for strictly determined games, that is, games that have saddle points, and for games in which mixed strategies are needed, that is, where several moves have to be alternated with a certain probability.
- *To apply* linear programming techniques to obtain optimal strategies.
- *To point out* how the ideas of game theory can be applied to a variety of situations in the social and management sciences.

THE PRESS SEPTEMBER 18, 1978

Computer: "I Give Up"

Toronto (Sept.)—David Levy, an international chess master from Great Britain, successfully won his 1968 bet that no chess computer would beat him in the next ten years.

A Control Data Corporation CYBER 176 computer, using a program called Chess 4.7 and located in Arden Hills, Minnesota was matched against David Levy, who sat in an air-conditioned booth in Toronto. The computer was programmed to analyze any new moves and to print out its answer on a TV screen. A playing board was arranged in front of Levy with lights flashing under the piece to be moved and under the position to which it should be moved.

After game 4 of the match, which took place over the Labor Day weekend, the score stood at $2\frac{1}{2}$–$1\frac{1}{2}$ and the outcome was in doubt. In the fifth and deciding game, Levy took no chances. The match ended $3\frac{1}{2}$–$1\frac{1}{2}$ in favor of Levy.

Within the past ten years, the idea of playing a serious game of chess with a computer has come a long way. Today computers that are capable of playing chess at all levels are readily available on the market.

The game of chess has been played for hundreds of years. What is most fascinating about the game is the staggering number of possible strategies that have to be analyzed in relation to any move. With the advent of the high-speed computer, the task was considerably simplified and the idea of chess computers arose. In 1949 Claude Shannon, a U.S. programmer, described how a computer could be programmed to play chess. In 1956 A. M. Turing designed a chess-playing machine called MADAM. Unfortunately his machine, which used fewer pieces and a smaller board than the standard game, was a poor chess player and made foolish moves. After only several moves, the machine would give up.

Despite this disheartening start, further improvements were added. Today computers are capable of playing a fairly advanced game of chess by analyzing the possible moves and their consequences. By doing this, the computer arrives at an optimal strategy (best move). Today's computers can beat many "good" players. As the newspaper article indicates, David Levy challenged a chess computer programmed with Chess 4.7. Levy had indicated in 1968 that no computer would beat him in the next ten years. Chess 4.7 almost did.

There are many games on the market that follow a pattern similar to the Chess 4.7 computer. Among such games are electronic backgammon, checkers, poker, and football simulation. The number of television games on the market is also growing constantly. In each case all the possible moves are analyzed; then the game or gadget is programmed with the optimal strategies.

How do we determine optimal strategies?

7.1 Introduction

As part of our normal way of living, we often play different types of games. Some games are played simply for fun, others to pass the time, and still others to make money. There are certain economic and social activities that so closely resemble games that we frequently use the terminology of game theory in discussing them. The connection between games and these other activities is more than just a casual one. Thus, we might hear one say "that was a good political deal," "he plays the market," "let's agree on this deal," and so on. Since these activities so closely resemble games, we can apply the mathematical theory of games to analyze them. It is for this reason that game theory has found wide applications in the fields of economics and sociology.

A typical example of this occurs when a group of workers (usually a union) sits down to negotiate a pay raise and a new contract with the bosses. The two groups have opposing interests. Ultimately they arrive at an agreement that each side must live with but that may turn out to be more advantageous to one side than to the other. In arriving at an agreement, each side will bluff, twist arms, and use a little "friendly" persuasion in the form of strikes, lockouts, or other tactics. Each side will attempt to discover its opponent's strategies and act accordingly, while at the same time prevent discovery of its own strategy. Under such circumstances, the theory of games can often be applied.

Similarly, when playing chess, poker, or tic-tac-toe, the two players have precisely opposite interests. Each player has several possible moves that he or she may make and that are unknown to his or her opponent. The main objective of each player is to win, and the moves are planned accordingly.

As we shall see in this chapter, game theory can be applied to analyzing countries at war, politicians competing to get elected, investment alternatives, and appropriate airport landing fees for airlines. Game theory has also been applied to the automobile industry when manufacturers have several strategies possible, such as the addition of new accessories, new body designs, and the best time to announce new model price increases. In this chapter we apply game theory to help companies determine where to open new factories when several alternate sites are possible.

In many games there is an optimal or pure strategy. Thus, a player who follows a particular sequence of moves will have the safest and best possible strategy no matter what his or her opponent does or even if the opponent knows his or her moves. A simple example of this is the game of tic-tac-toe. In a perfectly played game, a tie should always result. Unfortunately this does not always happen.

There are other games, however, that have no single best possible strategy. Most military situations are of this type. In such cases a combination of strategies is required and the player's main objective is to prevent any leak of information from himself to his opponent. Thus, rather than choose a precise course of action, the various alternatives are considered with different probabilities.

The first person to develop a method for determining best moves under these circumstances was John Von Neumann. It is with this theory and its applications that we concern ourselves in this chapter.

Historical Note

The first studies of game theory were conducted by three mathematicians, John Von Neumann, Émil Borel, and Oskar Morgenstern. Émil Borel (1871–1956) was a French mathematician who made important contributions to the field of measure theory. Oskar Morgenstern (1902–1977) was a German-born economist who coauthored the book *Theory of Games and Economic Behaviour* with John Von Neumann. He was a noted economist who worked at Princeton University.

John Von Neumann, one of the twentieth century's greatest mathematical geniuses, was born in Budapest, Hungary, on December 28, 1903. He studied at the University of Berlin, the Technical Institute of Zurich, Switzerland, and the University of Budapest. He was only 20 years old when he received his Ph.D. degree in mathematics in 1923 from the University of Budapest and a degree in chemical engineering from the Federal Institute of Technology of Zurich, Switzerland.

After lecturing in Europe, he left there just before World War II and came to the United States, where he became a professor of mathematics at the Institute for Advanced Study at Princeton University. He worked there from 1933 until his death in 1957. Von Neumann served on the U.S. Atomic Energy Commission and was one of the scientists who helped to develop and produce the first atomic bomb.

Together with Oskar Morgenstern in 1944, John Von Neumann wrote a book entitled *Theory of Games and Economic Behaviour*. In this book Von Neumann developed the theory that enables the economist to study economic behavior through the use of game theory methods. Von Neumann proved that there is a "rational course of action" or best strategy in all games, even those that are not strictly determined. This important idea paved the way for many new applications of game theory to the fields of economics and sociology.

7.2 Two-Person Zero-Sum Games

Let us look in on Ronald and Cynthia, who are playing the following game. At a given signal, each player will show one or two fingers. If the total number of fingers shown by both players is even, then Ronald wins. The amount of money won (in dollars) is the total number of fingers shown by both players. If the number of fingers shown by both players is odd, then Cynthia wins. Again the amount of money won (in dollars) is the total number of fingers shown by both players. For example, if Ronald shows two fingers and Cynthia shows two fingers, then the sum is even so that Ronald will win \$4 from Cynthia. On the other hand, if Ronald shows one finger and Cynthia shows two fingers, then the sum is odd so that Cynthia will win \$3 from Ronald. You will notice immediately that whenever Ronald wins any money this means that Cynthia loses that amount of money. Thus, if Ronald wins \$4, then Cynthia loses \$4. This situation results in a zero-sum game, since the sum of one player's gain and the other player's loss (a negative number) is zero.

Definition 7.1 A game is called a *zero-sum game* if on each move, one player wins what the other player loses.

Comment There are many games, such as Monopoly, that are not zero-sum games simply because there is a kitty or a bank. We will not concern ourselves here with such games.

In the game we are discussing, two players are involved, Ronald and Cynthia. While there are games that may involve many players, we will restrict ourselves to games involving two players only.

Definition 7.2 A *two-person game* is any game that involves only two players.

In order to study two-person zero-sum games properly, it is advisable to indicate all the possible plays and the corresponding payments from one player to another in the form of a table, which is called the *payoff matrix*. In this case we get a 2×2 payoff matrix, since each player has two options, namely, put out 1 finger or 2 fingers. We set up the payoff matrix in terms of Ronald's winnings and losses. The payoff matrix for this game is:

		Cynthia: Shows 1 finger	Cynthia: Shows 2 fingers
Ronald	Shows 1 finger	+2	−3
	Shows 2 fingers	−3	+4

The entries for this matrix are obtained as follows: If Ronald shows 1 finger and Cynthia shows 1 finger, then the sum is 2, an even number, so that Ronald wins \$2. This represents the entry in the first row, first column.

If Ronald shows 1 finger and Cynthia shows 2 fingers or if Ronald shows 2 fingers and Cynthia shows 1 finger, then the sum is 3, an odd number, so that Ronald loses \$3. This represents the entry in the first row, second column and also the entry in the second row, first column.

If Ronald shows 2 fingers and Cynthia shows 2 fingers, then the sum is 4, an even number, so that Ronald wins $4. This represents the entry in the second row, second column.

We can abbreviate the payoff matrix as

		C	
		1	2
R	1	+2	−3
	2	−3	+4

Comment The preceding payoff matrix represents Ronald's (R's) winnings and losses, which are the same as Cynthia's (C's) losses and winnings, respectively. We could have set up the payoff matrix in terms of C's winnings and losses; however, it is standard practice to set up the matrix in terms of R's winnings or losses.

Definition 7.3 In any game, the *payoff matrix* is a table of numbers indicating the amount of money won and lost by each player as a result of any move.

To further illustrate how to set up a payoff matrix, we give another example.

Example 1 Richie and Carol each have a nickel, dime, quarter, and half-dollar. At a given signal, each player will show one coin. If the sum represented by both coins is even, then Richie wins Carol's coin. Otherwise, Carol wins Richie's coin. Set up the payoff matrix.

Solution If Richie shows a nickel and Carol shows a nickel, then the sum is an even number, so that Richie will win a nickel from Carol. Thus, we enter +5 in the first row, first column of the payoff matrix. If Richie shows a nickel and Carol shows a dime, then the sum is an odd number, so that Richie will lose a nickel to Carol. Thus, we enter a −5 in the first row, second column of the payoff matrix. In a similar manner, we determine the other entries for the payoff matrix. The complete matrix is shown here:

		Carol shows			
		Nickel	Dime	Quarter	Half-dollar
Richie shows	Nickel	+5	−5	+25	−5
	Dime	−10	+10	−10	+50
	Quarter	+5	−25	+25	−25
	Half-dollar	−50	+10	−50	+50

Once we set up a payoff matrix for any game, two questions immediately arise:

1. Is the game fair? If not, which player does the game favor?
2. What is the best move for each player? By a best move, we mean a move that either wins the most amount of money or loses the least amount of money. Such a best move is called an *optimal strategy*.

The theory of games allows us to determine optimal strategies for each player. In the next few sections, we indicate how this is done.

7.3 Strictly Determined Games

Once we set up the payoff matrix, the next step is to determine the optimal strategies for each player. Suppose we have a card game involving two players, Robyn and Carl. Each player has been given three cards: a 6 card, an 8 card, and a 9 card. The rules of the game are as follows: At a given signal, each player will show a card. If the numbers on both cards match, then nobody wins. If the numbers do not match, then the player with the higher number wins. The amount of money won is the sum of the numbers shown. The payoff matrix then is:

		Carl		
		6 card	8 card	9 card
Robyn	6 card	0	−14	−15
	8 card	+14	0	−17
	9 card	+15	+17	0

What is the optimal strategy for each player? If Robyn plays the 6 card, then she can break even, lose \$14, or lose \$15. If she plays the 8 card, she can win \$14, break even, or lose \$17. If she plays the 9 card, she can never lose any money; she can break even, win \$15, or win \$17. Clearly her best move is always to play the 9 card, since then she can never lose any money. We circle this row horizontally in the payoff matrix to indicate Robyn's optimal move. This is done in Figure 7.1.

What about Carl? If he plays the 6 card, he can break even, lose \$14, or lose \$15. (Remember, the payoff matrix is set up in terms of Robyn's winnings. Thus, when Robyn wins \$14 or \$15, Carl loses \$14 or \$15.) If Carl plays the 8 card, he can win \$14, break even, or lose \$17.

		Carl		
		6 card	8 card	9 card
Robyn	6 card	0	−14	−15
	8 card	+14	0	−17
	9 card	+15	+17	0

Figure 7.1

If Carl plays the 9 card, he can win \$15, win \$17, or break even. Clearly his best move is to play the 9 card, since then he can never lose any money. We circle this column vertically in Figure 7.1 to indicate Carl's best move.

You will notice that one number has been circled twice. It is the number 0. Such a number is called a *saddle point* and represents the value of the game. A game that has a saddle point is called a *strictly determined game.* If the value of a game is 0, then we say that the game is *fair*. If the value of the game is a negative number, then the game favors C, and if the value of the game is a positive number, then the game favors R. Assuming no player makes any foolish moves, if a game is fair then no money will be won by either R or C in the long run.

Any game that has a saddle point is called a *strictly determined game* since each player knows which is his or her best move. These moves are known to both players. When a game has a saddle point, the optimal strategies are called *pure strategies*. This is because each player will always use the same strategy every time the game is played no matter what his or her opponent does.

For the 3×3 game we are discussing, we can find the saddle point in a more convenient way. We first determine the smallest number in row 1. It is -15. (Remember -15 is smaller than 0.) Is -15 the largest number of the column that it is in? Obviously not, since in the third column there is a 0, which is larger than -15.

What about the second row? The smallest number is -17. Is it the largest number of the column it is in? Again the answer is no, since there is a 0 in the third column, which is larger than -17.

For the third row of the payoff matrix, the smallest number is 0. Is 0 the largest number of the column that it is in? The answer is yes. Thus, the 0 in the third row is the minimum value in its row and at the same time the maximum value in its column. Such a point is called a saddle point.

We can summarize our discussion in the following definitions and rule.

Definition 7.4 A *saddle point* is a number in a payoff matrix that is at the same time the smallest number in its row and the largest number in its column.

Definition 7.5 A game is said to be *strictly determined* if it has a saddle point.

Definition 7.6 The number that represents the saddle point is called the *value of the game* and will be denoted by v.

Definition 7.7 A game is called a *fair game* if its value is zero.

Using these definitions, we can specify the optimal strategies for each player when the game is strictly determined.

Rule 1 For strictly determined games, the *optimal strategies* for the players are as follows:

Player R: Play the row that contains the saddle point. (If there is more than one such row, play either.)

Player C: Play the column that contains the saddle point. (If there is more than one such column, play either.)

Let is illustrate the use of this rule with several examples.

Example 2 Robyn and Carl are again playing cards. This time each player has four cards: a 2 card, a 6 card, an 8 card, and a 9 card. The rules of the game are the same as before. At a given signal, each player shows a card. If the numbers on the cards are the same, then nobody wins. If the numbers do not match, then the player with the higher number wins. The amount of money won is the difference between the numbers shown (larger minus smaller).

a) Find the payoff matrix.
b) Is the game strictly determined?
c) Find the value of the game.
d) Find the optimal strategy for each player.

Solution a) The payoff matrix is:

		Carl			
		2 card	6 card	8 card	9 card
Robyn	2 card	0	−4	−6	−7
	6 card	4	0	−2	−3
	8 card	6	2	0	−1
	9 card	7	3	1	0

b) The smallest number in the first row is −7. It is *not* the largest in its column; 0 is larger. The smallest number in the second row is −3. It is *not* the largest in its column; 0 is larger. The smallest number in the third row is −1. It is *not* the largest in its column; 0 is larger. The smallest number in the fourth row is 0. It *is* the largest in its column. Thus, 0 is a saddle point. It is at the same time the smallest number in its row and the largest number in its column. Since the game has a saddle point, it is strictly determined.

c) The saddle point is 0, so the value of the game is 0. Thus, the game is fair.

d) We use Rule 1. The optimal strategy for Robyn is to play row 4, or the card with a 9 on it. The optimal strategy for Carl is to play column 4, or the card with a 9 on it. This is illustrated in the following matrix.

		Carl			
		2	6	8	9 (Optimal for Carl)
Robyn	2	0	−4	−6	−7
	6	4	0	−2	−3
	8	6	2	0	−1
Optimal for Robyn →	9	7	3	1	0 ← Saddle point and value of game

Example 3 *More card games.* Rona and Cassandra are playing cards. Rona has the ace and 2 of spades, whereas Cassandra has the ace, 2, and 3 of diamonds. The rules of the game are as follows: At a given signal, each player will play one card. If Rona plays the ace of spades, then she will win \$7, \$3, or \$1 depending on whether Cassandra plays the ace, 2, or 3 of diamonds, respectively. If Rona plays the 2 of spades, then she will win \$15, lose \$6, or lose \$3 depending on whether Cassandra plays the ace, 2, or 3 of diamonds, respectively.

a) Find the payoff matrix.
b) Is the game strictly determined?
c) Find the value of the game.
d) Find the optimal strategy for each player.

Solution a) In this case, since Rona has only 2 cards and Cassandra has 3 cards, we get the following 2×3 payoff matrix:

		Cassandra		
		Ace of diamonds	2 of diamonds	3 of diamonds
Rona	Ace of spades	7	3	1
	2 of spades	15	−6	−3

b) The smallest number in the first row is 1. It is also the largest number in its column. Thus, 1 is a saddle point. The smallest number in the second row is -3. It is *not* the largest number in its column. Since the game has a saddle point, it is strictly determined.
c) The saddle point is 1, so the value of the game is 1. Since the value is $+1$, the game favors player R (Rona).
d) We use Rule 1. The optimal strategy for Rona is to play row 1, or the ace of spades. The optimal strategy for Cassandra is to play column 3, or the 3 of diamonds. This is illustrated in the following matrix.

		Cassandra		
		Ace of diamonds	2 of diamonds	3 of diamonds (Optimal for Cassandra)
Rona	Ace of spades (Optimal for Rona)	7	3	1 ← Saddle point and value of game
	2 of spades	15	−6	−3

Comment Perhaps you are wondering why one would want to play a game that has a value of zero since no money can be won. Furthermore, why would one want to play a game whose value is not zero? Obviously, such a game favors one of the players. It is only when a game is analyzed mathematically that we can determine whether it is fair or not. Furthermore, in the game of tic-tac-toe, if each player played his or her optimal strategy, then each game would always end in a tie. This does not always happen. Also, there are certain games such as chess, as discussed at the beginning of this chapter, that have so many possible strategies that even a computer has difficulties analyzing all of them and acting accordingly.

Comment An important feature of an optimal strategy is that it does not matter if your opponent knows what you are going to do.

Example 4 Marilyn and Mickey are playing a card game that has the following payoff matrix.

		Mickey			
		Ace of clubs	2 of clubs	3 of clubs	4 of clubs
Marilyn	Ace of spades	8	2	3	7
	2 of spades	6	−1	−4	4
	3 of spades	5	2	6	3
	4 of spades	3	−4	−7	9

a) Is the game strictly determined?
b) Find the value of the game.
c) Find the optimal strategy for each player.

Solution a) The smallest number in the first row is 2. It *is* also the largest number in the column. Thus, the 2 in the first row is a saddle point.

The smallest number in the second row is -4. It is *not* the largest number in its column. The number 6 is larger. The smallest number in the third row is 2. It is also the largest number in its column. Thus, the 2 in the third row is a saddle point. The smallest number in the fourth row is -7. It is not the largest number in its column. The number 6 is larger. In this game we have two saddle points. The game is strictly determined.

b) The saddle points are both 2, so the value of the game is 2. The game obviously favors Marilyn.

c) We use Rule 1. The optimal strategy for Marilyn is to play either row 1 or row 3, that is, the ace of spades or the 3 of spades. The optimal strategy for Mickey is to play row 2, that is, the 2 of clubs.

Example 5 *Paper, stone, scissors.* A familiar game played by children is the game of paper, stone, and scissors. The game is played as follows: At a given signal, each player says the word *paper*, *stone*, or *scissors*. A stone breaks scissors, scissors cut paper, and paper wraps stone. Assume that two children, Steven and Henry, are playing this game. If Steven says stone and Henry says scissors, Steven wins a nickel since a stone breaks scissors. If Steven says stone and Henry says paper, then Henry wins a nickel since paper wraps stone. Similarly, if Steven says paper and Henry says scissors, then Henry wins a nickel. If both players say the same word, then nobody wins.

a) Set up the payoff matrix.
b) Is the game strictly determined?
c) Find the optimal strategy for each player.

Solution a) The payoff matrix is

		Henry		
		Paper	Stone	Scissors
Steven	Paper	0	+5	−5
	Stone	−5	0	+5
	Scissors	+5	−5	0

b) The smallest number in the first row is -5. It is *not* the largest number in the column; 0 is larger. The smallest number in the second row is -5. It is *not* the largest number in the column; 0 is larger. The smallest number in the third row is -5. It is *not* the largest number in the column; 0 is larger. Since there is no saddle point, the game is *not* strictly determined.

c) In the next section we will discuss optimal strategies for players when the game is not strictly determined.

Comment The name *saddle* point is based on a comparison with a saddle. When a saddle is attached to a horse, there are two curves involved; one

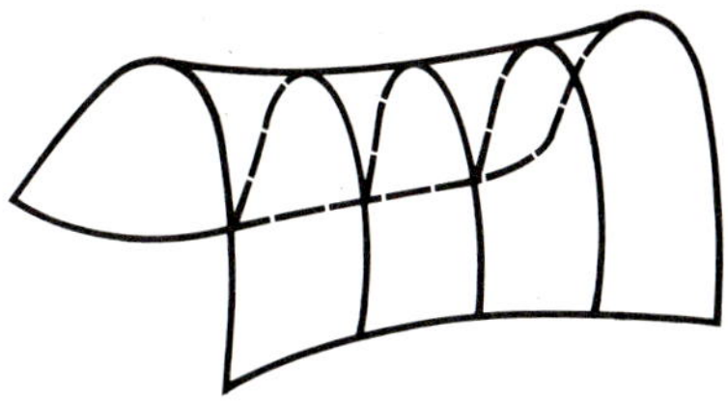

Figure 7.2 The saddle point.

is the saddle in which the rider sits and the other is the curvature of the horse's back to which the saddle is attached. The saddle point in the middle of the saddle is the intersection of both curves (see Figure 7.2). Similarly, the saddle point in game theory analysis represents the point of intersection of two opposing strategies.

In this section we have merely indicated how to find the optimal strategies for games that are strictly determined. In Section 7.6 we apply these ideas to many different situations in the social and management sciences.

Exercises

Each of the payoff matrices in Exercises 1–9 represents some game. For each,

a) Determine whether the game has a saddle point or not.
b) If the game is strictly determined, find the optimal strategy for each player and the value of the game.

1. C

R		
	2	3
	−2	1

2. C

R		
	7	2
	−1	5

3. C

R			
	3	1	4
	7	0	5
	2	−1	8

4. C

R			
	−3	7	0
	−5	2	9

5. C

R				
	2	3	−1	5
	3	0	−5	8
	4	6	−8	10
	7	6	1	9

6. C

R		
	−1	8
	−1	0

7.

C

R	8	−2
	1	−6
	5	3

8.

C

R	5	−1
	6	7
	−3	4

9.

C

R	3	7	0	5	4
	1	−2	−6	8	10
	2	3	0	8	1
	10	−1	−4	0	7

10. Renée and Calvin are choosing as follows: At a given signal, each player will show one or two fingers. Renée pays Calvin an amount equal to the product of the number of fingers shown, whereas Calvin pays Renée an amount equal to the sum of the number of fingers shown.

a) Find the payoff matrix.
b) Does the game have a saddle point?
c) Is the game strictly determined?
d) If yes, find the optimal strategy for each player.

11. *Three-finger Morra.* A popular game, three-finger Morra, is played as follows: At a given signal, each player shows one, two, or three fingers. If both players show the same number of fingers, then nobody wins. Otherwise, the player who shows the greater number of fingers wins. In each case the amount of money won (in dollars) is the sum of the number of fingers shown by both players.

a) Is the game strictly determined?
b) If yes, find the optimal strategy for each player.

12. *Boy or girl?* Two sisters, Ruth and Cindy, have just been admitted to Maternity Hospital to have their babies. Ruth promises to pay her sister \$100 if the babies born to her and her sister are of the same sex. Otherwise, Cindy will pay her \$100.

a) Set up the payoff matrix.
b) Find the saddle point for this game (if it exists).

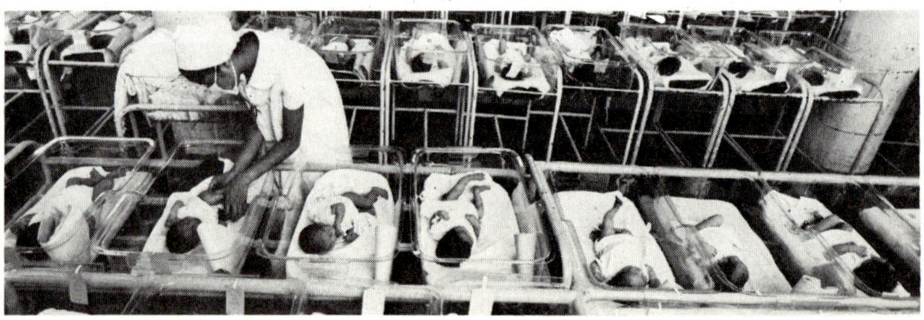

Photo: David Burnett, DeWys.

★13. Consider a game that has the following payoff matrix.

	C	
R	a	b
	a	d

a) Show that this game is strictly determined for all values of a, b, and d.
b) Find the saddle point(s).
c) Find the optimal strategy for each player.

★14. Consider a game that has the following payoff matrix.

	C	
R	a	a
	c	d

a) Show that this game is strictly determined for all values of a, c, and d.
b) Find the saddle point(s).
c) Find the optimal strategy for each player.

7.4 Games with Mixed Strategies

All the games we have discussed until now have been strictly determined. In that case we can apply Rule 1 to determine the optimal strategy for each player.

Unfortunately, many of the games we often encounter are not strictly determined. For example, consider the simple game of choosing in which each player must show one or two fingers. If we assume that the two players are Alice and Bob and that Alice wins when the number of fingers shown by both is the same or else Bob wins, then we have the following payoff matrix. (In this case the amount of money won by each player is the total number of fingers shown by both players.)

		Bob: Shows 1 finger	Bob: Shows 2 fingers
Alice	Shows 1 finger	2	-3
	Shows 2 fingers	-3	4

The smallest number in the first row is -3. It is not the largest number in its column; 4 is larger. The smallest number in the second row is -3. It also is not the largest number in its column. Thus, since there is no number that is both the smallest in its row and the largest in its column, the game is *not strictly determined.*

What is the best move for each player? Alice reasons as follows: "If I show one finger, then I either win \$2 or lose \$3. If I show two fingers, then I lose \$3 or win \$4. Since in either case I can lose \$3, my best move is to show two fingers. In that case I may win \$4 as opposed to \$2 when I show only one finger." Thus, it would seem that Alice should always show two fingers. Bob is sure to notice this, however, and will always show one finger, thereby guaranteeing that Alice will always lose \$3. What should Alice do?

It is obvious that Alice should mix her strategies by sometimes showing one finger and sometimes two fingers but according to no obvious pattern that Bob can discover. She should show one or two fingers randomly. Bob should also follow a similar mixed strategy. Just how should they accomplish this? Simply flipping a coin will not work. Why not?

When a game is not strictly determined, how do we determine the optimal strategy for each player? Must we go through a long chain of reasoning as was done in the preceding problem, or can we calculate optimal strategies in a more systematic manner? Fortunately, the answer to the latter is yes. John Von Neumann proved that there is a course of action that represents the best strategy or saddle point even when the games are not strictly determined. The idea involves choosing a particular strategy with a certain probability only. If a player uses a statistical strategy, then discovery by his or her opponent will be impossible. We analyze here optimal strategies for nonstrictly determined games that have a 2 × 2 payoff matrix.

It can be shown[1] that any nonstrictly determined game that has a 2 × 2 payoff matrix of the type.

		Player C	
		Column 1	Column 2
Player R	Row 1	a	b
	Row 2	c	d

has the following optimal strategies, which we state as a rule.

Rule 2 *Optimal strategies for nonstrictly determined games with a 2 × 2 payoff matrix.* Let

$$R_1 = \frac{d-c}{a+d-b-c} \qquad R_2 = \frac{a-b}{a+d-b-c}$$

$$C_1 = \frac{d-b}{a+d-b-c} \qquad C_2 = \frac{a-c}{a+d-b-c}$$

[1] See, for example, J. Kemeny, J. Snell, and G. Thompson, *Introduction to Finite Mathematics*, pp. 345–350 (Englewood Cliffs, N.J.: Prentice-Hall, 1966).

Then the optimal strategies for the players are:

Player R: Depending on the total number of games played, play row 1 R_1 of the time and row 2 R_2 of the time.

Player C: Depending on the total number of games played, play column 1 C_1 of the time and column 2 C_2 of the time.

Since the game is not strictly determined, there is no saddle point. Thus, we must define the value of the game by a formula.

Rule 3 For a game that is not strictly determined, we define the value as

$$v = \frac{ad - bc}{a + d - b - c}$$

These formulas may seem somewhat complicated, but they are easy to use as the following examples illustrate.

Example 6 *Which hand?* Mark and Betty are playing the following game: Mark hides a \$1 bill in one hand and a \$5 bill in his other hand. Betty must now guess which bill Mark has in his right hand. If she guesses correctly, she wins the bill. Otherwise, she loses a \$5 bill.

a) Find the payoff matrix for the game.
b) Is the game strictly determined?
c) Find the optimal strategy for each player.
d) Find the value of the game.

Solution a) The payoff matrix is

		Betty guesses that Mark's right hand contains	
		\$1 bill	\$5 bill
Mark's right hand contains	\$1 bill	−1	+5
	\$5 bill	+5	−5

b) The smallest number in the first row is −1. It is *not* the largest in its column; +5 is larger. The smallest number in the second row is −5. It is *not* the largest in its column; +5 is larger. Thus, there is no saddle point and the game is *not* strictly determined.

c) To find the optimal strategy for each player, we apply Rule 2. In this case we have $a = -1$, $b = +5$, $c = +5$, and $d = -5$, so that

$$R_1 = \frac{d-c}{a+d-b-c} = \frac{(-5)-(+5)}{(-1)+(-5)-(+5)-(+5)} = \frac{-10}{-16} = \frac{5}{8}$$

$$R_2 = \frac{a-b}{a+d-b-c} = \frac{(-1)-(+5)}{(-1)+(-5)-(+5)-(+5)} = \frac{-6}{-16} = \frac{3}{8}$$

$$C_1 = \frac{d-b}{a+d-b-c} = \frac{(-5)-(+5)}{(-1)+(-5)-(+5)-(+5)} = \frac{-10}{-16} = \frac{5}{8}$$

$$C_2 = \frac{a-c}{a+d-b-c} = \frac{(-1)-(+5)}{(-1)+(-5)-(+5)-(+5)} = \frac{-6}{-16} = \frac{3}{8}$$

Thus, player R should play row 1, $\frac{5}{8}$ of the time and row 2, $\frac{3}{8}$ of the time. This means that Mark should hide the \$1 bill in his right hand $\frac{5}{8}$ of the time and the \$5 bill in his right hand $\frac{3}{8}$ of the time. Betty should follow a similar strategy and guess that the right hand contains the \$1 bill $\frac{5}{8}$ of the time and that it contains the \$5 bill $\frac{3}{8}$ of the time.

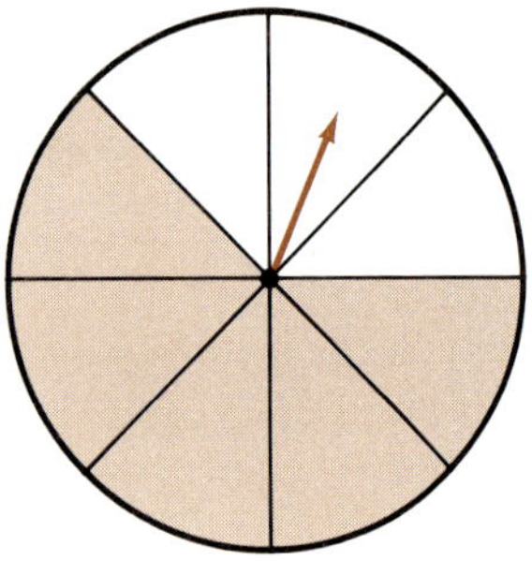

How can Mark accomplish this? If he hides the \$1 bill in his right hand five times in a row and then hides the \$5 bill in his right hand three times in a row, Betty is sure to discover this pattern and guess accordingly. Mark must therefore find a way of randomly hiding the \$1 bill $\frac{5}{8}$ of the time in his right hand and $\frac{3}{8}$ of the time in his left hand. One way he can do this is to use the spinner shown here. The pointer will land in the shaded portion $\frac{5}{8}$ of the time and in the unshaded portion $\frac{3}{8}$ of the time. Thus, Mark should spin the pointer. If it lands in the shaded portion, he should hide the \$1 bill in his right hand. Otherwise, he should hide it in his left hand. In the long run, using the spinner, he will end up hiding the \$1 bill in his right hand $\frac{5}{8}$ of the time.

What about Betty? How should she proceed? She should use the same spinner and guess that the right hand contains the \$1 bill when the pointer lands in the shaded portion.

d) To find the value of the game, we use Rule 3. We have

$$v = \frac{ad-bc}{a+d-b-c} = \frac{(-1)(-5)-(+5)(+5)}{(-1)+(-5)-(+5)-(+5)} = \frac{-20}{-16} = \frac{5}{4}$$

Since the value of v is not zero, the game is not fair. It favors player R, or Mark.

Comment As a quick check, the value of $R_1 + R_2$ should always be one. Can you see why? The value of $C_1 + C_2$ should also always be one. Can you see why?

Example 7 Calvin and Ruth are playing a card game that has the following payoff matrix:

		Calvin	
		Card 1	Card 2
Ruth	Card 1	2	−3
	Card 2	−1	4

a) Is the game strictly determined?
b) Find the optimal strategy for each player.
c) Find the value of the game.

Solution a) The smallest number in the first row is −3. It is *not* the largest in its column; 4 is larger. The smallest number in the second row is −1. It is *not* the largest in its column; 2 is larger. Thus, there is no saddle point and the game is *not* strictly determined.

b) To find the optimal strategy for each player, we apply Rule 2. In this case we have $a = +2$, $b = -3$, $c = -1$, and $d = +4$, so that

$$R_1 = \frac{d - c}{a + d - b - c} = \frac{(+4) - (-1)}{(+2) + (+4) - (-3) - (-1)} = \frac{5}{10} = \frac{1}{2}$$

$$R_2 = \frac{a - b}{a + d - b - c} = \frac{(+2) - (-3)}{(+2) + (+4) - (-3) - (-1)} = \frac{5}{10} = \frac{1}{2}$$

$$C_1 = \frac{d - b}{a + d - b - c} = \frac{(+4) - (-3)}{(+2) + (+4) - (-3) - (-1)} = \frac{7}{10}$$

$$C_2 = \frac{a - c}{a + d - b - c} = \frac{(+2) - (-1)}{(+2) + (+4) - (-3) - (-1)} = \frac{3}{10}$$

Thus, player R should play row 1 half the time and row 2 half the time. This means that Ruth should play card 1 half the time and card 2 half the time. One way to accomplish this is to flip a coin and play card 1 each time heads comes up. Otherwise, she should play card 2. In the long run, she will end up playing card 1 half the time and card 2 half the time.

Calvin should play card 1, $\frac{7}{10}$ of the time and card 2, $\frac{3}{10}$ of the time. One way he can do this is to use a spinner similar to the one given in Example 6 but with ten equal parts, seven of which have been shaded.

c) To find the value of the game, we use Rule 3:

$$v = \frac{ad - bc}{a + d - b - c} = \frac{(+2)(+4) - (-3)(-1)}{(+2) + (+4) - (-3) - (-1)} = \frac{5}{10} = \frac{1}{2}$$

Since the value of v is not zero, the game is not fair. It favors player R, or Ruth.

We apply the ideas of nonstrictly determined games to the social and management sciences in Section 7.6.

Exercises

Each of the payoff matrices in Exercises 15–22 is for a nonstrictly determined game. In each case find the following:

a) The values of R_1, R_2, C_1, C_2, and v
b) The optimal strategy for each player
c) Whether the game is fair or not

	C	
R	3	4
	5	3

16.

	C	
R	−2	3
	0	−1

17.

	C	
R	−5	8
	2	0

18.

	C	
R	0	2
	3	−1

19.

	C	
R	0	−2
	−3	0

20.

	C	
R	1	3
	3	2

21.

	C	
R	3	4
	4	3

22.

	C	
R	−2	0
	−1	−5

23. Bill and Mary are playing the following card game: Bill has two cards, a black 2 and a red 5. He places one of the cards face down on a table. Mary must now guess which card is on the table. If she guesses correctly, then she wins an amount of money (in dollars) equal to the number shown on the card. Otherwise, she loses \$3.

a) Set up the payoff matrix.
b) Is the game strictly determined?
c) Find the optimal strategy for each player.
d) Find the value of the game.
e) Is the game fair?

★ 24. In Exercise 23 assume that Mary loses \$$x$ to Bill when she guesses incorrectly. Find a value of x so that this game will be fair.

25. *Fleeing spy.* A spy has just obtained very valuable information from a government installation and he must deliver it to his superiors. However, a government agent is hot on his trail. The spy can escape either through the center of town or on a deserted road. If he selects the deserted road and the government agent catches him, he will be killed. If he selects the city road and is caught by the government agent, then he can only be arrested and detained for 24 hours. Assign the spy 100 points if he delivers the message. Assign the government agent 1000 points if he kills the spy and 50 points if he arrests the spy.

a) Set up the payoff matrix.
b) Is the game strictly determined?
c) Find the optimal strategy for the spy and for the government agent.

Note: The method discussed in this section works even for this exercise, although it is a nonzero sum game. Also, although the game is never repeated, we nevertheless can use the same analysis.

7.5 Use of Linear Programming to Solve Matrix Games[2]

When dealing with games that are not strictly determined, we can find the optimal strategy for each player by using the mixed strategies technique discussed in Section 7.4. However, we can also find the optimal strategy for each player by setting up the problem as a linear programming problem and then applying the *simplex method* discussed in Chapter 5.

This method involves finding strategies for both player R and player C so that player R will play row 1 R_1 of the time, row 2 R_2 of the time, row 3 R_3 of the time, and so on, where $R_1 + R_2 + R_3 + \cdots = 1$, since these are all the probabilities and they must add to 1. Similarly, player C should play column 1 C_1 of the time, column 2 C_2 of the time, and so on, where $C_1 + C_2 + C_3 + \cdots = 1$.

We shall assume that the value of the game is positive and greater than zero; that is, $v > 0$. If this is not the case, then we add a large number to each element of the payoff matrix so that the new payoff matrix will contain only positive values. It can be shown that if each element of the payoff matrix contains only positive numbers, then the value of the game is positive.[3]

We then introduce the variables $x_1, x_2, \ldots,$ such that

$$x_1 = \frac{C_1}{v} \qquad x_2 = \frac{C_2}{v} \qquad x_3 = \frac{C_3}{v} \qquad \text{etc.} \tag{1}$$

[2] This section assumes some background knowledge of algebra; it is slightly more difficult and may require some additional time and thought.

[3] See, for example, J. Kemeny, J. Snell, and G. Thompson, *Introduction to Finite Mathematics*, 3d ed., pp. 402–410 (Englewood Cliffs, N.J.: Prentice-Hall, 1974).

In view of the fact that $C_1 + C_2 + \cdots + C_n = 1$, we have:

$$x_1 + x_2 + \cdots + x_n = \frac{C_1 + C_2 + C_3 + \cdots + C_n}{v} = \frac{1}{v} \tag{2}$$

The objective function to be maximized then is

$$z = x_1 + x_2 + x_3 + \cdots + x_n \tag{3}$$

and the constraint inequalities are set up using the entries in the payoff matrix as coefficients of $x_1, x_2, \ldots$, and where each inequality is equal to or less than one.

The following examples illustrate the exact procedure.

Example 8 Roy and Cathy are choosing. At a given signal, each player will show one or two fingers. If both players show the same number of fingers, then Roy wins. Otherwise, Cathy wins. In each case the amount of money won is the sum of the number of fingers shown. Find the optimal strategy for each player and also the value of the game.

Solution The payoff matrix is

		Cathy	
		1 finger	2 fingers
Roy	1 finger	2	−3
	2 fingers	−3	4

Before proceeding any further, we first check to make sure that the game is not strictly determined. If it is strictly determined, then we apply Rule 1. In this case the game is not strictly determined since there are no saddle points. Applying Rule 3, we find that the value of the game is

$$v = \frac{ad - bc}{a + d - b - c} = \frac{(2)(4) - (-3)(-3)}{2 + 4 - (-3) - (-3)} = \frac{-1}{12}$$

Since the value is negative, we add $+3$ to each number in the payoff matrix so that the resulting entries will be nonnegative. (We could actually add any positive number that would make all the entries nonnegative.) We then have a new payoff matrix:

$2+3$	$-3+3$
$-3+3$	$4+3$

or

5	0
0	7

Since all the entries in this payoff matrix are positive, the value of the game will be positive. Thus, we can apply the simplex method discussed in Chapter 5. At this point it is suggested that the reader review that chapter before continuing.

Now we must maximize

$$z = x_1 + x_2$$

subject to

$$5x_1 + 0x_2 \le 1$$
$$0x_1 + 7x_2 \le 1$$
$$x_1 \ge 0, x_2 \ge 0$$

Applying the simplex method, we first add slack variables x_3 and x_4. We get the following linear programming problem:

$$\begin{aligned} 5x_1 + 0x_2 + x_3 \quad &= 1 \\ 0x_1 + 7x_2 \quad + x_4 &= 1 \\ \max z = x_1 + x_2 + 0x_3 + 0x_4 & \end{aligned}$$

where

$$x_1 \ge 0, x_2 \ge 0, x_3 \ge 0, x_4 \ge 0$$

The first tableau is:

$$\text{Tableau 1} \quad \begin{array}{c} \\ \end{array} \begin{pmatrix} a_1 & a_2 & a_3 & a_4 & b \\ \hline 5 & 0 & 1 & 0 & 1 \\ 0 & 7 & 0 & 1 & 1 \\ \hline -1 & -1 & 0 & 0 & 0 \end{pmatrix} \quad \begin{array}{c} \text{Vectors in basis} \\ a_3 \\ a_4 \end{array}$$

Replacing a_3 by a_1 in the "vectors in basis" column gives the second tableau:

$$\text{Tableau 2} \quad \begin{pmatrix} a_1 & a_2 & a_3 & a_4 & b \\ \hline 1 & 0 & \frac{1}{5} & 0 & \frac{1}{5} \\ 0 & 7 & 0 & 1 & 1 \\ \hline 0 & -1 & \frac{1}{5} & 0 & \frac{1}{5} \end{pmatrix} \quad \begin{array}{c} \text{Vectors in basis} \\ a_1 \\ a_4 \end{array}$$

Replacing a_4 by a_2 in the "vectors in basis" column gives the third tableau:

$$\text{Tableau 3} \quad \begin{pmatrix} a_1 & a_2 & a_3 & a_4 & b \\ \hline 1 & 0 & \frac{1}{5} & 0 & \frac{1}{5} \\ 0 & 1 & 0 & \frac{1}{7} & \frac{1}{7} \\ \hline 0 & 0 & \frac{1}{5} & \frac{1}{7} & \frac{12}{35} \end{pmatrix} \quad \begin{array}{c} \text{Vectors in basis} \\ a_1 \\ a_2 \end{array}$$

Since there are no negative numbers in the bottom row of the tableau, we are done. We can read our answer from the tableau. We have

$$x_1 = \frac{1}{5} \qquad x_2 = \frac{1}{7}$$

Now we use equation (2) to get

$$x_1 + x_2 = \frac{1}{v}$$

or

$$\frac{1}{5} + \frac{1}{7} = \frac{12}{35} = \frac{1}{v}$$

so that

$$v = \frac{35}{12}$$

Since we originally added 3 to each term, however, we must now subtract it, getting

$$v = \frac{35}{12} - 3 = -\frac{1}{12}$$

Thus, the value of the game is $-\frac{1}{12}$. It favors player C. This is the same answer we obtained before when we used Rule 3.

Now we use equation (1). We have

$$x_1 = \frac{C_1}{v} \quad \text{and} \quad x_2 = \frac{C_2}{v}$$

so that

$$C_1 = x_1 v \quad \text{and} \quad C_2 = x_2 v$$

Thus,

$$C_1 = \frac{1}{5} \cdot \frac{35}{12} = \frac{7}{12}$$

$$C_2 = \frac{1}{7} \cdot \frac{35}{12} = \frac{5}{12}$$

We now know Cathy's optimal strategy: Cathy should play C_1, or show one finger, $\frac{7}{12}$ of the time and play C_2, or show two fingers, $\frac{5}{12}$ of the time.

What is Roy's optimal strategy? In order to find his optimal strategy, we must find values of y_1 and y_2 that minimize

$$z = y_1 + y_2$$

subject to

$$5y_1 + 0y_2 \geq 1$$
$$0y_1 + 7y_2 \geq 1$$

where

$$y_1 \geq 0, y_2 \geq 0$$

This is the dual of the problem already solved. (We suggest that the reader review the discussion of duality given in Section 5.5.) The reader should verify that $y_1 = \frac{1}{5}$ and $y_2 = \frac{1}{7}$.

To find R_1 and R_2, we use a formula similar to equation (1). We have

$$y_1 = \frac{R_1}{v} \quad \text{and} \quad y_2 = \frac{R_2}{v}$$

so that

$$R_1 = y_1 v \quad \text{and} \quad R_2 = y_2 v$$

Thus,

$$R_1 = \frac{1}{5} \cdot \frac{35}{12} = \frac{7}{12}$$

$$R_2 = \frac{1}{7} \cdot \frac{35}{12} = \frac{5}{12}$$

Therefore, Roy's optimal strategy is to play row one, or show one finger, $\frac{7}{12}$ of the time and play row 2, or show two fingers, $\frac{5}{12}$ of the time.

Example 9 A certain game has the following payoff matrix:

	Column 1	Column 2	Column 3
Row 1	5	1	6
Row 2	8	9	0

Find the value of the game and the optimal strategy for each player.

Solution Since the entries in the payoff matrix are all positive numbers, the value of the game will be positive. Then we can apply the simplex method. We must maximize

$$z = x_1 + x_2 + x_3$$

subject to

$$5x_1 + x_2 + 6x_3 \leq 1$$
$$8x_1 + 9x_2 + 0x_3 \leq 1$$

where

$$x_1 \geq 0, x_2 \geq 0, x_3 \geq 0$$

Applying the simplex method, we first add the slack variables x_4 and x_5. We get the following linear programming problem:

$$\begin{aligned} 5x_1 + x_2 + 6x_3 + x_4 &= 1 \\ 8x_1 + 9x_2 + 0x_3 + x_5 &= 1 \\ \max z = x_1 + x_2 + x_3 + 0x_4 + 0x_5 & \end{aligned}$$

where

$$x_1 \geq 0, x_2 \geq 0, x_3 \geq 0, x_4 \geq 0, x_5 \geq 0$$

The first tableau is:

Tableau 1

a_1	a_2	a_3	a_4	a_5	b	Vectors in basis
5	1	6	1	0	1	a_4
8	9	0	0	1	1	a_5
−1	−1	−1	0	0	0	

Since the bottom row has negative numbers, we replace a_5 by a_1 in the "vectors in basis" column. Actually we could have replaced a_5 by a_1 or a_2. We then have the second tableau:

Tableau 2

a_1	a_2	a_3	a_4	a_5	b	Vectors in basis
0	$-\frac{37}{8}$	6	1	$-\frac{5}{8}$	$\frac{3}{8}$	a_4
1	$\frac{9}{8}$	0	0	$\frac{1}{8}$	$\frac{1}{8}$	a_1
0	$\frac{1}{8}$	−1	0	$\frac{1}{8}$	$\frac{1}{8}$	

Replacing a_4 by a_3 in the "vectors in basis" column gives the third tableau:

Tableau 3

a_1	a_2	a_3	a_4	a_5	b	Vectors in basis
0	$-\frac{37}{48}$	1	$\frac{1}{6}$	$-\frac{5}{48}$	$\frac{1}{16}$	a_3
1	$\frac{9}{8}$	0	0	$\frac{1}{8}$	$\frac{1}{8}$	a_1
0	$-\frac{31}{48}$	0	$\frac{1}{6}$	$\frac{1}{48}$	$\frac{3}{16}$	

Replacing a_1 by a_2 in the "vectors in basis" column gives the fourth tableau:

Tableau 4

a_1	a_2	a_3	a_4	a_5	b	Vectors in basis
$\frac{37}{54}$	0	1	$\frac{1}{6}$	$-\frac{1}{54}$	$\frac{4}{27}$	a_3
$\frac{8}{9}$	1	0	0	$\frac{1}{9}$	$\frac{1}{9}$	a_2
$\frac{31}{54}$	0	0	$\frac{1}{6}$	$\frac{5}{54}$	$\frac{7}{27}$	

Since there are no negative numbers in the bottom row of the tableau, we are done. We can read our answer from the tableau. As x_1 is not in the basis, $x_1 = 0$. Thus, our solution is

$$x_1 = 0 \qquad x_2 = \frac{1}{9} \qquad x_3 = \frac{4}{27}$$

Now we use equation (2) to get

$$x_1 + x_2 + x_3 = \frac{1}{v}$$

$$0 + \frac{1}{9} + \frac{4}{27} = \frac{1}{v}$$

$$\frac{7}{27} = \frac{1}{v}$$

so that

$$v = \frac{27}{7}$$

The game favors player R.

Applying equation (1) gives

$$x_1 = \frac{C_1}{v} \qquad x_2 = \frac{C_2}{v} \qquad x_3 = \frac{C_3}{v}$$

so that

$$C_1 = x_1 v \qquad C_2 = x_2 v \qquad C_3 = x_3 v$$

Thus,

$$C_1 = 0 \cdot \frac{27}{7} = 0$$

$$C_2 = \frac{1}{9} \cdot \frac{27}{7} = \frac{3}{7}$$

$$C_3 = \frac{4}{27} \cdot \frac{27}{7} = \frac{4}{7}$$

We now know player C's optimal strategy. Player C should never play C_1, play C_2 $\frac{3}{7}$ of the time, and play C_1 $\frac{4}{7}$ of the time.

To find player R's optimal strategy, we must find values of y_1 and y_2 that minimize

$$z = y_1 + y_2$$

subject to

$$5y_1 + 8y_2 \geq 1$$
$$y_1 + 9y_2 \geq 1$$
$$6y_1 + 0y_2 \geq 1$$

where

$$y_1 \geq 0, y_2 \geq 0$$

Again this is the dual of the original problem. Solving this dual problem gives $y_1 = \frac{1}{6}$ and $y_2 = \frac{5}{54}$.

To find R_1 and R_2, we use a formula similar to equation (1). We have

$$y_1 = \frac{R_1}{v} \quad \text{and} \quad y_2 = \frac{R_2}{v}$$

so that

$$R_1 = y_1 v \quad \text{and} \quad R_2 = y_2 v$$

Thus,

$$R_1 = \frac{1}{6} \cdot \frac{27}{7} = \frac{9}{14}$$

$$R_2 = \frac{5}{54} \cdot \frac{27}{7} = \frac{5}{14}$$

Therefore, player R's optimal strategy is to play row 1, $\frac{9}{14}$ of the time and play row 2, $\frac{5}{14}$ of the time.

Comment Whereas Rule 2 for finding optimal strategies for nonstrictly determined games works only when the payoff matrix is 2×2, the method discussed in this section is more general. It works for any game with a payoff matrix of dimension $m \times n$.

Exercises

In Exercises 26–39 use the methods discussed in this section to find the optimal strategies for each player.

26. C

R		
	5	−2
	−2	6

27. C

R		
	−3	5
	6	−7

28. C

R		
	2	3
	10	1

29. C

R		
	5	6
	9	−8

30.

R		
	1	−5
	−7	9

31. C

R		
	8	1
	2	3

32.

R		
	2	0
	1	2

33. C

R			
	1	0	−2
	−2	−3	5

34. C, R

8	4
12	6
1	10

35. C, R

3	6	−1	2
2	−8	0	1

36. C, R

1	2
2	5
3	−4

37. C, R

1	−3	0
3	0	4
−4	1	7

38. C, R

3	4	−1	0
8	1	2	5
−3	6	8	9

★ **39.** C, R

8	1	3	9
0	6	1	4
5	3	4	2
0	1	7	9

40. *Matching coins.* Rochelle and Cindy each have a dime, a quarter, and a half-dollar. At a given signal, each shows one coin. If the total value of the coins shown is even, then Rochelle wins that amount of money. Otherwise, Cindy wins the total value of the coins shown.

a) Set up the payoff matrix.
b) Find the value of the game. Is it fair?
c) Find the optimal strategy for each player.

41. *Deciding what to build.* Louis Pennington, a real estate developer, has acquired a large tract of land alongside State Highway 1. Louis can build an amusement park, a shopping center, and/or a sports stadium on the land. The county, which opposes Louis' development plans, is considering building a correctional facility nearby. This would affect Louis' profits directly as shown in the accompanying table. Assuming Louis can use the land for all three activities, find his optimal strategy.

		If county: Builds facility	If county: Does not build facility
Louis' profit if he builds	Amusement park	12% profit	48% profit
	Shopping center	18% profit	35% profit
	Sports stadium	19% profit	33% profit

42. *Paper, stone, and scissors again.* Refer to the game of paper, stone, and scissors discussed in Example 5. The accompanying payoff matrix was

		Henry		
		Paper	Stone	Scissors
Steven	Paper	0	+5	−5
	Stone	−5	0	+5
	Scissors	+5	−5	0

given, and we indicated that the game was not strictly determined.

a) Find the value of the game.
b) Find the optimal strategy for each player.

43. *Gambling.* Phil is at a gambling table where there is a \$3 charge to play the following game. The dealer hides a \$1 bill, a \$5 bill, or a \$10 bill. The player must guess which bill has been hidden. If the player guesses correctly, he wins that amount of money. Otherwise, he wins nothing.

a) Set up the payoff matrix.
b) Find the value of the game. Is it fair?
c) Find the optimal strategy for each player.

44. *War games.* An airplane pilot is planning to bomb a coastal town that is protected by antiaircraft guns. The pilot can fly either low or high while dropping the bombs. If he flies low, he will increase the effectiveness of the bombing as he will be able to see the target. However, the enemy has antiaircraft guns that are much more accurate when the plane flies low. The antiaircraft guns can be adjusted to one of two positions, for low-flying planes and for high-flying planes. Credit the pilot with +10 points if he avoids enemy gunfire and −6 points if he is within range of the enemy antiaircraft guns. Assign an additional +7 points to the pilot if he flies low during the bombing raid.

a) Set up the payoff matrix.
b) Find the value of the game.
c) Find the optimal strategies for the pilot and for the enemy antiaircraft gunners.

Photo: Wide World Photos

★45. *Six-finger Morra.* Rochelle and Sharon are playing six-finger Morra. The rules of the game are as follows: At a given signal, each player shows from one to six fingers. The total number of fingers shown by both players is then divided by 3. If the remainder is 1, then Rochelle wins. If the remainder is 2, then Sharon wins. If there is no remainder, nobody wins. In each case the amount of money won (in dollars) is the number of fingers shown by both players.

a) Set up the payoff matrix.
b) Find the value of the game.
c) Is the game fair?
d) Find the optimal strategy for each player.

46. *Cards.* Margaret and Walter are playing cards. Margaret has the ace of hearts, ace of clubs, 2 of hearts, and 2 of clubs. Walter has the ace of diamonds, ace of spades, 2 of diamonds, and 2 of spades. The rules of the game are as follows: At a given signal, each player shows one card. If both cards show the same number, then the black card beats the red card. Otherwise, the card with higher number wins. (An ace is counted as a 1.) The amount of money won (in dollars) is the sum of the numbers shown.

a) Set up the payoff matrix.
b) Find the value of the game.
c) Is it fair?
d) Find the optimal strategy for each player.

47. *How to invest?* Abe Gardner has just inherited a large sum of money that he wishes to invest for a period of time. His financial advisor has recommended several alternatives along with the possible profits to be made depending on the economic conditions that prevail. Assuming Abe can invest the money in all three possibilities, find his optimal investment strategy.

		Economic conditions	
		Normal	Abnormal
Investment possibilities	Real estate	15% profit	28% profit
	High-grade municipal bonds	12% profit	19% profit
	Stocks	24% profit	7% profit

7.6 Applications of Game Theory to the Social and Management Sciences

In this section we apply the ideas of game theory to many different problems in the social and management sciences. Some of the situations can be analyzed by applying Rule 1, which gives optimal strategies for strictly determined games. Others have to be analyzed by applying Rule 2, which gives optimal strategies for nonstrictly determined games. The reader should study each example carefully.

Example 10

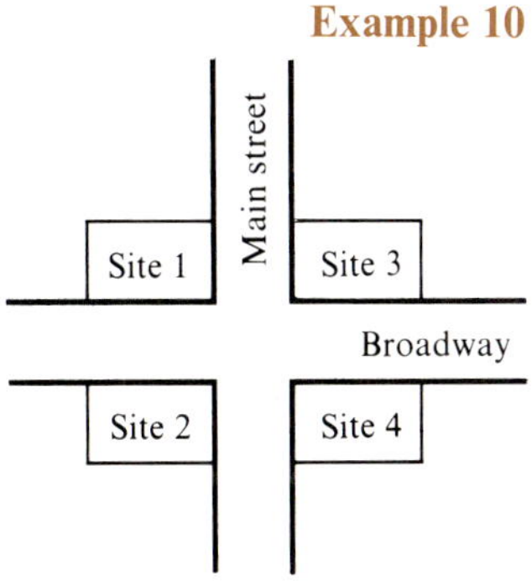

Where to build new gas stations? A recent change in the zoning laws permits the building of two gas stations at the intersection of Main Street and Broadway as shown in the accompanying diagram. Two oil companies are planning to build a new station on this busy intersection and have submitted bids on the exact location of the station. Company A will build on either site 1 or site 2, whereas company B will build on either site 3 or site 4. Because of the varying traffic flow patterns, there are advantages and disadvantages to selecting one particular corner. Thus, if both companies select corners that are diagonally opposite each other, then both companies will split the business, 50% to each company. If company A selects site 1 and company B selects site 3, then company A can expect to get 60% of the business. If company A selects site 2 and company B selects site 4, then company A can expect to get 40% of the business. Assume that all the business at this intersection will be shared only by these two companies and that when one company increases its share of the business, this is at the expense of the other company.

a) Set up the payoff matrix.
b) Is the game strictly determined?
c) Find the optimal strategy for each company.

Solution a) The payoff matrix in terms of possible business for company A is

		Company B opens at	
		Site 3	Site 4
Company A opens at	Site 1	60%	50%
	Site 2	50%	40%

or in simplified form

		B	
		3	4
A	1	0.60	0.50
	2	0.50	0.40

b) The smallest number in the top row is 0.50. This *is* also the largest number in its column. Thus, the 0.50 in the first row, second column is a saddle point. The smallest number in the second row is 0.40. It is *not* the largest number in its column; 0.50 is larger. Since we have a saddle point, the game is strictly determined.

c) As the game is strictly determined, we can apply Rule 1. The optimal strategies are: Company A should open at site 1 and company B should open at site 4.

Example 11 *Sherlock Holmes escaping death.* In Arthur Conan Doyle's famous story "The Final Problem," Sherlock Holmes was being pursued by his arch enemy Professor Moriarty. In the story Holmes boards a train traveling from London to Dover in an attempt to escape to the Continent. As the train speeds out of the London station, however, Holmes and Moriarty spot each other. Moriarty decides to charter a special train and pursue Holmes. There is only one stop between London and Dover. This is at Canterbury. What should each person do? Should Sherlock or Moriarty get off at Canterbury or go all the way to Dover? If both get off at Dover or both get off at Canterbury, then Moriarty is sure to kill Holmes. If Moriarty gets off at Canterbury and Holmes goes on to Dover, then he will escape the professor (at least temporarily). If Holmes gets off at Canterbury and Moriarty goes on to Dover, the chase is not over. It can be considered a temporary draw. Assign a value of 500 to Moriarty if he catches Holmes and kills him. Assign a value of 250 to Holmes if he escapes to the Continent and a value of 0 if he gets off at Canterbury while Moriarty gets off at Dover.

a) Set up the payoff matrix.
b) Find the value of the game? Is the game fair?
c) What should Sherlock Holmes do?

Solution a) The payoff matrix in terms of Sherlock Holmes's points is:

		Moriarty gets off at	
		Canterbury	Dover
Holmes gets off at	Canterbury	−500	0
	Dover	+250	−500

b) The smallest number in the top row is −500. It is *not* the largest in its column; +250 is larger. The smallest number in the second row is −500. It is *not* the largest in its column; 0 is larger. Thus, there are no saddle points and the game is not strictly determined. We must therefore apply Rule 3. We have $a = -500$, $b = 0$, $c = +250$, and $d = -500$, so that

$$v = \frac{ad - bc}{a + d - b - c} = \frac{(-500)(-500) - (0)(+250)}{(-500) + (-500) - (0) - (+250)} = -200$$

Since the value of the game is negative, the game is not fair. It favors Professor Moriarty.

c) In order to find the optimal strategy, we apply Rule 2. We have

$$R_1 = \frac{d - c}{a + d - b - c} = \frac{(-500) - (+250)}{(-500) + (-500) - (0) - (+250)} = 0.6$$

$$R_2 = \frac{a - b}{a + d - b - c} = \frac{(-500) - (0)}{(-500) + (-500) - (0) - (+250)} = 0.4$$

$$C_1 = \frac{d - b}{a + d - b - c} = \frac{(-500) - (0)}{(-500) + (-500) - (0) - (+250)} = 0.4$$

$$C_2 = \frac{a - c}{a + d - b - c} = \frac{(-500) - (+250)}{(-500) + (-500) - (0) - (+250)} = 0.6$$

Thus, if the same situation would be repeated many times, Holmes should get off at Caterbury 40% of the time and at Dover 60% of the time.

Comment According to the story, Holmes got off at Canterbury and watched as the train carrying Professor Moriarty sped by on its way to Dover. Sherlock thus escaped death.

Example 12 *War games.*[4] In 1943 the Japanese Army sent a large convoy of ships carrying supplies from the Port of Rabaul in New Britain to reinforce its troops at Lae in New Guinea. The Japanese commander could route the convoy over one of two routes; one route was north of the island of New Britain and the other was south of New Britain. In either case the trip would take three days. At that time the commander of the American forces was General Kinney. He was interested in using most of his available aircraft to bomb the convoy on either route. The remaining aircraft would conduct routine patrols on the other route. Clear weather was forecast for the southern route, whereas rain and poor visibility were forecast for the northern route. According to General Kinney, if both his planes and the Japanese went by the northern route, then the convoy would be sighted on the second day, thereby allowing two days of bombing before the convoy reached Lae. If both his planes and the Japanese went south, then the convoy would be spotted immediately. This would allow three days for bombing the convoy. If the convoy went by the southern route and his planes went by the northern route, then alternate planes would spot the convoy on the second day. This would allow two days for bombing the convoy. On the other hand, if the convoy went by the northern route and the planes went by the southern route, then the convoy would not be spotted for two days. This would allow only one day for bombing the convoy.

a) Set up this military problem in a payoff matrix.
b) Find the optimal strategy for the Japanese commander and for General Kinney.

Solution a) We set up the payoff matrix in terms of the number of possible bombing days as given in the problem.

		Japanese commander sends his convoy	
		North	South
General Kinney sends his planes	North	2	2
	South	1	3

b) The smallest number in the top row is 2. The 2 in the first row, first column is also the largest number in its column. Thus, it is a saddle point. The smallest number in the second row is 1. It is *not* the largest number in its column. Therefore the optimal strategies are for General Kinney to send his planes north and for the Japanese convoy to use the northern route.

[4] This example is based on O. G. Haywood's article "Military Decision and Game Theory," which appeared in the *Journal of the Operations Research Society of America* (vol. 2, pp. 365–385, 1954).

Comment According to history, this was the course of action actually taken by both commanders. General Kinney sent his planes north. Due to bad weather and poor visibility, they spotted the convey traveling by the northern route on the second day. This allowed them two days of bombing.

Example 13 *Aircraft landing fees.*[5] For years official of the Birmingham Airport in England have been concerned with determining an appropriate airport pricing and investment policy in view of constantly recurring congestion. The nearest competing airport to Birmingham is in Manchester. S. C. Littlechild and G. F. Thompson, in "Aircraft Landing Fees: A Game-Theory Approach," decided to apply linear programming and game theory techniques to help determine fair and efficient landing fees as well as optimal runway size. Specifically, they studied the investment and pricing policies of Birmingham Airport for the years 1968–1969. By appropriately analyzing the different types of aircraft using the airport, they concluded that it was optimal to extend the runway to accommodate all types of aircraft using the airport. Furthermore, their analyses (which are beyond the scope of this text) indicated how the game theory rules given in this chapter can be applied to determine the landing fees to be charged.

Exercises

48. *Health food store.* Two health food companies, A and B, are planning to sell their new 100% natural banana food mix in one of three neighborhoods. The accompanying payoff matrix indicates the differences in sales depending on which neighborhood each chooses. (A + sign indicates that company A sells more, and a − sign indicates that company B sells more.) Find the optimal strategy for each company.

		Company B		
		Poor neighborhood	Middle-class neighborhood	Upper-class neighborhood
Company A	Poor neighborhood	4	−8	3
	Middle-class neighborhood	−2	6	0
	Upper-class neighborhood	−1	0	1

49. *Where to build a new plant.* Two fast-food chains are considering opening new branches in one of the three towns of Morganville, Skyville, and Dollytown. The distances between these towns is shown in the

[5] This example is based on the article "Aircraft Landing Fees: A Game-Theory Approach" by S. C. Littlechild and G. F. Thompson, which appeared in the *Bell Journal of Economics* (vol. 8, no. 1, pp. 186–204, 1977).

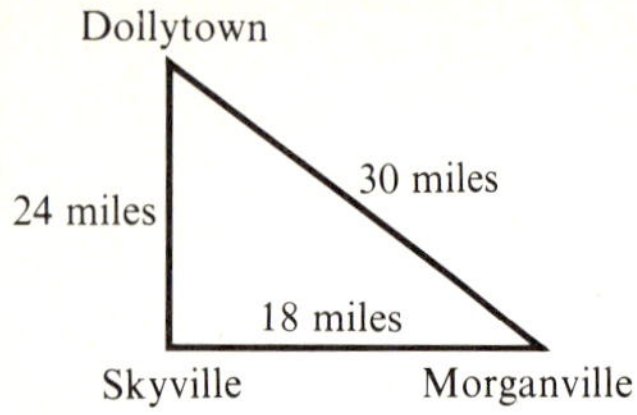

accompanying diagram. Various estimates indicate that the amount of business to be obtained by a food chain in each of these cities is approximately the same. If both food chains open in the same town, then each can expect to capture 50% of the market. If they open in different cities, however, then the company that is closer to any given city will get all of that town's business.

a) Set up the payoff matrix.
b) Find the optimal strategy for each company.

50. *Escaping bank robber.* A bank robber is being chased by a police car when he comes to the end of the road. He can turn either right or left. If he turns right and the pursuing police car turns right also, then there is a 60% chance that he will be captured. If he turns right and the police car turns left, then there is a 40% chance that he will be captured by other police cars. If he turns left and the police car turns right, then there is a 20% chance that he will be captured. If he turns left and the police car also turns left, then there is a 50% chance that he will evade being captured since that area is covered by heavy brush.

a) Set up the payoff matrix.
b) Find the optimal strategy for the escaping bank robber and for the pursuing police officer.

51. *Prisoner's dilemma.* Police are interrogating two suspects, Al and Joe, in an attempt to find out more information on the latest terrorist bombings that have plagued the city. The suspects are being questioned separately. The prosecuting attorney knows that he does not have enough evidence to convict either of them but nevertheless approaches each prisoner separately with the following deal: "If you confess, then you will be set free and your partner will be given a 35-year jail sentence. If both you and your partner do not confess, then both of you will be given a 10-year jail sentence. If both of you do confess, then each of you will get a light jail sentence of 2 years."

a) Set up the payoff matrix.
b) Find the optimal strategy for each prisoner.

52. Read the article "Cooperation in the Decomposed Prisoner's Dilemma Game: A Question of Revealing or Concealing Information" by J. Pincus and V. E. Bixenstine in *Journal of Conflict Resolution* (vol. 21, pp. 519–530, 1977) for a further discussion of the prisoner's dilemma.

★53. *Tic-tac-toe.* Consider the game of tic-tac-toe where a typical board has been numbered as shown:

Space 1	Space 2	Space 3
Space 4	Space 5	Space 6
Space 7	Space 8	Space 9

a) Write the payoff matrix for the game.
b) Does the game have a saddle point?
c) Find the optimal strategy for each player.

54. *Escaping from the pursuer.* Roosevelt Smith is traveling in the first car of a 15-car passenger train going from New York to Canada. His archenemy Clayton Jones has just boarded the last car of the train but is unable to change cars. There is one stop between New York and Canada; this is at Albany. Neither person can see whether the other gets off the train until after the train pulls out. Each person can either get off at Albany or continue on to Canada. If both Roosevelt and Clayton get off at Albany or in Canada, then Clayton is sure to see Roosevelt and kill him. If this happens, credit Clayton with 400 points. If Roosevelt gets off at Albany and Clayton goes on to Canada, then Roosevelt will escape from Clayton only temporarily. In this case credit Roosevelt with 50 points. If Roosevelt goes to Canada and Clayton gets off at Albany, then Roosevelt will escape from Clayton permanently. In this case credit Roosevelt with 200 points.
a) Set up the payoff matrix.
b) Find the value of the game. Is the game fair?
c) What are the optimal strategies; that is, where should Roosevelt get off and where should Clayton get off?
d) What device can each person use to accomplish this?

★55. *Alternate getaway plans.* Refer to Exercise 54. Assume that the train will make two stops between New York and Canada. These are at Binghamton and at Albany. The other conditions remain the same. If both Roosevelt and Clayton get off at Binghamton, at Albany, or in Canada, then Clayton is sure to see Roosevelt and kill him. If this happens, credit Clayton with 400 points. If Clayton gets off at Binghamton or Albany and Roosevelt goes on to Canada, then Roosevelt will escape from Clayton permanently. In this case credit Roosevelt with 200 points. In any other case Roosevelt will escape from Clayton only temporarily. If this happens, credit Roosevelt with 50 points.
a) Set up the payoff matrix.
b) Find the value of the game. Is the game fair?
c) What is the optimal strategy for each player; that is, where should Roosevelt get off and where should Clayton get off?

7.7 Summary

In this chapter we discussed the basic ideas of game theory and how they can be applied to many different situations. If the game is strictly determined, then the optimal strategies are easy to determine. If the game is not strictly determined, then we apply Von Neumann's theorem to help us determine optimal strategies for each player.

Additionally we indicated how the simplex method used in solving linear programming problems can be used to find optimal strategies for nonstrictly determined games. Unlike the Von Neumann theorem discussed in this chapter, however, which can be applied only when the nonstrictly determined game has a 2×2 payoff matrix, the simplex method can be applied even to the more general game that has an $m \times n$ payoff matrix. It should also be noted that we can apply graphical methods to solve game theory problems in a manner similar to that used for linear programming problems.

All the games discussed in this chapter were very simple. Thus, we were able to analyze them mathematically. There are more complicated games such as poker, chess, and bridge that mathematicians have attempted to analyze also. As the newspaper article at the beginning of this chapter indicates, chess 4.7 is one such attempt.

Today computers have been programmed to play all kinds of games. One such game is tic-tac-toe. If the computer is programmed with the optimal strategies, it can never lose. Either the computer wins or the game ends in a tie.

Study Guide

You should be able to demonstrate your knowledge of the following ideas presented in this chapter by giving definitions, descriptions, or specific examples. Page references are given for each term so that you can check your answer.

Zero-sum game (p. 311)
Payoff matrix (p. 312)
Pure strategies (p. 314)
Strictly determined game (p. 314)
Fair game (p. 314)
Nonstrictly determined game (p. 321)
Use of linear programming for matrix games (p. 327)
Two-person game (p. 311)
Optimal strategy (p. 313)
Saddle point (p. 314)
Value of game (p. 314)
Optimal strategies for strictly determined games (p. 315)
Optimal strategies for nonstrictly determined games (p. 322)
Simplex method for matrix games (p. 327)

Formulas to Remember

The following list summarizes all the formulas given in this chapter.

Rule 1: For strictly determined games, the optimal strategies are:
Player R: Play the row that contains the saddle point.
Player C: Play the column that contains the saddle point.

Rule 2: If the payoff matrix is 2×2 and of the form

a	b
c	d

then optimal strategies for nonstrictly determined games are:

$$R_1 = \frac{d - c}{a + d - b - c} \qquad R_2 = \frac{a - b}{a + d - b - c}$$

$$C_1 = \frac{d - b}{a + d - b - c} \qquad C_2 = \frac{a - c}{a + d - b - c}$$

Player R: Play row 1 R_1 of the time and row 2 R_2 of the time.

Player C: Play column 1 C_1 of the time and column 2 C_2 of the time.

Rule 3: For nonstrictly determined games, value

$$v = \frac{ad - bc}{a + d - b - c}$$

When using the simplex method to solve game theory problems,

$$x_1 = \frac{C_1}{v} \qquad x_2 = \frac{C_2}{v} \qquad x_3 = \frac{C_3}{v} \qquad \text{etc.}$$

$$x_1 + x_2 + x_3 + \cdots + x_n = \frac{C_1 + C_2 + \cdots + C_n}{v} = \frac{1}{v}$$

$$z = x_1 + x_2 + x_3 + \cdots + x_n$$

$$y_1 = \frac{R_1}{v} \qquad y_2 = \frac{R_2}{v} \qquad y_3 = \frac{R_3}{v} \qquad \text{etc.}$$

$$y_1 + y_2 + y_3 + \cdots + y_n = \frac{R_1 + R_2 + \cdots + R_n}{v} = \frac{1}{v}$$

$$z = y_1 + y_2 + \cdots + y_n$$

Mastery Tests

FORM A

1. What is the name for a table indicating the amount of money won or lost by each player as a result of any move?
2. What do we call a best move?
3. Can a strictly determined game have more than one saddle point? Explain your answer.
4. What type of games have pure strategies?
5. A game with a value that is a negative number is unfair. Which player does it favor?
6. The minimax theorem of Von Neumann is used to find optimal strategies for what type of games?

7. Find the value of a game whose payoff matrix is

2	7
0	4

8. What are the optimal strategies for a game with the accompanying payoff matrix?

1	−1
−1	1

9. Which of the following payoff matrices are strictly determined?

a)

2	7
4	3

b)

1	0
5	6

c)

−1	2
0	5

d)

5	4
−1	6

10. Which of the following games are fair?

a)

−1	2
0	5

b)

1	5
0	−3

c)

4	1
−3	0

d)

4	5
1	3

FORM B

■ ★ 1. *Rolling dice.* John and Roland are throwing dice. The rules of the game are as follows: At a given signal, both players roll a die at the same time and must guess whether the sum of the number of dots appearing on both dice will be even or odd. If it is even, then Roland wins. Otherwise, John wins. In each case the amount of money won (in dollars) is the sum of the number of dots appearing on both dice.

a) Set up the payoff matrix.
b) Find the value of the game. Is it fair?
c) Find the optimal strategy for each player.

2. *Choosing.* Robin and Carolyn are choosing as follows: At a given signal, each player will show three or four fingers. If the sum of the fingers shown by both players is even, then Robin wins. Otherwise, Carolyn wins. In each case the amount of money won (in dollars) is the sum of the fingers shown.
   a) Set up the payoff matrix.
   b) Find the value of the game. Is it fair?
   c) Find the optimal strategy for each player.

3. *Defending military installations.* Colonel Smith has several divisions of marines, which he can use to defend a radar installation on one island or a military airfield on another island. The enemy commander can use all of his army to attack either target. Each officer must commit his entire force to either operation. If the enemy attacks the radar installation with his army and all divisions of marines are there defending it, then the attacking enemy force will be destroyed. In this case credit the defenders with 200 points. If the enemy attacks the airfield but the marines are defending the radar installation, the airfield will be completely destroyed. In this case credit the enemy with 150 points. If the enemy attacks the radar installation but the marines are defending the airfield, then the radar installation will be destroyed. In this case credit the enemy with 100 points. If the enemy attacks the airfield and the marines are there defending it, then the attacking enemy will be destroyed. In this case credit the defenders with 300 points.
   a) Set up the payoff matrix.
   b) Find the value of the game.
   c) Find the optimal strategy for each commander.

★ 4. *Dominance.* Consider a game that has the following payoff matrix.

$$\begin{pmatrix} 1 & 14 & -4 \\ -3 & 6 & 7 \\ 0 & -2 & -5 \end{pmatrix}$$

We notice that each element of row 1 is larger than the corresponding elements of row 3. In this case we say that row 1 *dominates* row 3 or that row 3 is *dominated* by row 1. We can then eliminate it from consideration since R is never going to select it. The resulting payoff matrix is

$$\begin{pmatrix} 1 & 14 & -4 \\ -3 & 6 & 7 \end{pmatrix}$$

Similarly, each element of column 2 is larger than the corresponding element of column 1; that is, column 2 dominates column 1. In terms of player C, we can then eliminate column 2. The resulting payoff matrix is

$$\begin{pmatrix} 1 & -4 \\ -3 & 7 \end{pmatrix}$$

   a) Find the optimal strategies for this new payoff matrix.
   b) Find the optimal strategies for the original payoff matrix.

★5. Using the procedure outlined in question 4, find the optimal strategies for each of the following games by eliminating dominated rows and columns.

a) $\begin{pmatrix} 5 & 1 & 2 \\ 2 & 3 & 1 \\ -4 & 0 & 1 \end{pmatrix}$ b) $\begin{pmatrix} 1 & 0 & 5 \\ -2 & 3 & 1 \\ -6 & -1 & 4 \end{pmatrix}$

6. Read each of the following articles to see how game theory analysis is used in applied situations.

a) "How We Play Our War Games with Soviet Pieces" by J. Galub in *Popular Mechanics*, vol. 149, pp. 100–103, April 1978.
b) "Reasoned Strategy Prevails As You Battle the Odds" by F. Peterson and J. R. Kesselman in *Science Digest*, vol. 83, pp. 52–56, April 1978.
c) "Computer Chess: A Grandmaster Analyzes Play" by David Levy in *Science Digest*, vol. 83, pp. 48–53, August 1978.

Suggested Further Reading

Bellman, R., and D. Blackwell, "Red Dog, Blackjack, Poker," *Scientific American*, pp. 44–47, October 1951.

Farquharson, Robin, *Theory of Voting*, New Haven, Conn.: Yale University Press, 1969.

Kuhn, H. W., and A. W. Tucker, "Theory of Games," *Encyclopedia Britannica*, 1973.

Luce, Robert D., and Howard Raiffa, *Games and Decisions*, New York: Wiley, 1975.

MacDonald, John, *Strategy in Poker, Business and War*, New York: Norton, 1950.

McKinsey, John C., *Introduction to the Theory of Games*, New York: McGraw-Hill, 1952.

Morgenstern, Oskar, "The Theory of Games," *Scientific American*, vol. 180, pp. 294–308, 1950.

Rapoport, Anatol, *Fights Games and Debates*, Ann Arbor: University of Michigan Press, 1960.

Rapoport, Anatol, *Two-Person Game Theory: The Essential Ideas*, Ann Arbor: University of Michigan Press, 1966.

Shubik, Martin, *Strategy and Market Structure*, New York: Wiley, 1959.

Thompson, G. L., "Game Theory," *McGraw-Hill Encyclopedia of Science and Technology*, New York: McGraw-Hill, 1971.

Von Neumann, John, and Oskar Morgenstern, *Theory of Games and Economic Behaviour*, rev. ed., Princeton, N.J.: Princeton University Press, 1953.

Williams, John D., *The Compleat Strategyst*, rev. ed., New York: McGraw-Hill, 1965.

CHAPTER 8 Decision Analysis

Objectives to be Achieved in this Chapter

- *To discuss* the problems encountered when making a decision in the face of uncertainties. This is decision analysis.
- *To analyze* how to find the expected value in a variety of situations.
- *To indicate* how to arrive at an optimal decision. Our approach is to calculate the expected maximum profit from among all the possible alternatives.
- *To study* mortality tables, which are tables used by the insurance industry in determining the probability that an individual of a given age will die.
- *To apply* the ideas of expected value to indicate how insurance companies determine the premium to be charged for life insurance policies.

"Wildcatters" Search for Oil off Jersey Coast

DAILY PRESS MARCH 25, 1979

New York (March 25)—In an effort to determine whether the Baltimore Canyon trough contains recoverable quantities of oil or gas, the Glomar Pacific began exploratory drilling yesterday 95 miles off the coast of New Jersey. The search for oil in the U.S. waters of the Atlantic Ocean—one of the nation's last petroleum frontiers—has important implications. If oil in sufficient quantities is found in the area, this may eventually help reduce our nation's dependence on foreign oil and help turn the Northeast—the country's most energy-poor region—into a hub of petroleum activity.

The Exxon Corporation, one of the companies that is financing the operation, estimates that the first hole in the Baltimore Canyon will cost \$9 million. According to an official of the United States Geological Survey, there is only a 20% probability of finding significant but hardly staggering quantities of hydrocarbons in the area.

When the U.S. government allowed the oil companies to drill for oil off the East Coast of the United States, the oil companies invested huge amounts of money in this venture. As the newspaper article indicates, the costs are enormous. After all, if no oil is found, then the oil companies will have to absorb huge losses. A decision to explore was made even though geologists estimate that there is only a 20% probability of finding significant but hardly staggering amounts of hydrocarbons in the area. Of course, if oil is found in significant quantities, then the oil companies stand to make enormous profits.

All of us are faced with similar situations where we must make important decisions in the face of uncertainties. How are such decisions reached?

8.1 Introduction

Let us look in on Mary Lou, who is the purchasing agent and manager of the student cafeteria at State University. One of the more popular items that the cafeteria sells is fresh glazed doughnuts, which it orders daily. Unfortunately the doughnuts are highly perishable and must be sold by the end of the day. If they are not sold by the end of the day, they must be thrown away. This, of course, results in a considerable financial loss.

Mary Lou pays \$5 for a box of doughnuts and sells them at a profit of \$2 per box. Each morning she must decide how many boxes of doughnuts to order. To help her decide how many boxes to order, she might calculate the *expected number* of boxes that will be sold (assuming that the selling conditions for the doughnuts remain the same). In the previous weeks the frequencies with which she sold different quantities of doughnuts are shown in Table 8.1.

Table 8.1

Number of boxes of doughnuts	Number of times this quantity of doughnuts was sold	Probability
5	11	$\frac{11}{60} = .18$
6	13	$\frac{13}{60} = .22$
7	14	$\frac{14}{60} = .23$
8	12	$\frac{12}{60} = .20$
9	7	$\frac{7}{60} = .12$
10	3	$\frac{3}{60} = .05$
	60	

Before making a decision, however, other factors have to be considered. For example, if too many boxes of doughnuts are ordered, the cafeteria will lose $5 for each box not sold. On the other hand, if too few boxes are ordered, then the demand will be greater than the stock available. The cafeteria will then lose a possible profit of $2 on each box of doughnuts not in stock.

Mary Lou must make a decision in the face of many uncertainties. Should she gamble and order more in order to realize a greater profit, or should she play it safe and order less?

Often in our daily lives we are confronted with similar situations where we must make a decision in the face of uncertainties. In this chapter we discuss methods of analyzing these uncertainties to help us arrive at an optimal decision. It is important to note that *decision analysis* does not remove the uncertainty or risk present. It merely helps us improve our decision making under these circumstances.

Historical Note

The mathematical theory used in decision analysis is still primarily in the research stage. With the ever-growing need to make decisions in the face of increasing uncertainties, mathematicians are often called upon to formulate decision-making processes and to develop techniques for choosing the best possible alternative even when the number of alternatives is large. Several approaches are currently being used. These include the minimax criterion discussed in Chapter 7 (game theory) and decision trees. In each case, no matter which technique is used, all the possible alternatives must be clearly defined and enumerated. Hopefully, once the subject is on a firm foundation and the theory sufficiently developed, the problem of decision making under uncertainty should be realistically solvable.

8.2 Expected Value

Let us return to Mary Lou's problem discussed in the previous section. In recent weeks she knows that she has sold between 5 and 10 boxes of doughnuts per day. The frequencies with which she has sold these quantities are shown in Table 8.1. Also given in the table are the probabilities that given quantities are sold. These probabilities are obtained by dividing the frequencies with which the different numbers of boxes were sold by 60, the total frequency.

We can now use the information in Table 8.1 to calculate the expected number of boxes of doughnuts that will be sold. We multiply each possibility by its probability and add the results. We have

$$5(.18) + 6(.22) + 7(.23) + 8(.20) + 9(.12) + 10(.05) = 7.01$$

Thus, assuming no unusual circumstances, the cafeteria can expect to sell 7.01 boxes of doughnuts per day. Before ordering 7.01 boxes of doughnuts, however, Mary Lou has to consider other factors as well. Some of these are analyzed in Section 8.3.

In this doughnut problem, the number of boxes ordered was either 5, 6, 7, 8, 9, or 10. In determining the expected value, we multiplied each of these possibilities by its probability and added the results.

Definition 8.1 If a quantity can take the values $x_1, x_2, \ldots, x_n$ and if the probabilities of $x_1, x_2, x_3, \ldots, x_n$ are, respectively, $p_1, p_2, \ldots, p_n$, then the *expected value* or *mathematical expectation* of the quantity is defined as

$$x_1p_1 + x_2p_2 + x_3p_3 + \cdots + x_np_n$$

The subscripts have no special significance. We use them only to avoid using too many different letters.

We illustrate the use of expected value with several examples.

Example 1 *Expected number of sick days.* Union officials of the Northville Policemen's Benevolent Association are negotiating a new contract with the city. They are interested in increasing the number of paid sick days per year that each police officer is allowed. Records indicate that the number of paid sick days taken by police personnel in the last contract year was as shown in the table. Find the expected number of sick days used by a police officer in Northville.

Number of sick days used	3	4	5	6	7	8	9	10	11	12
Percentage of police officers using these days	.03	.05	.06	.07	.09	.18	.21	.11	.13	.07

Solution In this case the number of sick days taken was 3, 4, 5, . . ., 12 and the probabilities are .03, .05, .06, . . ., .07. Applying Definition 8.1, we have

$$\begin{aligned}\text{Expected number of sick days}\\ \text{used by a police officer} &= 3(.03) + 4(.05) + 5(.06) + 6(.07)\\ &\quad + 7(.09) + 8(.18) + 9(.21)\\ &\quad + 10(.11) + 11(.13) + 12(.07)\\ &= 8.34\end{aligned}$$

Thus, the expected number of sick days used by a police officer in Northville is 8.34.

Example 2 *Selecting a plant site.* The Mark Manufacturing Company is expanding its operations and is considering opening two new factories, one on the East Coast in New York and one on the West Coast in San Diego. Market research indicates that if it opens the new factory in San Diego, it can expect to make a profit of $100,000 per year with a probability of .62. On the other hand, if it opens in New York, it can expect to make a profit of $175,000 with a probability of .38. What is the company's mathematical expectation?

Solution We use Definition 8.1. We have

$$\begin{aligned}100{,}000(.62) + 175{,}000(.38) &= 62{,}000 + 66{,}500\\ &= 128{,}500\end{aligned}$$

Thus, the company's mathematical expectation is $128,500.

Example 3 *Oil exploration.* A small oil company is willing to spend $10 million to search thoroughly for oil on a particular offshore site. If oil is found in sufficient quantities, the company can expect to make a profit of $500 million (over a 15-year period). Geologists estimate that the probability of finding oil in the area is only .16, and the probability of not finding oil in sufficient quantities is .84. What is the company's mathematical expectation?

Photo: Wide World Photos

Solution We use Definition 8.1. If no oil is found, then the company will lose the \$10 million it invests. If we let a + sign denote a gain and a − sign a loss, then the company's mathematical expectation is

$$\begin{aligned} -10{,}000{,}000(.84) + 500{,}000{,}000(.16) &= -8{,}400{,}000 + 80{,}000{,}000 \\ &= 71{,}600{,}000 \end{aligned}$$

Thus, the company's mathematical expectation is \$71.6 million.

Example 4 *Expected family size.* Based on extensive research, a sociologist has collected the following data on family size in Sharon Lakes. If a family

Number of people in family	Probability
2	.14
3	.31
4	.26
5	.15
6	.08
7	.04
8	.02

is selected at random, what is the expected family size?

Solution We use Definition 8.1. We have

$$\begin{aligned} &2(.14) + 3(.31) + 4(.26) + 5(.15) + 6(.08) + 7(.04) + 8(.02) \\ &\quad = 0.28 + 0.93 + 1.04 + 0.75 + 0.48 + 0.28 + 0.16 \\ &\quad = 3.92 \end{aligned}$$

Thus, the expected family size is 3.92 people.

Example 5 *Winning lottery ticket.* In a local charity drive, 5000 lottery tickets are sold at \$5 apiece. Helen buys one lottery ticket. If there will be only one winning prize of \$8000, what is Helen's mathematical expectation.

Solution We apply Definition 8.1. Since Helen bought one \$5 ticket, her loss is \$5 if she does not win and her profit is \$8000 − \$5 or \$7995 if she does win. Thus, her mathematical expectation is

$$\begin{aligned} -5\left(\frac{4999}{5000}\right) + 7995\left(\frac{1}{5000}\right) &= -4.999 + 1.599 \\ &= -3.40 \end{aligned}$$

Thus, Helen's mathematical expectation is −\$3.40.

In the next section we indicate how to interpret expected value.

Exercises

1. *Sewer construction.* A contractor is bidding on a sewer construction project that promises a profit of \$500,000 with a probability of .68. Furthermore, there is a .32 probability that the contractor will not be awarded the construction project. The contractor will then lose the \$75,000 spend in preparing construction plans. What is the contractor's mathematical expectation?

2. *Rolling dice.* Bill and Martin are playing a game. Martin rolls a pair of dice. Bill agrees to pay Martin \$1 for each dot that comes up on both dice. Find Martin's mathematical expectation.

3. *Planning a family.* Mr. and Mrs. Hans are planning to have five children. Find the expected number of girls in the family. (Assume that the birth of a boy or a girl is equally likely.)

4. *Operating a ferry.* Marcelle operates a ferry between two islands. Based on past experience, he has determined that the number of cars using the ferry between the hours of 11:00 A.M to 2:00 P.M. is as follows:

Number of cars using ferry	0	1	2	3	4	5	6	7	8
Probability	.07	.19	.22	.15	.08	.12	.09	.06	.02

What is the expected number of cars using the ferry during these hours?

5. *Raising funds for charity.* The town of Waterloo is sponsoring a bazaar to raise funds to buy a new police car. If the bazaar is held outdoors, then \$3800 can be raised. If it rains, the bazaar will have to be held indoors and only \$2000 can be raised. If the probability of rain is $\frac{3}{7}$ and the probability of nice weather is $\frac{4}{7}$, how much money can the town expect to raise?

6. *Suicide prevention.* The local chapter of the suicide prevention unit receives the following number of calls for assistance on the night before Thanksgiving Day. What is the expected number of calls to the suicide prevention unit during the indicated time period?

Number of calls	0	1	2	3	4	5	6	7	8	9	10	11
Probability	.05	.10	.15	.17	.12	.08	.09	.06	.05	.05	.04	.04

7. *Fast-food store.* A company that operates fast-food stores is planning to open a new store in one of two locations. Market research indicates that if it opens the new store in location A, its annual profit will be \$80,000 if successful and its annual loss will be \$10,000 if it is not. If the company opens in locations B, its annual profit will be \$60,000 if

successful but an annual loss of $8,000 if not successful. If the probability of success at either location is $\frac{1}{2}$, where should the company open its new store?

8. *Ski equipment.* Sherry operates a ski shop and sells different items used by skiers. The probability with which a skier will buy several pieces of equipment as well as the profit earned are shown in the accompanying table. Find Sherry's expected profit.

Equipment	Probability	Profit earned
Skis	.41	$40.00
Boots	.28	30.00
Poles	.08	3.00
Goggles	.04	1.50
Bindings	.19	15.00

9. *Which contract is best?* A small oil company is considering bidding on four different tracts of land. The following chart indicates the expected incomes, estimated costs of drilling, and estimated probabilities of finding oil in the different tracts.

Tract	Estimated cost of drilling	Estimated yearly profit	Probability of finding oil
A	$10 million	$300 million	.26
B	15 million	412 million	.19
C	4 million	188 million	.32
D	21 million	281 million	.23

a) Find the expected profit from each tract of land.
b) On which tract of land should the oil company bid?

10. *Emergency call for ambulance service.* The city of Newville is analyzing the possibility of decreasing the number of ambulances on call during the hours from 12:00 P.M. to 6:00 A.M. Based on past experience, the numbers of emergency calls for ambulance service during these hours are as shown in the table. What is the expected number of emergency calls for ambulance service during these hours?

Number of calls for service	0	1	2	3	4	5	6	More than 6
Probability	.03	.08	.19	.31	.26	.09	.04	0

11. *Baseball bets.* Two baseball fans are discussing the upcoming baseball world series between the American and National leagues. Mel has made

the following bet with Pete. If the American League wins the series in four or five games, then Mel will pay Pete $50. If the American League wins in six or seven games, then he will pay Pete $25. On the other hand, if the National League wins in four or five games, then Pete will pay him $60, and if the National League wins in six or seven games, then Pete will pay him $20. Mel believes that the probabilities associated with these outcomes are as shown in the table. Find Mel's expected winnings.

Winner	Probability
National League in 4 or 5 games	.41
National League in 6 or 7 games	.27
American League in 4 or 5 games	.19
American League in 6 or 7 games	.13

12. *Selecting the best bid.* The Hoffman Construction Company is bidding on four different jobs. The possible profits and losses associated with each job and the corresponding probabilities are shown in the accompanying table. (The cost associated with submitting a bid becomes a loss if the company does not win the contract on the job.)

Construction job	Profit if company wins contract	Probability of winning contract	Loss if company does not win contract
Paper recycling plant	$100,000	.18	$11,000
Animal shelter	70,000	.27	9,000
Commuter parking lot	13,000	.43	4,000
Shopping center	90,000	.12	8,000

a) Find the company's expected profit to be made from each possible construction job.
b) Which construction job has the greatest expected profit?

13. *Roll the coin.* Six coins are thrown. For each heads that comes up, Jack will pay Jill $1. How much money can Jill expect to win if this game is repeated 50 times?

★14. *Playing craps.* Mike and John are playing craps. A pair of dice is rolled and the number of dots that comes up is observed. If the number of dots showing is 7 or 11, then John will pay Mike $20. If the number of dots showing is 2, 3, or 12, then Mike pays John $20. If the number of dots showing is 4, 5, 6, 8, 9, or 10, then the dice are rolled again until either a 7 comes up or the number of dots showing on two successive throws is the same. If the latter happens, then Mike will win $20. If the

7 comes up, however, then Mike will pay John \$20. What is the expected amount of money that Mike can win?

8.3 Decision Making

Until now we have been merely analyzing the expected number of items to be sold. In a realistic situation, however, before a decision is made, other factors such as possible profits and lossess have to be analyzed also. For example, in the doughnut example discussed at the beginning of the previous section, Mary Lou knows that if too many boxes are ordered, then the cafeteria will lose \$5 for each unsold box of doughnuts. Furthermore, the unsold boxes may or may not have some salvage value. On the other hand, if the demand for the doughnuts exceeds the supply, then the cafeteria will lose \$2 for each box not in stock. In the process of arriving at a decision, all these factors have to be considered.

Mary Lou has prepared a chart indicating the possible profits that can be made depending on the number of doughnuts ordered. Such a chart is given in Table 8.2. In each case an entry represents the profit minus the loss from unsold stock. For example, if Mary Lou orders 5 boxes and sells 5 boxes, then the profit is

$$5(2) - 5(0) = \$10$$

However, if she orders 10 boxes and sells only 5 boxes, the profit is

$$5(2) - 5(5) = -\$15$$

This represents a loss of \$15 to the cafeteria.

We can now combine the data in Table 8.2 with the probabilities of selling the indicated numbers of boxes of doughnuts as given in Table 8.1. Then we calculate the expected profit depending on the number of boxes of doughnuts ordered.

If 5 boxes are ordered, then the expected profit is

$$\begin{aligned}&10(.18) + 10(.22) + 10(.23) + 10(.20) + 10(.12) + 10(.05)\\&\quad = 1.80 + 2.20 + 2.30 + 2.00 + 1.20 + 0.50 = \$10.00\end{aligned}$$

Table 8.2 The Possible Profits That Can Be Earned Depending on the Number of Boxes Ordered and Sold

		Number of boxes sold					
		5	6	7	8	9	10
Number of boxes ordered	5	10	10	10	10	10	10
	6	5	12	12	12	12	12
	7	0	7	14	14	14	14
	8	−5	2	9	16	16	16
	9	−10	−3	4	11	18	18
	10	−15	−8	−1	6	13	20

If 6 boxes are ordered, then the expected profit is

$$5(.18) + 12(.22) + 12(.23) + 12(.20) + 12(.12) + 12(.05)$$
$$= 0.90 + 2.64 + 2.76 + 2.40 + 1.44 + 0.60 = \$10.74$$

If 7 boxes are ordered, then the expected profit is

$$0(.18) + 7(.22) + 14(.23) + 14(.20) + 14(.12) + 14(.05)$$
$$= 0 + 1.54 + 3.22 + 2.80 + 1.68 + 0.70 = \$9.94$$

If 8 boxes are ordered, then the expected profit is

$$-5(.18) + 2(.22) + 9(.23) + 16(.20) + 16(.12) + 16(.05)$$
$$= -0.90 + 0.44 + 2.07 + 3.20 + 1.92 + 0.80 = \$7.53$$

If 9 boxes are ordered, then the expected profit is

$$-10(.18) + (-3)(.22) + 4(.23) + 11(.20) + 18(.12) + 18(.05)$$
$$= -1.80 - 0.66 + 0.92 + 2.20 + 2.16 + 0.90 = \$3.72$$

If 10 boxes are ordered, then the expected profit is

$$-15(.18) + (-8)(.22) + (-1)(.23) + 6(.20) + 13(.12) + 20(.05)$$
$$= -2.70 + (-1.76) + (-0.23) + 1.20 + 1.56 + 1.00 = -\$0.93$$

The maximum expected profit of $10.74 occurs when 6 boxes of doughnuts are ordered. Since Mary Lou wishes to maximize the expected profit, she should order 6 boxes.

We can summarize the results of the previous paragraphs in the following rule.

Rule 1 *Optimal decision-making rule:* In determining how much stock to purchase, the best decision to obtain the most profit is to purchase the quantity of stock that results in the *greatest expected profit.*

Comment In the preceding problem the maximum expected profit is $10.74. How do we interpret this result? Does it mean that the cafeteria will earn $10.74 profit when it orders six boxes of doughnuts? Obviously not. Table 8.2 indicates that when six boxes of doughnuts are ordered, then the profit is either $5 or $12, depending on how many boxes are sold. The $10.74 is a long-range prediction of the expected profit. If Mary Lou uses the same procedure discussed each day to determine how many boxes of doughnuts to order, then in the long run the cafeteria will earn $10.74 profit.

We illustrate the use of Rule 1 with another example.

Example 6 *How many newspapers?* Sam operates a newspaper stand in a busy bus terminal. Each morning he must decide how many newspapers to order. Sam pays 16¢ for a paper that he, in turn. sells for 20¢. At the end of the day, any unsold newspapers have salvage value of 3¢ each. Based on past experience, he has calculated the probabilities shown in the accompanying table. How many newspapers should he order?

Number of newspapers ordered	Probability of selling this quantity of newspapers
75	.03
100	.11
125	.26
150	.32
175	.15
200	.07
225	.06

Solution In order to apply the optimal decision-making rule, we must first calculate the possible profits from the different numbers of newspapers ordered. We have:

$$\text{Profit} = \text{income} - \text{cost} + \text{salvage value}$$

so we get the possible profit chart shown below.

Now we calculate the expected profit by multiplying the possible profits by the associated probabilities. If 75 newspapers are ordered, then the expected profit is

$$\begin{aligned} &3.00(.03) + 3.00(.11) + 3.00(.26) + 3.00(.32) + 3.00(.15) \\ &\quad + 3.00(.07) + 3.00(.06) \\ &= 0.09 + 0.33 + 0.78 + 0.96 + 0.45 + 0.21 + 0.18 \\ &= \$3.00 \end{aligned}$$

Photo: DeWys

Possible Profit with Salvage Value

		Number of newspapers sold						
		75	100	125	150	175	200	225
Number of newspapers ordered	75	3.00	3.00	3.00	3.00	3.00	3.00	3.00
	100	−0.25	4.00	4.00	4.00	4.00	4.00	4.00
	125	−3.50	0.75	5.00	5.00	5.00	5.00	5.00
	150	−6.75	−2.50	1.75	6.00	6.00	6.00	6.00
	175	−10.00	−5.75	−1.50	2.75	7.00	7.00	7.00
	200	−13.25	−9.00	−4.75	−0.50	3.75	8.00	8.00
	225	−16.50	−12.25	−8.00	−3.75	0.50	4.75	9.00

If 100 newspapers are ordered, then the expected profit is

$$\begin{aligned} &-0.25(.03) + 4.00(.11) + 4.00(.26) + 4.00(.32) + 4.00(.15) \\ &\quad + 4.00(.07) + 4.00(.06) \\ &= -0.0075 + 0.44 + 1.04 + 1.28 + 0.60 + 0.28 + 0.24 \\ &= \$3.87 \end{aligned}$$

If 125 newspapers are ordered, then the expected profit is

$$\begin{aligned}
&-3.50(.03) + 0.75(.11) + 5.00(.26) + 5.00(.32) \\
&\quad + 5.00(.15) + 5.00(.07) + 5.00(.06) \\
&= -0.11 + 0.08 + 1.30 + 1.60 + 0.75 + 0.35 + 0.30 \\
&= \$4.27
\end{aligned}$$

If 150 newspapers are ordered, then the expected profit is

$$\begin{aligned}
&-6.75(.03) + (-2.50)(.11) + 1.75(.26) + 6.00(.32) \\
&\quad + 6.00(.15) + 6.00(.07) + 6.00(.06) \\
&= -0.2025 + (-0.275) + 0.455 + 1.92 + 0.90 + 0.42 + 0.36 \\
&= \$3.58
\end{aligned}$$

If 175 newspapers are ordered, then the expected profit is

$$\begin{aligned}
&-10.00(.03) + (-5.75)(.11) + (-1.50)(.26) + 2.75(.32) \\
&\quad + 7.00(.15) + 7.00(.07) + 7.00(.06) \\
&= -0.30 + (-0.6325) + (-0.39) + 0.88 + 1.05 + 0.49 + 0.42 \\
&= \$1.52
\end{aligned}$$

If 200 newspapers are ordered, then the expected profit is

$$\begin{aligned}
&13.25(.03) + (-9.00)(.11) + (-4.75)(.26) + (-0.50)(.32) \\
&\quad + 3.75(.15) + 8.00(.07) + 8.00(.06) \\
&= -0.3975 + (-0.99) + (-1.235) + (-0.16) \\
&\quad + 0.5625 + 0.56 + 0.48 \\
&= -\$1.18
\end{aligned}$$

If 225 newspapers are ordered, then the expected profit is

$$\begin{aligned}
&-16.50(.03) + (-12.25)(.11) + (-8.00)(.26) + (-3.75)(.32) \\
&\quad + 0.50(.15) + 4.75(.07) + 9.00(.06) \\
&= -0.495 + (-1.3475) + (-2.08) + (-1.20) \\
&\quad + 0.075 + 0.3325 + 0.54 \\
&= -\$4.18
\end{aligned}$$

Thus, to maximize his profit, Sam should order 125 newspapers daily. If he orders 125 newspapers daily, his expected profit is \$4.27.

Comment The preceding calculations do not allow for the losses that occur when the number of newspapers demanded is greater than the number of newspapers available. Such a loss is called an *opportunity loss*. It is obtained by calculating the greatest possible profit that could be made if there had been sufficient newspapers to meet the demand minus the profit that was actually made. In most cases, however, the greatest expected profit will occur for the same amount of stock that results in the smallest expected loss.

Comment It is also possible that when Sam does not order a sufficient supply of newspapers, then the customers will go to another newspaper stand to buy a paper. These customers may then continue to buy their newspapers at the competing newsstand and never return to Sam even when he does have enough newspapers.

Exercises

For Exercises 15–20 use the following information: Martha owns and operates a drive-up milk store. She receives deliveries twice a week, on Monday and Friday. Any milk purchased on Monday that is not sold by Friday must be dumped out as it spoils. Similarly any milk that is purchased on Friday must be sold by Monday. Martha would like to know how many cases of milk she should order each Monday. In the past, she has been ordering 40 cases of milk every Monday morning at a cost of \$7 a case. She makes a profit of \$2 on each case of milk sold. The accompanying table shows the number of cases of milk sold by the store between Monday and Friday over a 50-week period.

Cases of milk sold	Frequency (number of times this happened)
29	3
30	6
31	8
32	9
33	10
35	7
38	4
39	2
40	1
	50

15. Find the probability of selling 29, 30, 31, 32, 33, 35, 38, 39, or 40 cases of milk.

16. Find the expected number of cases of milk that Martha will sell.

17. Compute the possible profits to be made by selling the different numbers of cases of milk.

18. How many cases of milk should Martha order to maximize her expected profit?

19. Assume that Martha raises the price so that she now makes a profit of \$2.50 on each case of milk she sells. Compute the possible profits to be made now by selling the different numbers of cases of milk. (Assume that the demand for milk will remain the same despite the increase in price.)

20. How many cases of milk should Martha now order so as to maximize her expected profit?

For Exercises 21–27 use the following information: Refer to the previous exercises. Martha also sells loaves of bread, which she orders daily. Each day Martha orders 100 loaves of bread for which she pays 45¢ apiece. She sells them at a profit of 7¢ each. Unlike the milk, however, any bread not sold by the end of two days does not have to be thrown away. The old bread can be used for animal feed, bread crumbs, and other purposes. The salvage value is 13¢ a loaf. The accompanying table shows the numbers of loaves of bread sold by the store daily over a three-month period.

Loaves of bread sold	Frequency (number of times this happened)
89	10
92	14
93	16
96	18
97	11
98	10
99	7
100	4
	90

21. Find the probability of selling 89, 92, 93, 96, 97, 98, 99, or 100 loaves of bread.

22. Find the expected number of loaves of bread that Martha will sell.

23. Compute the possible profits to be made by selling the different numbers of loaves of bread.

24. How many loaves of bread should Martha order to maximize her expected profit?

25. Assume that the wholesale supplier lowers the price of a loaf of bread to 43¢. Martha will not lower her prices. Compute the possible profits to be made now by selling the different numbers of loaves of bread.

26. Assume that Martha raises the price of bread so that she now makes a profit of 9¢ a loaf. Compute the possible profits to be made now by selling the different numbers of loaves of bread. (Assume that the demand for bread will remain the same despite the increase in price.)

27. Compare the answers obtained in Exercises 25 and 26. Comment.

8.4 Applications to Life Insurance

The ideas of probability and statistics play an important role in the insurance business. Today there are companies that issue policies providing health insurance, liability insurance, cancer (only) insurance, fire insurance, and life insurance, to name a few of the different types available to the individual. Since insurance companies are not in business to lose money, the ideas of expected value discussed in the preceding section are crucial in helping the company decide what premium to charge.

Let us see how this is done by analyzing one particular type of insurance available to an individual—life insurance. The National Association of (State) Insurance Commissioners collected data on the experience of life insurance companies with ordinary life insurance policies for the years 1950–1954. The resulting data were organized and arranged in the form of *mortality tables*, one of which is given in Table 8.3. This table gives the number of individuals living at the various ages, the average number of deaths per 1000 people at these ages, and the average number of years people of a given age will live (*life expectancy*). These tables are periodically reviewed and revised by

Table 8.3 Commissioners 1958 Standard Ordinary Mortality Table (1950–1954)

Age	Number living	Deaths per 1000	Life expectancy (in years)	Age	Number living	Deaths per 1000	Life expectancy (in years)
0	10,000,000	1.08	68.30	25	9,575,636	1.93	45.82
1	9,929,200	1.76	67.78	26	9,557,155	1.96	44.90
2	9,911,725	1.52	66.90	27	9,538,423	1.99	43.99
3	9,896,659	1.46	66.00	28	9,519,442	2.03	43.08
4	9,882,210	1.40	65.10	29	9,500,118	2.08	42.16
5	9,868,375	1.35	64.19	30	9,480,358	2.13	41.25
6	9,855,053	1.30	63.27	31	9,460,165	2.19	40.34
7	9,842,241	1.26	62.35	32	9,439,447	2.25	39.43
8	9,829,840	1.23	61.43	33	9,418,208	2.32	38.51
9	9,817,749	1.21	60.51	34	9,396,358	2.40	37.60
10	9,805,870	1.21	59.58	35	9,373,807	2.51	36.69
11	9,794,005	1.23	58.65	36	9,350,279	2.64	35.78
12	9,781,958	1.26	57.72	37	9,325,594	2.80	34.88
13	9,769,633	1.32	56.80	38	9,299,482	3.01	33.97
14	9,756,737	1.39	55.87	39	9,271,491	3.25	33.07
15	9,743,175	1.46	54.95	40	9,241,359	3.53	32.18
16	9,728,950	1.54	54.03	41	9,208,737	3.84	31.29
17	9,713,967	1.62	53.11	42	9,173,375	4.17	30.41
18	9,698,230	1.69	52.19	43	9,135,122	4.53	29.54
19	9,681,840	1.74	51.28	44	9,093,740	4.92	28.67
20	9,664,994	1.79	50.37	45	9,048,999	5.35	27.81
21	9,647,694	1.83	49.46	46	9,000,587	5.83	26.95
22	9,630,039	1.86	48.55	47	8,948,114	6.36	26.11
23	9,612,127	1.89	47.64	48	8,891,204	6.95	25.27
24	9,593,960	1.91	46.73	49	8,829,410	7.60	24.45

each of the insurance companies using them to reflect changing (favorable or unfavorable) company experiences, medical and technological advances, and other factors.

When an individual applies for life insurance, the company must determine the probability that the individual will die (or live) while the insurance policy is in effect. This can be calculated easily using the data in Table 8.3. We illustrate the technique with several examples.

Example 7 Find the probability that a 40-year-old individual will *not* live to be 41 years old.

Solution From Table 8.3 we find that there were 9,241,359 people alive at age 40 and 9,208,737 alive at age 41. Thus, 9,241,359 − 9,208,737 or 32,622 people died between their fortieth and forty-first birthdays. The probability that a 40-year-old individual will not live to be 41 years

Table 8.3 Mortality Table (*Continued*)

Age	Number living	Deaths per 1000	Life expectancy (in years)	Age	Number living	Deaths per 1000	Life expectancy (in years)
50	8,762,306	8.32	23.63	75	4,129,906	73.37	7.81
51	8,689,404	9.11	22.82	76	3,826,895	79.18	7.39
52	8,610,244	9.96	22.03	77	3,523,881	85.70	6.98
53	8,524,486	10.89	21.25	78	3,221,884	93.06	6.59
54	8,431,654	11.90	20.47	79	2,922,055	101.19	6.21
55	8,331,317	13.00	19.71	80	2,626,372	109.98	5.85
56	8,223,010	14.21	18.97	81	2,337,524	119.35	5.51
57	8,106,161	15.54	18.23	82	2,058,541	129.17	5.19
58	7,980,191	17.00	17.51	83	1,792,639	139.38	4.89
59	7,844,528	18.59	16.81	84	1,542,781	150.01	4.60
60	7,698,698	20.34	16.12	85	1,311,348	161.14	4.32
61	7,542,106	22.24	15.44	86	1,100,037	172.82	4.06
62	7,374,370	24.31	14.78	87	909,929	185.13	3.80
63	7,195,099	26.57	14.14	88	741,474	198.25	3.55
64	7,003,925	29.04	13.51	89	594,477	212.46	3.31
65	6,800,531	31.75	12.90	90	468,174	228.14	3.06
66	6,584,614	34.74	12.31	91	361,365	245.77	2.82
67	6,355,865	38.04	11.73	92	272,552	265.93	2.58
68	6,114,088	41.68	11.17	93	200,072	289.30	2.38
69	5,859,253	45.61	10.64	94	141,191	316.66	2.07
70	5,592,012	49.79	10.12	95	97,165	351.24	1.80
71	5,313,586	54.15	9.63	96	63,037	400.56	1.51
72	5,025,855	58.65	9.15	97	37,787	488.42	1.18
73	4,731,089	63.26	8.69	98	19,331	668.15	0.83
74	4,431,800	68.12	8.24	99	6,415	1000.00	0.50

old is

$$\frac{32{,}622}{9{,}241{,}359} = .00353$$

The same probability can be obtained by referring to the "Deaths per 1000" column in Table 8.3. According to the table, there were 3.53 deaths per 1000 individuals aged 40.

Example 8 Find the probability that a 35-year-old individual will live to be 50 years old.

Solution From Table 8.3 we find that there were 9,373,807 people of the original 10,000,000 people alive at age 35 and 8,762,306 alive at age 50. Thus, 8,762,306 of the original 9,373,807 remained alive. The probability that a 35-year-old individual will live to be 50 years old is

$$\frac{8{,}762{,}306}{9{,}373{,}807} = .93476$$

Example 9 Etham Wilson is 75 years old.

a) Find Etham's life expectancy.
b) What is the probability that Etham will live to be 90 years old?

Solution

a) By referring to the fourth column of Table 8.3, we find that the life expectancy of a 75-year-old individual is 7.81 years. Thus, a person of Etham's age can expect to live, on the average, 75 + 7.81, or 82.81 years.
b) From Table 8.3 we find that there were 4,129,906 people living at age 75 and 468,174 living at age 90. Thus, the probability that 75-year-old Etham will live to be 90 years old is

$$\frac{468{,}174}{4{,}129{,}906} = .11336$$

Comment It is important to interpret the data in Table 8.3 properly. For example, according to Table 8.3, the life expectancy of a 30-year-old individual is 41.25 years. This does *not guarantee* that a 30-year-old individual will definitely live an additional 41.25 years. On the contrary, the individual may die within a year. In the long run, however, when many individuals are considered, a 30-year-old person will live on the average an additional 41.25 years.

We can summarize the results of the previous examples in the following formula.

Formula 8.1 Let N_a be the number of individuals alive who are a years old and let N_b be the number of individuals who are b years old.

Then the probability that an individual who is a years old will live to be b years old is

$$\frac{N_b}{N_a}$$

The probability that this individual will not live to be b years old is

$$\frac{N_a - N_b}{N_a}$$

where the values of N_a and N_b can be obtained by referring to Table 8.3.

Once an insurance company knows the probability that an individual who is a years old will live to be b years old, the next thing to determine is the premium to charge for life insurance. Here is where expected value plays an important role. In our discussion we will neglect any overhead costs, any costs associated with building up a reserve, and any other incidental costs. In reality the premium charged to an individual must cover both mortality costs and all other costs.

To calculate the premium charges, let us assume that Hiram Kirkland, who is 47 years old, applies to a life insurance company for a one-year, \$10,000 life insurance policy. Thus, if he dies within the year, the company will pay his beneficiary \$10,000. How much should the company charge for such a policy? It should be obvious that if Hiram *does not* die within the year, then the premium paid will be profit to the company. On the other hand, if Hiram does die within the year, then the company will have to pay \$10,000 to his beneficiary. Obviously the company will then lose money, as the premium will be considerably less than the amount the company must pay out.

Since Hiram will live or die within the year, we can calculate the company's expected profit as follows: Let p be the premium charged. Then we have the following table.

Profit to the company if Hiram lives	Profit to the company if Hiram dies within a year
p	$p - 10{,}000$

Comment If Hiram dies within a year, then $p - 10{,}000$ will be a negative number. This represents the company's loss.

Let us now calculate the probability that a 47-year-old person will live to be 48 years old. Using Formula 8.1 and Table 8.3, we have

$$\text{Probability that a 47-year-old person will live to be 48 years old} = \frac{N_{48}}{N_{47}}$$

$$= \frac{8{,}891{,}204}{8{,}948{,}114} = .99364$$

and

$$\text{Probability that a 47-year-old person will } \textit{not} \text{ live to be 48 years old} = \frac{N_{47} - N_{48}}{N_{47}}$$

$$= \frac{8{,}948{,}114 - 8{,}891{,}204}{8{,}948{,}114}$$

$$= .00636$$

Then the expected profit for the insurance company is

$$\begin{pmatrix}\text{Profit if}\\ \text{Hiram lives}\end{pmatrix} \cdot \begin{pmatrix}\text{probability}\\ \text{that}\\ \text{Hiram will}\\ \text{live}\end{pmatrix} + \begin{pmatrix}\text{profit if}\\ \text{Hiram dies}\end{pmatrix} \cdot \begin{pmatrix}\text{probability}\\ \text{that}\\ \text{Hiram will not}\\ \text{live}\end{pmatrix}$$

$$= p(.99364) + (p - 10{,}000)(.00636)$$
$$= .99364p + .00636p - (10{,}000)(.00636)$$
$$= p - 63.60$$

If the company wants to break even, then its expected profit is zero. Thus, we solve this equation for p:

$$0 = p - 63.60$$
$$p = 63.60$$

This means that if the company charges \$63.60 as its premium for every 47-year-old person that applies for a one-year, \$10,000 life insurance policy, then in *the long run* the company will break even. In order to make a profit, the premium charged by the company must be more than \$63.60.

Comment As we indicated earlier, our calculations do not take into consideration overhead costs, costs associated with building up a reserve fund, and any other incidental costs. These costs are added to the premium an individual must pay.

Example 10 *Living to a ripe old age.* Marilyn Black is 88 years old and has applied to an insurance company for a \$5000, one-year life insurance policy. Based on mortality tables only, what premium should the company charge Marilyn?

Solution As in the earlier illustration about Hiram Kirkland, we let p be the premium the company should charge Marilyn. Then we have the following table.

Profit to the company if Marilyn lives	Profit to the company if Marilyn dies within a year
p	$p - 5000$

Using Formula 8.1 and Table 8.3, we calculate the probability that Marilyn will live or die within the year.

$$\begin{aligned}\text{Probability that an 88-year-old person will live to be 89 years old} &= \frac{N_{89}}{N_{88}} \\ &= \frac{594{,}477}{741{,}474} = .80175\end{aligned}$$

and

$$\begin{aligned}\text{Probability that an 88-year-old person will } \textit{not} \text{ live to be 89 years old} &= \frac{N_{88} - N_{89}}{N_{89}} \\ &= \frac{741{,}474 - 594{,}477}{741{,}474} \\ &= .19825\end{aligned}$$

The expected profit for the insurance company is

$$\begin{aligned}&\begin{pmatrix}\text{Profit if}\\ \text{Marilyn lives}\end{pmatrix}\cdot\begin{pmatrix}\text{probability}\\ \text{that}\\ \text{Marilyn}\\ \text{will live}\end{pmatrix} + \begin{pmatrix}\text{profit if}\\ \text{Marilyn dies}\end{pmatrix}\cdot\begin{pmatrix}\text{probability}\\ \text{that}\\ \text{Marilyn will}\\ \text{not live}\end{pmatrix}\\ &= p(.80175) + (p - 5000)(.19825)\\ &= .80175p + .19825p - (5000)(.19825)\\ &= p - 991.25\end{aligned}$$

Then we solve the equation

$$0 = p - 991.25$$

for p and get

$$p = 991.25$$

Thus, the company should charge Marilyn at least \$991.25 for the one-year, \$5000 life insurance policy.

Exercises

28. Jack Nelson is 58 years old. Assuming no unusual circumstances, find the probability that he will live to be 62 years old.

29. *Planning a wedding.* Betty Washington just turned 40 years old. She has a 10-year-old daughter. If her daughter will marry when she is 23 years old, find the probability that Betty will be alive then to help plan the wedding.

30. *Living to pay off a mortgage.* Mark Ellis owns a factory on which there is a mortgage of 20 years remaining. Mark is 56 years old. Find the probability that Mark will be alive when the final payment on the mortgage is made in 20 years.

31. *Maturing bonds.* George Barr owns some noncallable New York City municipal bonds that mature on January 1, 1992. On January 1, 1979,

George celebrated his sixty-eighth birthday. Find the probability that George will die before the bonds mature.

32. *Beating the rap.* Law enforcement officials are encountering considerable difficulty in gathering evidence to prosecute Fingers Nelson for a crime committed during 1978. At the time of the crime Fingers was 62 years old. There is a seven-year statute of limitations, after which time law enforcement officials cannot prosecute Fingers Nelson.
    a) Find the probability that Fingers Nelson dies within the seven-year period.
    b) What is the probability that he will be alive at the end of the seven-year period?

33. *Happy birthday.* On his thirty-fifth birthday, Randy Sullivan wishes to buy a one-year, $50,000 life insurance policy. What premium should the insurance company charge?

34. Jill Kennedy purchased a five-year term life insurance policy for $20,000. Jill, who is a 21-year-old healthy woman, agrees to pay the insurance company $120 at the beginning of each year for this coverage. Find the insurance company's expected profit.

35. On her forty-eighth birthday, Joyce Williamson wants to buy a five-year insurance policy that will pay her beneficiary $25,000 if she dies within the five years. What annual premium should the insurance company charge for the first year of the policy to make a profit? (Assume that the annual premium will be the same for each of the five years.)

36. Mollie Goldfarb is 55 years old and wishes to buy a life insurance policy to pay $40,000 to her beneficiaries if she dies before she begins to collect social security at age 65. What premium should the insurance company charge for the first year of the policy to make a profit?

37. *Proud father celebrates.* Isaac Lichtenfeld has just become the proud father of a little boy. Isaac, who is 32 years old, wishes to obtain life insurance for himself to provide $35,000 to his wife, Hilda, and son if he dies before his son reaches his eighteenth birthday.
    a) Find Isaac's life expectancy.
    b) Find Hilda's life expectancy if she was 29 years old when the son was born.
    c) What is the son's life expectancy at birth?
    d) What premium should the insurance company charge Isaac for the first year of the policy so as to make a profit?
    e) What premium should the insurance company charge Hilda for the first year of the policy if she wishes to obtain a similar life insurance policy?

8.5 Summary

In this chapter we attempted to analyze the difficulties involved in making decisions in the face of ever-present uncertainties. Our approach was to calculate the expected profit to be attained under all the possible alternatives and then to select the alternative with the maximum expected profit. It is for this reason that we analyzed expected value.

It should be pointed out that the mathematical theory involved in decision analysis is only in the development stage. Our approach was to calculate the greatest expected profit. An alternate approach that results in the same answer is to use marginal analysis and determine the minimum marginal loss or the maximum marginal profit.

We also studied mortality tables compiled by insurance companies. We indicated how expected value can be used by insurance companies to determine the minimum premium to charge to at least break even. Our discussion was intended merely to introduce the reader to some applications of expected value to life insurance. For a more thorough analysis, the reader can consult the references given at the end of this chapter.

Study Guide

You should be able to demonstrate your knowledge of the following ideas presented in this chapter by giving definitions, descriptions, or specific examples. Page references are given for each term so that you can check your answer.

Decision analysis (p. 351)
Optimal decision-making rule (p. 359)
Opportunity loss (p. 361)
Life expectancy (p. 364)
Expected value or mathematical expectation (p. 352)
Greatest expected profit (p. 359)
Mortality tables (p. 364)

Formulas to Remember

The following list summarizes all the formulas given in the chapter.

1. **Optimal decision-making rule:** Purchase the quantity of stock that results in the greatest expected profit.
2. The probability that an individual who is a years old now will be alive when he or she is b years old is

   $$\frac{N_b}{N_a}$$

   The probability that the individual will not be alive when he or she is b years old is

   $$\frac{N_a - N_b}{N_a}$$

Mastery Tests

FORM A

1. Find the probability that an individual who is alive at age 12 will still be alive at age 70.

2. Find the probability that an individual who is alive at age 12 will *not* be alive at age 80.

3. Find the probability that a newborn will be alive at age 17.

4. True or false: Expected profit can never be a negative number.

5. Define opportunity loss.

6. Mr. and Mrs. Carter are the proud parents of a family of eight children. What is the expected number of girls in the family? (Assume that the probability that a child is a girl is $\frac{1}{2}$.)

7. A fair coin will be tossed until six heads are obtained. Find the expected number of tosses.

8. What do we call the average number of additional years that people of a given age will live?

9. The best decision to obtain the most profit is to purchase the quantity of stock that results in the ____________.

10. Joyce buys a lottery ticket for \$8. If there are 7000 tickets sold, and if only one \$20,000 prize is to be awarded, what is her mathematical expectation?

FORM B

1. *Family income.* A sociologist has determined the family incomes of the 4000 families in Silver Springs as shown in the table. If a family is selected at random, what is the expected family income?

Family income	Number of families with this income
Under \$4,000	161
\$4,000–\$5,000	219
\$5,000–\$7,000	528
\$7,000–\$9,000	764
\$9,000–\$11,000	842
\$11,000–\$13,000	657
\$13,000–\$16,000	542
More than \$16,000	287

2. *Insurance premiums.* What premium should an insurance company charge an individual who is 49 years old and wants to buy a \$100,000, one-year life insurance policy?

3. A coin will be thrown continuously until heads comes up for the first time or until five tails come up in a row. Find the expected number of times that the coin must be thrown.

For questions 4–10 use the following information: Pete operates the lunchroom in the Apex Manufacturing Company. Each day he sells prepackaged lunches in a vending machine. The lunches are purchased from a local vendor and consist of a tuna sandwich, a salad, and a dessert, which are individually packaged. Any lunch that is not sold must be thrown away. Pete pays 80¢ for a lunch, which he sells for 90¢. In the past month Pete has ordered 200 lunches a day and has sold the number of lunches over this period as shown in the table.

Number of lunches sold	Frequency (number of times this happened)
185	1
188	3
192	2
196	4
197	3
198	4
199	8
200	6
	31

4. Find the probability of selling 185, 188, 192, 196, 197, 198, 199, or 200 lunches.

5. Find the expected number of lunches that Pete will sell.

6. Compute the possible profits to be made by selling the different numbers of lunches.

7. How many lunches should Pete order to maximize his expected profit?

8. Pete decides to increase the price of a lunch by 5¢. Assume customer demand will remain the same despite the increase in price. Compute the possible profits to be made now by selling the different numbers of lunches.

9. After carefully researching the problem of unsold lunches, Pete finds a customer who will purchase any unsold lunches for 8¢ each. Compute the possible profits to be made under these new conditions of selling the different quantities of lunches.

10. Refer to question 9. How many lunches should Pete order under these new conditions to maximize his expected profit?

★11. Refer to the newspaper clipping at the beginning of this chapter. After carefully reading the article about the company's costs associated with offshore oil drilling and the probability of finding oil, can you calculate the company's expected profit?

Suggested Further Reading

MacDonald, J., *Strategy in Poker, Business and War*, New York: Norton, 1950.

McAdams, A. K., *Mathematical Analysis for Management Decisions*, New York: Macmillan, 1970.

Shubik, Martin, *Strategy and Market Structure*, New York: Wiley, 1959.

CHAPTER 9 Statistics

Objectives to be Achieved in this Chapter

- *To discuss* three different arithmetic ways of analyzing a set of numbers. These measures of central tendency are the mean, the median, and the mode.
- *To present* a notation, summation notation, that is used to indicate the operation of adding numbers.
- *To discuss* two different ways of indicating how spread out the numbers in a set are. These measures of dispersion are the range and the standard deviation.
- *To study* frequency distributions, which are graphical techniques that can be used along with bar graphs to analyze a large list of numbers.
- *To draw* circle graphs, line graphs, and pictographs, which can also be used to interpret data visually.
- *To analyze* one very important distribution in statistics. This is the familiar bell-shaped normal curve.
- *To indicate* how to determine whether a relationship exists between several variables and then to find the equation expressing this relationship. This is linear correlation and regression.
- *To point out* several common ways in which statistics are misused.

Population Projections

	1975 population (estimated)	2000 population (projected)	25-Year change 1975–2000
Northeast			
Maine	1,058,000	1,273,000	Up 20%
New Hampshire	812,000	1,121,000	Up 38%
Vermont	472,000	614,000	Up 30%
Massachusetts	5,814,000	6,787,000	Up 17%
Rhode Island	931,000	1,107,000	Up 19%
Connecticut	3,100,000	3,713,000	Up 20%
New York	18,076,000	18,816,000	Up 4%
New Jersey	7,333,000	8,958,000	Up 22%
Pennsylvania	11,860,000	12,465,000	Up 5%
Midwest			
Ohio	10,735,000	11,999,000	Up 12%
Indiana	5,313,000	6,069,000	Up 14%
Illinois	11,198,000	12,491,000	Up 12%
Michigan	9,111,000	10,970,000	Up 20%
Wisconsin	4,589,000	5,476,000	Up 19%
Minnesota	3,921,000	4,637,000	Up 18%
Iowa	2,861,000	3,058,000	Up 7%
Missouri	4,767,000	5,506,000	Up 16%
North Dakota	637,000	631,000	Down 1%
South Dakota	681,000	679,000	Down 0.3%
Nebraska	1,544,000	1,755,000	Up 14%
Kansas	2,280,000	2,516,000	Up 10%
South			
Delaware	579,000	742,000	Up 28%
Maryland	4,122,000	5,612,000	Up 36%
District of Columbia	712,000	697,000	Down 2%
Virginia	4,981,000	6,414,000	Up 29%
West Virginia	1,799,000	1,912,000	Up 6%
North Carolina	5,441,000	6,830,000	Up 26%
South Carolina	2,816,000	3,644,000	Up 29%
Georgia	4,931,000	6,625,000	Up 34%
Florida	8,277,000	12,924,000	Up 56%
Kentucky	3,387,000	4,032,000	Up 19%
Tennessee	4,173,000	5,085,000	Up 22%
Alabama	3,615,000	4,148,000	Up 15%
Mississippi	2,341,000	2,653,000	Up 13%
Arkansas	2,110,000	2,545,000	Up 21%
Louisiana	3,806,000	4,471,000	Up 17%
Oklahoma	2,715,000	3,347,000	Up 23%
Texas	12,238,000	16,654,000	Up 36%

Population Projections

	1975 population (estimated)	2000 population (projected)	25-Year change 1975–2000
West			
Montana	746,000	862,000	Up 16%
Idaho	813,000	1,069,000	Up 31%
Wyoming	376,000	450,000	Up 20%
Colorado	2,541,000	3,615,000	Up 42%
New Mexico	1,144,000	1,409,000	Up 23%
Arizona	2,212,000	3,452,000	Up 56%
Utah	1,203,000	1,643,000	Up 37%
Nevada	590,000	851,000	Up 44%
Washington	3,559,000	4,759,000	Up 34%
Oregon	2,284,000	3,066,000	Up 34%
California	21,198,000	28,083,000	Up 32%
Alaska	365,000	474,000	Up 30%
Hawaii	868,000	1,157,000	Up 33%
Total	213,032,000	259,869,000	Up 22%

Source: U.S. Department of Commerce.
Source: *U.S. News & World Report*, November 20, 1978, p. 80.

Interest Rate Paid on Short-Term Loans

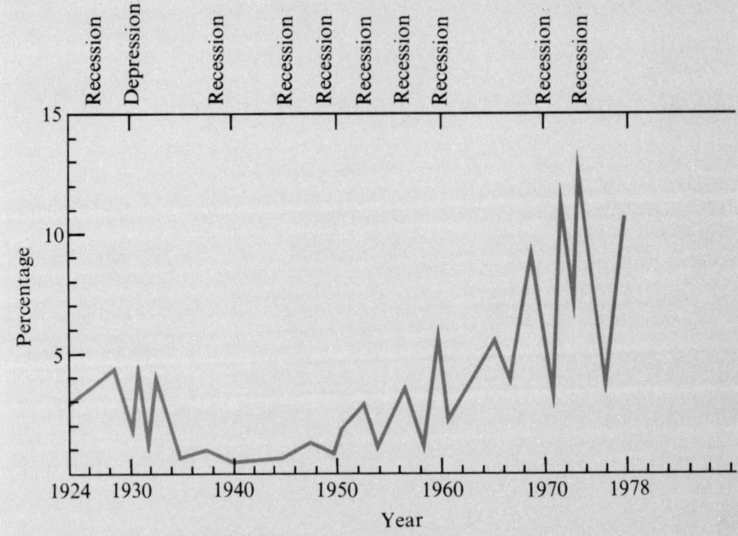

Source: Federal Reserve Board, U.S. Dept. of Commerce

Jobs Are Few for New Ph.D.s

Half of the people who obtain doctorates through 1985 will not get jobs suitable to their training, according to projections by the U.S. Bureau of Labor Statistics. This continues a trend that began with the end of the boom of the 1960s and reflects the continuing inability of our present economy to fully absorb these highly educated people.

Statistics are used in making all kinds of projections and predictions. In the first table statistics are used to determine what the U.S. population will look like, state by state, in the year 2000. By carefully examining each state's projected population in the year 2000, one immediately notices that the South and the West will grow twice as fast as the Northeast and the Midwest. Such projections have important implications, as the South and West stand to gain considerable political power to go with their larger populations.

Now consider the second illustration. Here a statistical graph from the year 1924 on is used to predict that a recession is inevitable, since interest rates are high.

In the third article we have an extremely important projection from the student point of view.

In all these articles predictions and projections are made based on past statistics. How are such projections made? What is the probability that they are accurate?

9.1 Introduction

The average person tends to think of statistics as having something to do with charts, tables, and graphs. While this is not entirely inaccurate, the word *statistics* means more than this to the mathematician. Generally speaking, the subject of statistics is concerned with two main ideas. The first involves the collection of numerical facts called *data* and the organization and arrangement of them usually in the form of charts, graphs, or tables. This branch of statistics is called *descriptive statistics.* Before the advent of the modern high-speed computer or calculator, much of the statistician's time was spent with this simple task.

The second branch of statistics is concerned with interpreting the data that have been organized and with making predictions based on the data. This branch of statistics, which is known as *inferential statistics*, requires careful and thorough analysis by the statistician.

As a simple example, consider a group of agents from the state's Environmental Protection Agency who are monitoring the level of mercury in fish caught from a polluted river. Suppose a sample of 1000 fish from different parts of the river is taken. The mercury content within these fish is then thoroughly analyzed in the laboratory and a summary of these findings is made. If the survey indicates that the level of mercury in the sample fish is too high, then the agents will conclude that all the fish in the river contain mercury levels that are too high. As a result, fishing in the river will be banned. This decision involves statistical inference and must be made only after thoroughly analyzing the data. A decision that is based on only a casual analysis of the data can often result in great monetary losses and the eventual ruin of the fishing industry in a particular area.

In this example both descriptive statistics and inferential statistics are used. Descriptive statistics is used to collect the sample fish and arrange the data obtained from the sample in some meaningful form such as charts, graphs, or tables. Then statistical inference is used in making predictions about the remaining fish in the lake.

It should be immediately apparent that statistics involves taking *samples* from the entire group, which is called the *population*. Thus, in this example a sample of 1000 fish is taken and tested as opposed to testing *all* the fish in the river.

This is but one use of statistics. Today statistics is applied to many different areas including medicine, insurance, population studies, agriculture, education, psychology, sociology, and business, to mention but a few. The following examples of the use of statistics should be quite familiar to you.

1. Statistics indicate that people who work in factories that manufacture asbestos products stand a greater chance of getting cancer than do other industrial workers.
2. Statistics gathered by a large insurance company show that most automobile accidents occur within 25 miles of the driver's home.
3. Statistics indicate that the crime rate increased last year by 12%.
4. Statistics show that a nationally advertised boycott against coffee prices was effective in lowering these prices.
5. Statistics show that more people tune in one television network during the prime hour of 7:00 P.M. to 8:00 P.M. than any other network.
6. Statistics show that people are now living longer than ever before.

As indicated earlier, great care must be exercised when interpreting statistics. It has been said that statistics can be made to say what you want them to. This is because people have misused statistics to their advantage. To understand how statistics can be misused, consider the following data: The State Thruway Authority reported that there were 70 accidents recorded on a particular road during ice storms as opposed to 491 accidents recorded when the weather was nice. Can we then conclude that it is much safer to drive on this particular road during an ice storm than during nice weather?

In this chapter we discuss several aspects of descriptive statistics, that is, how to organize the data, and several aspects of statistical inference, that is, how to interpret the data. We will also indicate how statistics can be misused.

Historical Note

The pioneer in the development of the field of statistics was John Graunt (1620–1674). He was a prosperous London haberdasher who dealt with buttons, needles, and other small items. His wealth afforded him much free time. During the seventeenth century there was a weekly publication put out by the local government called the *Bills of Mortality* listing the births and deaths in each region and their causes. To satisfy his curiosity, Graunt analyzed the data on the causes of death given in the *Bills of Mortality*. He published his findings in a book called *Natural and Political Observations Made Upon the Bills of Mortality*. Graunt studied the causes of death in different cities and noticed that the percentage of deaths from different causes was about the same and did not change considerably from year to year. For example, deaths from suicide, accidents, and certain diseases occurred not only with surprising regularity but also in approximately the same percentages from year to year. Furthermore, Graunt's statistical analysis led him to discover that there were more male than female births. Since men were more subject to death from occupational hazards, diseases, and war, however, it turned out that at marriageable age, the numbers of men and women were about equal. Graunt believed that this was nature's way of assuring monogamy.

After Graunt published his book, many other mathematicians became interested in statistics and made important contributions. Among these were Pierre Simon de Laplace (1749–1827), Abraham De Moivre (1667–1754), and Carl Friedrich Gauss (1777–1855), who studied and applied the normal distribution, and Karl Pearson (1857–1936) and Sir Francis Galton (1822–1911), who studied the correlation coefficient. Both of these ideas are discussed in this chapter.

With the advent of the computer and its ability to perform long and difficult calculations in a short time, the statistician was afforded the opportunity of developing new theory for use in statistical inference.

9.2 Measures of Central Tendency

To understand how to analyze data, let us consider two fire companies located in the northern part of town. As an economy move, city officials have decided to close down one of these fire stations. In an effort to determine which one to close, the information in the following table on the number of responses to fires made by each company has been obtained for a 13-week period.

Week	Number of responses to fires Company 1	Company 2
1	12	10
2	19	12
3	13	8
4	14	12
5	22	6
6	19	37
7	4	12
8	9	11
9	19	49
10	10	7
11	7	4
12	19	12
13	15	15
Total	182	195

If you were making the decision, which fire company would you close?

In terms of the total number of responses to fire, it would seem that fire company 2 is more important since it responded to 195 fires as compared with fire company 1, which responded to only 182 fires. Let us analyze the data more carefully, however, by taking a simple average. We add the total number of responses by each company and divide the sum by 13, the total number of weeks. For company 1, the average is

$$\frac{12 + 19 + 13 + 14 + 22 + 19 + 4 + 9 + 19 + 10 + 7 + 19 + 15}{13}$$

$$= \frac{182}{13} = 14$$

For company 2, the average is

$$\frac{10 + 12 + 8 + 12 + 6 + 37 + 12 + 11 + 49 + 7 + 4 + 12 + 15}{13}$$

$$= \frac{195}{13} = 15$$

Thus, it would appear that fire company 2 is more important since it responded to an average of 15 fires per week as compared with fire company 1, which responded to an average of only 14 fires per week.

Responses by company 1	Responses by company 2
4	4
7	6
9	7
10	8
12	10
13	11
(14)	(12)
15	12
19	12
19	12
19	15
19	37
22	49

Further analysis of the data indicates that fire company 1 reponded to 19 fires most often (4 weeks), whereas fire company 2 responded to 12 fires most often. Thus, it would now seem that fire company 1 is more important in terms of consistent performance.

Let us arrange the number of responses by each fire company in order from smallest to largest to determine a middle number, or a number that cuts the group in half. We have the accompanying table.

You will notice that two numbers have been circled. For fire company 1 this number is 14, and for fire company 2 this number is 12. Thus, for fire company 1 the number 14 cuts the group in half; six numbers are above it and six numbers are below it. For fire company 2 the number 12 cuts the group in half. Therefore, it would seem that fire company 1 is more important.

Our discussion indicates that there are many different ways of finding one typical number that can be used to represent a whole group of numbers. Which one should city officials use?

We can summarize the previous problem with the following definitions.

Definition 9.1 The *mean* or *average* of a set of numbers is found by adding them together and dividing the sum by the total number of entries added.

Notation If we are dealing with an entire set of numbers, then we denote the mean by the Greek letter μ, read as mu. This is called the *population mean.* On the other hand, when we are dealing with only a sample from the population, it is standard to use $\bar{x}$ to represent the *sample mean.* In this book we will always be dealing with the entire population and will therefore use the symbol μ to represent the mean.

We can rewrite this definition by using *summation notation.* In this notation the Greek symbol Σ, read as sigma, stands for the operation of adding a sequence of numbers. Thus, suppose we have a group of numbers $x_1, x_2, x_3, \ldots, x_n$ where n is the total number in the group. Then we have Formula 9.1.

Formula 9.1

$$\mu = \frac{\Sigma x}{n} = \frac{x_1 + x_2 + x_3 + \cdots + x_n}{n}$$

Definition 9.2 The *mode* of a set of numbers is the number that occurs most often. If every number occurs only once, there is no mode. A set of numbers may have more than one mode.

Definition 9.3 The *median* of a set of numbers is the number that is in the middle when the numbers are arranged in order from smallest to largest.

To find the median of a set of n numbers, we first arrange the numbers in order from smallest to largest. Then if n is odd, the median is the middle number, and if n is even, the median is halfway between the two middle numbers.

When these definitions are applied to our fire station example, we get the following table.

	Company 1	Company 2
Mean	14	15
Mode	19	12
Median	14	12

Again, we repeat the question raised before. If you were making the decision, which fire company would you close?

Example 1 In 1969 nine students graduated from Poly University with their master's degrees. Their names and ages at graduation are shown in the accompanying table. Find the mean, median, and mode for the ages of these graduates.

Name	Age at graduation
Nancy Rodgers	21
Hugh Maxfield	28
Bridgitte Nelson	19
Jeremiah Clarkson	34
Patricia Callighan	22
Nelson Thompson	24
Cary Schleifer	22
Paul Kemeny	25
Sigmund Owens	23

Solution Let x represent the age of each student at graduation. The mean age then is

$$\mu = \frac{\Sigma x}{n} = \frac{x_1 + x_2 + \cdots + x_n}{n}$$

$$= \frac{21 + 28 + 19 + 34 + 22 + 24 + 22 + 25 + 23}{9}$$

$$= \frac{218}{9} = 24.22 \text{ years}$$

Solution Let us now arrange the ages in order from lowest to highest. We get

19, 21, 22, 22, (23), 24, 25, 28, 34

The number in the middle has been circled. This number is the median. Thus, the median age of a graduating student is 23 years. The mode is the number that occurs most often. In this case the mode is 22 years. Therefore,

Mean = 24.22 years
Median = 23 years
Mode = 22 years

Example 2 According to the district tax office, there are eight houses on Park Place that are assessed for real estate taxes as shown in the accompanying table. Find the mean, median, and mode for these assessed values.

Address	Assessed value
1 Park Place	\$100,000
5 Park Place	91,000
9 Park Place	120,000
17 Park Place	110,000
25 Park Place	95,000
33 Park Place	88,000
39 Park Place	140,000
51 Park Place	275,000

Solution The mean assessed value is

$$\mu = \frac{\Sigma x}{n} = \frac{100{,}000 + 91{,}000 + 120{,}000 + 110{,}000 + 95{,}000 + 88{,}000 + 140{,}000 + 275{,}000}{8}$$

$$\mu = \frac{1{,}019{,}000}{8} = \$127{,}375$$

To find the median, we rearrange the numbers in order from lowest to highest:

88,000, 91,000, 95,000, (100,000, 110,000,) 120,000, 140,000, 275,000

The median is midway between the middle two numbers (the circled ones). Thus, the median is

$$\frac{100{,}000 + 110{,}000}{2} = \$105{,}000$$

Since no number occurs more than once, there is no mode.

Example 3 There are 19 employees in the math department of Bologa University. Their salaries are determined by their rank as shown in the table. What are the mean, median, and modal annual salaries of these employees?

Rank	Number of people with this rank	Annual salary
Professor	2	\$35,000
Associate professor	7	28,600
Assistant professor	4	23,500
Instructor	2	19,400
Lecturer	3	16,100
Secretary	1	10,305

Solution To find the mean, we must find the sum of the salaries of all 19 employees. We have

Number of employees	Annual salary	Total payment
2	\$35,000	\$70,000
7	28,600	200,200
4	23,500	94,000
2	19,400	38,800
3	16,100	48,300
1	10,305	10,305
		\$461,605

Thus, the mean annual salary is

$$\frac{\$461{,}605}{19} = \$24{,}295$$

The modal annual salary is \$28,600 and the median annual salary is \$23,500. Can you see why?

Comment In Example 3 the mean annual salary is \$24,295 even though nobody actually earns that amount.

Example 4 Refer to Example 2. The district tax office has decided to raise everybody's assessed value by 10% for the coming year.

a) Find the new assessed values.
b) Find the mean, median, and mode for these new assessed values.
c) What changes in the mean, median, and mode have taken place?

Solution a) If each house's assessed value is increased by 10%, then the new assessed value will be as shown in the accompanying table.

Address	Old assessed value	Increase	New assessed value
1 Park Place	$100,000	$10,000	$110,000
5 Park Place	91,000	9,100	100,100
9 Park Place	120,000	12,000	132,000
17 Park Place	110,000	11,000	121,000
25 Park Place	95,000	9,500	104,500
33 Park Place	88,000	8,800	96,800
39 Park Place	140,000	14,000	154,000
51 Park Place	275,000	27,500	302,500

b) The new mean is $140,112.50. There is no mode. The new median is $115,500. The reader should verify these results.

c) The mean and median are both increased by 10%. Since there was no mode originally, there again is no mode.

Comment The mean, median, and mode are known as *measures of central tendency*, since each measures some central or general trend, that is, some location of the data.

Exercises

1. Bill Waverly is a stockbroker. Yesterday he completed six stock transactions. The values of the stock involved in these transactions were $495, $1215, $780, $1140, $1245, and $975. Find the mean, median, and modal prices of a stock transaction.

2. A letter carrier made a survey of the number of monthly magazines received by 12 families in a large apartment house. The results are shown in the accompanying table. Find the mean, median, and mode for the number of magazines received.

Family	Magazincs received	Family	Magazines received
Blau	4	Kosco	4
Blois	3	Lee	2
Caleandra	2	Maye	8
Frost	7	Prenowitz	3
Hirth	1	Smith	4
Jackson	6	Thompson	16

3. There are 70 employees on the payroll of the Warren Bus Company as shown in the table. Find the mean, median, and modal salaries.

Category	Number of workers in this category	Annual salary
Driver	25	$19,000
Mechanic	14	21,000
Ticket agent	8	14,000
Porter	6	12,000
Secretary	3	9,000
Other	14	10,500

4. Jan loves to jog. Over the past ten days he has jogged the indicated numbers of miles. Find the mean, median, and modal number of miles jogged.

Day	Number of miles jogged
Sunday	12
Monday	14
Tuesday	11
Wednesday	17
Thursday	13
Friday	14
Saturday	10
Sunday	8
Monday	14
Tuesday	17

Photo: Everett Johnson, DeWys

5. A shoe salesman claims that the average size of men's shoe sold is 9. To which average is the salesman referring: the mean, median, or mode?

6. Herbert received the grades of 71, 83, 66, 49, 92, and 87 on his first six quizzes. What grade must he receive on the next quiz so that his average for the seven quizzes will be 77?

7. The following data are available on the salaries of the workers of the Printex Corporation: mean salary $21,000, median salary $17,000, modal salary $11,000. The management proudly boasts that the "average" salary is quite high, whereas the union claims that it is very low. To which "average" is each referring? Explain your answer.

8. Professor Black has told her class that a student must have an average of 88 on the eight tests given in order to receive an A for the course. Wes has received the scores of 91, 61, 97, 94, 90, 87, and 95 on his first seven tests. What is the lowest score that Wes can get on his eighth test to get an A?

★9. There are 48 people on the payroll of Gene's Department Store. The mean salary for a week is \$257, the median salary for a week is \$220, and the modal salary for a week is \$184.30. What is the total weekly payroll?

10. A set of numbers has a mean of 17. What happens to the mean if the following occur?
   a) Each number in the set is increased by 2.
   b) Each number in the set is decreased by 2.
   c) Each number in the set is multiplied by 2.
   d) Each number in the set is divided by 2.

11. Is it safe for a nonswimmer who is 6 feet tall to wade in a pool whose average depth is 4 feet? Explain your answer.

12. Consider the set of numbers 1, 5, 8, 3, 9, and 4. Compute the following.
   a) Σx
   b) μ
   c) $\Sigma(x - \mu)$
   d) $\Sigma(x - \mu)^2$

13. In Exercise 12 compute
   a) Σx^2

   Are the answers the same? Explain why or why not.

14. A new photocopying machine is to be installed on the tenth floor of a large office building to serve the needs of the numerous offices located on the floor. In an effort to determine exactly where the machine should be placed, a list is made of the distances from each office to one particular spot. The mean, median, and modal distances are then computed. Which of these *average* distances would be more helpful in deciding where the machine should be placed? Explain your answer.

15. The Ajax Construction Corporation is negotiating a new labor contract with the union that represents its 621 workers. Management claims that the *average* hourly rate of pay is \$8.29, which actually exceeds the pay of similar workers in other companies. The union claims that the *average* hourly rate of pay is only \$6.71, which is lower than the pay of other workers. A labor negotiator believes that both union and management claims are accurate. How can this be? Explain your answer.

16. Consider the table, which shows how taxes compare for 15 countries. The table indicates that the average total tax revenue is 38.9% of the total national output. Is 38.9% the actual average of these numbers? Explain the use of average here.

Tax Revenue as a Share of National Output for 1977

Country	Percentage
Sweden	53.3
Norway	47.4
Netherlands	46.7
Denmark	45.0
Belgium	42.8
Finland	41.5
Austria	40.4
France	39.4
West Germany	38.2
Italy	37.8
Britain	36.6
Canada	32.7
Switzerland	31.0
United States	30.4
Japan	20.5
Average	38.9

Source: Organization for Economic Cooperation and Development

17. There are six members in the Putter family as shown in the table. The average age of these family members is 25 years, which is older than all but one of the ages. What can you conclude?

Name	Age
Ned (father)	23
Mary (mother)	21
Kate (daughter)	3
Melissa (daughter)	2
Salvatore (son)	1
Frances (grandmother)	100

18. Consider the accompanying table on housing costs. Explain why the median is used when we compare housing costs across the United States.

Median Amounts Paid Monthly by Home Buyers (1977)

Metropolitan area	Mortgage payment	Real estate taxes	Utilities	Hazard insurance	Total
Atlanta	$298	$40	$62	$20	$420
Chicago	291	64	60	14	429
Denver	281	63	50	12	406
Detroit	235	73	70	11	389
Hartford	240	76	70	15	401
Houston	291	48	74	26	439
Los Angeles	403	99	50	15	567
Miami	297	50	80	16	443
Minneapolis-St. Paul	302	54	50	15	421
New York	291	111	70	25	497
Philadelphia	250	63	70	10	393
Portland	240	56	70	10	376
St. Louis	226	43	51	13	333
San Francisco	445	99	50	20	614
Washington, D.C.	388	85	91	14	578
Nationwide	273	54	60	13	400

Source: United States League of Savings Associations.

19. A new air conditioner was recently installed in a lecture hall at State University. Several students have complained to the teacher that it is now too cold, whereas others have complained that it is still too warm. In an attempt to satisfy everybody, the students are polled as to the ideal temperature for the lecture hall. The following data are obtained: mean 70°, median 72°, mode 68°. At what temperature should the air conditioner be set to satisfy as many students as possible?

9.3 Measures of Dispersion

Although the mean is a fairly good indicator of the middle or central tendency of a set of numbers, it is usually used with another indicator that measures the dispersion or spread of the data. To see why some measure of dispersion is needed, consider the salaries of the top five executives of two competing companies as shown in the accompanying table.

Company A	Company B
$35,000	$18,000
39,000	19,000
41,000	41,000
57,000	42,000
68,000	120,000
Mean: $48,000	Mean: $48,000
Median: $41,000	Median: $41,000
Mode: none	Mode: none

For both companies the mean salary is $48,000 and the median salary is $41,000, but beyond that the companies are quite different. For company A the salaries are not so spread out as they are for company B. Thus, we need some way of measuring this *dispersion.*

One common way of measuring the spread of a set of numbers is by means of the range.

Definition 9.4 The *range* of a set of numbers is the difference between the largest number and the smallest number in the set.

For the two companies we are discussing, the range is as follows:

$$\begin{aligned}\text{Range for company A} &= \text{highest salary} - \text{lowest salary}\\ &= \$68{,}000 - \$35{,}000 = \$33{,}000\\ \text{Range for company B} &= \text{highest salary} - \text{lowest salary}\\ &= \$120{,}000 - \$18{,}000 = \$102{,}000\end{aligned}$$

Thus, it is immediately apparent that company B's salary range is about three times as large as company A's.

The range is used frequently by stockbrokers to describe the prices of certain stocks. Statements like "stock x had a price range of $20 to $85 or $65 during the year 1979" are quite common. Although the range is by far the simplest measure of dispersion to calculate (since only two numbers are needed to calculate it), it is not used commonly because it does not tell us how the other numbers vary. Furthermore, if there is one extreme value in the group of numbers, then the dispersion or range will appear quite large. If we remove the extreme value, the range may be quite small. For example, the time it took Donna to drive to school during one week is shown in the accompanying table. On Friday there was an accident on the highway and it took Donna considerably longer to drive to school. If Friday *is* included, then the range of the time needed to drive to school is 88 – 21, or 67 minutes. When Friday is *not* included, then the range is 31 – 21, or 10 minutes, which is considerably less. Because of this sensitivity, the range is not used as often as other measures of dispersion.

Day	Driving time (minutes)
Monday	25
Tuesday	31
Wednesday	28
Thursday	21
Friday	88

One of the most popular measures of dispersion used today is the *standard deviation.* To see how to calculate it, let us consider the numbers 5, 8, 10, 13, and 24. The mean of these numbers is found by adding them together and dividing the sum by 5. The sum of these numbers is 60, so the mean is $\frac{60}{5} = 12$. Let us now subtract the mean from each of these numbers. We will arrange our work in tabular form as shown in the margin. (As a check, we can add all these differences. The sum of these differences, or deviations, from the mean will always be zero.)

Number	Difference from mean
5	$5 - 12 = -7$
8	$8 - 12 = -4$
10	$10 - 12 = -2$
13	$13 - 12 = 1$
24	$24 - 12 = 12$
	0

We now square each of these deviations as shown in the next table. If we do not square these differences, then the sum will always be zero, as the negative deviations will cancel out the positive deviations.

Number	Difference from mean	Square of difference from mean
5	-7	$(-7)(-7) = 49$
8	-4	$(-4)(-4) = 16$
10	-2	$(-2)(-2) = 4$
13	1	$(1)(1) = 1$
24	12	$(12)(12) = 144$
		214

The average of these squares is $\frac{214}{5} = 42.8$.

The square root of the average of these squares is defined as the standard deviation and is denoted by the Greek letter σ. Thus, in our example the standard deviation[1]

$$\sigma = \sqrt{42.8} \approx 6.54$$

We can summarize the procedure for finding the standard deviation of a set of numbers by the following rule.

Rule The *standard deviation* of a set of numbers is found by calculating (in order)

a) The mean or average of the numbers
b) The difference between each number and the mean
c) The square of each of these differences
d) The average of these squares
e) The square root of the average of these squares

We illustrate the use of this rule with two examples.

Example 5 *Cost of having a baby.* In an attempt to determine the cost of having a baby, a news reporter asked ten new mothers what their obstetricians had charged for delivering the baby. The costs were \$650, \$725, \$550, \$475, \$800, \$830, \$750, \$740, \$700, and \$680. Find the mean and the standard deviation.

Solution We arrange the data in tabular form as shown on top of next page and then calculate the mean and the standard deviation. The mean is \$690 and the standard deviation of the cost is $\sqrt{10{,}665}$, or \$103.27.

[1] A knowledge of how to compute square roots is not needed. Such values can be obtained by using a hand-held calculator.

Cost	Difference from mean	Square of difference from mean
\$650	650 − 690 = −40	(−40)(−40) = 1,600
725	725 − 690 = 35	(35)(35) = 1,225
550	550 − 690 = −140	(−140)(−140) = 19,600
475	475 − 690 = −215	(−215)(−215) = 46,225
800	800 − 690 = 110	(110)(110) = 12,100
830	830 − 690 = 140	(140)(140) = 19,600
750	750 − 690 = 60	(60)(60) = 3,600
740	740 − 690 = 50	(50)(50) = 2,500
700	700 − 690 = 10	(10)(10) = 100
680	680 − 690 = −10	(−10)(−10) = 100
\$6900	0	106,650
Mean = $\frac{6900}{10}$ = 690		Mean = $\frac{106,650}{10}$ = 10,665

Example 6 According to police department records, the number of burglaries reported daily for the first eight days of June was 23, 35, 17, 16, 19, 12, 31, and 27. Find the mean and the standard deviation.

Solution We arrange the data in tabular form as shown and then calculate the mean and the standard deviation. The mean is 22.5 crimes per day and the standard deviation is $\sqrt{55.5}$, or 7.45 rounded off.

Burglaries reported	Difference from mean	Square of difference from mean
23	23 − 22.5 = 0.5	(0.5)(0.5) = 0.25
35	35 − 22.5 = 12.5	(12.5)(12.5) = 156.25
17	17 − 22.5 = −5.5	(−5.5)(−5.5) = 30.25
16	16 − 22.5 = −6.5	(−6.5)(−6.5) = 42.25
19	19 − 22.5 = −3.5	(−3.5)(−3.5) = 12.25
12	12 − 22.5 = −10.5	(−10.5)(−10.5) = 110.25
31	31 − 22.5 = 8.5	(8.5)(8.5) = 72.25
27	27 − 22.5 = 4.5	(4.5)(4.5) = 20.25
180	0	444
Mean = $\frac{180}{8}$ = 22.5		Mean = $\frac{444}{8}$ = 55.5

Comment You may feel that the standard deviation is a rather complicated number to calculate, so why bother. As we shall see, however, it is extremely important when working with the normal distribution discussed in Section 9.5.

Comment Some books denote a number in a set of numbers by the symbol x. The difference from the mean is then denoted by $x - \mu$ and the square of this difference by $(x - \mu)^2$. The formula for the standard deviation can then be written as

$$\sigma = \sqrt{\frac{\Sigma\,(x - \mu)^2}{n}}$$

Exercises

20. The annual salaries of the eight mechanics at Joe's Repair Shop are \$12,000, \$17,000, \$13,000, \$19,500, \$14,600, \$18,400, \$21,300, and \$20,400. Find the range and the standard deviation for these salaries.

21. The number of miles driven per day by a bus on a cross-country drive are 420, 580, 622, 701, 482, 375, 572, and 655. Find the range and the standard deviation for the number of miles driven.

22. The number of people who volunteer to give blood daily over a ten-day period at a local unit of the American Red Cross are 21, 18, 39, 17, 25, 34, 29, 30, 47, and 16. Find the range and the standard deviation for the number of daily volunteers.

23. The number of calls to the police for emergency assistance during a 12-hour period are shown in the accompanying table. Find the range and the standard deviation for the number of calls received.

Time	Calls
Midnight–1:00 A.M.	23
1:00–2:00 A.M.	17
2:00–3:00 A.M.	69
3:00–4:00 A.M.	42
4:00–5:00 A.M.	19
5:00–6:00 A.M.	22
6:00–7:00 A.M.	31
7:00–8:00 A.M.	46
8:00–9:00 A.M.	42
9:00–10:00 A.M.	51
10:00–11:00 A.M.	32
11:00 A.M.–noon	29

24. The number of rocks collected daily by a geology class expedition during the first two weeks of the expedition were 58, 71, 69, 82, 47, 55, 69, 77, 80, and 50. Find the range and the standard deviation for the number of rocks collected.

25. In 1978 the prices of a pound of ground meat at five supermarkets were 79¢, 84¢, 65¢, 89¢, and 95¢. Find the mean and the standard deviation of the price of ground meat in 1978.

26. *Inflation.* Refer to Exercise 25. In 1979 the same five supermarkets were surveyed again. This time it was found that the price of a pound of ground meat at each of these supermarkets had doubled.

a) Calculate the mean and the standard deviation for these new prices.

b) How do these answers compare with the answers obtained in Exercise 25?

27. The weekly earnings of two salespeople, Matty and Harriet, over the past eight weeks are as follows:

Matty: \$212, \$218, \$197, \$201, \$219, \$207, \$200, and \$210

Harriet: \$211, \$219, \$191, \$222, \$199, \$212, \$218, and \$216

a) Find the range and the standard deviation of the salary for each employee.

b) Who is a more consistent employee?

★**28.** Verify that the sum of the differences from the mean, that is $\Sigma(x - \mu)$, is 0.

★**29.** Consider a set of numbers $x_1, x_2, x_3, \ldots, x_n$. Subtract a constant amount c from each number, thereby producing the new set of numbers $x_1 - c, x_2 - c, x_3 - c, \ldots, x_n - c$. What effect does this operation have on the standard deviation? Can you generalize?

30. Each year many television stations conduct fund-raising drives through telethons for various charities. For the past seven years one television station has kept records on the time that passed before the first \$25,000 was pledged. The times are given in the accompanying table. Find the range and the standard deviation.

Year	Minutes passed
1972	19
1973	17
1974	16
1975	14
1976	10
1977	9
1978	8

31. *Rising cost of health insurance.* The average cost of medical health insurance last year in a certain city was \$840 with a standard deviation of \$32. This year, because of inflation, the rates will be increased by \$15. What will the new mean and standard deviation be?

9.4 Frequency Distributions and Graphs

Consider the following situation: A psychologist is performing experiments with rats running through a maze. He has conducted the experiment with many rats and has recorded the number of minutes required by each rat to find its way through the maze. The results of the experiment are shown here.

5	6	9	12	7	8	12	10	14	12
11	8	6	7	13	9	7	11	13	13
13	5	7	6	15	12	8	15	17	17
17	11	12	13	19	17	9	13	19	14
6	13	15	14	5	14	8	11	10	11
8	17	18	18	6	16	9	18	8	12
14	19	7	17	10	15	11	8	6	10
19	14	8	14	14	14	19	6	5	6
11	16	9	13	11	10	11	19	12	5
15	8	14	11	7	18	12	12	11	8

The only thing we can say for sure from these numbers is that some rats needed as long as 19 minutes to get through the maze while some needed only 5 minutes to find their way. Since the numbers are not arranged in any particular order, it is difficult to conclude anything else. Obviously, the data must be reorganized into a more meaningful form. To accomplish this, we use a frequency distribution.

Definition 9.5 A *frequency distribution* is a convenient way of grouping data so that meaningful patterns can be found. The word *frequency* will mean how often some number occurs.

A frequency distribution is easy to construct. We first make a list of numbers starting at 5 and going up to 19 to indicate the number of minutes required for a rat to find its way through the maze. This list is the first column. In the second column we make tally marks for each number. We go through the original numbers and put a tally mark in the appropriate space for the number of minutes needed by a rat to find its way through the maze. Finally, in the third column we enter the total number of tally marks for each line. This is the frequency column. When this procedure is applied to our example, we get the distribution shown in Table 9.1

Once we have done this, we can come up with some meaningful interpretations. Since 14 occurred most often, this means that most rats require 14 minutes to find their way through the maze. Also the frequency fluctuated as the time increased beyond 14 minutes.

The previous frequency distribution can be shortened somewhat as shown in Table 9.2. Although Table 9.2 is more compact, some information is lost in this version. Thus, we can tell that 11–12 or 13–14 minutes is the most frequent time needed to complete the maze. However, we cannot tell how many rats actually needed 11 minutes to complete the maze.

Table 9.1

Minutes required	Tally	Frequency
5	𝍸	5
6	𝍸 𝍷𝍷𝍷	8
7	𝍸 𝍷	6
8	𝍸 𝍸	10
9	𝍸	5
10	𝍸	5
11	𝍸 𝍸	10
12	𝍸 𝍷𝍷𝍷𝍷	9
13	𝍸 𝍷𝍷𝍷	8
14	𝍸 𝍸 𝍷	11
15	𝍸	5
16	𝍷𝍷	2
17	𝍸 𝍷	6
18	𝍷𝍷𝍷𝍷	4
19	𝍸 𝍷	6
		100

Table 9.2

Minutes required	Tally	Frequency
5–6	𝍸 𝍸 𝍷𝍷𝍷	13
7–8	𝍸 𝍸 𝍸 𝍷	16
9–10	𝍸 𝍸	10
11–12	𝍸 𝍸 𝍸 𝍷𝍷𝍷𝍷	19
13–14	𝍸 𝍸 𝍸 𝍷𝍷𝍷𝍷	19
15–16	𝍸 𝍷𝍷	7
17–18	𝍸 𝍸	10
19–20	𝍸 𝍷	6
		100

Comment It is customary to use about ten categories (or *classes*) in a frequency distribution. With too few classes, too much information is lost. When the number of classes is too large, then nothing is gained by using a frequency distribution.

Once we have a frequency distribution, we can present the information it contains in the form of a graph. To do this we first draw two lines, one vertical and one horizontal. We indicate frequency along the vertical line and the number of minutes along the horizontal line. We then draw rectangles over each interval, with the height of each rectangle equal to the frequency of that class. Such a graph is known as a *bar graph*. For the data of Table 9.2, the bar graph is shown in Figure 9.1.

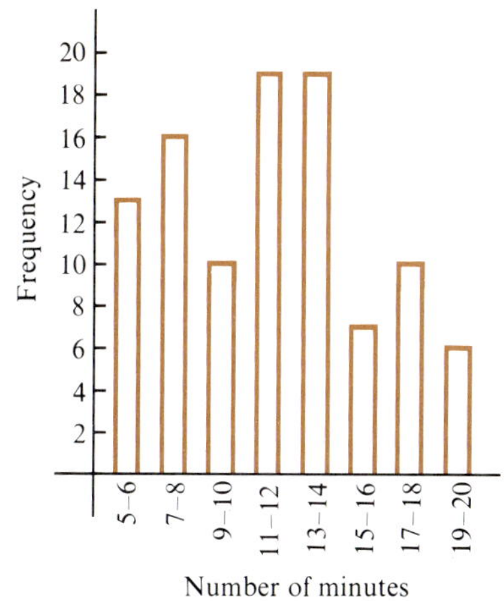

Figure 9.1

Example 7 The weights of 100 people in a health club are as follows:

130	135	138	156	160	162	162	158	179	171
131	148	132	142	140	158	178	140	170	162
146	162	148	139	135	142	132	132	140	172
169	173	141	178	162	147	147	153	138	155
179	151	170	179	173	146	169	162	147	142
171	142	167	156	142	131	173	173	148	137
168	153	166	140	157	172	165	152	156	148
171	168	142	133	137	138	148	134	158	153
142	172	143	138	165	162	154	158	161	157
134	175	158	177	142	157	159	162	172	166

a) Set up a frequency distribution using ten intervals.

b) Draw the bar graph for the frequency distribution obtained in part a.

Solution a) A frequency distribution using ten intervals is shown here.

Weight	Tally	Frequency
130–134	卌 ||||	9
135–139	卌 ||||	9
140–144	卌 卌 ||||	14
145–149	卌 卌	10
150–154	卌 |	6
155–159	卌 卌 |||	13
160–164	卌 卌	10
165–169	卌 ||||	9
170–174	卌 卌 |||	13
175–179	卌 ||	7
		100

b) The bar graph is shown in Figure 9.2.

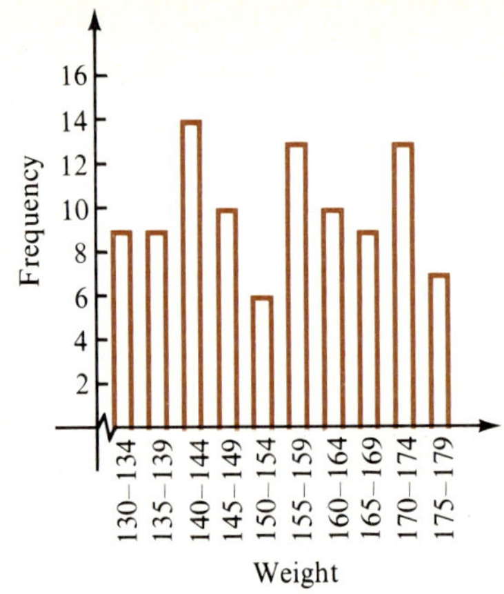

Figure 9.2

Example 8 An IQ test was given to 80 students. The following scores were obtained:

121	111	101	114	97	104	108	118
111	114	97	110	87	102	104	93
103	79	83	108	93	124	105	105
111	114	107	121	101	99	113	84
79	74	83	91	121	87	119	97
94	88	91	92	132	111	106	122
76	84	94	107	127	109	119	87
139	141	97	102	116	122	118	104
137	144	101	107	93	127	103	83
131	115	103	111	99	101	78	96

a) Set up a frequency distribution using ten intervals.

b) Draw the bar graph for the frequency distribution obtained in part a.

Solution a) A frequency distribution using ten intervals is shown here.

IQ scores	Tally	Frequency
74–81	𝍸	5
82–88	𝍸 ||||	9
89–95	𝍸 |||	8
96–102	𝍸 𝍸 |||	13
103–109	𝍸 𝍸 𝍸	15
110–116	𝍸 𝍸 ||	12
117–123	𝍸 ||||	9
124–130	|||	3
131–137	|||	3
138–144	|||	3
		80

b) The bar graph is shown in Figure 9.3.

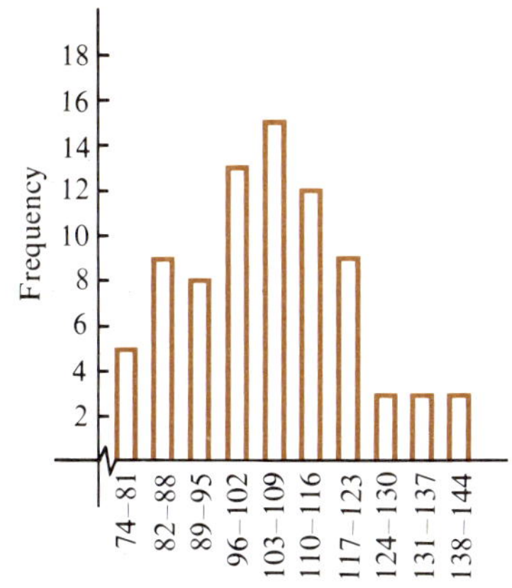

Figure 9.3

Bar graphs are often used to show comparative data.

Example 9 The bar graphs given in Figure 9.4 show the composition of the student body at six colleges in the East.

a) Which college has the most men?
b) Which college has the most women?
c) How many more men are in college E than in college A?

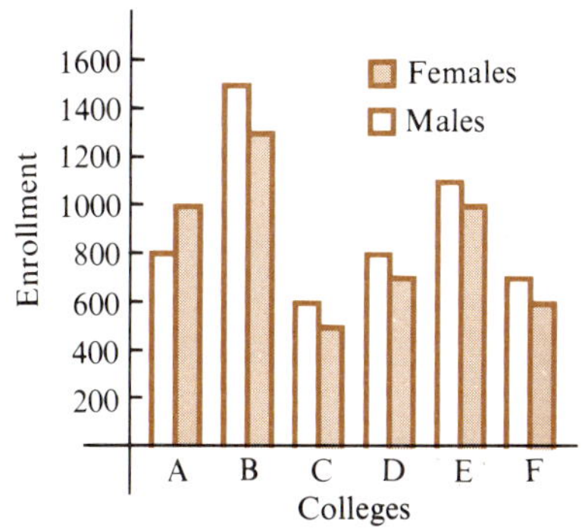

Figure 9.4

Solution a) College B has the most men.
b) College B has the most women.
c) College E has 1100 men and college A has 800 men, so college E has 300 more men then college A.

Bar graphs can also be drawn horizontally as in Example 10.

Example 10 The total number of math courses offered by the six colleges discussed in Example 9 during the spring 1979 and spring 1980 semesters are shown in the accompanying table. Draw a bar graph picturing this information. Use horizontal bars.

College	Spring 1979	Spring 1980
A	19	16
B	15	16
C	17	12
D	13	18
E	14	14
F	12	17

Solution The bar graph is shown in Figure 9.5.

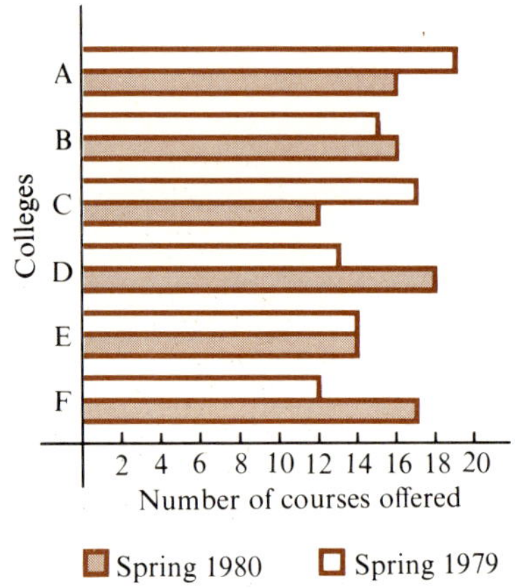

Figure 9.5

In recent years the bars of a bar graph have often been replaced by a series of equally spaced identical pictures, where each picture or symbol represents a specified quantity. Such graphs, which are called *pictographs* or *pictograms*, are used to catch the reader's eye and to make the graph more appealing.

Example 11 *The price of gas.* The pictograph in Figure 9.6 shows how the price of gas has changed over the past few years in one particular city.

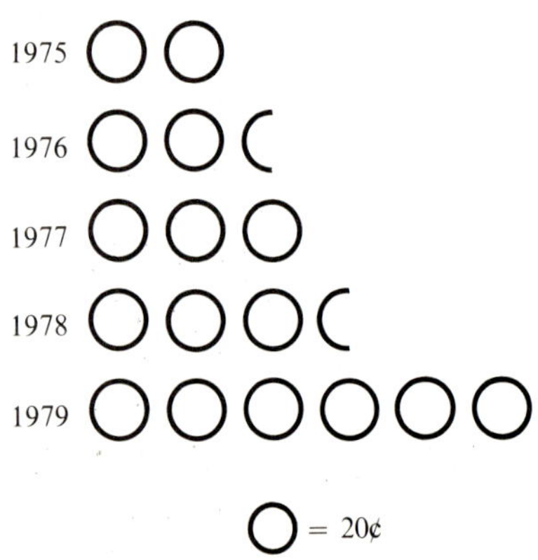

Figure 9.6 Average selling price of a gallon of gasoline.

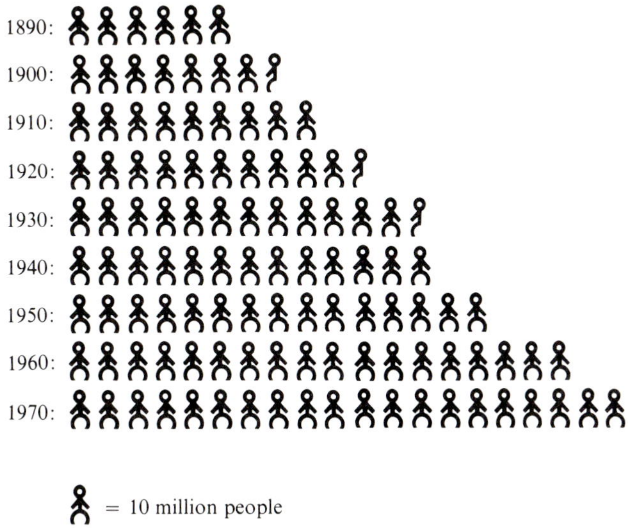

Figure 9.7 (*Source:* U.S. Bureau of the Census)

Example 12 *Population changes.* The pictograph in Figure 9.7 indicates the population change of the United States from 1890 to 1970.

a) How many people were there in 1930?

b) By approximately how many people did the population of the United States change from 1920 to 1970?

Solution a) In 1930 there were 125 million people in the United States.

b) The population of the United States increased from 105 million in 1920 to 200 million in 1970, or by 95 million people.

Another variation of the bar graph is obtained by taking the middle of the top of each bar of a bar graph and then erasing the bars. When these points are joined together, we get a *line graph.* Line graphs are used primarily to show how things have changed over a period of time; that is, it shows trends.

Example 13 Consider the line graph given in Figure 9.8. In which year were housing starts at the lowest point?

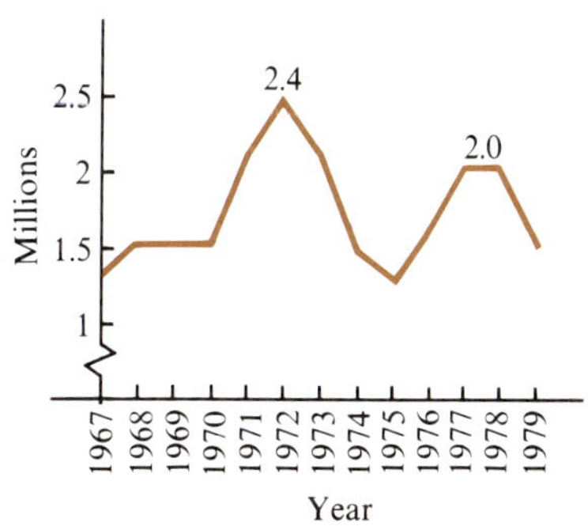

Figure 9.8 New starts for private and public housing. (*Source:* U.S. Dept. of Commerce)

Example 14 The graph given at the beginning of this chapter concerning interest rates is another example of a line graph. Based on this graph, in which year was the rate of interest the highest?

Another type of graph that is often used to picture information is the *circle graph* or *pie chart*. This type of graph is particularly helpful when parts of a total are compared with other parts or with the total itself, as shown in Figure 9.9 and the accompanying table.

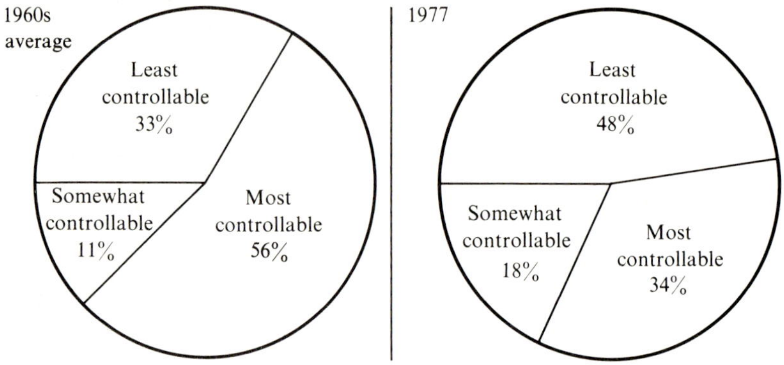

Figure 9.9 Share of total federal spending. (*Source:* Federal Reserve Board)

Programs	Spending 10 years ago	Spending this year	Increase
Social security	\$26.8 billion	\$103.1 billion	284.7%
Additional retirement programs	\$6.3 billion	\$26.8 billion	325.0%
Unemployment assistance	\$2.9 billion	\$11.9 billion	310.3%
Veterans' benefits	\$5.7 billion	\$12.6 billion	121.1%
Medicare	\$6.6 billion	\$30.0 billion	354.5%
Medicaid	\$2.3 billion	\$11.6 billion	404.3%
Public assistance and related programs	\$3.9 billion	\$23.0 billion	489.7%
Net interest on debt	\$12.7 billion	\$39.8 billion	213.4%
Farm price supports	\$4.1 billion	\$4.6 billion	12.2%

Source: Office of Management and Budget

In a circle graph, the size of each central angle has the same proportion of the 360 degrees in a circle that the item has to the total. Thus, to make a circle graph, we first set up a chart indicating what percentage each item is of the total. Then we multiply each of these percentages by 360 to obtain the number of degrees in the central angle for each category. Then we draw a circle and measure off the appropriate angles; that is, we construct successive central angles containing the appropriate number of degrees.

To illustrate the procedure for drawing a circle graph, consider Example 15.

Example 15 Draw a circle graph to indicate the family budget of a typical resident in Charlesville.

Item	Amount budgeted per month
Rent	$400
Food	300
Clothing	200
Entertainment	100
Education	200
Savings	100
Miscellaneous	200
Total	$1500

Solution We first determine the percentage of the budget allowed for each item.

Item	Amount budgeted per month	Percentage of total budget
Rent	$400	$\frac{400}{1500} = 0.2667$, or 26.67%
Food	300	$\frac{300}{1500} = 0.20$, or 20%
Clothing	200	$\frac{200}{1500} = 0.1333$, or 13.33%
Entertainment	100	$\frac{100}{1500} = 0.667$, or 6.67%
Education	200	$\frac{200}{1500} = 0.1333$, or 13.33%
Savings	100	$\frac{100}{1500} = 0.0667$, or 6.67%
Miscellaneous	200	$\frac{200}{1500} = 0.1333$, or 13.33%

Now

$$\begin{aligned}
26.67\% \text{ of } 360^\circ &= 0.2667(360^\circ) = 96^\circ \\
20\% \text{ of } 360^\circ &= 0.20(360^\circ) = 72^\circ \\
13.33\% \text{ of } 360^\circ &= 0.1333(360^\circ) = 48^\circ \\
6.67\% \text{ of } 360^\circ &= 0.0667(360^\circ) = 24^\circ \\
13.33\% \text{ of } 360^\circ &= 0.1333(360^\circ) = 48^\circ \\
6.67\% \text{ of } 360^\circ &= 0.0667(360^\circ) = 24^\circ \\
13.33\% \text{ of } 360^\circ &= 0.1333(360^\circ) = 48^\circ
\end{aligned}$$

We now construct the circle graph. It is shown in Figure 9.10.

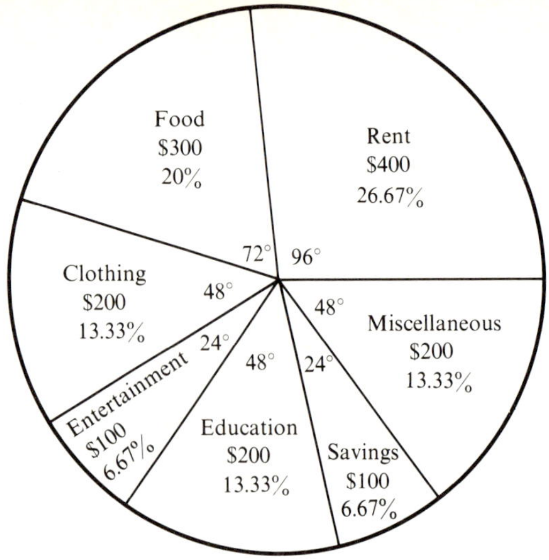

Figure 9.10

Example 16 The cause of hospitalization for 10,000 patients last year in a particular city is shown in Figure 9.11. How many people were hospitalized as a result of the following?

a) Cancer
b) Automobile or industrial accidents
c) Fires

Solution

a) 14% of 10,000 = (0.14)(10,000) = 1400. Thus, 1400 people were hospitalized as a result of cancer.
b) 13% + 5% or 18% of 10,000 = (0.18)(10,000) = 1800. Thus, 1800 people were hospitalized as a result of automobile or industrial accidents.
c) 7% of 10,000 = (0.07)(10,000) = 700. Thus, 700 people were hospitalized as a result of fire.

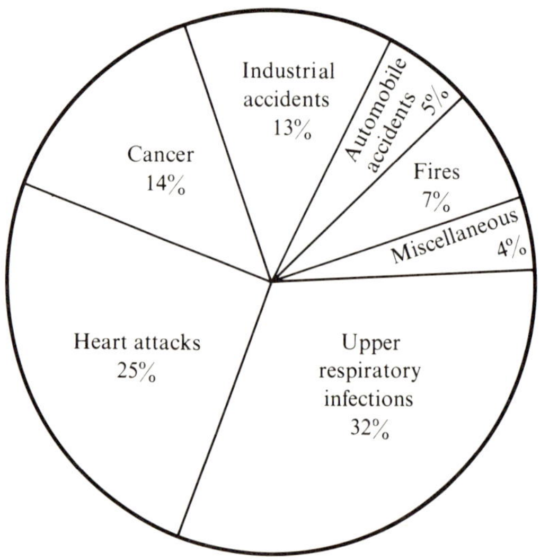

Figure 9.11

32. Draw a bar graph to picture the following information: The weights of the brains of a particular group of people were determined and the results were grouped according to the distributions in the following table.

Weight (grams)	Frequency (old person)	Frequency (young person)
800–949	3	0
950–1049	6	2
1050–1099	8	14
1100–1149	19	22
1150–1199	11	45
1200–1249	14	54
1250–1299	14	55
1300–1349	3	23
1350–1399	6	13
1400–1449	1	10

33. The amounts of pollution (on a particular scale) over the past 100 days in the air of a large city were as follows:

261	258	264	245	271	273	263	245	239	257
257	258	278	277	252	278	275	277	253	264
261	269	256	271	243	239	245	261	255	273
254	274	255	263	279	239	258	263	244	262
254	254	266	267	256	278	263	239	262	244
264	258	268	269	271	258	272	258	257	275
258	266	262	277	244	261	263	273	267	274
275	274	239	255	277	275	269	264	274	268
261	275	239	266	264	243	242	251	271	241
276	277	264	255	266	255	239	259	274	248

Draw a bar graph to picture this information.

34. Fifty college students were asked to indicate how much money each had spent for supplies for the current semester. Their answers (rounded to the nearest dollar) were as follows:

16	19	14	23	21	27	17	20	17	16
18	35	24	33	26	33	35	23	26	20
28	21	18	21	19	18	15	21	18	22
23	32	22	25	27	25	19	27	40	27
26	20	21	15	20	26	18	24	19	24

Construct the frequency distribution for these data and then draw the bar graph.

35. The ages of the students in Phil's statistics class are as follows:

28	25	18	33	17	27	15	20	21	28
29	34	16	30	31	17	19	34	32	22
34	25	19	18	19	26	18	32	24	32
31	36	24	22	21	18	29	27	36	31

Construct the frequency distribution for these data and then draw the bar graph.

36. The birth and death rates for several countries are given in the accompanying table. Draw the bar graph to picture this information.

Country	Birth rate	Death rate
Egypt	34.8	13.1
Israel	27.2	7.1
Japan	19.4	6.6
France	16.4	10.7
Philippines	24.8	7.3
Mexico	44.7	9.1
United States	15.0	9.4
USSR	17.7	8.7
Puerto Rico	24.1	6.7

37. Although the United States has 6% of the world population, it consumes very high percentages of the world production of certain materials as shown in the accompanying table. Draw a bar graph to picture this information.

Material	Percentage
Natural gas	57
Silver	42
Aluminum	36
Petroleum	32
Tin	32
Nickel	30
Copper	27
Steel	19

38. Bill has made a study on how he spends his typical day. The results are as follows:

School: 7 hours
Sleeping: 6 hours
Homework: 2 hours
Television: 3 hours
Meals: 2 hours
Miscellaneous: 4 hours

Draw a circle graph to picture this information.

39. There are seven continents. The areas contained within each continent are given in the accompanying table. Draw a circle graph to picture this information.

Continent	Area (square miles)
Africa	11,506,000
Antarctica	5,500,000
Asia	16,988,000
Australia	2,975,000
Europe	3,745,000
North America	9,390,000
South America	6,795,000

40. A sample is taken of 80 workers to see how many photocopies each makes per week (on the average) on the machine located on the third floor. The following results are obtained:

24	18	7	4	16	12	26	22
25	8	26	18	11	15	25	20
9	4	12	8	5	29	14	30
14	20	16	25	15	17	11	14
7	22	21	15	16	27	31	7
23	15	31	30	12	1	21	17
25	10	11	34	18	13	16	33
2	26	14	25	16	10	7	22
20	23	25	2	19	19	26	23
11	16	26	16	20	5	12	8

Construct the frequency distribution for these data and then draw the bar graph.

41. Make a line graph to show how the demand for electricity varies according to the month for a particular community.

Month	Demand (megawatt-hours)	Month	Demand (megawatt-hours)
January	894	July	1550
February	1162	August	1625
March	1272	September	1570
April	975	October	1300
May	1384	November	1140
June	1400	December	1050

42. The pictograph in Figure 9.12 shows the 1974 automobile production in the United States.

a) How many cars were produced by General Motors in 1974?
b) How many cars were produced by Chrysler Corporation in 1974?

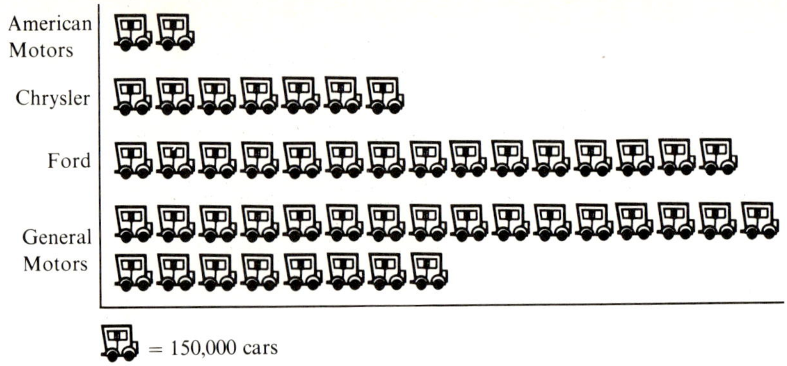

Figure 9.12

43. The information in the accompanying table is available on the number of *new* life insurance policies sold by a large company during the past seven years. Draw a line graph to indicate the changes in the number of policies sold.

Year	Policies sold
1972	1742
1973	827
1974	1328
1975	1962
1976	1500
1977	1400
1978	1850

44. *Affirmative action.* From Figure 9.13 compare the percentage of black women in the labor force with the percentage of white women in the work force for the years 1960 and 1974. Comment.

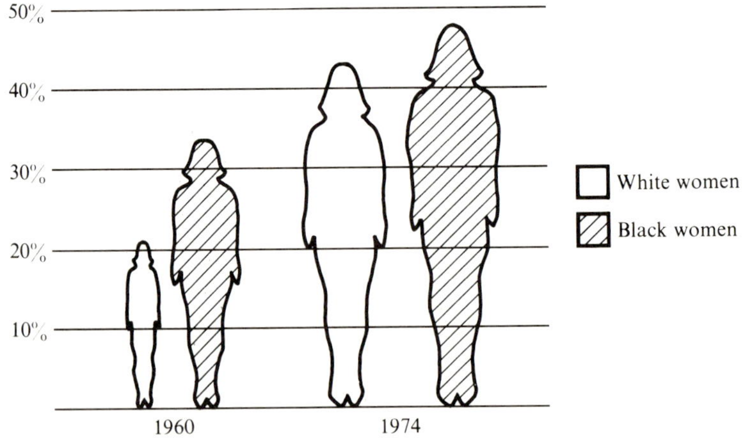

Figure 9.13 Percentage of white and black women in the labor force. (*Source:* Bureau of Labor Statistics)

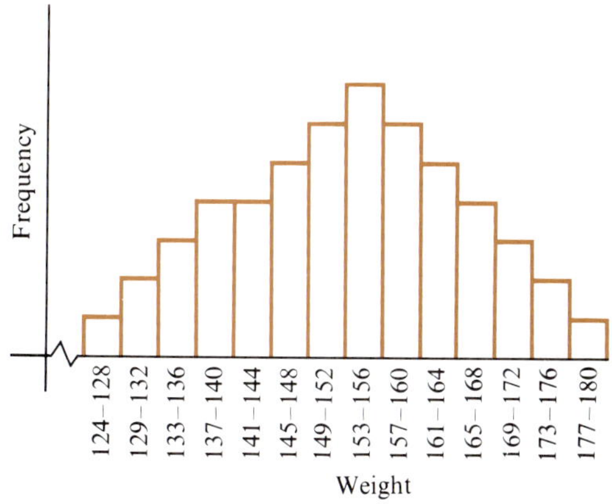

Figure 9.14

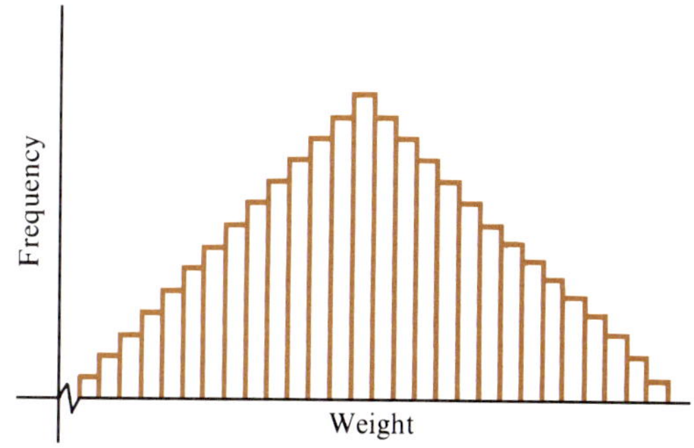

Figure 9.15

9.5 The Normal Distribution

Recently a survey was taken of the weights of the 800 students at Trixy College. After the data were arranged into a frequency distribution containing 14 intervals, the bar graph shown in Figure 9.14 was obtained. When the data were grouped into a frequency distribution containing 28 intervals, the bar graph shown in Figure 9.15 was obtained. If we were to continue this process of adding additional intervals, the graph would tend to smooth out into a continuous curve similar to the one shown in Figure 9.16. Such a curve is called a *normal curve*, and the distribution that gives rise to it is called a *normal distribution.*

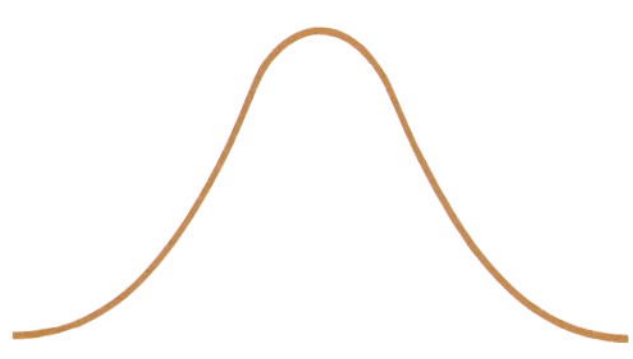

Figure 9.16 The normal curve.

Since the normal distribution has wide-ranging applications, we need a careful description of a normal curve and some of its properties. The graph of a normal distribution is a bell-shaped curve that extends in both directions. Although the curve gets closer and closer to the horizontal axis, it never really crosses it, no matter how far it is extended.

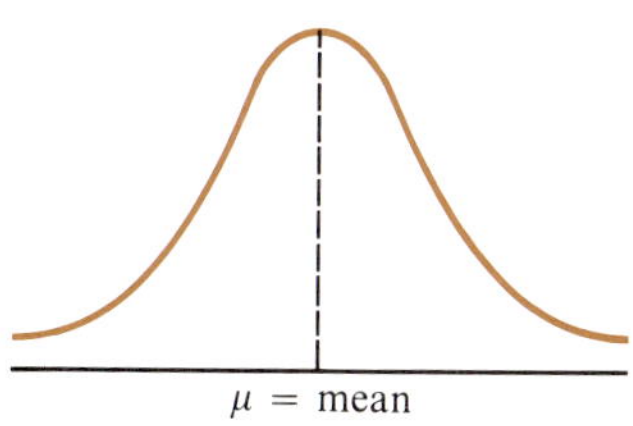

Figure 9.17

The mean of the normal distribution is at the center of the curve, and the curve is symmetric about the mean. This tells us that we can fold the curve along the dotted line in Figure 9.17 and either portion of the curve will correspond with the other portion.

For a normal distribution, the mean, the median, and the mode are all equal.

The data that make up the normal distribution tend to cluster around the middle with very few values more than three standard deviations from the mean on either side. As a matter of fact, about 68.3% of the data will fall within one standard deviation of the mean on either side (see Figure 9.18). Also approximately 95.4% of the data will fall within two standard deviations of the mean on either side (see Figure 9.19), and approximately 99.75% of the data will fall within three standard deviations of the mean on either side (see Figure 9.20).

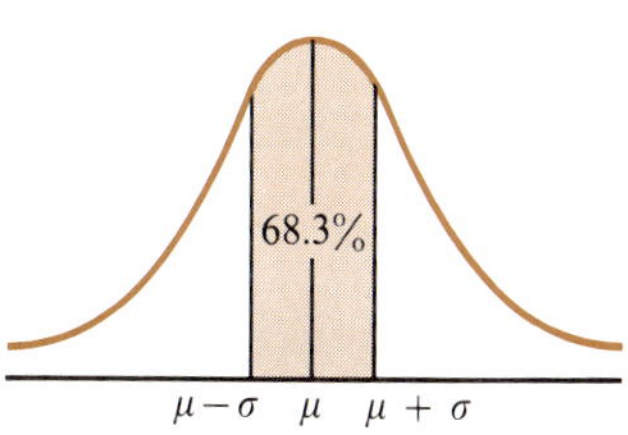

Figure 9.18

Thus, in our case if the weights of the college students are normally distributed with a mean of 150 pounds and a standard deviation of

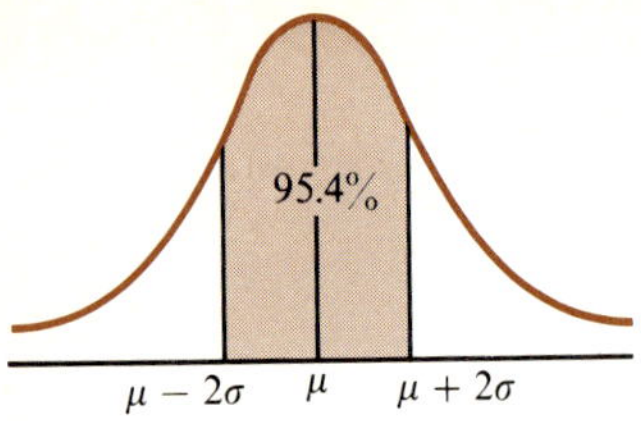

Figure 9.19

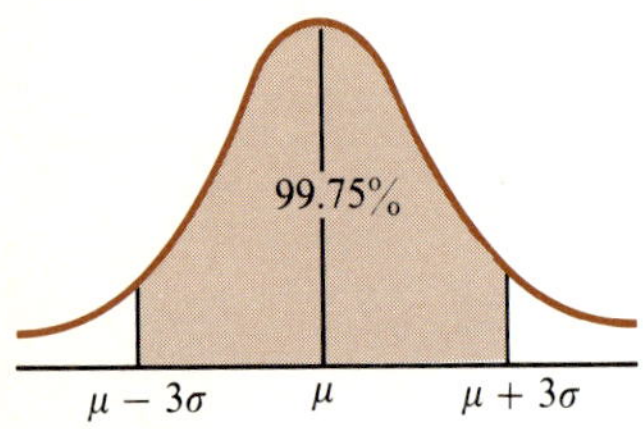

Figure 9.20

10 pounds, then we would expect approximately 68.3% of the students to weigh between

$$\mu - \sigma \quad \text{and} \quad \mu + \sigma$$
$$150 - 10 \quad \text{and} \quad 150 + 10$$

or between 140 and 160 pounds. Similarly, we would expect approximately 95.4% of the students to weigh between

$$\mu - 2\sigma \quad \text{and} \quad \mu + 2\sigma$$
$$150 - 2(10) \quad \text{and} \quad 150 + 2(10)$$

or between 130 and 170 pounds. Also 99.75% of the students should weigh between

$$\mu - 3\sigma \quad \text{and} \quad \mu + 3\sigma$$
$$150 - 3(10) \quad \text{and} \quad 150 + 3(10)$$

or between 120 and 180 pounds.

A normal distribution is completely specified by its mean and standard deviation. Thus, although all normal distributions are basically mound-shaped, different means and different standard deviations will describe different bell-shaped curves (see Figure 9-21). It is possible, however, to convert each of these different normal distributions into one standard form. You may be wondering, why bother? The answer is rather simple as we shall see.

The total area under the normal curve is exactly one (one square unit). Since the normal curve is symmetric about the mean, we can immediately conclude that $\frac{1}{2}$ (or 0.5000) of *all* the area is to the right of the mean and $\frac{1}{2}$ (or 0.5000) of the area is to the left of the mean (see Figure 9.22).

Area under the normal curve is associated with probability. Thus, if a measurement x, the weights of college students in our case, is normally distributed, then the probability that x will fall between the values of a and b is equal to the *area* under the normal curve between a and b.

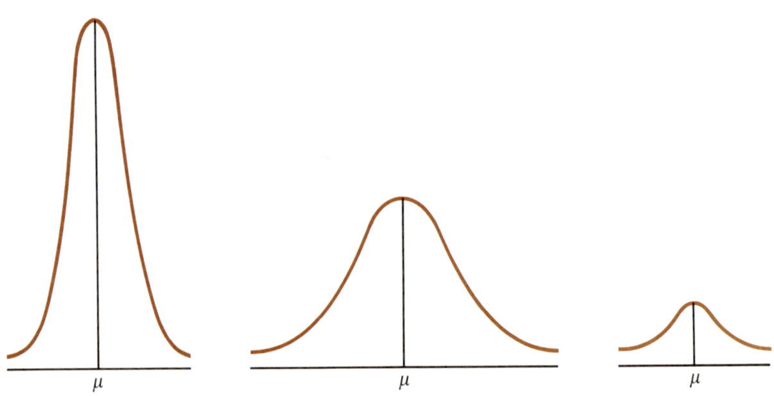

Figure 9.21 Different normal distributions.

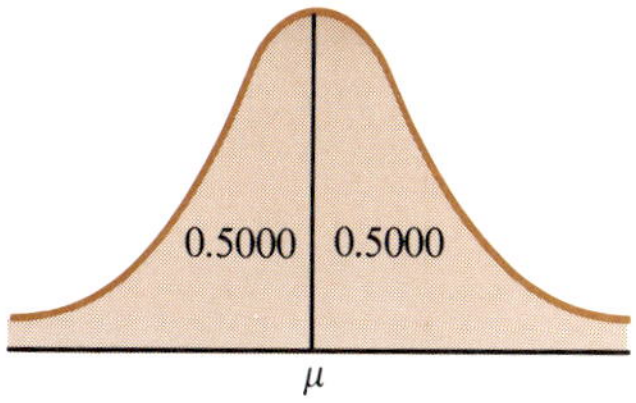

Figure 9.22

Since areas under a normal distribution are related to probability, we can use special normal distribution tables for calculating probabilities. Since the mean and the standard deviation can be any values, however, it would seem that we need an endless number of tables. Fortunately, this is not the case; we need only one standardized table. (See Table 3 in the appendix.) Thus, the area under the curve between 40 and 60 of a normal distribution with a mean of 50 and a standard deviation of 10 will be the same as the area between 70 and 80 of another normally distributed variable with mean 75 and standard deviation 5. They are both within standard deviation unit from the mean. It is for this reason that statisticians use a standard normal distribution.

Definition 9.6 A *standard normal distribution* is a normal distribution with a mean of 0 and a standard deviation of 1.

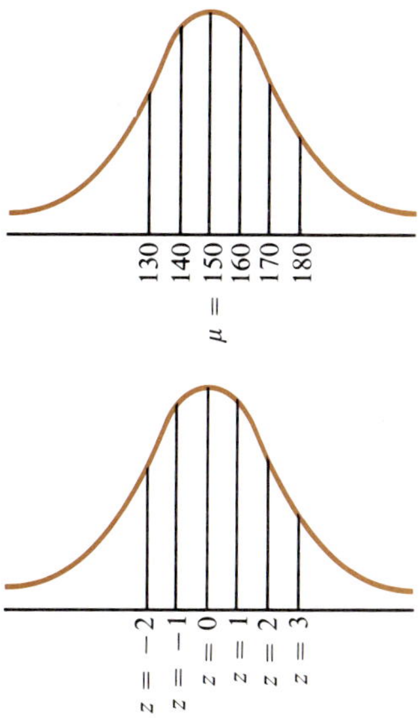

Figure 9.23 The correspondence between the original data and the standard normal curve.

Thus, what we attempt to do is to convert any normal curve with mean μ and standard deviation σ into a normal curve with mean 0 and standard deviation 1. For example, suppose we consider the weights of college students (discussed earlier) whose mean $\mu = 150$ and whose standard deviation $\sigma = 10$. From the original data, one standard deviation to the right of the mean corresponds to $150 + 10$, or 160. On the standard normal curve, this simply corresponds to $+1$. Also, from the original data, two standard deviations to the right of the mean corresponds to $150 + 2(10)$, or 170. On the standard normal curve, this simply corresponds to $+2$. The corresponding value on the standard normal curve is usually expressed in units of z (see Figure 9.23). Similarly, one standard deviation to the left of the mean corresponds to a weight of $150 - 10$, or 140 pounds. On the standard normal curve, this corresponds to a negative z-value of -1. In a similar manner, a weight of 151 pounds corresponds to $z = 0.1$, a weight of 152 corresponds to $z = 0.2$, and so on.

More generally, we can convert any raw score x into a z-score by using Formula 9.2.

Formula 9.2 The *z-score* or *z-value* of any number x in a normal distribution is given by

$$z = \frac{x - \mu}{\sigma}$$

where x = original score, μ = mean, and σ = standard deviation.

Comment A z-score tells us how many standard deviation units a particular number is away from the mean as well as whether this number is above or below the mean.

Example 17 A certain brand of flashlight battery has a mean life μ of 40 hours and a standard deviation σ of 5 hours. Find the z-score of a battery that lasts for

a) 50 hours b) 35 hours c) 40 hours

Solution Since $\mu = 40$ and $\sigma = 5$, we use Formula 9.2.

a) The z-score of 50 is

$$z = \frac{x - \mu}{\sigma} = \frac{50 - 40}{5} = 2$$

Since the z-score is a positive number, we know that $x = 50$ lies to the right of the mean.

b) The z-score of 35 is

$$z = \frac{x - \mu}{\sigma} = \frac{35 - 40}{5} = -1$$

Since the z-score is a negative number, we know that $x = 35$ lies to the left of the mean.

c) The z-score of 40 is

$$z = \frac{40 - 40}{5} = \frac{0}{5} = 0$$

Since the z-score is 0, we know that $x = 40$ coincides with the mean.

Comment Since σ is always a positive number, z will be a negative number whenever x is less than μ, as $x - \mu$ is then a negative number. A z-score of 0 implies that the term has the same value as the mean.

Let us now apply the idea of z-score to find the areas under a normal curve and thus probability calculations.

Table 3 in the Appendix gives the areas of a standard normal distribution between $z = 0$ and $z = 3.09$. We read the table as follows: The first two digits of the z-score are under the column headed by z; the third digit heads the other columns. Thus, to find the area from $z = 0$ to $z = 2.59$, we first look under z to 2.5 and then across from $z = 2.5$ to the column headed by 0.09. The area is 0.4952 or 49.52%.

Similarly, to find the area from $z = 0$ to $z = 1.94$, we first look under $z = 1.9$ and then move across to the column headed by 0.04. The area is 0.4738.

Example 18 Find the area between $z = 0$ and $z = 1.13$ in a standard normal curve.

Solution We first draw a sketch as shown in Figure 9.24. Then using Table 3 in the Appendix for $z = 1.13$, we find that the area between $z = 0$ and $z = 1.13$ is 0.3708. This means that the probability of a score with this normal distribution falling between $z = 0$ and $z = 1.13$ is .3708.

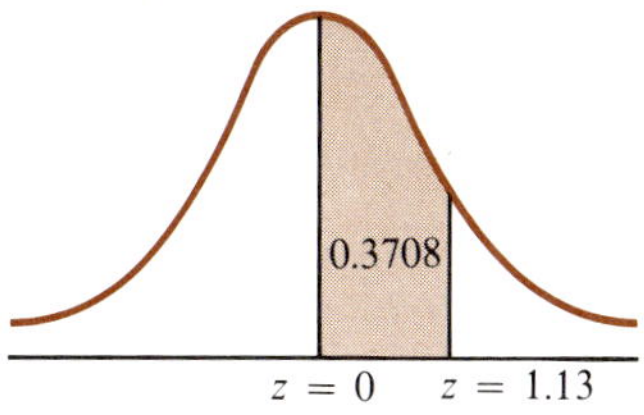

Figure 9.24

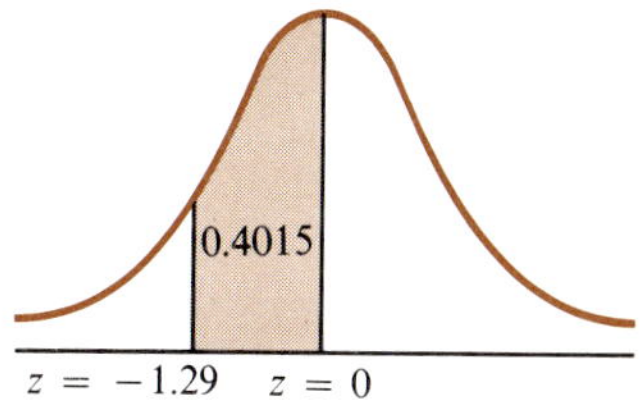

Figure 9.25

Example 19 Find the area between $z = -1.29$ and $z = 0$ in a standard normal distribution.

Solution We first draw a sketch as shown in Figure 9.25. Then using Table 3, we look up the area between $z = 0$ and $z = 1.29$. The area is 0.4015, not -0.4015. A negative value of z tells us that the value of z is to the left of the mean. The area under the curve (and the resulting probability) is *always* a positive number. Thus, the probability of getting a z-score between 0 and -1.29 is .4015.

Example 20 Find the area between $z = -1.37$ and $z = 1.78$ in a standard normal distribution.

Solution We draw the sketch as shown in Figure 9.26. Since Table 3 gives the area only from $z = 0$ on, we first look under the normal curve from $z = 0$ to $z = 1.78$. We get 0.4625. Then we look up the area between $z = 0$ and $z = 1.37$. We get 0.4147. Finally we add these two together and get $0.4625 + 0.4147 = 0.8772$. Thus, the probability that a z-score is between $z = -1.37$ and $z = 1.78$ is .8772.

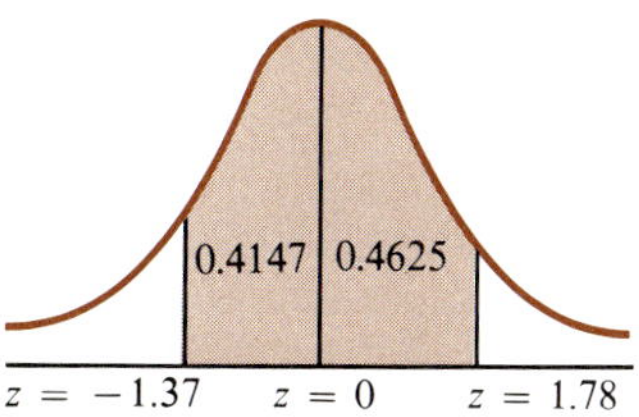

Figure 9.26

By following a procedure similar to that used in Example 20, you should verify the following:

1. The probability that a z-score falls within one standard deviation of the mean on either side, that is, between $z = -1$ and $z = 1$, is approximately 68%.
2. The probability that a z-score falls within two standard deviations of the mean, that is, between $z = -2$ and $z = 2$, is approximately 95%.
3. The probability that a z-score falls within three standard deviations of the mean is 99.7%. Thus, approximately 99.7% of the z-scores fall between $z = -3$ and $z = 3$.

In many cases we have to find areas between two given values of z or areas to the right or left of some value of z. Finding these areas is an easy task provided we remember that the area under the entire normal distribution is 1. Thus, since the normal distribution is symmetrical about $z = 0$, we conclude that the area to the right of $z = 0$ and the area to the left of $z = 0$ are both equal to 0.5000.

Example 21 Find the area between $z = 0.96$ and $z = 2.83$ in a standard normal distribution.

Solution We cannot look this up directly since the chart starts at 0, not at $z = 0.96$. However, we can look up the area between $z = 0$ and $z = 2.83$ and get 0.4977 and then the area between $z = 0$ and $z = 0.96$ and get 0.3315 (see Figure 9.27). We then take the difference between the two and get $0.4977 - 0.3315 = 0.1662$.

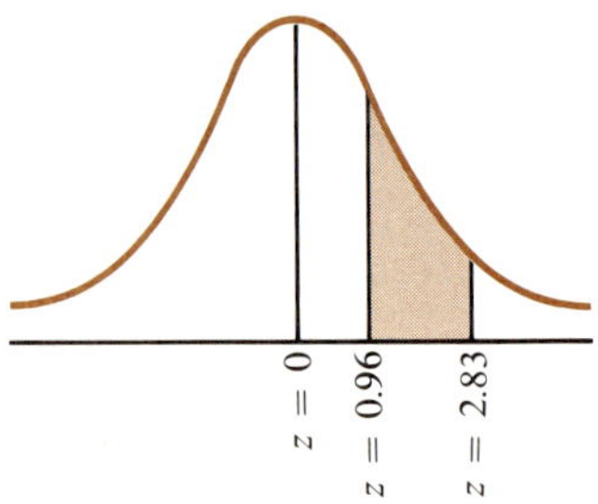

Figure 9.27

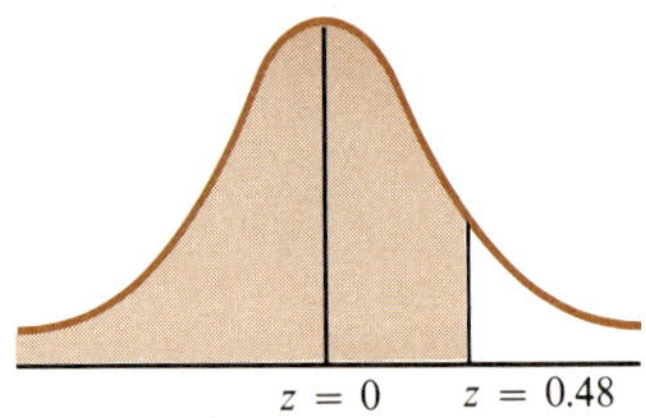

Figure 9.28

Example 22 Find the probability of getting a z-value less than 0.48 in a standard normal distribution.

Solution The probability of getting a z-value less than 0.48 really refers to the area under the curve to the left of $z = 0.48$. This represents the shaded portion of Figure 9.28. We look up the area from $z = 0$ to $z = 0.48$ and get 0.1844. We add this to 0.5000, getting $0.1844 + 0.5000 = 0.6844$. Thus, the probability of getting a z-value less than 0.48 is .6844.

If we are given a normal distribution whose mean is different from 0 and whose standard deviation is different from 1, we can convert this normal distribution into a standard normal distribution by converting each of its scores into a standard score by using Formula 9.2.

Expressing the scores of a normal distribution as standard scores allows us to calculate different probabilities, as Example 23 shows.

Example 23 In a normal distribution with $\mu = 40$ and $\sigma = 5$, find the probability of obtaining the following.

a) A value greater than 50
b) A value less than 25

Solution a) We use Formula 9.2. We have $\mu = 40$, $x = 50$, and $\sigma = 5$, so that

$$z = \frac{x - \mu}{\sigma} = \frac{50 - 40}{5} = \frac{10}{5} = 2$$

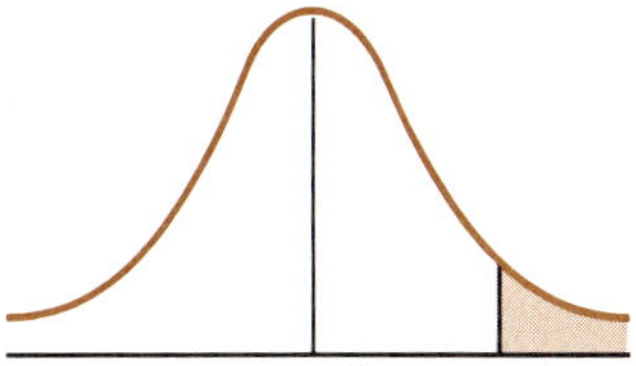

Figure 9.29

See Figure 9.29. Thus, we are really interested in the area to the right of $z = 2$ of a standard normal curve. The area from $z = 0$ to $z = 2$ is 0.4772. The area to the right of $z = 2$ is then $0.5000 - 0.4772 = 0.0228$. Therefore, the probability of obtaining a value greater than 50 is .0228.

b) We use Formula 9.2. We have $\mu = 40$, $x = 25$, and $\sigma = 5$, so that

$$z = \frac{x - \mu}{\sigma} = \frac{25 - 40}{5} = -\frac{15}{5} = -3$$

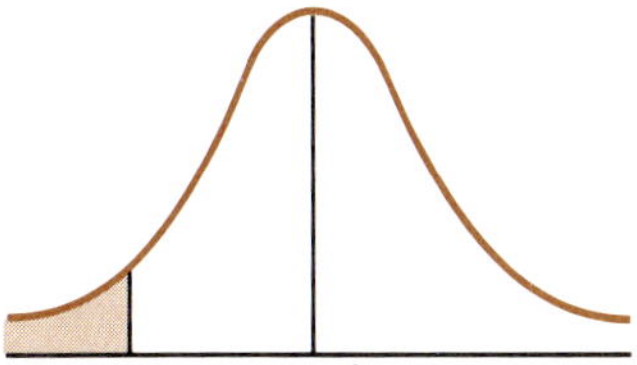

Figure 9.30

See Figure 9.30. Thus, we are interested in the area to the left of $z = -3$. The area from $z = 0$ to $z = -3$ is 0.4987. Thus, the area to the left of $z = -3$ is $0.5000 - 0.4987 = 0.0013$. The probability of obtaining a value less than 25 is therefore .0013.

Let us now apply the normal distribution to some examples.

Examples 24 From past experience it has been found that the weight of a newborn at a maternity hospital is normally distributed with a mean of $6\frac{1}{2}$ pounds (104 ounces) and a standard deviation of 21 ounces. If a newborn baby is selected at random, find the probability that the baby weighs more than 8 pounds (128 ounces)?

Solution We use Formula 9.2. Here $\mu = 104$, $\sigma = 21$, and $x = 128$, so that

$$z = \frac{x - \mu}{\sigma} = \frac{128 - 104}{21} = \frac{24}{21} = 1.14$$

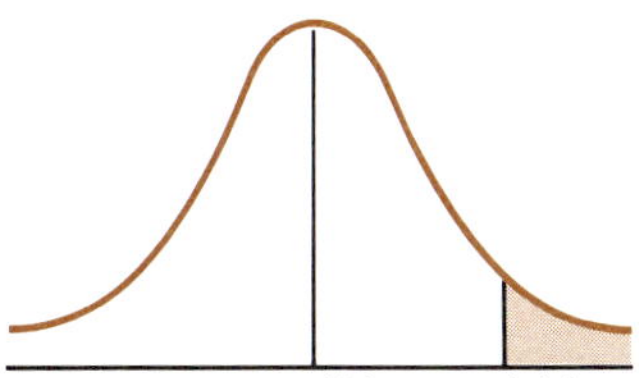

Figure 9.31

Thus, we are interested in the area to the right of $z = 1.14$. The area from $z = 0$ to $z = 1.14$ is 0.3729, so the area to the right of $z = 1.14$ is $0.5000 - 0.3729 = 0.1271$. See Figure 9.31. Therefore, the probability that a randomly selected baby weighs more than 8 pounds is .1271.

Example 25 In a recent study in a certain town it was found that the number of hours that a typical ten-year-old child watches television per week is normally distributed with a mean of 7 hours and a standard deviation of 0.83 hour. Lester is a typical ten-year-old child in this town. What is the probability that he watches between 5 and 8 hours of television per week?

Solution We first find the probability that Lester will watch television between 7 and 8 hours and add to this the probability that he will watch television between 5 and 7 hours per week. Using Formula 9.2, we get

$$z = \frac{x - \mu}{\sigma} = \frac{8 - 7}{0.83} = 1.20$$

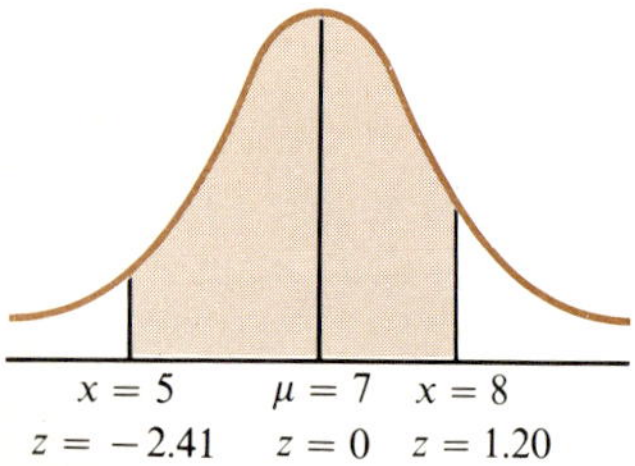

Figure 9.32

The area between $z = 0$ and $z = 1.20$ is 0.3849. See Figure 9.32. Also,

$$z = \frac{x - \mu}{\sigma} = \frac{5 - 7}{0.83} = -2.41$$

The area between $z = 0$ and $z = -2.41$ is 0.4920. Adding these two probabilities, we get $.4920 + .3849 = .8769$. Thus, the probability[2] that Lester watches between 5 and 8 hours of television per week is .8769.

Exercises

45. In a standard normal distribution, find the following areas.

a) Between $z = 0$ and $z = 1.76$
b) Between $z = -0.59$ and $z = 0$
c) To the right of $z = 0.88$
d) To the left of $z = -1.72$
e) To the right of $z = 2.18$
f) Between $z = -1.69$ and $z = 1.23$
g) Between $z = -2.71$ and $z = 1.09$
h) Between $z = -0.86$ and $z = -1.55$

46. Find the percentage of z-scores in a standard normal distribution that are in the following areas.

a) Above $z = -1.58$
b) Below $z = 2.23$
c) Between $z = 0.82$ and $z = 3.09$
d) Abovc $z - 1.76$
e) Below $z = -3.06$
f) Between $z = -1.13$ and $z = -1.58$
g) Between $z = -2.63$ and $z = 1.93$

47. In a recent bowling match the average score was 230 points with a standard deviation of 10. Find the z-score of an individual with these scores:

a) 240 points
b) 207 points
c) 290 points
d) 230 points

48. The z-scores in the table on page 417 were obtained by a consumer's testing group for five different brands of mayonnaise.

[2] For a more detailed discussion of the applications and uses of the normal distribution, see *Statistics and Probability in Modern Life* by Joseph Newmark, 2d ed., chap. 8 (New York: Holt, Rinehart and Winston, 1977).

Brand	z-score
A	2.63
B	−1.02
C	−1.36
D	0.97
E	2.02

a) Rank these brands from highest to lowest.
b) Which brands were above average?
c) Which brands were below average?

49. Find the percentage of z-scores in a normal distribution with $\mu = 28$ and $\sigma = 3$ that are in the following areas.
a) Between 24 and 32
b) Greater than 29
c) Less than 21
d) Between 25 and 30

50. It has been found that the amount of soda dispensed by a particular vending machine is normally distributed with a mean of 7 ounces and a standard deviation of 0.3 ounce. If a cup that has been filled by this machine is randomly selected, find the probability that it contains less than 6.6 ounces.

51. Mortgage statistics collected by a bank indicate that the number of years that the average new homeowner will occupy a house before moving or selling is normally distributed with a mean of 6.7 years and a standard deviation of 2.06 years. If a homeowner is randomly selected, find the probability that the owner will sell the house after living in it for at least 4.5 years.

52. Grades on a particular scholarship test are normally distributed with a mean of 220 and a standard deviation of 27. Find the probability that a randomly selected student will score higher than 250 on this exam.

53. The number of minutes that it takes Martha to drive to work is normally distributed with a mean of 36 and a standard deviation of 7. What percentage of the time will it take Martha more than 49 minutes to drive to work?

54. According to college officials, the amount of time needed to complete registration in the fall semester is normally distributed with a mean of 5 hours and a standard deviation of 0.93 hour. Find the probability that Maria will complete the registration procedure in less than 4 hours.

9.6 Linear Correlation and Regression[3]

Many colleges require students to take a mathematics placement test before allowing them to enroll in any calculus courses. Presumably a student who scores well on such a placement test is more likely to do better in a calculus course than a student who scores poorly on such an exam. College officials may be interested in determining the reliability of such tests. Furthermore, they may be interested in being able to predict a student's performance in the calculus course when his or her performance on the placement is known.

Similarly, many colleges administer vocational aptitude tests. The officials may be interested in knowing whether there is any relationship

[3] Examples and Exercises in this section should be solved with the aid of a hand-held calculator.

between the math aptitude score and the business aptitude score. Do students who score well on the math part of the aptitude exam also do well on the business part? If we know a student's math score, can we predict the student's business score? Questions of this nature frequently arise when we have many variables and are interested in determining relationships between them.

Sir Francis Galton, a cousin of Charles Darwin, undertook a detailed study of human characteristics. He was interested in determining whether a relationship exists between the heights of fathers and the heights of their sons. Do tall parents have tall children? Do intelligent parents or successful parents have intelligent or successful children? In *Natural Inheritance*, Galton introduced the idea that we today refer to as *correlation*. This mathematical idea allows us to measure the closeness of the relationship between two variables. Galton found that there exists a very close relationship between the heights of fathers and the heights of their sons. On the question of whether intelligent parents have intelligent children, it has been found that the correlation is .55. As we shall see, this means that it is not necessarily true that intelligent parents have intelligent children.

To understand what is meant by correlation, let us consider the different aptitude scores that were obtained by ten students at State Tech College as shown in the accompanying table. Is there any relationship between math aptitude scores and business aptitude scores? If a student scores well in math, will the student also score well in business or in language?

Student	Math score	Language score	Music score	Business score
A	44	75	20	40
B	70	31	9	72
C	32	22	50	31
D	49	11	17	50
E	51	19	24	49
F	63	67	13	59
G	28	31	54	24
H	26	48	57	27
I	49	53	23	49
J	52	26	22	48

We can analyze the situation pictorially by means of a *scatter diagram*. We simply draw two lines (called *axes*), one vertical and one horizontal. On one of these axes we indicate the math scores and on the other we indicate the business scores. After both axes are labeled, we use a point to represent each student's score. The point is placed directly above the student's math score and directly to the right of the business score. Thus, the point for student A's scores is placed directly above the 44 score on the math axis and to the right of 40 on the business axis.

This is pictured in Figure 9.33. In a similar manner, we plot the other scores. You will notice that these points form an approximate straight line. When this happens, we say that there is a *linear correlation* between the two variables. Notice that the higher the math score, the higher is the business score. The line moves in a direction from lower left to upper right. When this happens, we say that there is a *positive correlation* between the math scores and the business scores. This means that a student with a higher math score will also have a higher business score.

Now let us draw the scatter diagram for the business aptitude scores and the music aptitude scores. It is given in Figure 9.34. In this case you will notice that the higher the business score, the lower the music score. Again the points arrange themselves in the form of a line, but this time the line moves in a direction from upper left to lower right. When this happens, we say that there is a *negative correlation* between the business scores and the music scores. This means that a student with a high business score will have a low music score.

Now let us draw the scatter diagram for the language scores and the music scores. It is given in Figure 9.35. In this case the points do not form a straight line. When this happens, we say that there is little or no correlation between the language scores and the music scores.

Once we have determined that there is a linear correlation between two variables, we may be interested in determining the strength of the linear relationship. Karl Pearson developed a *coefficient of linear*

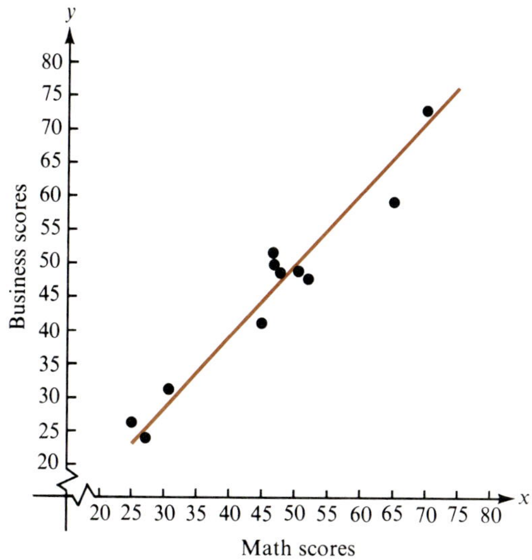

Figure 9.33 Scatter diagram for the math and business scores.

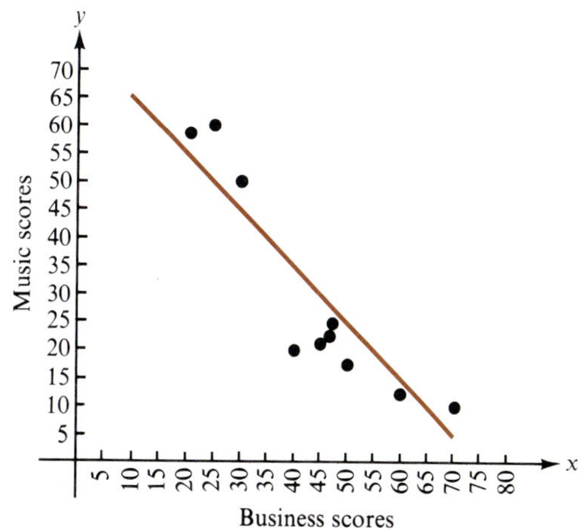

Figure 9.34 Scatter diagram for the business and music scores.

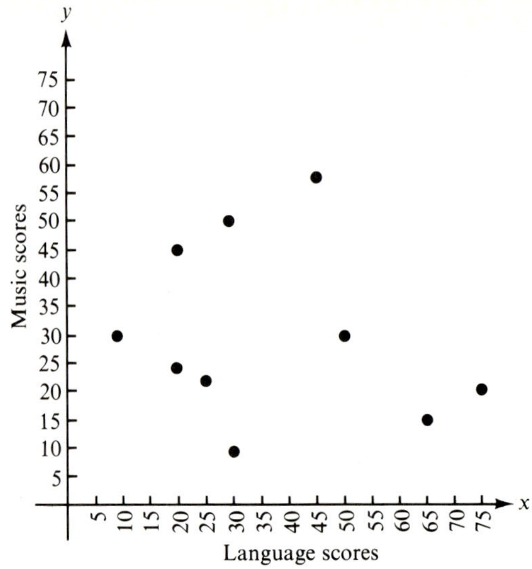

Figure 9.35 Scatter diagram for the language and music scores.

correlation, which measures the strength of the relationship between two variables. The value of the coefficient of linear correlation is calculated by means of a formula.

Formula 9.3 The *coefficient of linear correlation* is given by

$$r = \frac{n(\Sigma\, xy) - (\Sigma\, x)(\Sigma\, y)}{\sqrt{n(\Sigma\, x^2) - (\Sigma\, x)^2}\sqrt{n(\Sigma\, y^2) - (\Sigma\, y)^2}}$$

where

x = label for one of the variables

y = label for the other variable

n = number of pairs of scores

When using Formula 9.3 the coefficient of correlation will always have a value between $+1$ and -1. A value of $+1$ means perfect positive correlation and corresponds to the situation where all the dots lie exactly on a straight line. A value of -1 means perfect negative correlation and again corresponds to the situation where all the points lie exactly on a straight line. Correlation is considered high when it is close to $+1$ or -1 and low when it is close to 0. If the coefficient of linear correlation is 0, we say that there is no linear correlation. These possibilities are indicated in Figures 9.36 and 9.37.

Although Formula 9.3 looks complicated, it is rather easy to use. The only new symbol that appears is $\Sigma\, xy$. This value is found by

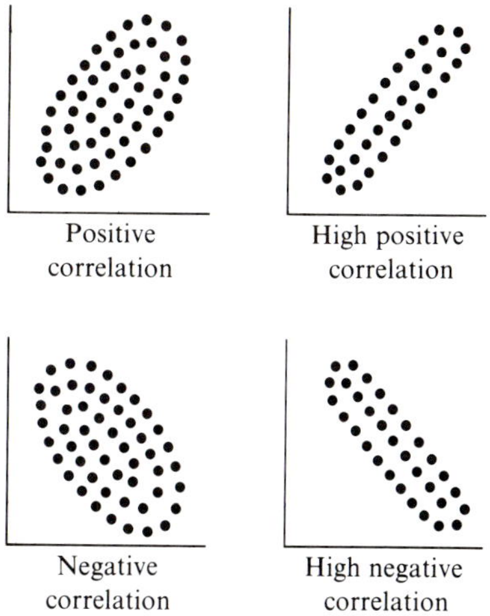

Figure 9.36 Possible correlations.

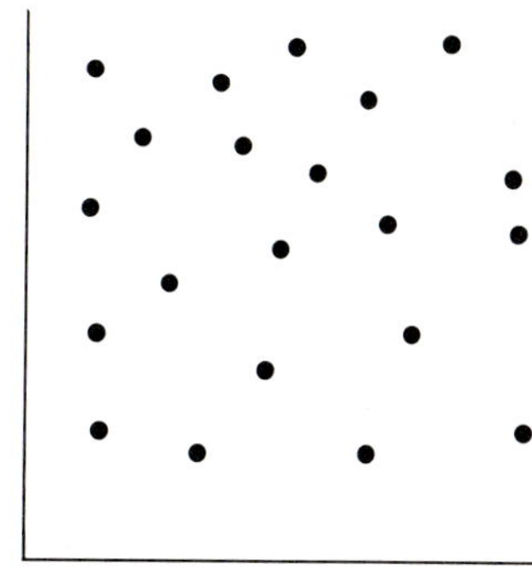

Figure 9.37 No correlation.

multiplying the corresponding values of x and y and then adding all the products. The following examples illustrate the procedure.

Example 26 Find the coefficient of linear correlation between the business and music scores given here.

Business score	40	72	31	50	49	59	24	27	49	48
Music score	20	9	50	17	24	13	54	57	23	22

Solution We let x represent the business score and y the music score. Then we arrange the data in tabular form.

x	y	x^2	y^2	xy
40	20	1600	400	800
72	9	5184	81	648
31	50	961	2500	1550
50	17	2500	289	850
49	24	2401	576	1176
59	13	3481	169	767
24	54	576	2916	1296
27	57	729	3249	1539
49	23	2401	529	1127
48	22	2304	484	1056
$\Sigma x = 449$	$\Sigma y = 289$	$\Sigma x^2 = 22{,}137$	$\Sigma y^2 = 11{,}193$	$\Sigma xy = 10{,}809$

Now we use Formula 9.3. We have

$$
\begin{aligned}
r &= \frac{n(\Sigma xy) - (\Sigma x)(\Sigma y)}{\sqrt{n(\Sigma x^2) - (\Sigma x)^2}\sqrt{n(\Sigma y^2) - (\Sigma y)^2}} \\
&= \frac{10(10{,}809) - (449)(289)}{\sqrt{10(22{,}137) - 449^2}\sqrt{10(11{,}193) - 289^2}} \\
&= \frac{-21{,}671}{\sqrt{19{,}769}\sqrt{28{,}409}} = \frac{-21{,}671}{(140.60)(168.55)} = \frac{-21{,}671}{23{,}698.13} \\
&= -.9145
\end{aligned}
$$

Thus, the coefficient of correlation is $-.9145$. Since this value is close to -1, we say that there is a high degree of negative correlation. Figure 9.34 also indicated this.

Example 27 A fire department official believes that as the temperature decreases, the number of fires increases. To support this claim, he has gathered the statistics shown in the accompanying table. Compute the coefficient of linear correlation and comment.

Temperature, x	40°	35°	30°	25°	20°	15°	10°	5°
Number of fires, y	33	37	40	44	56	60	61	71

Photo: Louise E. Jafferson, Monkmeyer

Solution We arrange the data in tabular form.

x	y	x^2	y^2	xy
40	33	1600	1089	1320
35	37	1225	1369	1295
30	40	900	1600	1200
25	44	625	1936	1100
20	56	400	3136	1120
15	60	225	3600	900
10	61	100	3721	610
5	71	25	5041	355
$\Sigma x = 180$	$\Sigma y = 402$	$\Sigma x^2 = 5100$	$\Sigma y^2 = 21{,}492$	$\Sigma xy = 7900$

Now we use Formula 9.3. We have, with $n = 8$,

$$\begin{aligned} r &= \frac{n(\Sigma\, xy) - (\Sigma\, x)(\Sigma\, y)}{\sqrt{n(\Sigma\, x^2) - (\Sigma\, x)^2}\sqrt{n(\Sigma\, y^2) - (\Sigma\, y)^2}} \\ &= \frac{8(7900) - (180)(402)}{\sqrt{8(5100) - 180^2}\sqrt{8(21{,}492) - 402^2}} \\ &= \frac{-9160}{\sqrt{8400}\sqrt{10{,}332}} = \frac{-9160}{(91.65)(101.65)} \\ &= -.9832 \end{aligned}$$

Thus, the coefficient of correlation is $-.9832$, a rather high negative correlation. Therefore, the data would seem to support the fire official's claim.

Although the coefficient of correlation is usually the first number that is calculated when we are given several sets of scores, great care must be used in interpreting the results. It can undoubtedly be said that among all the statistical measures discussed in this chapter, the correlation coefficient is the one that is most misused. One reason for this is the assumption that because the two variables are related, a change in one will result in a change in the other. Frequently two variables may appear to have a high correlation, even though they are not directly associated with each other. There may be a third variable that is highly correlated to these two variables.

Once we determine that there is a correlation between two variables, we might be interested in finding the equation of the line that best fits the set of points. If a relationship in the form of an equation can be found between the two variables, we can then use this equation to *predict* the value of one of the variables if the value of the other is known.

Fitting a line to a set of numbers is by no means an easy task. Nevertheless, methods have been designed to handle such prediction problems. The *least squares method* determines the line in such a way that the sum of the squares of the y distances between the given points and the line is a minimum. Such a line is called a *regression line* of y on x.

Formula 9.4 The equation of the regression line is given by

$$y = m_y + b(x - m_x)$$

where m_x and m_y are the means of x and y, respectively, and

$$b = \frac{n(\Sigma\, xy) - (\Sigma\, x)(\Sigma\, y)}{n(\Sigma\, x^2) - (\Sigma\, x)^2}$$

and n is the number of pairs of scores.

Let us use Formula 9.4 in the next examples to find the regression connecting two variables.

Example 28 Fifteen students have been receiving special instruction prior to taking a state civil service exam. In an effort to determine the effectiveness of the program, a comparison is made between the grade on the exam and the number of weeks each student received the special instruction. The results are given in the table.

Number of weeks, x	0.50	0.75	1.00	1.25	1.50	1.75	2.00	2.25	2.50	2.75	3.00	3.25	3.50	3.75	4.00
Grade, y	57	64	59	68	74	76	79	83	85	86	88	89	90	94	96

a) Compute the correlation coefficient between x and y.
b) Find the regression equation that will predict a student's score if we know how many weeks of special instruction he or she received.
c) If a student receives 0.65 week of special instruction, what is the student's predicted score on the exam?

Solution We arrange the data in tabular form.

x	y	x^2	y^2	xy
0.50	57	0.2500	3249	28.50
0.75	64	0.5625	4096	48.00
1.00	59	1.0000	3481	59.00
1.25	68	1.5625	4624	85.00
1.50	74	2.2500	5476	111.00
1.75	76	3.0625	5776	133.00
2.00	79	4.0000	6241	158.00
2.25	83	5.0625	6889	186.75
2.50	85	6.2500	7225	212.50
2.75	86	7.5625	7396	236.50
3.00	88	9.0000	7744	264.00
3.25	89	10.5625	7921	289.25
3.50	90	12.2500	8100	315.00
3.75	94	14.0625	8836	352.50
4.00	96	16.0000	9216	384.00
$\Sigma x = 33.75$	$\Sigma y = 1188$	$\Sigma x^2 = 93.4375$	$\Sigma y^2 = 96{,}270$	$\Sigma xy = 2863.00$

a) To compute the coefficient of correlation, we use Formula 9.3:

$$r = \frac{n(\Sigma\, xy) - (\Sigma\, x)(\Sigma\, y)}{\sqrt{n(\Sigma\, x^2) - (\Sigma\, x)^2}\sqrt{n(\Sigma\, y^2) - (\Sigma\, y)^2}}$$

$$= \frac{15(2863) - (33.75)(1188)}{\sqrt{15(93.4375) - 33.75^2}\sqrt{15(96{,}270) - 1188^2}}$$

$$= \frac{2850}{\sqrt{262.5}\sqrt{32{,}706}} = \frac{2850}{(16.20)(180.85)} = .9728$$

Thus, the coefficient of correlation is .9728.

b) To find the regression equation, we must first calculate the values of m_y, m_x, and b. We have

$$m_y = \frac{\Sigma y}{n} = \frac{1188}{15} = 79.2$$

$$m_x = \frac{\Sigma x}{n} = \frac{33.75}{15} = 2.25$$

$$b = \frac{n(\Sigma xy) - (\Sigma x)(\Sigma y)}{n(\Sigma x^2) - (\Sigma x)^2} = \frac{15(2863) - (33.75)(1188)}{15(93.4375) - 33.75^2}$$

$$= \frac{2850}{262.5} = 10.86$$

Thus, the regression equation predicting a student's score when we know how many weeks of special instruction he or she received is, using Formula 9.4

$$\begin{aligned} y &= m_y + b(x - m_x) \\ &= 79.2 + 10.86(x - 2.25) \end{aligned}$$

Therefore, the regression equation is $y = 79.2 + 10.86(x - 2.25)$.

c) If the student receives 0.65 week of instruction, then $x = 0.65$. Substituting this value of x into the regression equation gives

$$\begin{aligned} y &= 79.2 + 10.86(0.65 - 2.25) \\ &= 79.2 + 10.86(-1.6) \\ &= 79.2 - 17.38 \\ &= 61.82 \end{aligned}$$

Thus, the student's predicted score on the exam is 61.82.

Example 29 As we indicated earlier, Galton believed that there exists a very close relationship between the heights of fathers and the heights of their sons. To test this claim, a scientist selects eight men at random and records their heights and the heights of their sons (in inches) as shown in the table.

Father, x	66	68	71	72	69	69	73	70
Son, y	63	66	70	74	70	68	73	70

a) Find the regression equation.
b) If a father is 74 inches tall, what is the predicted height of his son?

Solution We arrange the data in tabular form.

x	y	x^2	xy
66	63	4356	4158
68	66	4624	4488
71	70	5041	4970
72	74	5184	5328
69	70	4761	4830
69	68	4761	4692
73	73	5329	5329
70	70	4900	4900
$\Sigma x = 558$	$\Sigma y = 554$	$\Sigma x^2 = 38{,}956$	$\Sigma xy = 38{,}695$

a) To find the regression equation, we calculate the values of m_y, m_x, and b. We have

$$m_y = \frac{\Sigma y}{n} = \frac{544}{8} = 69.25$$

$$m_x = \frac{\Sigma x}{n} = \frac{558}{8} = 69.75$$

$$b = \frac{n(\Sigma xy) - (\Sigma x)(\Sigma y)}{n(\Sigma x^2) - (\Sigma x)^2} = \frac{8(38{,}695) - (558)(554)}{8(38{,}956) - 558^2}$$

$$= \frac{428}{284} = 1.51$$

Thus, the regression equation is

$$y = m_y + b(x - m_x)$$
$$= 69.25 + 1.51(x - 69.75)$$

b) If a father is 74 inches tall, then $x = 74$. Substituting this value of x into the regression equation gives

$$y = 69.25 + 1.51(74 - 69.75)$$
$$= 69.25 + 1.51(4.25)$$
$$= 69.25 + 6.42$$
$$= 75.67$$

Thus, the predicted height of the son is 75.67 inches.

After determining the least squares regression equation that predicts a value of y when x has a particular value, we may be interested in determining how the predicted value of y and the observed value of y compare. Quite often there may be large differences between the two.

Fortunately, statisticians have devised a method for measuring the differences between the predicted and the observed values. This is the *standard error of the estimate* and is given by the formula

$$\sqrt{\frac{\Sigma(Y - Y_p)^2}{n - 2}}$$

where Y_p is the predicted value, Y is the observed value, and n is the number of pairs of scores.[4]

Exercises

55. A forest ranger keeps records on the number of inches of rainfall reported in a certain area and the number of fires reported during the past eight years.

Inches of rainfall, x	50	55	40	20	71	68	60	37
Fires reported, y	53	47	57	80	30	32	35	71

a) Draw the scatter diagram for these data.
b) Compute the coefficient of correlation.
c) Find the regression equation.
d) If 45 inches of rain were reported this year, find the predicted number of fires.

56. A butcher claims that as the price of a pound of ground meat goes up, then the number of pounds he sells per week decreases. To support his claim, he has gathered the following data.

Price per pound (cents), x	100	120	130	140	150	160	170	200
Pounds sold per week, y	427	380	350	320	300	260	237	175

a) Draw the scatter diagram.
b) Compute the coefficient of correlation.
c) Find the regression equation.
d) If the price of a pound of ground meat rises to \$2.25, what is the predicted number of pounds the butcher will sell?

[4] A complete discussion of this formula is beyond the scope of this book. The interested reader can find a detailed discussion in *Statistics and Probability in Modern Life* by Joseph Newmark, 2 ed., chap. 9 (New York: Holt, Rinehart and Winston, 1977).

57. The following table gives the retail value of a particular car as it ages.

Age of car (years), x	0	$\frac{1}{2}$	1	$1\frac{1}{2}$	2	$2\frac{1}{2}$	3
Value (dollars), y	4800	4000	3700	3200	2900	2500	1800

a) Draw the scatter diagram.
b) Compute the coefficient of correlation.
c) Find the regression equation.
d) What is the predicted value of a car after four years?

58. The number of deer shot in a particular area and the number of hunting permits issued for the area over the last few years are shown in the table.

Year	Number of permits issued, x	Number of deer shot, y
1974	27	312
1975	24	300
1976	28	320
1977	31	340
1978	25	301
1979	29	325

a) Draw the scatter diagram.
b) Compute the coefficient of correlation.
c) Find the regression equation.
d) In 1980 the conservation department plans to issue 25 hunting permits. What is the predicted number of deer to be shot?

59. The number of speeding tickets issued and the number of radar traps set up by police officials during the past eight weekends are shown in the table.

Number of speed traps, x	10	8	12	6	9	4	11	10
Number of tickets issued, y	112	89	120	80	94	60	115	110

a) Draw the scatter diagram.
b) Compute the coefficient of correlation.
c) Find the regression equation.
d) If the police plan to operate 15 speed traps next weekend, what is the predicted number of tickets to be issued?

60. The amount of money that a particular publisher is willing to invest in the production of a book is determined by the number of good reviews received as shown in the table.

Number of good reviews received, x	0	1	2	3	4	5
Amount of money to be invested, y	\$0	\$10,000	\$15,000	\$25,000	\$35,000	\$40,000

a) Find the regression equation.
b) If a book has received seven good reviews, what is the predicted amount of money that the publisher will invest in the production of the book?

61. According to one supermarket manager, the number of quarts of milk returned by customers because the milk is sour is determined by the outside temperature as shown in the table.

Outside temperature, x	70°	75°	80°	85°	90°	95°
Quarts of milk returned, y	16	23	26	30	32	39

a) Find the regression equation.
b) If the outside temperature rises to 100°, what is the predicted number of quarts of milk that will be returned?

62. According to one study, the average salary of an individual in a certain town is determined by the number of years of schooling beyond high school obtained by the individual as shown in the table.

Years of schooling beyond high school, x	0	1	2	3	4	5	6
Average salary, y	\$12,280	\$14,170	\$16,275	\$18,140	\$19,000	\$21,000	\$23,000

a) Find the regression equation.
b) Find the predicted salary of an individual who has obtained eight years of schooling beyond high school.

63. According to a real estate agent, the number of requests for information that he receives per day is influenced by the number of times his commercial is broadcast on radio or television as shown in the table.

Times commercial is broadcast, x	0	1	2	3	4	5	6
Requests for information, y	26	35	39	43	48	50	56

a) Find the regression equation.
b) If the agent has contracted to have the commercial broadcast eight times per day, what is the predicted number of requests for information that will be received?

■ **64.** A large department store is planning to hold its annual spring sale. It has found that the demand for a particular AM-FM radio that is placed on sale is determined by the sale price as shown in the table.

Sale price, x	\$12.99	\$12.49	\$11.99	\$11.49	\$10.99	\$10.49	\$9.99
Units sold, y	82	97	120	150	190	250	320

a) Find the regression equation.
b) If the store plans to sell the radio for \$7.99, what is the predicted number of radios that will be sold?
c) Why would the store manager be interested in obtaining such information?

9.7 Misuses of Statistics

As indicated earlier, statistics can be very useful and informative when applied properly. If statistics are not used properly, however, they can be made to say anything you want. The following examples on the misuse of statistics illustrate the nature of the difficulty involved.

1. *Wrong sampling techniques.* The greatest misuse of statistics occurs when the data are obtained from samples that are not random. An improperly selected sample can lead to the wrong conclusion. Over the years many incorrect predictions have been made on the basis of nonrandom samples. For example, in 1936 *the Literary Digest* was interested in determining who would win the coming presidential election. It decided to poll the voters by mailing 10 million ballots. On the basis of the approximately 2 million ballots returned, it predicted that Alfred E. Landon would be elected. An October 31 headline read

   Final return in the *Digest's* poll of 10 million voters
   Landon 1,293,669
   Roosevelt 972,897

   Actually Franklin Roosevelt carried 46 of the 48 states and many of them by a landslide. The 10 million people to whom the *Digest* had sent ballots were selected from telephone listings and from the list of its own subscribers. The year 1936 was a depression year and many people could not afford telephones or magazine subscriptions. Thus, the *Digest* did not select a random sample of voters of the United States. *The Literary Digest* soon went out of business.

   On April 6, 1976, both the ABC and NBC television networks projected that Morris Udall would win the Democratic primary in Wisconsin. Their predictions were based on samples from selected precincts and did not take certain districts into consideration. When all the rural votes were counted, Jimmy Carter came out on top.

Many newspapers were so confident of their predictions that they incorrectly printed the morning editions of their newspapers with the headline "Carter Upset by Udall."

In both examples the reason for the incorrect prediction is that it was based on information obtained from poor samples. It is for this reason that statisticians insist that samples be selected randomly.

2. *Different horizontal and/or vertical scales.* Consider the two graphs shown in Figure 9.38. Both indicate how the crime rate has changed in a particular city over the past six years. One paints a more alarming picture than the other. Can you see why?

3. *Correlation.* Many people have applied a positive correlation to prove a cause-and-effect relationship that may not even exist. To illustrate the point, it has been shown that there is a high positive correlation between teacher's salaries and the use of drugs on campus. Does this mean that the more money a teacher earns, the more prevalent will be the use of drugs on campus?

4. *Disregarding unfavorable data.* Often statistics are used in the area of consumer products. For example, a drug manufacturer may be interested in claiming that 90% of the doctors surveyed recommend his particular product. He then has his advertising agent poll groups of 100 doctors until he finds a group in which the overwhelming majority use the product. The unfavorable results are then discarded.

5. *Units not indicated.* Consider the following headline that appeared in a local newspaper: "In 1978 Murder in Our City Increased by 100%." Upon careful analysis it was found that in 1978 there were two murders committed in a city of 5 million and in 1979 there were four murders committed. Thus, although the headline may be mathematically accurate, it definitely does give one the impression that crime is widespread in the city.[5]

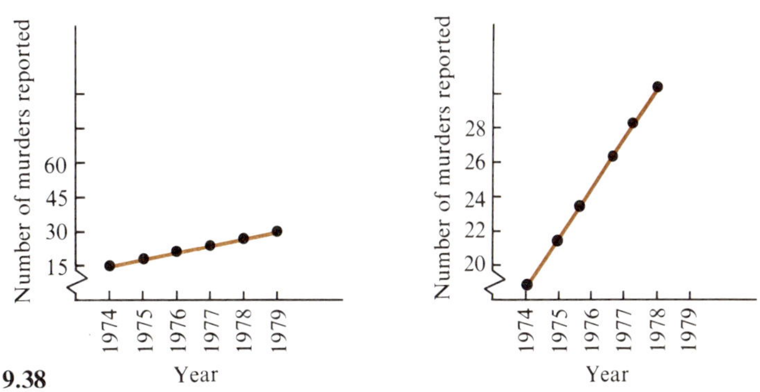

Figure 9.38

[5] For further examples of how statistics can be misused or misinterpreted, see *How to Lie with Statistics* by Darrell Huff (New York: Norton, 1954).

9.8 Summary

In this chapter we discussed different ways of analyzing a mass of data so that meaningful interpretations can be made. One method we mentioned is to use one of the measures of central tendency along with some measure of the dispersion involved. We indicated that the most commonly used measure of dispersion is the standard deviation. In discussing measures of central tendency, we introduced summation notation, a notation used to indicate the operation of a sum of numbers.

A large quantity of numbers can also be analyzed by graphical techniques. We simply set up a frequency distribution and then draw the bar graph to picture the information contained in the frequency distribution. We also indicated how pictographs, circle graphs, and line graphs can be used, each in its own right, to picture information visually.

One particular distribution that we discussed in detail is the normal distribution. We indicated how it can be applied to a variety of problems.

Finally, we discussed a method for determining whether a relationship exists among several variables and, if so, how strong this relationship is. We indicated how to find the equation of the regression line so that we can make predictions.

It should be apparent that our discussion was by no means a thorough analysis of all the ideas in the field of statistics. Our discussions were merely intended to introduce the basic ideas of statistics and how they can be misused.

A more detailed discussion of the topics covered in this chapter can be found in any elementary statistics book. See the references at the end of this chapter.

Study Guide

You should be able to demonstrate your knowledge of the following ideas presented in this chapter by giving definitions, descriptions, or specific examples. Page references are given for each term so that you can check your answer.

Statistics (p. 377)
Descriptive statistics (p. 377)
Sample (p. 378)
Mean or average (p. 381)
Mode (p. 381)
Measures of central tendency (p. 385)
Range (p. 390)
Frequency distribution (p. 395)
Classes (p. 396)
Pictographs or pictograms (p. 400)
Normal curve (p. 409)
Data (p. 377)
Inferential statistics (p. 377)
Population (p. 378)
Summation notation (p. 381)
Median (p. 382)
Measures of dispersion (p. 390)
Standard deviation (p. 390)
Frequency (p. 395)
Bar graph (p. 396)
Line graph (p. 401)
Circle graph or pie chart (p. 402)
Standard normal distribution (p. 411)

Normal distribution (p. 409)
Correlation (p. 418)
Scatter diagram (p. 418)
Linear correlation (p. 419)
Negative correlation (p. 419)
Least squares method (p. 423)
Standard error of the estimate (p. 427)
z-score or z-value (p. 411)
Axes (p. 418)
Positive correlation (p. 419)
Coefficient of linear correlation (p. 420)
Regression line (p. 423)

Formulas to Remember

The following list summarizes the formulas discussed in this chapter.

Mean or average: $\mu = \dfrac{\Sigma x}{n}$

Range = largest number − smallest number

Standard deviation: $\sigma = \dfrac{\sqrt{\Sigma (x - \mu)^2}}{n}$ or the rule on page 391

z-score or z-value: $z = \dfrac{x - \mu}{\sigma}$

Coefficient of linear correlation:

$$r = \frac{n(\Sigma xy) - (\Sigma x)(\Sigma y)}{\sqrt{n(\Sigma x^2) - (\Sigma x)^2}\sqrt{n(\Sigma y^2) - (\Sigma y)^2}}$$

Regression equation: $y = m_y + b(x - m_x)$ where

$$m_y = \frac{\Sigma y}{n}$$

$$m_x = \frac{\Sigma x}{n}$$

$$b = \frac{n(\Sigma xy) - (\Sigma x)(\Sigma y)}{n(\Sigma x^2) - (\Sigma x)^2}$$

Standard error of the estimate: $\sqrt{\dfrac{\Sigma (Y - Y_p)^2}{n - 2}}$

Mastery Tests

FORM A

1. What type of correlation exists between the incidence of heart attacks and being overweight?
   a) Positive correlation
   b) Negative correlation
   c) Zero
   d) Depends on the individual
   e) None of these

2. When $r = 0$, what is the value of this expression?

$$\sqrt{\frac{\Sigma(Y - Y_p)^2}{n - 2}}$$

a) 1
b) -1
c) 0
d) A very large number
e) None of these

3. The distribution of scores on a certain physical fitness test is normally distributed with a mean of 79 and a standard deviation of 10. Find the z-score of a girl who scored 74 on the exam.

a) -0.5000
b) 5
c) 0.5000
d) -5
e) None of these

4. Find the area under the normal curve from $z = -1$ to $z = 0.75$.

a) 0.1587
b) 0.2266
c) 0.3413
d) 0.6147
e) None of these

5. The graduating averages at a certain high school were alleged to be normally distributed with a mean of 75 and a standard deviation of 5. What is the probability that a graduating senior had an average below 65?

a) .9972
b) .4772
c) .0228
d) $-.0228$
e) None of these

6. Transistors are manufactured by a certain process that gives them a lifetime of 600 hours with a standard deviation of 45 hours. What percentage will have a lifetime of between 580 and 620 hours?

a) 97.36%
b) 48.68%
c) 95.76%
d) 0%
e) None of these

7. The standard deviation of the standard normal distribution is:

a) 0
b) 1
c) -1
d) 4
e) None of these

8. Which of the following z-scores has the smallest probability of occurrence for a normal distribution?

a) 0.23
b) -1.2
c) 4.5
d) 0
e) None of these

9. The proportion of the area under the normal curve from $z = 0$ to $z = 0.5$ is 0.1915. What is the proportion of the area under the normal curve between $z = 0.5$ and $z = 1$?

a) 0.3413
b) 0.1915
c) 0.1498
d) 0.3830
e) None of these

10. In a normal distribution, the mean is at:

a) $z = -1$
b) $z = 0$
c) $z = 1$
d) $z = 0.5$
e) None of these

11. The number of hats and the number of pairs of shoes owned by five female employees of the BAC Corporation are shown in the accompanying table. The coefficient of correlation for these data is:

a) .18
b) .93
c) .57
d) .39
e) None of these

Number of hats, x	Number of pairs of shoes, y
4	7
7	9
5	6
6	8
9	11

12. The least squares prediction equation for the data of question 11 is:

a) $y = 8.2 + 0.93(x - 6.2)$ b) $y = 0.93 + 8.2(x - 6.2)$
c) $y = 6.2 + 0.93(x - 8.2)$ d) $y = 8.2 + 6.2(x - 0.93)$
e) None of these

13. From the data in question 11, what is the predicted number of pairs of shoes owned by a female employee who has eight hats?

a) 9.87 b) 2.73 c) 6.01
d) 52.03 e) None of these

FORM B

1. If $x_1 = 15$, $x_2 = 18$, $x_3 = 24$, $x_4 = 12$, and $x_5 = 23$, find the following.

a) Σx^2 b) $(\Sigma x)^2$

2. When a manufacturer claims that the average size dress he sells is size 10, to which average is he referring?

3. In a certain class the scores are normally distributed with a mean of 75 and a standard deviation of 10. Furthermore, the top 15% of the students will get a grade of A. What is the lowest score needed for an A?

For questions 4 and 5 use the following information: At Tech College of Engineering there were 500 entering freshmen in 1978 and 600 in 1979. They were divided into major fields as shown in the accompanying graphs.

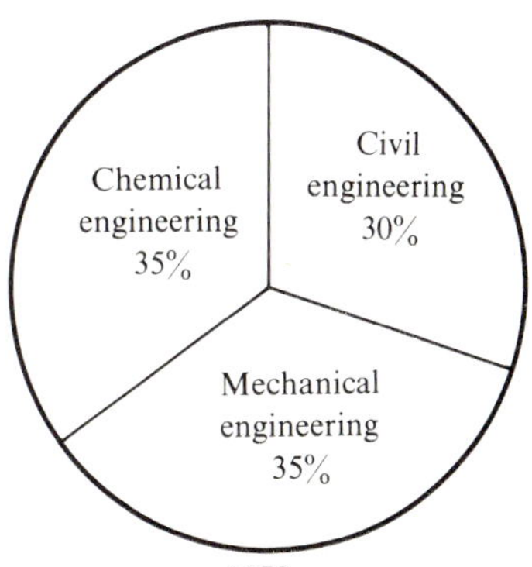

1978
500 students

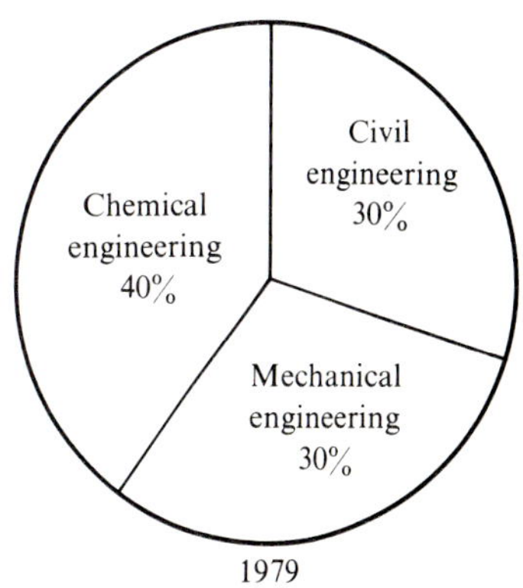

1979
600 students

4. How many students were in civil engineering in 1978?

5. From 1978 to 1979 what happended to the *number* of freshman students in mechanical engineering?
   a) Decreased by 25
   b) Decreased by 5
   c) Increased by 5
   d) Doubled
   e) None of these

6. Consider the accompanying graphs. Is it true that because there are more lawyers, the number of civil cases filed has increased? What type of correlation would you expect? Explain your answer.

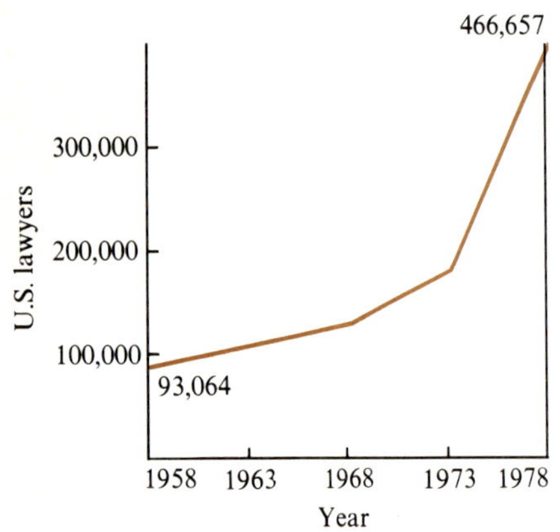

Source: American Bar Association

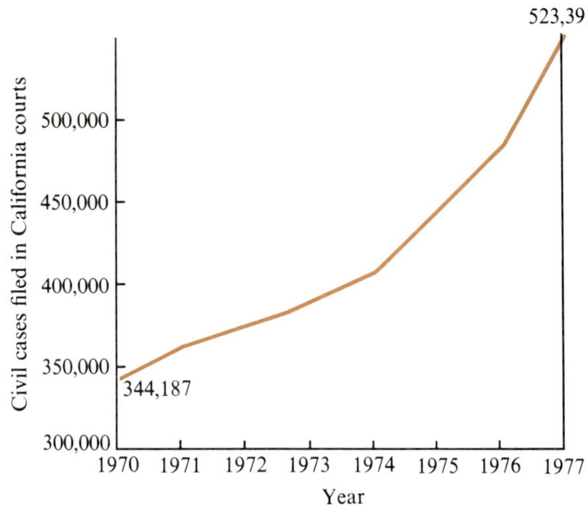

Source: Administrative Office of California Courts

Suggested Further Reading

Adler, I., *Probability and Statistics for Everyone*, New York: New American Library, 1966.

Cooke, W. P., "Beginning Statistics at the Track," *Mathematics Magazine* Vol. 46, pp. 250–255, 1973.

Huff, D., *How to Lie with Statistics*, New York: Norton, 1954.

Los Angeles Times, "Nielsen raters' views decide your TV fare," June 9, 1974.

Mathematics in the Modern World (readings from *Scientific American*), article 25, San Francisco: Freeman, 1968.

New York Times, "On the Nielsen families," September 15, 1974.

Newmark, J., *Statistics and Probability in Modern Life*, 2d ed., New York: Holt, Rinehart and Winston, 1977.

Senter, R. J., *Analysis of Data: Introductory Statistics for the Behavioral Sciences*, Glenview, Ill.: Scott Foresman, 1969.

Tanur, J. M., ed., *Statistics: A Guide to the Unknown*, San Francisco: Holden-Day, 1972.

Weaver, W., *Lady Luck*, New York: Doubleday (Anchor Books), 1963.

CHAPTER 10

The Mathematics of Finance

Objectives to be Achieved in this Chapter

- *To distinguish* between simple and compound interest and to study the different types of simple interest and when each type is used.
- *To analyze* what is meant by continuously compounded interest and to calculate the interest earned under these conditions.
- *To understand* what an annuity is and how to calculate the total value of its accumulations by introducing special annuity and present value charts and learning how to interpret them.
- *To distinguish* between amortization and sinking funds and when and how each is used to pay off a debt now or in the future.
- *To apply* the ideas of interest not only to business situations but to other fields as well.

We Pay You the Highest Interest on Savings Certificates Allowed by Law!

Maturity	Guaranteed rate	Annual yield	Minimum deposit
8 years	8.00% per annum compounded daily credited quarterly	8.45%	\$1000
6 years	7.75% per annum compounded daily credited quarterly	8.17%	\$1000
4 years	7.50% per annum compounded daily credited quarterly	7.90%	\$1000
$2\frac{1}{2}$ years	6.75% per annum compounded daily credited quarterly	7.08%	\$500
1 year	6.50% per annum compounded daily credited quarterly	6.81%	\$500
90 days	5.75% per annum compounded daily credited quarterly	6.00%	\$500

Our special six-month savings certificate pays $\frac{1}{4}$% more than the U.S. treasury bill six-month discount rate! Each week our interest rate will be based on the U.S. treasury bill six-month discount rate for that week. (Our rate will always be $\frac{1}{4}$% higher than this and the commercial bank rate for the same week!) Savings certificates issued during that week will be paid at this established rate for the 26-week term (182 days) of the certificate.

Minimum \$10,000

The effective yield on treasury bills is higher than the discount rate.

Federal regulations provide that premature withdrawals from certificate accounts receive the passbook rate on the withdrawn amount. In addition, a penalty of up to 90 days interest is forfeited.

And, we pay the best interest available on day-in/day-out accounts too!

5.47% effective annual yield on 5.25% per annum compounded daily credited quarterly

No bank offers you both these features in a regular savings account.

Interest paid from day of deposit to day of withdrawal, and compounded daily on balances of \$10 or more

Ten extra interest days on your money means even greater earnings for you! Deposits made by the tenth earn from the first, provided they remain till the end of the quarter.

And, a reminder, all accounts are legal for corporate, trust, and pension funds.

Yields on all of the above rates are effective when principal and interest are left on deposit for a full year

The table indicates that interest rates at this bank are compounded daily but credited quarterly. Furthermore, the annual yield and the effective annual yield are not the same. Does this seem reasonable? Shouldn't the rates be equivalent? Why is the effective rate on treasury bills higher than the discount rate?

There are some banks in the United States that compound interest continuously. How is this possible?

Obviously a knowledge of interest and how it is calculated is necessary to understand fully the nature and intent of bank advertising.

10.1 Introduction

At one time or another in our lives we must either borrow money or lend money. We may borrow money to help pay for college tuition, to finance a car that has been bought on "time," or to buy a house that sells for $50,000 when we have only $5,000 for the down payment. In these situations we need money and we are willing to pay *interest* to someone or some business for the use of the money.

On the other hand, we may lend money to someone or to some business or government agency. In this case the business or government agency may sell bonds, which in essence promise to repay the loan at a specified date. In the meantime interest will be paid, usually at six-month intervals, for the use of the money.

Many of us have department store credit cards or general bank credit cards such as Master Card or Visa. When we do not repay the monthly charges by the date specified, we are charged interest. This is usually labeled "finance charges" on the monthly statement we receive.

As a matter of fact, a large part of the U.S. economy is based on borrowing and interest. It is for this reason that a knowledge of interest, whether on money borrowed or on money saved, is a valuable tool. In this chapter we analyze some of the ideas of interest and more generally the mathematics of finance. This is done from both the consumer point of view and the business point of view.

It is strongly suggested that the reader use a calculator to work many of the exercises in this chapter.

Historical Note

Interest on money borrowed has had an unusual and often tainted history. Over the years scoundrels have charged enormous interest rates for the use of their money. There also have been "legitimate" businesses that purposely disguised the true amount or percentage of interest charged. An interest charge of 6% may have been an actual rate of 18%.

In 1969 the U.S. Congress passed the Truth in Lending Law for the benefit of consumers who use credit. Under the provisions of the law, full disclosure must be made of any interest charged (finance charges) and of the method used to calculate the interest. Interest must be expressed as an annual percentage rate on the monthly statement received. Although these actual charges vary from state to state, great progress has been made in making the ideas of interest understandable by the everyday person. It should be noted, however, that the Truth in Lending Law is a consumer protection act and does not apply to business loans.

10.2 Simple and Compound Interest

As indicated earlier, *interest* is the money paid for the use of money. The amount of money borrowed when an individual applies for a loan from a bank or when an individual lends money either to a bank in the form of a savings account or to a government agency in the form of a bond is called *principal.*

When money is borrowed, the borrower agrees to pay the lender interest at a specified *rate* over a period of time. For example, 6% over three years or 10% over four years. In each case the *rate of interest* is specified as a percentage. This is accomplished by using the % symbol or by writing it in decimal form. Thus, we can write an interest rate of 18 percent as 18% or as 0.18. Also we can write an interest rate of 5 percent as 5% or as 0.05.

Comment Many savings banks pay $5\frac{1}{4}$% interest on savings accounts. This is written in decimal form as 0.0525, *not* as 0.525. The latter represents 52.5%, a high interest rate.

The interest charged on a loan may be simple or compound. *Simple interest* is usually used for short-term loans of a year or less and is computed on the whole amount of money borrowed for the entire period of the loan. On the other hand, when a loan is for a longer period of time or for interest paid by a bank, the interest is added to the original loan at specified times (quarterly, semiannually, monthly, etc.), thereby increasing the amount of interest paid. This is known as *compound interest* and will be discussed later in this section.

Simple Interest

Suppose \$4000 is borrowed for 1 year. If the rate of interest is 7% per year, then the interest is 7% of \$4000, or

$$0.07 \cdot \$4000 = \$280$$

If the money is borrowed for 2 years, then the total interest paid is

$$0.07 \cdot \$4000 \cdot 2 = \$560$$

where 0.07 is the rate of interest, \$4000 is the amount of money borrowed (the principal), and 2 is the time period.

More generally we have the following simple interest formula:

Formula 10.1 *Simple interest formula.* The total simple interest, I, on a loan of P dollars for n years, where the interest rate is $r\%$ per year, is given by

$$I = P \cdot r \cdot n$$

Example 1 Find the simple interest on a loan of \$700 for 3 years if the rate of interest is 5% a year.

Solution We use Formula 10.1. Here $P = 700$, $r = 0.05$, and $n = 3$. Then

$$\begin{aligned} I &= P \cdot r \cdot n \\ &= 700 \cdot 0.05 \cdot 3 \\ &= 105 \end{aligned}$$

Thus, the total simple interest is \$105.

When working with Formula 10.1, both r(rate) and n(time) must be expressed in the same unit of time, for example, days, months, or years. If this is not the case, then one of them must be converted so that they are both in the same unit of time. This is illustrated in the following examples.

Example 2 *Friendly loans.* Bill arranges to borrow \$500 from his "friend" for 8 months. The simple interest rate is 30% per year. How much money does Bill have to repay?

Solution We use Formula 10.1 with $P = 500$ and $r = 0.30$. Since the loan is only for 8 months, we must express this as a fraction of the year. Thus, $n = 8$ months $= \frac{8}{12}$ year (n is usually expressed in years). Then

$$\begin{aligned} I &= P \cdot r \cdot n \\ &= 500 \cdot 0.30 \cdot \frac{8}{12} \\ &= 100 \end{aligned}$$

Thus, the interest is \$100. Since the loan was for \$500, Bill must repay \$600 to his friend.

Example 3 *Off-track betting (OTB).* Harry is in an OTB office in New York when he receives a tip on a particular horse. He borrows \$840 from a friend and agrees to repay \$900 to his friend in a month. Find the rate of interest.

Photo: Rhoda Sidney, DeWys

Solution Again we use Formula 10.1 with $P = 840$ and $n = 1$ month $= \frac{1}{12}$ year. Since Harry borrowed \$840 and will repay \$900, the interest is $900 - 840 = \$60$. Then

$$I = P \cdot r \cdot n$$

$$60 = 840 \cdot r \cdot \frac{1}{12}$$

$$60 = 70r$$

Now we divide both sides of the equation by 70 and get

$$r = \frac{60}{70} = 0.8571$$

Thus, the rate of interest is 85.71%.

Example 4 *Short-term loan.* Marie Cartright arranges for a short-term loan with her bookie at an annual interest rate of 12%. The total interest charged was \$100. If Marie repays the loan in 2 months, how much money did she borrow?

Solution Here we are interested in finding the amount of money borrowed, so we must find the value of P. We use Formula 10.1 with $r = 0.12$, $n = 2$ months $= \frac{2}{12}$ year, and $I = 100$. We have

$$I = P \cdot r \cdot n$$

$$100 = P \cdot 0.12 \cdot \frac{2}{12}$$

$$100 = 0.02P$$

Now we divide both sides of this equation by 0.02. We get

$$\frac{100}{0.02} = \frac{\cancel{0.02}P}{\cancel{0.02}}$$

$$5000 = P$$

Thus, Marie borrowed \$5000.

Table 10.1 Types of Simple Interest

Type	Days in month	Days in year
Exact	Actual number of days	365 (assuming no leap year)
Ordinary	30	360
Banker's rule	Actual number of days	360

Often a loan is made for a specific number of days rather than for a month or for a year. It would seem reasonable to use 365 days as the denominator of the time fraction just as we used 12 as the denominator when the time was given in months (see Example 3). However, this is usually not done. Instead, it is assumed that the year contains 360 days and that a month contains 30 days. Interest calculated using these numbers is called *ordinary interest* or *banker's rule interest* as opposed to *exact interest*, which is calculated using 365 days in the year. This is illustrated in Table 10.1.

Comment Most lending institutions use both the exact and the banker's rule for computing interest. Banker's rule interest is used for large loans, whereas exact interest is used for small loans of usually less than \$75,000. Ordinary interest is used for long-term transactions such as real estate deals.

Example 5 *Paying real estate taxes.* On February 10 Mike Sullivan arranges for a 3-month loan for \$5000 to pay his company's quarterly real estate taxes to the city's tax collector. The interest rate is 8%. Find the simple interest using the following rules.

a) Exact interest rule
b) Ordinary interest rule
c) Banker's rule

Solution a) If the loan is made on February 10 and has to be repaid in 3 months, then it has to be repaid on May 10. Now we must calculate the number of days between February 10 and May 10. To simplify this computation, we can use Table 10.2, which gives the number of each day of the year. February 10 is the 41st day of the year and May 10 is the 130th day. Thus, there are 130 − 41 or 89 days between February 10 and May 10. For exact interest, the year is assumed to contain 365 days. Now we can use Formula 10.1 with $P = 5000$, $r = 0.08$, and $n = 89$ days $= \frac{89}{365}$ year. We have

$$\begin{aligned} I &= P \cdot r \cdot n \\ &= 5000 \cdot 0.08 \cdot \frac{89}{365} \\ &= 97.53 \qquad \text{(rounded off)} \end{aligned}$$

Thus, the interest using the exact rule is \$97.53.

Table 10.2 The Number of Each Day of the Year

Day of month	Jan.	Feb.	Mar.	Apr.	May	June	July	Aug.	Sept.	Oct.	Nov.	Dec.	Day of month
1	1	32	60	91	121	152	182	213	244	274	305	335	1
2	2	33	61	92	122	153	183	214	245	275	306	336	1
3	3	34	62	93	123	154	184	215	246	276	307	337	3
4	4	35	63	94	124	155	185	216	247	277	308	338	4
5	5	36	64	95	125	156	186	217	248	278	309	339	5
6	6	37	65	96	126	157	187	218	249	279	310	340	6
7	7	38	66	97	127	158	188	219	250	280	311	341	7
8	8	39	67	98	128	159	189	220	251	281	312	342	8
9	9	40	68	99	129	160	190	221	252	282	313	343	9
10	10	41	69	100	130	161	191	222	253	283	314	344	10
11	11	42	70	101	131	162	192	223	254	284	315	345	11
12	12	43	71	102	132	163	193	224	255	285	316	346	12
13	13	44	72	103	133	164	194	225	256	286	317	347	13
14	14	45	73	104	134	165	195	226	257	287	318	348	14
15	15	46	74	105	135	166	196	227	258	288	319	349	15
16	16	47	75	106	136	167	197	228	259	289	320	350	16
17	17	48	76	107	137	168	198	229	260	290	321	351	17
18	18	49	77	108	138	169	199	230	261	291	322	352	18
19	19	50	78	109	139	170	200	231	262	292	323	353	19
20	20	51	79	110	140	171	201	232	263	293	324	354	20
21	21	52	80	111	141	172	202	233	264	294	325	355	21
22	22	53	81	112	142	173	203	234	265	295	326	356	22
23	23	54	82	113	143	174	204	235	266	296	327	357	23
24	24	55	83	114	144	175	205	236	267	297	328	358	24
25	25	56	84	115	145	176	206	237	268	298	329	359	25
26	26	57	85	116	146	177	207	238	269	299	330	360	26
27	27	58	86	117	147	178	208	239	270	300	331	361	27
28	28	59	87	118	148	179	209	240	271	301	332	362	28
29	29		88	119	149	180	210	241	272	302	333	363	29
30	30		89	120	150	181	211	242	273	303	334	364	30
31	31		90		151		212	243		304		365	31

Note: This table assumes that there is no leap year. In leap years, add 1 to the number given for each day after February 28.

b) The ordinary interest rule assumes that each month has 30 days and that the year has 360 days. Since the loan was for 3 months, we have $P = 5000$, $r = 0.08$, and $n = \frac{90}{360} = \frac{1}{4}$ year. Using Formula 10.1, we get

$$\begin{aligned} I &= P \cdot r \cdot n \\ &= 5000 \cdot 0.08 \cdot \frac{1}{4} \\ &= 100 \end{aligned}$$

Thus, the interest using the ordinary rule is \$100.00.

c) The banker's rule uses the actual number of days and assumes that the year contains 360 days. Thus, $n = 89$ days $= \frac{89}{360}$ year. Using Formula 10.1 with $P = 5000$, $r = 0.08$, and $n = \frac{89}{360}$, we get

$$
\begin{aligned}
I &= P \cdot r \cdot n \\
&= 5000 \cdot 0.08 \cdot \frac{89}{360} \\
&= 98.89 \qquad \text{(rounded off)}
\end{aligned}
$$

Thus, with the banker's rule, the interest is \$98.89.

Example 6 Gary Peters arranges for a \$7500 loan from the Commercial State Bank. The loan is for 60 days. The annual rate of interest is 6%.

a) Find the ordinary interest charge.

b) Find the total amount of money that Gary has to repay to the bank.

Solution a) We use Formula 10.1 with $P = 7500$, $r = 0.06$, and $n = 60$ days $= \frac{60}{360}$ year. Using the ordinary interest rule, we have

$$
\begin{aligned}
I &= P \cdot r \cdot n \\
&= 7500 \cdot 0.06 \cdot \frac{60}{360} \\
&= 75
\end{aligned}
$$

Thus, the interest is \$75.

b) Since Gary has to repay the principal along with interest, he will repay \$7500 + \$75 or \$7575 to the bank.

In Example 6 the interest rate was 6% and the time period was 60 days. Since this particular combination occurs quite often, there is a special 6% 60-day shortcut rule that allows us to compute the ordinary interest quickly. This rule is particularly helpful when the principal is a small number and involves decimals, for example, when the principal is \$1836.96. We state this shortcut rule as a formula.

Formula 10.2 *6% 60-Day rule.* The ordinary interest on any principal at 6% for 60 days can be found by moving the decimal point in the principal two places to the left.

Example 7 Using the 6% 60-day rule, find the ordinary interest on the following loans.

a) \$543.78 at 6% for 60 days

b) \$543.78 at 3% for 180 days

Solution We use the 6% 60-day rule.

a) Since the loan of \$543.78 is at 6% for 60 days, the ordinary interest can be obtained (using Formula 10.2) by moving the decimal point two places to the left. Thus, the interest is

\$543.78 or \$5.44 (rounded off)

b) Since the interest rate is 3%, we must first find the interest at 6% and then divide by 2. From part a, the interest for a \$543.78 loan at 6% for 60 days is \$5.44. Thus, at 3% the interest is half this amount, or \$2.72. This is the interest charge for 60 days. Since the loan is for 180 days, we must multiply this amount by 3. Therefore, the interest charge is

$$\$2.72 \cdot 3 = \$8.16$$

Compound Interest Suppose a person has some money that he or she wishes to deposit in a savings bank that pays compound interest. Since some banks in the United States now pay interest *compounded continuously*, let us analyze what is meant by *compound interest.*

To get started, suppose we deposit \$1000 (the principal) in a bank that pays 6% interest a year compounded annually. Then at the end of one year we would have \$1060. This amount represents the \$1000 principal plus the 1000(0.06) or \$60 interest earned on the money.

During the second year the bank will pay 6% interest on the \$1060, so that we will earn \$1060(0.06) or an additional \$63.60 in interest. Thus, at the end of the second year we would have a total accumulation of \$1123.60. The same thing will happen in succeeding years. Table 10.3 indicates the amount of money accumulated after many years assuming the bank pays 6% interest compounded annually.

Table 10.3 Amount of Money Accumulated after Six Years Assuming a 6% Interest Rate Compounded Annually

Year	Amount of money on deposit at beginning of year	Interest earned during year	Amount of money on deposit at end of year	Simplified form
1	\$1000.00	1000.00(0.06)	$1000.00 + 1000.00(0.06) = 1000.00(1 + 0.06)$	\$1060.00
2	\$1060.00	1060.00(0.06)	$1060.00 + 1060.00(0.06) = 1000.00(1 + 0.06)^2$	\$1123.60
3	\$1123.60	1123.60(0.06)	$1123.60 + 1123.60(0.06) = 1000.00(1 + 0.06)^3$	\$1191.02
4	\$1191.02	1191.02(0.06)	$1191.02 + 1191.02(0.06) = 1000.00(1 + 0.06)^4$	\$1262.48
5	\$1262.48	1262.48(0.06)	$1262.48 + 1262.48(0.06) = 1000.00(1 + 0.06)^5$	\$1338.23
6	\$1338.23	1338.23(0.06)	$1338.23 + 1338.23(0.06) = 1000.00(1 + 0.06)^6$	\$1418.52

Table 10.4 Amount of Money Accumulated After One Year at 6% Interest Using Different Compounding Periods

Principal	Compounded annually	Compounded semiannually	Compounded quarterly	Compounded monthly	Compounded continuously
\$1000.00	\$1060.00	\$1060.90	\$1061.36	\$1061.68	\$1061.84
\$5000.00	\$5300.00	\$5304.50	\$5306.82	\$5308.39	\$5309.18

On the other hand, if the bank pays 6% interest compounded semiannually, then it is really paying only 3% interest for each six-month period. Similarly, when the bank pays 6% interest compounded quarterly, it is really paying only $1\frac{1}{2}$% interest for every three-month period. What happens when the bank pays 6% interest compounded continuously? Table 10.4 gives the amount of money that can be accumulated when different compounding periods are used. The values given in the table were obtained by using what is called the compound interest formula, to be discussed shortly.

Tables 10.3 and 10.4 suggest that as the number of compounding periods increases, the amount of money that we accumulate also increases. Yet, the amount of money we accumulate does not increase without bounds (unfortunately!). There is a limit. To find this limit, we can use the compound interest formula, or Formula 10.3, which follows directly from the fourth column of Table 10.3.

Formula 10.3 *Compound interest formula.* The compound amount A (principal plus interest) that results when P dollars (the principal) are invested at r% per *period* for n periods is given by the formula

$$A = P(1 + r)^n$$

When the money is compounded continuously, we have Formula 10.4.

Formula 10.4 If P dollars are invested in a bank that pays r% interest per year compounded continuously, the amount of money accumulated, A, after n years is

$$A = Pe^{rn}$$

where the values of e to the appropriate power can be determined from Table 4 in the Appendix.

Comment The letter e given in Formula 10.4 is used often in mathematics to represent a number with a value of 2.71828. . . . The values of e to the appropriate power can be found in Table 4 in the Appendix.

Example 8 John Miser has \$1000 that he can deposit in one of three banks, all of which pay 8% interest yearly. However, one bank compounds the interest annually, one compounds it quarterly, and one compounds it continuously. Compute the different amounts of interest that can be earned in each of these banks in 3 years.

Solution We will use Formula 10.3 for compound interest, $A = P(1 + r)^n$. For the bank that compounds interest annually, there will be 3 paying periods (one for each year), so that $n = 3$. Then

$$\begin{aligned} A &= 1000(1 + 0.08)^3 \\ &= 1000(1.08)^3 \\ &= \$1259.71 \end{aligned}$$

For the bank that compounds interest quarterly, there will be 12 paying periods (4 for each year), so that $n = 12$. Also the rate of interest is 8% per year or $\frac{8}{4} = 2\%$ per paying period. Then

$$\begin{aligned} A &= \$1000(1 + 0.02)^{12} \\ &= 1000(1.02)^{12} \\ &= \$1268.24 \end{aligned}$$

For the bank that compounds interest continuously, we use Formula 10.4, $A = Pe^{rn}$. The money will remain in the bank for 3 years, so that $n = 3$. (When using Formula 10.4, n is expressed in years. The same is true for the interest rate, r.) Then

$$\begin{aligned} A &= 1000e^{0.08(3)} \\ &= 1000e^{0.24} \end{aligned}$$

Using Table 4 in the Appendix, we find that the value of $e^{0.24}$ is 1.27125. Thus,

$$\begin{aligned} A &= 1000e^{0.24} \\ &= 1000(1.27125) \\ &= \$1271.25 \end{aligned}$$

In this problem the difference between compounding continuously and compounding annually amounts to \$1271.25 − \$1259.71 or \$11.54 per \$1000 invested.

Many banks now quote interest rates with the expression "effective interest rate." This is calculated by using Formula 10.5.

Formula 10.5 The *effective interest rate* corresponding to an interest rate of $r\%$ compounded continuously is given by

$$e^r - 1$$

Thus, if a bank pays 6% interest compounded continuously, the effective annual interest rate is

$$\begin{aligned} e^r - 1 &= e^{0.06} - 1 \\ &= 1.0618 - 1 \\ &= 0.0618, \text{ or } 6.18\% \end{aligned}$$

This means that a 6% interest rate compounded continuously will yield the same amount of money to the depositor as a 6.18% interest rate compounded annually.

Example 9 What is the effective annual yield of a 7% interest rate compounded continuously?

Solution We use Formula 10.5. We have $r = 0.07$, so that

$$\begin{aligned} e^r - 1 &= e^{0.07} - 1 \\ &= 1.0725 - 1 \\ &= 0.0725, \text{ or } 7.25\% \end{aligned}$$

Thus, the effective annual interest rate is 7.25% or $7\frac{1}{4}\%$.

The function $f(n) = Pe^{rn}$, where r is the rate and n is the number of years, has many other interesting applications outside of finance. For example, the function is often used when studying such topics as population and bacteria growth, wealth and capital growth, and so forth.[1]

Example 10 *Decreasing purchasing power.* In recent years the purchasing power of the U.S. dollar has been decreasing constantly because of inflation. Assuming a 6% yearly rate of inflation, how much will \$1000 be worth (in present purchasing power) 5 years from now?

Solution Since purchasing power has been decreasing, $r = -0.06$. We then apply the compound interest formula with $P = 1000$ and $n = 5$. We have

$$\begin{aligned} A &= P(1 + r)^n \\ &= 1000(1 - 0.06)^5 \\ &= 1000(0.94)^5 \\ &= 733.90 \end{aligned}$$

Thus, in 5 years \$1000 will be worth only \$733.90 in purchasing power. It will lose approximately \$266.10 in purchasing power due to inflation.

[1] A more detailed discussion of such functions can be found in many elementary calculus books for liberal arts students under the heading *growth and decay problems.* See, for example, *The Usefulness of Calculus for the Behavioural, Life and Managerial Sciences* by Joseph Newmark (New York: Harper & Row, 1978).

Example 11 *Current value.* Trudy purchased a U.S. treasury note that pays 6% annual interest and is redeemable for \$10,000 after 1 year. If Trudy wishes to sell the note after 6 months, what is the (current) value? (Assume that the treasury note increases in value continuously and that the rate of change is proportional to its value.)

Solution Since the value of the treasury note is increasing continuously, we use Formula 10.4. As the note will be worth \$10,000 after 1 year, $A = 10{,}000$ when $n = 1$. Also $r = 0.06$. Substituting these values into Formula 10.4 gives

$$A = Pe^{rn}$$
$$10{,}000 = Pe^{0.06(1)}$$
$$10{,}000 = Pe^{0.06}$$

Now we solve for P, getting

$$P = \frac{10{,}000}{e^{0.06}}$$
$$= \frac{10{,}000}{1.061836547}$$
$$= 9417.65$$

Therefore,

$$A = 9417.65e^{0.06n}$$

represents the value of the treasury note at any time n. After 6 months (that is, when $n = \frac{1}{2}$ year), the value of the note is

$$A = 9417.65e^{0.06(1/2)}$$
$$= 9417.65e^{0.03}$$
$$= 9417.65(1.0305)$$
$$= 9704.89$$

Thus, the current value of the note (after 6 months) is \$9704.89.

Example 12 *Getting bald.* Upon reaching his thirtieth birthday, a movie star is told that he will lose his hair continuously at a rate of 4% annually. What percentage of his hair will the movie star have remaining when he reaches his forty-fifth birthday?

Solution Again we apply the compound interest formula with $P = 100$, $r = -0.04$, and $n = 45 - 30$, or 15 years. Thus, we have

$$A = P(1 + r)^n$$
$$= 100(1 - 0.04)^{15}$$
$$= 100(0.96)^{15}$$
$$= 54.21$$

Thus, the movie star will have approximately 54% of his hair remaining when he reaches his forty-fifth birthday.

Comment When using the compound interest formula, the value of $(1 + P)^n$ can also be obtained from compound interest charts. These values can be obtained easily using any hand-held calculator, however; and it is for this reason that compound interest charts are not given in this book.

Exercises

1. Find the simple interest on a loan of \$4500 for seven years at an annual rate of $4\frac{1}{2}\%$.
2. Find the simple interest on a loan of \$6200 for six years at an annual rate of $7\frac{3}{4}\%$.
3. Alex borrows \$800 from a finance agency and agrees to repay the money in ten months. The simple interest rate is 18% per year. How much money does Alex have to repay the finance company?
4. What is the finance charge on a loan of \$820 for $3\frac{1}{2}$ months if the simple interest is 14% per year?
5. Cary borrows \$700 from friends and repays \$800 to them in five months. Find the simple rate of interest.
6. *Finance charges.* Frances purchases a television set with a down payment of \$400. She agrees to pay the \$100 balance in six months with a check of \$150. Find the rate of interest.
7. Jim borrows money from his aunt in order to purchase stereo equipment. The annual simple interest rate is 15%. Jim repays his aunt in four months. The total interest charge was \$175. How much money did Jim borrow from his aunt?
8. On February 1 Mike borrows \$10,000 from the Hamilton Bank for nine months so that he can install central air conditioning in his factory. The interest rate is 11%. Find the simple interest using the following rules.
   a) Exact interest rule
   b) Ordinary interest rule
   c) Banker's rule
9. *Farming.* On June 1 Cassandra borrows \$8000 from the Farmer's National Bank. Cassandra, who is a farmer, expects to repay the loan on September 1 when she harvests her crop. The simple interest rate is 9%. How much does Cassandra have to repay the bank if the bank computes the interest charges using these rules?
   a) Exact interest rule
   b) Ordinary interest rule
   c) Banker's rule
10. Find the simple ordinary interest on a 60-day loan of \$8426 if interest is 6% a year.

11. Find the simple ordinary interest on a 60-day loan of \$5418 if interest is 3% a year.

12. Find the simple ordinary interest on a 60-day loan of \$7958.16 if interest is 9% a year.

13. Find the simple ordinary interest on a 30-day loan of \$5618 if interest is 9% a year.

14. *Revenue sharing.* A large city in the Northeast is having difficulty meeting its monthly payroll. It arranges for a loan of \$2.5 million for 45 days, at which time it will receive funds from the federal government under the revenue-sharing program. If interest is 8% a year and is calculated using the banker's rule, find the total interest charge.

15. Find the amount of money that will accumulate in three years when \$800 is invested in a bank that pays 5% interest compounded annually.

16. Fay invests \$2000 at 7% compounded quarterly. How much money will accumulate after four years?

17. *Car loans.* Rochelle purchases a \$5000 car from a dealer and agrees to repay the \$5000 loan after a 36-month period. Interest is 10% a year compounded continuously. How much money will Rochelle have to repay the dealer?

18. Richie deposits \$30,000 in a bank that pays dividends at an annual interest rate of 7.5% compounded quarterly. If no other transactions are made, how much money will Richie have in his account in six years?

19. What is the effective annual interest rate of an 8% interest rate compounded continuously?

20. *Parking ticket.* Paul receives a \$25 summons for double parking. Interest on the fine is 9% a year compounded monthly. If it takes Paul two years to pay the fine, how much money will he end up paying?

21. *Individual Retirement Accounts.* Sharon is $39\frac{1}{2}$ years old and has just started an Individual Retirement Account (IRA) with the Community National Bank. The bank pays $6\frac{1}{2}\%$ annual interest compounded quarterly. According to government regulations, she may not withdraw her \$1250 deposit until she reaches age $59\frac{1}{2}$. How much money will she have in the bank by age $59\frac{1}{2}$?

22. *Husband and wife banking.* Chris deposits \$2000 in a bank that pays 6% interest compounded continuously. His wife Sonya deposits \$2000 in a bank that pays 6% interest compounded monthly. How much money will each have in three years?

23. Roy White repays a \$4000 loan to his "friends" in two years. He repaid a total of \$8000. Interest was compounded annually. Find the rate of interest.

24. In how many years will \$5000 invested at 5% yearly interest compounded continuously amount to \$8243.61?

★25. Carl Sanchez borrowed \$8000 and agreed to repay the loan in seven years with interest at 9% compounded monthly. How much does he still owe after four years?

10.3 Annuities

All of us have some idea of what an annuity is. When we make the weekly, monthly, or quarterly installment payments on the new car we bought, these payments represent an annuity. When we buy a life insurance policy and make equal semiannual or annual payments, then these payments represent an annuity. More generally, we have the following definitions.

Definition 10.1 A sequence of equal payments made at equal periods of time is called an *annuity*.

Definition 10.2 A *payment period* is the time between successive payments.

Definition 10.3 The *term* of an annuity is the time from the beginning of the first payment period to the end of the last payment period.

While there are many different types of annuities, we concern ourselves with only *ordinary annuities*. In this type of annuity the regular payments are made at the end of a period, at which time the interest is credited.

Example 13 Edmund decides to set up an annuity for his newborn child's college education by depositing \$200 every six months into an account that pays 6% interest compounded annually. Payments will be made for 20 years.

In this case we have an ordinary annuity where the payment period is six months, the interest is 6% per year or 3% per payment period, and the term of the annuity is 20 years.

At any point in time, an annuity may have accumulated a substantial amount of money from the payments made and from the interest earned.

Definition 10.4 The *amount of an annuity* is the sum of all the payments made plus all the interest accumulated.

How do we find the amount of an annuity? The amount of an ordinary annuity can be calculated in much the same way that we calculated compound interest in the preceding section. We illustrate the technique with an example.

Example 14 Find the amount of an ordinary annuity of \$100 per year after 5 years if interest is paid at 6% compounded annually.

Solution The first payment of \$100 is received at the end of year 1. Thus, it will earn interest for only 4 years. Similarly, the second payment will earn interest for only 3 years, and so forth. Now let A represent the amount of the annuity. Then from the compound interest formula of the preceding section, the value of the first \$100 payment after 4 years is

$$A_1 = 100(1 + 0.06)^4 = 100(1.262477) = \$126.25$$

The value of the second \$100 payment after 3 years is

$$A_2 = 100(1 + 0.06)^3 = 100(1.191016) = \$119.10$$

The value of the third \$100 payment after 2 years is

$$A_3 = 100(1 + 0.06)^2 = 100(1.1236) = \$112.36$$

The value of the fourth \$100 payment after 1 year is

$$A_4 = 100(1 + 0.06)^1 = 100(1.06) = \$106.00$$

The fifth \$100 payment, A_5, has a value of \$100 since it is deposited at the end of the fifth year. Thus, the amount of the annuity is

$$\begin{aligned} A &= A_1 + A_2 + A_3 + A_4 + A_5 \\ &= \$126.25 + \$119.10 + \$112.36 + \$106.00 + \$100 \\ &= \$563.71 \end{aligned}$$

In Example 14 we calculated the amount of the annuity after five years. Typically, annuities run for long periods of time. The calculations involved in finding the value of the annuity over a long period can be quite time-consuming. Fortunately, there are charts that enable us to determine the amount of an annuity after n payments. One such chart is given in Table 5 of the Appendix. This table gives the amount of an annuity at a particular interest rate for n periods assuming the amount deposited at each payment period is \$1. If the equal payments are more than \$1, then we simply multiply the chart value by the amount of the equal payment. We have the following formula.

Formula 10.6 If P represents the amount of money deposited into an annuity (at each payment period) that pays $r\%$ interest compounded per payment period, then the *amount A of the annuity* after n payments is

$A = P$ multiplied by the appropriate chart value given in Table 5

Example 15 *College education.* Lorraine set up an annuity for her child's college education. She will deposit \$150 every 6 months for 18 years into a fund that pays interest at the rate of 6% compounded semiannually. Find the amount of the annuity.

Solution Since payments into the annuity are made every 6 months for 18 years, there is a total of 36 payments. Thus, the number of periods, n, is 36. Also interest is 6% compounded semiannually, or for each period it is $\frac{6}{2} = 3\%$. Now we look in Table 5 to find the appropriate chart value. We first find the correct interest rate by moving across the top of the chart until we reach the 3% column. Then we move down this column until we reach the period 36 line. The chart value is 63.275944. Now we apply Formula 10.6 to find the amount of the annuity. We have

$$\begin{aligned} A &= P \text{ multiplied by appropriate chart value} \\ &= 150(63.275944) \\ &= 9491.39 \qquad \text{(rounded off)} \end{aligned}$$

Thus, the amount of the annuity will be \$9491.39.

Comment In Example 15 the amount of the annuity is \$9491.39. This represents Lorraine's contributions of \$150 · 36, or \$5400, plus the interest earned of \$9491.39 − \$5400, or \$4091.39.

Example 16 Randolph Scott deposits \$225 every three months into a savings account that pays 5% interest compounded quarterly. After making 41 deposits, how much money does Randolph have?

Solution Since Randolph made 41 payments, the number of periods $n = 41$. Also interest is 5% compounded quarterly. Thus, the interest rate is $\frac{5}{4} = 1\frac{1}{4}\%$ for each payment period. Now we look in Table 5 of the Appendix to find the appropriate chart value. It is found by locating the $1\frac{1}{4}\%$ interest column and then moving down the period column until we arrive at the 41 line. The value given is 53.133176. Now we apply Formula 10.6. We have

$$\begin{aligned} A &= P \text{ multiplied by appropriate chart value} \\ &= 225(53.133176) \\ &= 11{,}954.96 \qquad \text{(rounded off)} \end{aligned}$$

Thus, after making 41 deposits, the amount of the annuity will be \$11,954.96.

Example 17 *Vacation money.* Martha Chaney works for the Printex Corp. The company has a vacation plan under which any employee can contribute an equal amount to the plan at the end of each month. The plan pays 6% interest compounded monthly. At the end of a year, a check for the full accumulated amount is issued. The employee then has a large amount of vacation money. Martha has signed up for the plan and has agreed to contribute \$55 at the end of each of 12 months. After a year, how much vacation money will Martha have from the plan?

Solution In this case there are 12 equal payments, each of \$55. The interest rate is 6% a year compounded monthly, or $\frac{6}{12} = \frac{1}{2}\%$ for each month. From Table 5, the appropriate chart value is 12.335562. Then, from Formula 10.6, Martha will have 55(12.335562) or \$678.46 (rounded off) for vacation money.

10.4 Present Value of an Annuity

In the previous section we analyzed how to find the value of an annuity after n payment periods assuming equal deposits of P dollars are made. Thus, in Example 15 we found that if Lorraine deposits \$150 every six months into an annuity paying 6% interest compounded semiannually, then after 18 years, it will be worth \$9491.39.

Often we are interested in the reverse question. How much money, M, must be deposited today as a lump sum into an account paying $r\%$ interest compounded annually to have some given amount after n years? This amount of money, M, that must be deposited is called the *present value of the annuity.*

The present value of an annuity plays an important role in retirement plans. Typically when a person works, he or she contributes to an annuity plan. When the person retires, the annuity account contains all the worker's contributions (with the employer's contribution, if any) plus the accumulated interest. The account contains its largest balance. Thereafter, the person receives monthly or weekly payments from the account. Of course, the amount of money remaining in the account continues to earn interest until there is no more money left. The amount of money that the annuity must have before any payout begins, that is, before the first check is received, is the present value of the annuity. As each payout is made, the present value of the annuity decreases. Nevertheless, when the payouts first begin, the present value should be the same as the amount of the annuity (assuming all payments were made on time).

To find the present value of an annuity, we can use a chart similar to the one we used for finding the amount of an annuity. Such a chart is given in Table 6 in the Appendix. We use this chart exactly as we used the table for the amount of an annuity.

We have the following formula:

Formula 10.7 If P represents the amount of money to be received from an annuity at each payment period that pays $r\%$ interest compounded per payment period, then the *present value of the annuity* is

P multiplied by the appropriate chart value given in Table 6.

The following examples illustrate how this formula is used.

Example 18 *Retirement income.* Chuck has just retired and will receive \$50 a month for 7 years from an annuity earning 6% interest compounded monthly.

a) How much money is Chuck's annuity account before the first payout is made; that is, what is the present value of the annuity?
b) How much money will Chuck receive in the 7 years from the annuity?

Solution a) Chuck will receive 12 monthly payments a year for 7 years, so there is a total of $12 \cdot 7$ or 84 payment periods. Also the interest rate is 6% compounded monthly, or $\frac{6}{12} = \frac{1}{2}\%$ interest per payment period. Now we look in Table 6 in the Appendix to find the appropriate chart value. The value listed is 68.453042. This is the present value of \$1. To find the present value of \$50, simply multiply the chart value by 50. We get

$$50(68.453042) = 3422.65 \qquad \text{(rounded off)}$$

Thus, the present value of the annuity is \$3422.65. This means that there is \$3422.65 accumulated in Chuck's annuity account before he receives his first annuity payout check.

b) Chuck will receive 84 checks, each for \$50. Thus, he will receive a total of $84 \cdot 50$, or \$4200. Of this \$4200, the amount of \$3422.65 represents the present value, and \$4200 − \$3422.65 or \$777.35 represents the interest that continues to be earned while payments are being made. Payments will be made and interest earned until there is nothing left in the annuity account.

Example 19 *Dentist's office.* Dr. Bali, a dentist, has just renovated and refurbished her office. She has agreed to pay the supplier \$250 a month for 6 years to repay the loan. Dr. Bali lives in New York State where the (maximum) annual interest rate is 18% or $1\frac{1}{2}\%$ per month on any outstanding balance.

a) How much did Dr. Bali borrow from the supplier?
b) How much interest does Dr. Bali have to pay?

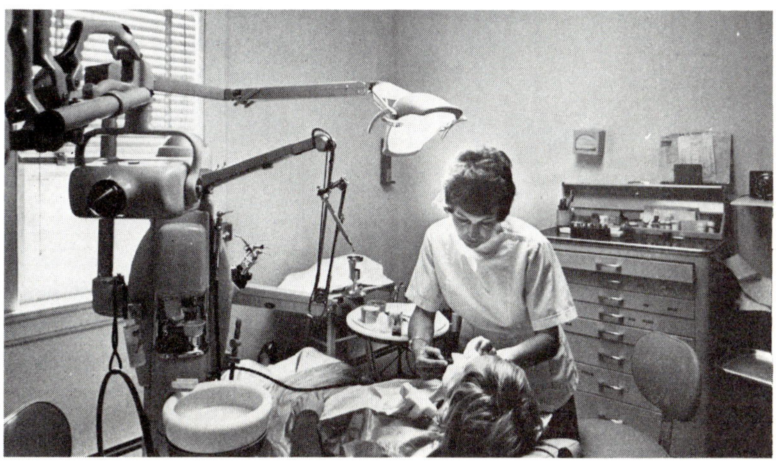

Photo: Mimi Forsyth, Monkmeyer

Solution a) The amount of money Dr. Bali borrowed from the supplier really represents the present value of an annuity paying \$250 per month for 6 years where the interest rate is $1\frac{1}{2}\%$ per month. Since Dr. Bali will repay the supplier \$250 a month for 6 years, there are $12 \cdot 6$ or 72 payment periods. Also the interest rate is $1\frac{1}{2}\%$ per period. Now we look in Table 6 to find the appropriate present value. The chart value is 43.844666, which is the present value of \$1. To find the present value of \$250, we simply multiply this by 250. We get

$$250(43.844666) = 10{,}961.17 \qquad \text{(rounded off)}$$

Thus, Dr. Bali borrowed \$10,961.17 from the supplier.

b) Dr. Bali will make 72 payments of \$250 each to the supplier, or a total of $250 \cdot 72$, or \$18,000. Of this \$18,000 repaid, \$10,961.17 is in repayment for the original loan, and \$18,000 − \$10,961.17 or \$7,038.83 is for interest charges.

Example 20 *Endowment funds.* Paul Conklin donated a large sum of money to his alma mater, State University, to provide for a yearly scholarship of \$1500 to a needy student who shows promise of high scholarship. College officials estimate that if they invest the money in a savings certificate paying 8% interest compounded annually, the fund will be able to provide scholarships for 25 years. How much money did Paul donate to the college?

Solution In this case we are interested in the present value of the annuity. Since the fund will last for 25 years, there are 25 payment periods. Also interest is 8% per period. Now we look in Table 6 to find the appropriate present value. The chart value is 10.674776. We then multiply the chart value by 1500, getting

$$1500(10.674776) = 16{,}012.16$$

Thus, Paul donated \$16,012.16 to the college. The college plans to award one \$1500 scholarship in each of 25 years, or a total of $25 \cdot \$1500$, or \$37,500. Therefore, the interest will be \$37,500 − \$16,012.16, or \$21,487.84. More than half of the amount to be awarded as scholarships will be from the interest earned.

Example 21 *How to buy a used car.* Roland McAllister has shopped around for a car. He has finally found one he likes. He can pay for it either with a check of \$5000 or with a down payment of \$1000 and 24 monthly payments of \$180 each. If he buys it with a down payment of \$1000, he can invest the rest of the money at 9% interest compounded monthly. Assuming he has the \$5000 cash available and that he would be able to make the monthly payments on time, which way will he end up saving money?

Solution In order to solve the problem, we must find the present value of an annuity at 9% interest where the monthly payment is \$180 for 2 years.

There are $12 \cdot 2$ or 24 monthly payments. The interest rate is 9% per year, or $\frac{9}{12} = \frac{3}{4}\%$ per month. From Table 6, the chart value is 21.889146. This we multiply by 180, getting

$$180(21.889146) = 3940.05$$

With the down payment of $1000 added to this amount, Roland will end up paying $4940.05 as opposed to paying with a check of $5000. Whether it is worthwile for Roland to buy the car on the installment plan or to pay for it in one payment really depends on the individual circumstances.

Example 22 *Setting retirement goals.* Martha Chaney, who is 35 years old, would like to retire from the Printex Corp. mentioned in Example 17 when she reaches her fifty-fifth birthday. When she retires, she wishes to receive $300 every 3 months for 25 years. The company annuity plan pays 6% interest a year compounded quarterly. How much money must Martha contribute as a lump sum now to realize her goals?

Solution We first find the present value of an annuity at 6% interest a year compounded quarterly where there are 4 payments per year for 25 years or 100 payment periods. The interest rate is 6% a year, or $\frac{6}{4} = 1\frac{1}{2}\%$ per payment period. From Table 6, the chart value is 51.624703. Thus, the present value of $300 is

$$300(51.624703) = 15{,}487.41 \qquad \text{(rounded off)}$$

Martha will have to accumulate $15,487.41 in her annuity account before she receives any payouts. Since Martha is now 35 years old, she has 20 years in which to do this. To find the amount of money that Martha must contribute now as a lump sum, we use the compound interest formula from Section 10.2, but in reverse. The amount of the annuity is to be $15,487.41, so that $A = 15{,}487.41$. The money will earn interest for a period of 20 years, and the rate of interest is 6% per year compounded quarterly. Thus, there will be 80 paying periods at $1\frac{1}{2}\%$ per period. From Formula 10.3, we have

$$\begin{aligned} A &= P(1 + r)^n \\ 15{,}487.41 &= P(1 + 0.015)^{80} \\ 15{,}487.41 &= P(1.015)^{80} \\ 15{,}487.41 &= 3.29066P \end{aligned}$$

Now we divide both sides by 3.29066, getting

$$4706.48 = P$$

Thus, Martha has to deposit $4706.48 now as a lump sum.

Comment There are charts available that would enable us to calculate how Martha can spread out these payments over a 20-year period to accumulate $15,487.41. However, these are beyond the scope of this book.

26. *Wise investment*. Jeremiah Wilson has agreed to invest \$120 per month into a fund for a period of seven years. The fund pays 9% interest a year compounded monthly. How much is the investment worth at the end of seven years?

27. *Company pension*. Charles Bren contributes \$125 every three months to his company pension plan, which pays 7% interest a year compounded quarterly. Charles plans to retire after 17 years of service. How much money will Charles have accumulated in his annuity account when he decides to retire?

28. *New car charges*. José and Maria Sanchez bought a new car. They made a down payment of \$1200 and agreed to pay the balance in installments of \$115 a month for three years. Interest is 18% a year compounded monthly. How much did the new car cost?

29. *Insurance paid up at 65*. Florence Ludlow pays an insurance company \$177.40 per year, which is to earn interest at the rate of 6% a year compounded annually. Florence, who is 41 years old, will make these payments until she is 65 years of age. How much will her insurance policy be worth (in terms of accumulations) when she is 65 years old?

30. *Moonlighting*. Shimon Del Hoyo is a police officer in Holtzville. On his days off, Shimon works as a security guard in a department store at a monthly take-home salary of \$193.78. Each month Shimon deposits the entire check into a savings account paying 6% interest a year compounded monthly. How much money will Shimon accumulate after seven years?

31. *Stock option plan*. Helen Kemeny invests \$135 every three months in a stock plan that pays a guaranteed 10% interest a year compounded quarterly. If all the gains and dividends are reinvested quarterly, how much will Helen's investment be worth at the end of three years?

32. *Lottery winning*. Sam Hoffman wins \$10,000 in the state lottery. He deposits the \$10,000 in a savings account paying 6% interest per year compounded monthly. The bank agrees to send Sam a monthly check of \$175 for living expenses. For how many years will Sam be able to receive the monthly checks before there is no more money left in the account?

★33. *Sweepstakes prize*. General Noodles Inc. is conducting a sweepstakes. The first-prize winner will receive \$100 a month for five years, the second-prize winner will receive \$50 a month for three years, and the third-prize winner will receive \$25 a month for two years. General Noodles Inc. has set up a special fund in a bank from which to distribute the prizes. If the bank pays 6% interest a year compounded monthly, what lump sum must General Noodles deposit in the account now?

34. *Paying off a loan.* Robert Adler bought a new car for \$5300. He paid \$1300 as a down payment and agreed to pay off the balance plus interest by making monthly installment payments of \$217.48 each. If the rate of interest is $1\frac{1}{2}\%$ per month, how long will it take Robert to pay off the loan?

35. *Rebuilding playgrounds.* The Peter Fan Nursery School recently rebuilt its playground at a cost of \$23,134.72 (including interest over payment period). The school has agreed to pay off the debt by making monthly payments of \$200 to the construction company over a period of six years. Find the yearly rate of interest.

★ 36. *Loan consolidation.* The Zent Corporation has to pay off two loans to the Prudential Bank. One loan is for \$25,000 and is payable in eight years. The other loan is for \$32,000 and is payable in six years. Interest is 10% a year compounded quarterly. The company wishes to consolidate these loans into one loan that will be due in seven years. If the bank approves the company's application, how much will be due after seven years?

★ 37. *Comparison shopping for a mortgage.* Ed Furness has applied to two banks for a \$5441.76 mortgage. Both banks are willing to offer Ed the mortgage but under different conditions as follows:

   Second National City Bank: Quarterly mortgage payments of \$136.92 must be made over a period of 20 years.

   People's National Bank: Quarterly mortgage payments of \$129.83 must be made over a period of 23 years.

   Which of the two banks is offering the mortgage at a lower yearly rate of interest?

10.5 Further Applications: Amortization and Sinking Funds

Most businesses and government agencies borrow money in order to do business. For example, a company may borrow money from a bank to finance the installation of antipollution devices. The company usually agrees to repay the loan by making equal periodic payments to the bank. Similarly, when an individual obtains a long-term mortgage from a bank, he or she agrees to repay the mortgage by making equal periodic payments to the bank. In both cases each periodic payment includes partial repayment on the principal plus payment of interest on the declining balance of the principal. The process of making payments under the conditions described is known as *amortization.*

How do we determine the periodic payments necessary to amortize (pay off under the conditions described) a loan? Fortunately, mathematicians have compiled many charts that simplify the computations considerably. One such chart is given in Table 7 in the Appendix. We illustrate the use of this chart with several examples.

Example 23 *Paying your mortgage.* Theresa Borgen arranged for a 15-year mortgage with the Second National City Bank. The mortgage was for \$34,000. She has decided to amortize the loan by making equal payments to the bank every 3 months. Interest is 8% compounded quarterly. How much will Theresa have to pay the bank every 3 months?

Solution Since the mortgage is to run for 15 years and there are to be 4 payments per year, there will be a total of $15 \cdot 4$ or 60 payment periods. Interest is 8% compounded quarterly, or $\frac{8}{4} = 2\%$ per payment period. Now we use Table 7 with 60 payment periods and an interest rate of 2%. The chart value is 0.028768. We multiply the chart value by the amount of the mortgage, getting

$$34{,}000(0.028768) = 978.11$$

Thus, Theresa will have to pay the bank \$978.11 every 3 months.

Example 24 *Buying a television on installment.* Gwendolyn bought a color television for \$489. She made a down payment of \$89 and agreed to repay the \$400 balance in equal monthly payments over a 3-year period. Interest is 18% a year compounded monthly. How much money does Gwendolyn have to pay monthly to amortize the loan over the 3-year period?

Solution Since Gwendolyn will repay the loan over a 3-year period, there will be $3 \cdot 12$ or 36 payment periods. Also the interest rate is 18% a year, or $\frac{18}{12} = 1\frac{1}{2}\%$ a month. Now we use Table 7. The chart value for 36 payment periods and a $1\frac{1}{2}\%$ rate of interest is 0.036152. We multiply the chart value by the amount of the balance, getting

$$400(0.036152) = 14.46 \qquad \text{(rounded off)}$$

Thus, her monthly payment is \$14.46. Over the 3 years, Gwendolyn will pay $36 \cdot \$14.46$, or \$520.56. The original loan was for \$400. Therefore she will pay \$120.56 in interest.

Example 25 *Closed-circuit television.* To combat increased shoplifting, the Rochelle Department Store purchased \$200,000 of closed-circuit television monitors for its several stores. The department store chain agreed to pay for the equipment by making equal monthly payments to the manufacturer over a 5-year period. Interest is 9%. What are the monthly payments?

Solution Since the loan will be repaid over a 5-year period, there will be a total of $12 \cdot 5$ or 60 payment periods. The interest rate is $\frac{9}{12}$ or $\frac{3}{4}\%$ per month. Now we use Table 7. The chart value is 0.020758. Since the loan was for \$200,000, the monthly payments will be

$$200{,}000(0.020758) = \$4151.60$$

Government agencies or corporations often borrow money by issuing bonds. A *bond* represents a promise by the corporation or government agency to repay a certain amount of money on a specified day with fixed rate of interest. For example, suppose a city issues 20-year bonds totaling \$50 million to finance sewer construction programs. Suppose further that interest is 8% a year payable semiannually.

In order to pay the semiannual interest and to have the \$50 million needed to pay off the bonds at the end of 20 years, the city may set up a *sinking fund*. The city will agree to make equal periodic payments into a sinking fund account so that at the end of 20 years, the amount of money accumulated plus interest will equal \$50 million. Again the question is: How much money must be deposited into the sinking fund at the end of each payment period to accumulated (by the specified date) the amount of money needed to pay off the bonds?

As with amortization, mathematicians have compiled sinking fund charts that enable us to determine the amount of the periodic payments. One such chart is given in Table 8 of the Appendix. Its use is illustrated in the following examples.

Example 26 *Sewer construction.* A city issues \$50,000,000 worth of 20-year bonds to finance various sewer construction projects. Interest is 6% compounded quarterly. How much money must be deposited into a sinking fund to have sufficient funds to redeem the bonds in 20 years?

Solution Since the bonds are to be repaid in 20 years and the interest is payable quarterly, there is a total of $20 \cdot 4$ or 80 payment periods. Also the interest rate is 6% per year, or $\frac{6}{4} = 1\frac{1}{2}\%$ per quarter. Now we use Table 8 in the Appendix. This chart lists different values depending on the number of payment periods and the interest rate. The chart value for 80 payment periods and an interest rate of $1\frac{1}{2}\%$ is 0.006548. Now we multiply this chart value by the maturity value of the sinking fund. We get

$$50{,}000{,}000(0.006548) = 327{,}400$$

Thus, the city must deposit \$327,400 into the sinking fund every 3 months.

Example 27 *The cost of a college education.* Phil Anderson estimates that in 8 years he will need \$20,000 for his daughter's college education. He intends to set up a sinking fund by making monthly contributions into the account. Interest is 6% compounded yearly. How much money must he deposit into the sinking fund every month?

Solution Since Phil will need \$20,000 in 8 years and plans to make monthly payments into the sinking fund, there will be a total of 96 payment periods. Also the interest rate is 6% per year, or $\frac{6}{12} = \frac{1}{2}\%$ per month.

Now we use Table 8. The chart value is 0.008141. We multiply this value by \$20,000, the maturity value of the sinking fund, getting

$$20{,}000(0.008141) = 162.82$$

Thus, Phil must deposit \$162.82 each month into the sinking fund.

Exercises

38. *Down payments.* Deborah Andropoli wishes to accumulate \$20,000 in four years from now so that she will have enough money for the down payment on her dream house. She decides to make monthly payments into a savings account paying 6% interest a year compounded monthly. How much should Deborah deposit each month to realize her goals?

39. *Energy conservation.* In order to conserve energy, Roxana Vasquez has just installed a new heater pump and an energy-efficient thermostat. The total cost was \$800. Roxana has agreed to repay the \$800 in 24 monthly installments. Interest is 12% a year compounded monthly. What are Roxana's monthly installment payments?

40. *Burning garbage.* The city of Roxburg has been ordered by the Environmental Protection Agency to install, within two years, antipollution devices over the city's incinerators when it burns its garbage. The mayor estimates that the city will need \$100,000 to pay for the project when it is completed in two years. The city decides to make quarterly payments into an account paying 6% interest a year compounded quarterly to accumulate the necessary funds. How much money should be deposited quarterly into this account?

41. *Parking lots.* The Printex Corp. sold 10-year bonds totaling \$400,000 to raise money so that it could build an underground parking lot for its workers. Interest is 8% a year compounded semiannually. The company has set up a sinking fund to have sufficient funds to redeem the bonds on their due date. How much money should the company deposit every six months into the sinking fund?

42. *Buying new machines.* A company estimates that it will need \$50,000 in seven years to replace its machines as they wear out. The company plans to make deposits every three months into a sinking fund account paying 8% interest a year compounded quarterly. How much money should the company deposit into the account every three months?

43. *Hospital bills.* Isaac Merrick borrowed \$3000 from a finance company to pay for his wife's hospital bills. He will repay the loan by making 36 monthly payments. Interest is 12% a year compounded monthly. What are his monthly payments?

44. *False teeth.* It cost Gordon Mejia \$1200 for a pair of special dentures. He agrees to pay the dentist's bill by making 24 monthly payments. Interest is 9% a year compounded monthly. What are his monthly payments?

45. *Diamond engagement ring.* Irving Jackier buys a $4000 diamond engagement ring for his financée, Venus. He agrees to pay for the ring by making 18 monthly payments. Interest is 18% a year compounded monthly. What are his monthly payments?

46. *Buying snowmobiles.* Lynne Casolino purchased six snowmobiles for $20,000 for use in her winter amusement park. The total amount is to be paid off by making monthly payments over a period of 48 months. Interest is 18% a year compounded monthly. What monthly payments will pay off the loan?

47. *Mortgage options.* Joyce and Tom Napoli have applied for a $30,000 mortgage to purchase a new home. The bank has approved their application and has offered them the following options for the mortgage.

    Option A: Interest rate of 8% per year compounded quarterly over a period of 20 years

    Option B: Interest rate of 10% per year compounded quarterly over a period of 23 years

    Option C: Interest rate of 12% per year compounded quarterly over a period of 24 years

    a) Which option results in the lowest monthly payments?
    b) Find the total interest paid under each of these options.

★ 48. *Swimming pool.* Alex Tufano installed a heated pool in his backyard. The total cost of $20,000 is to be paid off over a 25-year period. Alex estimates that he can afford quarterly payments of $1600.72 and is negotiating the interest rate. What must be the annual rate of interest to pay off the loan under the conditions described?

★ 49. *Gas heat.* Charlotte Miller wishes to accumulate $10,000 so that she can convert from oil heat to gas heat. She will make $500 quarterly deposits into a sinking fund account that pays 10% interest a year compounded quarterly. How many deposits are necessary before Charlotte accumulates the $10,000?

★ 50. *Firemen's fund.* The Firemen's Pension Fund sold $100,000 of 15-year bonds with compound interest at 8% a year payable semiannually. It then deposited the $100,000 into a sinking fund of a bank paying compound interest at 9% a year compounded semiannually. How much money must the fund have available every six months to make the interest payments and the payments into the sinking fund?

10.6 Summary

In this chapter we discussed the difference between simple and compound interest. The three different types of simple interest in use today are mostly for short-term loans. Interest on long-term loans is usually compounded. Similarly, banks pay compound interest on money deposited.

We also analyzed what is meant by an annuity. When equal periodic payments are made into an account at the same time that the interest is credited, we have an ordinary annuity. After making a specified number of payments into the account, the annuity contains a certain amount of money from the cash deposits and from the interest earned. This represents the amount of the annuity.

On the other hand, when a specific amount of money is to be paid out monthly, quarterly, or at some other interval, then a lump sum of money must be deposited to allow such payments. This is typically the case with a person's pension account, which contains its maximum amount just prior to retirement before the first pension check is mailed out. This lump sum deposit is called the present value of an annuity. We introduced several charts to facilitate the computation of the amount and the present value of an annuity.

Finally we discussed amortization and sinking funds. These are used when a loan has to be paid off. Companies or government agencies may contribute a specified amount of money into a sinking fund so as to have sufficient money to pay interest and to pay back the principal when its bonds become due. Amortization involves making periodic payments so that after a specified number of payments, the loan plus the interest charges will be entirely repaid. Again we introduced several charts to make the computations easier.

Our discussions in this chapter were not intended to be a complete treatment of the mathematical ideas of finance. They were intended merely to introduce these ideas. The interested reader can find further discussion of the topics in the reference given at the end of this chapter.

Study Guide

You should be able to demonstrate your knowledge of the following ideas presented in this chapter by giving definitions, descriptions, or specific examples. Page references are given for each term so that you can check your answer.

Interest (p. 440)
Rate of interest (p. 440)
Compound interest (p. 440)
Banker's rule interest (p. 443)
6% 60-Day rule (p. 445)
Effective interest rate (p. 448)
Payment period (p. 453)
Ordinary annuity (p. 453)
Present value of an annuity (p. 456)
Bond (p. 463)

Principal (p. 440)
Simple interest (p. 440)
Ordinary interest (p. 443)
Exact interest (p. 443)
Compounded continuously (p. 446)
Annuity (p. 453)
Term (p. 453)
Amount of an annuity (p. 453)
Amortization (p. 461)
Sinking fund (p. 463)

Formulas to Remember

The following list summarizes all the formulas given in this chapter.

Simple interest: $I = P \cdot r \cdot n$ where P is the principal, n is the number of years, and r is the rate of interest per year

6% 60-day rule: Move the decimal point in the principal amount two places to the left when ordinary interest is 6% and the loan is for 60 days.

Compound interest: $A = P(1 + r)^n$

Continuously compounded interest: $A = Pe^{rn}$

Effective interest rate: $e^r - 1$

Amount of an annuity: $A = P$ multiplied by chart value in Table 5

Present value of an annuity: P multiplied by chart value in Table 6

Periodic payments to amortize loan: Use Table 7

Periodic payments to a sinking fund: Use Table 8

Mastery Tests

FORM A

1. What do we call money paid for the use of money?
2. Find the simple interest on a loan of \$500 for $3\frac{1}{2}$ years if the rate of interest is 7% a year.
3. Exact interest is calculated using how many days in the year?
4. Find the ordinary interest charge on a \$754.78 loan for 60 days if the rate of interest is 6%.
5. If \$10,000 is deposited in a bank that pays 7% interest compounded continuously, how much money will be in the account after five years?
6. A bank pays 7.25% per year interest compounded continuously. Find the effective interest rate.
7. What do we call a sequence of equal payments made at equal periods of time?
8. What is added to the sum of all the payments made to arrive at the amount of an annuity?
9. True or false: As each payout is made from an annuity, the present value remains the same.
10. True or false: It is quite possible that most of the payments received from an annuity over a period of years will be from the interest earned rather than from the principal.

FORM B

1. *Manhattan Island and the Indians.* The Indians sold Manhattan Island to Peter Minuit in 1626 for \$24. If the Indians had deposited the \$24 in a bank paying 5% interest compounded continuously, how much would it be worth today?

2. *Inflation.* A man retires on a \$10,000 annual pension in 1979. The annual rate of inflation in his city is 9%. After how many years will the purchasing power of his annual pension be cut in half?

★3. *The compound interest formula.* The compound interest formula, $P(1 + r)^n$, can also be evaluated by using *combinations* discussed in Chapter 6 and the binomial expansion theorem. The advantage in using this technique is that it gives the answer to any degree of accuracy with a minimum number of multiplications. For example, suppose \$1 is deposited into a savings account paying 5% interest a year compounded annually. Then after 10 years, the amount of money accumulated is

$$(1 + 0.05)^{10} = 1 + {}_{10}C_1(0.05)^1 + {}_{10}C_2(0.05)^2 + {}_{10}C_3(0.05)^3 + \cdots + {}_{10}C_{10}(0.05)^{10}$$

where

$${}_{10}C_1 = \frac{10!}{9!1!} = 10$$

$${}_{10}C_2 = \frac{10!}{8!2!} = 45$$

$${}_{10}C_3 = \frac{10!}{7!3!} = 120$$

and so on so that

$$(1 + 0.05)^{10} = 1 + 0.5 + 0.1125 + 0.015 + \cdots = 1.6275\ldots$$

More generally, we have

$$(1 + r)^n = 1 + {}_nC_1(r) + {}_nC_2(r^2) + {}_nC_3(r^3) + \cdots + {}_nC_n(r^n)$$

Using this formula, find the amount of money that will accumulate after six years assuming a \$1000 deposit and a 6% rate of interest compounded annually. Compare your answer with the results given in Table 10.3.

★4. Refer to question 3. Suppose \$2000 is deposited into an account paying 5% interest a year compounded quarterly. How much money will accumulate after five years? (Use the technique discussed in the previous question.)

5. *Refinancing a mortgage.* Douglas Lake has an outstanding balance of \$14,000 remaining on his mortgage. His monthly payments are \$400.

Due to changing financial conditions, Douglas finds that he is no longer able to make the \$400 monthly payments. He asks the bank to spread out the mortgage over a longer period of time at an interest rate of 9% a year compounded monthly. Douglas estimates that he can now afford monthly payments of only \$205.10. Under this new plan, how long will it take Douglas to repay the mortgage?

★ **6.** *Lease or buy.* The A.B. Montire Corp. is planning to expand its operations. It is considering two alternatives. Possibility 1: Build a new building on vacant land that the company now owns at a cost of \$750,000. Possibility 2: Lease an equivalent building at an annual rental fee of \$60,000 for 25 years with the option of buying the building for \$100,000 at the end of 25 years. The company estimates that it can earn 8% interest per year compounded annually with the money if it decides not to build the building. Which plan is more advantageous to the company?

7. *Parking ticket.* Ray Samson has accumulated 18 parking tickets. The judge fines Ray \$840. Interest is at the rate of 12% per year compounded monthly. It takes Ray one year to pay off the fine plus interest. How much does Ray end up paying?

8. *Fighting inflation.* The rate of inflation in Dodgetown is 9% a year. The Dodgetown Savings Bank offers its customers savings certificates with interest compounded quarterly. What rate of interest must Dodgetown Savings Bank offer on these certificates to enable the value of its customers' deposits to keep pace with inflation?

9. *Lottery winner.* Wanda Ward was the big winner in the state lottery. She deposited the money in a savings account paying 6% interest compounded every six months. Wanda, who is 32 years old, plans to leave the money in the bank until she is 55 years old, when she will retire and "live a little." She would like to receive \$10,000 a year income from the fund for 20 years. How much did Wanda win in the lottery?

10. *Changing rates.* Martin Cullinghan received \$1000 in gifts when he was born. His father deposited the money into a stock plan paying 6% interest per year compounded quarterly. When Martin was 5 years old, the stock plan increased the interest rate to 7% per year compounded quarterly. When he was 12 years old, the interest rate was again increased, this time to 8% per year compounded monthly. If all the dividends are left to accumulate interest, how much money will Martin have when he reaches 18 years?

11. *Rich aunt.* Refer to question 10. When Martin was 13 years old, his rich aunt Rose died and left him \$3000. This money was added to the existing stock plan account. How much money will Martin now have when he reaches 18 years?

12. *College scholarships.* The Martha Cortese Scholarship Fund was recently established at State University to provide scholarships for needy and deserving students. Two \$500 scholarships are to be awarded every six months to eligible students. How much money must be deposited now into an account paying 8% interest a year compounded semiannually if these scholarships are to be awarded for 30 years?

13. *Tennis courts.* Kenneth Deedy obtained a bank mortgage of $22,000 to build a tennis court. The mortgage is to be repaid by making quarterly payments over an eight-year period. The rate of interest is 12% compounded quarterly.
    a) Find Kenneth's quarterly payments.
    b) How much interest will Kenneth end up paying the bank?

14. *Settlements.* Mitchell Broad still owes a bank $8000 payable in eight years. However, he wishes to pay off the debt now. The current bank rate of interest is 6% a year compounded monthly. How much money must Mitchell give the bank now to pay off the debt completely?

15. *Inheritance.* Gerald Martin left his entire estate to his four-year-old grandson Dennis. Tax officials estimate that the estate will be worth $45,000 when Dennis reaches his eighteenth birthday. In the meantime, money for the estate will be deposited in a bank paying 6% interest a year compounded quarterly. How much was Gerald Martin's estate worth when he died?

Suggested Further Readings

Highland, Esther, *Business Mathematics*, Reston, Va: Reston Publishing Company, 1978.

CHAPTER 11

Markov Chains: Stochastic Analysis

Objectives to be Achieved in this Chapter

- *To define* what a Markov chain is and to give several illustrations of situations that give rise to Markov chains.
- *To study* transition matrices, which are matrices that give the probability of being in a particular state on the next trial of an experiment when we are in a given state now.
- *To understand* probability vectors, which are the rows of a transition matrix.
- *To analyze* regular Markov chains, which are Markov chains in which some power of the transition matrix contains only positive elements.
- *To discuss* absorbing Markov chains, which are Markov chains that contain at least one absorbing state. Once we enter an absorbing state, we cannot leave it.
- *To apply* Markov chains to a variety of situations in the social and management sciences.

Are People More Educated Today?

MORGANVILLE PRESS APRIL 21, 1978

Morganville (April 21)—According to the findings of the Mayor's Educational Commission released April 19, more and more people are being educated today than ever before. The findings revealed that 85% of the children of parents in Morganville who had only an elementary school education went on to complete high school and 25% of them even completed college studies. Only 15% of these children dropped out of school after completing elementary school. For parents who had completed high school, 84% of their children also completed high school, with 48% of them completing their college education. Only 16% of them dropped out of school after completing elementary school. On the other hand, if the parents were college graduates, then 98% of their children completed high school, with 81% of them completing college. A mere 2% of them dropped out after completing only elementary school.

In commenting on these statistics, the beaming Mayor Bailey siad, "The people of Morganville can be proud of their schools and of their educational heritage. The prospects of more and more college graduates can only result in a well-educated populace."

The newspaper article indicates that the educational level attained by a child in Morganville is influenced (at least to some extent) by his or her parents' highest educational level. What effect, if any, will a person's educational level have on the educational level attained by his or her children? Is it true that the percentage of parents who are college graduates and have children who are also college graduates is likely to increase in future generations? What about their grandchildren? How do we determine such probabilities?

11.1 Introduction

In Chapter 6 we discussed the binomial probability distribution. For this distribution we assumed that the outcomes for each trial were independent of the outcome of any preceding trial. Moreover, the probability of success remained the same for each trial of the experiment.

Let us now remove these restrictions and assume that the outcome of each experiment depends at most on the outcome of the experiment immediately preceding it. We also permit more than two possible outcomes per trial. These more general conditions allow us to consider many different types of problems that arise in the social and management sciences; it is these trials that we consider in this chapter. Such a series of trials is called a Markov chain in honor of the outstanding Russian mathematician A. A. Markov, who introduced them.

The following are two typical problems that can be solved using the ideas of Markov chains.

Example 1 *Spy message.* A spy is transmitting a coded message to an enemy country through a communications system involving enemy agents using only the digits 0 and 1. The system is not functioning properly. There is a probability of $\frac{3}{4}$ that a digit that is received by an agent will be transmitted unchanged by that agent. Each digit that enters the system must pass through two agents before it is received by the enemy country.

a) Find the probability that a digit transmitted by the spy as a 0 will be received and transmitted by each agent as a 0.
b) Find the probability that a digit transmitted by the spy as a 0 will be received by the enemy country as a 0.

Example 2 *Kicking child.* A child psychologist is observing a baby in a crib kick at a mobile that is suspended from the sides of the crib. It is observed that if the baby kicks at it with his left foot, then the probability that the next kick at it will be with his left foot is $\frac{2}{7}$. On the other hand, if he kicks at it with his right foot, then the probability that the next kick at it will be with his right foot is $\frac{4}{7}$. Find the probability that the fifth kick from now will be with his right foot if it is observed that he just kicked it with his left foot.

Historical Note

The modern theory of Markov chains (or stochastic processes) can be traced back to the works of the Russian mathematician Andrei Andreevich Markov (1856–1922). Markov studied at the University of St. Petersburg in Russia. In 1886 he was appointed professor at that institution. In 1896 he became a member of the Soviet Academy of Sciences.

In the early part of his life Markov was interested in number theory (an important branch of mathematics). His interests soon turned to probability and statistics and he was able to prove formally the central limit theorem. Pursuing his interest in probability theory, Markov studied experiments in which the outcome of any trial was dependent on the outcome of the immediately preceding trial. His pioneering works in the field laid the foundation for Markov chains.

11.2 Finite Markov Chains

As we indicated, Markov was interested in experiments in which the outcome of each trial depends at most on the outcome of the immediately preceding trial. Such an experiment is called a Markov chain.

Definition 11.1 A *Markov chain* is an experiment that consists of a finite number of outcomes called *states* in which the outcome of each trial depends at most on the outcome of the preceding trial.

Let us illustrate the idea of a Markov chain with an example.

Example 3 *Jogging or tennis.* In order to maintain his physical fitness, Chris either jogs or plays tennis every day. The indoor tennis club to which Chris belongs does not allow any member to use the courts two days in a row. (This is done so that everybody gets a chance.) On the other hand, if Chris jogs today, he is as likely to jog tomorrow as he is to play tennis. Is this situation a Markov chain?

Solution In this case we have a Markov chain where there are two states: {jogs, plays tennis}. If we let a_1 represent jogs and let a_2 represent plays tennis, then the states of the Markov chain are $\{a_1, a_2\}$.

In Example 3 we can arrange the information given in the form of a matrix as follows:

	Probability he jogs the next day	Probability he plays tennis the next day
Chris jogs one day	$\frac{1}{2}$	$\frac{1}{2}$
Chris plays tennis one day	1	0

Using the symbols introduced in Example 3, we can rewrite the problem as

$$\begin{array}{ll} & \text{Probability that tomorrow's state is} \\ & \quad\; a_1 \quad\; a_2 \\ \text{Today's state} \begin{array}{l} a_1 \\ a_2 \end{array} & \begin{pmatrix} \frac{1}{2} & \frac{1}{2} \\ 1 & 0 \end{pmatrix} \end{array}$$

How do we interpret these entries?

The entry of 1 in the second row, first column tells us that if Chris plays tennis today, then the probability that he jogs tomorrow is 1. In other words, if Chris is in state a_2 today, then the probability that he will be in state a_1 tomorrow is 1.

The entry of 0 in the second row, second column tells us that if Chris plays tennis today, then the probability that he will play tennis tomorrow is 0. In other words, if Chris is in state a_2 today, then the probability that he will be in state a_2 tomorrow is 0.

In a similar manner, we interpret the other entries in the matrix. Thus, each entry gives us the probability of being in a particular state tomorrow when we know today's state. Such a matrix is called a *transition matrix*. The rows of this matrix are called *probability vectors*.

In this case the probability vectors are

$$(\tfrac{1}{2} \quad \tfrac{1}{2}) \quad \text{and} \quad (1 \quad 0)$$

Notice that the sum of the probabilities of each row is one.

We can summarize our discussion of the previous paragraphs in the following definition.

Definition 11.2 Let $a_1, a_2, a_3 \ldots, a_n$ be the states of a Markov chain and let P_{ij} be the probability that an experiment that is in state a_i now will next move to state a_j. (Of course, i and j do not necessarily have to be different.) Then the row vectors

$$\begin{matrix} (P_{11} & P_{12} & P_{13} & \cdots & P_{1n}) \\ (P_{21} & P_{22} & P_{23} & \cdots & P_{2n}) \\ \vdots & \vdots & \vdots & & \vdots \\ (P_{n1} & P_{n2} & P_{n3} & \cdots & P_{nn}) \end{matrix}$$

that have elements that are probabilities and are such that each sum is one are called *probability vectors.* The matrix that consists of these probability vectors

$$\begin{pmatrix} P_{11} & P_{12} & P_{13} & \cdots & P_{1n} \\ P_{21} & P_{22} & P_{23} & \cdots & P_{2n} \\ \vdots & \vdots & \vdots & & \vdots \\ P_{n1} & P_{n2} & P_{n3} & \cdots & P_{nn} \end{pmatrix}$$

is called the *transition matrix.*

Let us return to the previous problem. Suppose we know that Chris jogged today and we wish to find the probability that he will jog two days from now. We can use the tree diagram shown in Figure 11.1 to help us arrive at the answer.

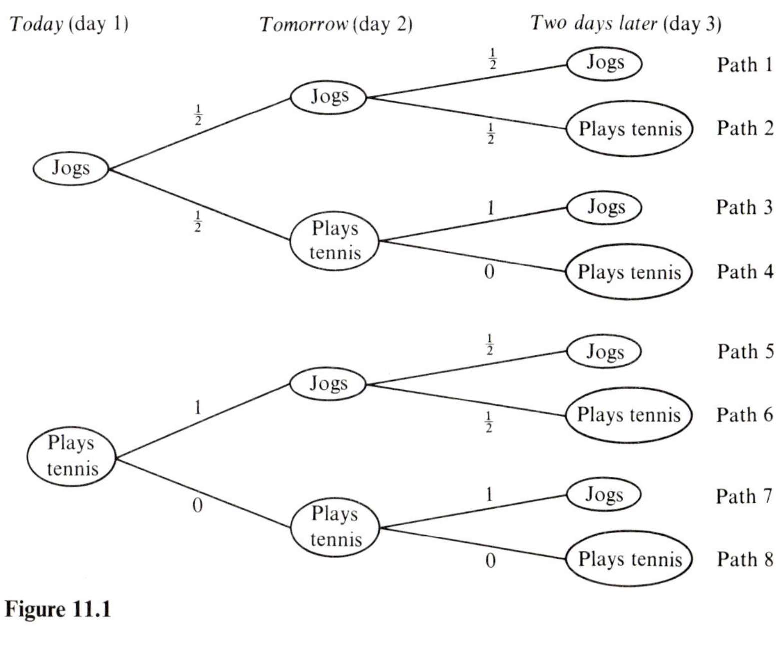

Figure 11.1

Suppose J denotes jogging and T denotes playing tennis. Then from Path 1

$$\text{Prob}(\text{J on day 2}|\text{J on day 1})\cdot\text{Prob}(\text{J on day 3}|\text{J on day 2})$$
$$=\frac{1}{2}\cdot\frac{1}{2}=\frac{1}{4}$$

and from Path 3

$$\text{Prob}(\text{T on day 2}|\text{J on day 1})\cdot\text{Prob}(\text{J on day 3}|\text{T on day 2})$$
$$=\frac{1}{2}\cdot 1=\frac{1}{2}$$

Thus, the probability that Chris will jog two days from now if he jogged today is the sum of the probabilities of these two cases, or

$$\frac{1}{4}+\frac{1}{2}=\frac{3}{4}$$

Let us analyze the transition matrix for this problem. It is

$$\mathbf{M}=\begin{pmatrix}\frac{1}{2} & \frac{1}{2}\\ 1 & 0\end{pmatrix}$$

If we multiply this matrix by itself, we get

$$\mathbf{M}\cdot\mathbf{M}=\mathbf{M}^2=\begin{pmatrix}\frac{1}{2} & \frac{1}{2}\\ 1 & 0\end{pmatrix}\cdot\begin{pmatrix}\frac{1}{2} & \frac{1}{2}\\ 1 & 0\end{pmatrix}=\begin{pmatrix}\frac{3}{4} & \frac{1}{4}\\ \frac{1}{2} & \frac{1}{2}\end{pmatrix}$$

Notice that the entry in the first row, first column of matrix $\mathbf{M}^2$ is $\frac{3}{4}$. This is exactly the probability we obtained when we used the tree diagram. As a matter of fact, the entries in matrix $\mathbf{M}^2$ give us the following probabilities:

$$\mathbf{M}^2=\begin{pmatrix}\frac{3}{4} & \frac{1}{4}\\ \frac{1}{2} & \frac{1}{2}\end{pmatrix}$$
$$=\begin{pmatrix}\text{Prob(jogs 2 days later}|\text{jogs today)} & \text{Prob(plays tennis 2 days later}|\text{jogs today)}\\ \text{Prob(jogs 2 days later}|\text{plays tennis today)} & \text{Prob(plays tennis 2 days later}|\text{plays tennis today)}\end{pmatrix}$$

This suggests that the entries obtained by squaring the transition matrix $\mathbf{M}$ give us the probabilities that an experiment now in state a_i will be in state a_j after the second trial.

Rule 1 If a transition matrix $\mathbf{M}$ is raised to the nth power, then the entry P_{ij} in the matrix $\mathbf{M}^n$ gives the probability that an experiment now in state i will be in state j after n trials.

Comment Many books refer to $\mathbf{M}^n$ as the *n-step transition matrix*, since it gives us the probability that an experiment starting in state i will be in state j after n trials (steps).

Example 4 *Future jogging.* Refer to the previous problem. The transition matrix was

$$\mathbf{M} = \begin{pmatrix} \frac{1}{2} & \frac{1}{2} \\ 1 & 0 \end{pmatrix}$$

If Chris jogs on day 1, find the following probabilities:

a) He will play tennis three days later.
b) He will jog three days later.

Solution Since the transition matrix is

$$\mathbf{M} = \begin{pmatrix} \frac{1}{2} & \frac{1}{2} \\ 1 & 0 \end{pmatrix}$$

we can apply Rule 1. Here we are interested in determining what Chris will do 3 days later, so $n = 3$. We have $\mathbf{M}^3 = \mathbf{M} \cdot \mathbf{M} \cdot \mathbf{M}$, so that

$$\begin{aligned}\mathbf{M}^3 &= \begin{pmatrix} \frac{1}{2} & \frac{1}{2} \\ 1 & 0 \end{pmatrix}^3 = \begin{pmatrix} \frac{1}{2} & \frac{1}{2} \\ 1 & 0 \end{pmatrix} \cdot \begin{pmatrix} \frac{1}{2} & \frac{1}{2} \\ 1 & 0 \end{pmatrix} \cdot \begin{pmatrix} \frac{1}{2} & \frac{1}{2} \\ 1 & 0 \end{pmatrix} \\ &= \begin{pmatrix} \frac{3}{4} & \frac{1}{4} \\ \frac{1}{2} & \frac{1}{2} \end{pmatrix} \cdot \begin{pmatrix} \frac{1}{2} & \frac{1}{2} \\ 1 & 0 \end{pmatrix} \\ &= \begin{pmatrix} \frac{5}{8} & \frac{3}{8} \\ \frac{3}{4} & \frac{1}{4} \end{pmatrix}\end{aligned}$$

We interpret the entries in this matrix as follows:

The entry P_{11} is the probability that if Chris jogs today, then he will jog three days later. This probability is $\frac{5}{8}$.

The entry P_{12} gives us the probability that if Chris jogs today, then he will play tennis three days later. This probability is $\frac{3}{8}$.

The entry P_{21} is the probability that if Chris plays tennis today, then he will jog three days later. This probability is $\frac{3}{4}$.

The entry P_{22} gives us the probability that if Chris plays tennis today, then he will play tennis three days later. This probability is $\frac{1}{4}$.

Example 5 *Long-range predictions.* In Example 4 if Chris jogs today, find the probabilities that he will jog and that he will play tennis 1 week later.

Solution Again we apply Rule 1. Here we are interested in determining what Chris will do 1 week later or 7 days later, so $n = 7$. We have

$$\mathbf{M}^7 = \mathbf{M}^3 \cdot \mathbf{M}^3 \cdot \mathbf{M}$$

From Example 4 we know that

$$\mathbf{M}^3 = \begin{pmatrix} \frac{5}{8} & \frac{3}{8} \\ \frac{3}{4} & \frac{1}{4} \end{pmatrix} \quad \text{and} \quad \mathbf{M} = \begin{pmatrix} \frac{1}{2} & \frac{1}{2} \\ 1 & 0 \end{pmatrix}$$

so that

$$\mathbf{M}^7 = \begin{pmatrix} \frac{5}{8} & \frac{3}{8} \\ \frac{3}{4} & \frac{1}{4} \end{pmatrix} \cdot \begin{pmatrix} \frac{5}{8} & \frac{3}{8} \\ \frac{3}{4} & \frac{1}{4} \end{pmatrix} \cdot \begin{pmatrix} \frac{1}{2} & \frac{1}{2} \\ 1 & 0 \end{pmatrix}$$

$$= \begin{pmatrix} \frac{43}{64} & \frac{21}{64} \\ \frac{21}{32} & \frac{11}{32} \end{pmatrix} \cdot \begin{pmatrix} \frac{1}{2} & \frac{1}{2} \\ 1 & 0 \end{pmatrix}$$

$$= \begin{pmatrix} \frac{85}{128} & \frac{43}{128} \\ \frac{43}{64} & \frac{21}{64} \end{pmatrix}$$

The entry P_{11} gives us the probability that if Chris jogs today, then he will jog 7 days later. This probability is $\frac{85}{128}$. The entry P_{12} gives us the probability that if Chris jogs today, then he will play tennis 7 days later. This probability is $\frac{43}{128}$.

Example 6 *Dating*. Bruce is dating Lynette, whom he met at a Saturday night rock party. As a matter of practice, Bruce knows that Lynette tries not to be late for their dates too often. Nevertheless, on occasion she is late for the date. Furthermore, if she is late on one date, then the probability is .95 that she will be on time for the next date. On the other hand, if she is on time for a date, then the probability is .70 that she will be late for the next date. Assume that Bruce will date Lynette for at least four dates. If Lynette is late on the first date, find the probabilities that she will be late and that she will be on time for the fourth date.

Solution In order to solve this problem using Markov chains, we must first set up a transition matrix. We have

		Probability on date 2	
		Late	On time
Date 1	Late	.05	.95
	On time	.70	.30

Thus, the transition matrix **M** is

$$\mathbf{M} = \begin{pmatrix} .05 & .95 \\ .70 & .30 \end{pmatrix}$$

Now we apply Rule 1. Here we are interested in determining the probabilities for the fourth date, so $n = 4$. We have

$$\mathbf{M}^4 = \begin{pmatrix} .05 & .95 \\ .70 & .30 \end{pmatrix} \cdot \begin{pmatrix} .05 & .95 \\ .70 & .30 \end{pmatrix} \cdot \begin{pmatrix} .05 & .95 \\ .70 & .30 \end{pmatrix} \cdot \begin{pmatrix} .05 & .95 \\ .70 & .30 \end{pmatrix}$$

$$= \begin{pmatrix} .6675 & .3325 \\ .245 & .755 \end{pmatrix} \cdot \begin{pmatrix} .6675 & .3325 \\ .245 & .755 \end{pmatrix}$$

$$= \begin{pmatrix} .52701875 & .47298125 \\ .3485125 & .6514875 \end{pmatrix}$$

The entry P_{11} gives the probability that if Lynette is late on the first date, then she will be late on the fourth date. This probability is .53.

The entry P_{12} gives the probability that if Lynette is late on the first date, then she will be on time on the fourth date. This probability is approximately .47.

11.3 Regular and Absorbing Markov Chains

In this section we consider two important kinds of Markov chains. These are the regular and the absorbing Markov chains. To get started, suppose we are given a transition matrix and multiply it by itself numerous times. What happens? Specifically, let the transition matrix be

$$\mathbf{M} = \begin{pmatrix} \frac{1}{2} & \frac{1}{2} \\ \frac{3}{5} & \frac{2}{5} \end{pmatrix}$$

or in decimal form

$$\mathbf{M} = \begin{pmatrix} 0.5 & 0.5 \\ 0.6 & 0.4 \end{pmatrix}$$

Then

$$\mathbf{M}^2 = \mathbf{M} \cdot \mathbf{M} = \begin{pmatrix} 0.5 & 0.5 \\ 0.6 & 0.4 \end{pmatrix} \cdot \begin{pmatrix} 0.5 & 0.5 \\ 0.6 & 0.4 \end{pmatrix} = \begin{pmatrix} 0.55 & 0.45 \\ 0.54 & 0.46 \end{pmatrix}$$

$$\mathbf{M}^3 = \mathbf{M}^2 \cdot \mathbf{M} = \begin{pmatrix} 0.55 & 0.45 \\ 0.54 & 0.46 \end{pmatrix} \cdot \begin{pmatrix} 0.5 & 0.5 \\ 0.6 & 0.4 \end{pmatrix} = \begin{pmatrix} 0.545 & 0.455 \\ 0.546 & 0.454 \end{pmatrix}$$

$$\mathbf{M}^4 = \mathbf{M}^2 \cdot \mathbf{M}^2 = \begin{pmatrix} 0.55 & 0.45 \\ 0.54 & 0.46 \end{pmatrix} \cdot \begin{pmatrix} 0.55 & 0.45 \\ 0.54 & 0.46 \end{pmatrix}$$

$$= \begin{pmatrix} 0.5455 & 0.4545 \\ 0.5454 & 0.4546 \end{pmatrix}$$

$$\mathbf{M}^5 = \mathbf{M}^3 \cdot \mathbf{M}^2 = \begin{pmatrix} 0.545 & 0.455 \\ 0.546 & 0.454 \end{pmatrix} \cdot \begin{pmatrix} 0.55 & 0.45 \\ 0.54 & 0.46 \end{pmatrix}$$

$$= \begin{pmatrix} 0.54545 & 0.45455 \\ 0.54546 & 0.45454 \end{pmatrix}$$

$$\mathbf{M}^6 = \mathbf{M}^3 \cdot \mathbf{M}^3 = \begin{pmatrix} 0.545 & 0.455 \\ 0.546 & 0.454 \end{pmatrix} \cdot \begin{pmatrix} 0.545 & 0.455 \\ 0.546 & 0.454 \end{pmatrix}$$

$$= \begin{pmatrix} 0.545455 & 0.454545 \\ 0.545454 & 0.454546 \end{pmatrix}$$

$$\vdots \qquad\qquad \vdots$$

$$\mathbf{M}^{10} = \mathbf{M}^5 \cdot \mathbf{M}^5 = \begin{pmatrix} 0.54545 & 0.45455 \\ 0.54546 & 0.45454 \end{pmatrix} \cdot \begin{pmatrix} 0.54545 & 0.45455 \\ 0.54546 & 0.45454 \end{pmatrix}$$

$$= \begin{pmatrix} 0.54545455 & 0.45454545 \\ 0.54545454 & 0.45454545 \end{pmatrix}$$

It can be shown that the decimal 0.54545455 . . . represents the fraction $\frac{6}{11}$ and the decimal 0.45454545 . . . represents the fraction $\frac{5}{11}$. This can be easily verified by converting $\frac{6}{11}$ and $\frac{5}{11}$ to their decimal forms.

Thus, you will notice that as the transition matrix is raised to higher powers, $\mathbf{M}^n$ approaches

$$\begin{pmatrix} \frac{6}{11} & \frac{5}{11} \\ \frac{6}{11} & \frac{5}{11} \end{pmatrix}$$

Let us now multiply the original transition matrix $\mathbf{M}$ by the probability vector

$$\left(\tfrac{6}{11} \quad \tfrac{5}{11}\right)$$

We get

$$\left(\tfrac{6}{11} \quad \tfrac{5}{11}\right) \cdot \begin{pmatrix} \frac{1}{2} & \frac{1}{2} \\ \frac{3}{5} & \frac{2}{5} \end{pmatrix} = \left(\tfrac{6}{11} \quad \tfrac{5}{11}\right)$$

In other words, multiplication of the original transition matrix $\mathbf{M}$ by the probability vector

$$\left(\tfrac{6}{11} \quad \tfrac{5}{11}\right)$$

gives the probability vector again. Such a vector is called a fixed probability vector and is usually denoted by $\mathbf{v}$.

Definition 11.3 A probability vector $\mathbf{v}$ that has the property $\mathbf{v} \cdot \mathbf{M} = \mathbf{v}$ for any matrix $\mathbf{M}$ is called a *fixed probability vector.*

How do we find such fixed probability vectors? In the following examples we illustrate the technique.

Example 7 Find the fixed probability vector (if it exists) for the transition matrix

$$\mathbf{M} = \begin{pmatrix} \frac{1}{2} & \frac{1}{2} \\ \frac{3}{5} & \frac{2}{5} \end{pmatrix}$$

Solution Let the fixed probability vector $\mathbf{v} = (x_1 \quad x_2)$. Then by Definition 11.3,

$$\mathbf{v} \cdot \mathbf{M} = \mathbf{v}$$

or

$$(x_1 \quad x_2) \cdot \begin{pmatrix} \frac{1}{2} & \frac{1}{2} \\ \frac{3}{5} & \frac{2}{5} \end{pmatrix} = (x_1 \quad x_2)$$

If we multiply the two matrices on the left side of the equation, we get

$$\left(\frac{1}{2}x_1 + \frac{3}{5}x_2 \qquad \frac{1}{2}x_1 + \frac{2}{5}x_2\right) = (x_1 \quad x_2)$$

For these two matrices to be equal, we must have

$$\frac{1}{2}x_1 + \frac{3}{5}x_2 = x_1$$

and

$$\frac{1}{2}x_1 + \frac{2}{5}x_2 = x_2$$

If we solve these two equations for x_1 and x_2, we get

$$x_1 = \frac{6}{5}x_2 \tag{1}$$

Since $\mathbf{v} = (x_1 \quad x_2)$ is a probability vector, $x_1 + x_2 = 1$ (the sum of the probabilities in a probability vector must be 1) or

$$x_1 = 1 - x_2 \tag{2}$$

Substituting this value for x_1 into equation (1) and solving for x_1 and x_2 gives

$$1 - x_2 = \frac{6}{5}x_2$$

$$1 = \frac{11}{5}x_2$$

so that

$$x_2 = \frac{5}{11} \quad \text{and} \quad x_1 = \frac{6}{11}$$

Therefore,

$$\left(\tfrac{6}{11} \quad \tfrac{5}{11}\right)$$

is the fixed probability vector.

Comment Example 7 illustrates that there is no need for us to raise the transition matrix to higher and higher powers to obtain the fixed probability vector. We can simply obtain the fixed probability vector by applying Definition 11.3.

Comment Perhaps you are wondering why we need fixed probability vectors. The answer is in their application. As we show in the next section, fixed probability vectors are often used in making long-range predictions.

Example 8 Find the fixed probability vector (if it exists) for the transition matrix

$$\mathbf{M} = \begin{pmatrix} \frac{1}{2} & \frac{1}{4} & \frac{1}{4} \\ \frac{1}{3} & \frac{1}{6} & \frac{1}{2} \\ \frac{1}{8} & \frac{1}{4} & \frac{5}{8} \end{pmatrix}$$

Solution Let the fixed probability vector $\mathbf{v} = (x_1 \quad x_2 \quad x_3)$. Then by Definition 11.3, we must have

$$\mathbf{v} \cdot \mathbf{M} = \mathbf{v}$$

or

$$(x_1 \quad x_2 \quad x_3) \cdot \begin{pmatrix} \frac{1}{2} & \frac{1}{4} & \frac{1}{4} \\ \frac{1}{3} & \frac{1}{6} & \frac{1}{2} \\ \frac{1}{8} & \frac{1}{4} & \frac{5}{8} \end{pmatrix} = (x_1 \quad x_2 \quad x_3)$$

Multiplying the two matrices on the left side of the equation gives

$$\left(\frac{1}{2}x_1 + \frac{1}{3}x_2 + \frac{1}{8}x_3 \qquad \frac{1}{4}x_1 + \frac{1}{6}x_2 + \frac{1}{4}x_3 \qquad \frac{1}{4}x_1 + \frac{1}{2}x_2 + \frac{5}{8}x_3\right)$$
$$= (x_1 \quad x_2 \quad x_3)$$

Thus, we have the following system of equations

$$\frac{1}{2}x_1 + \frac{1}{3}x_2 + \frac{1}{8}x_3 = x_1$$

$$\frac{1}{4}x_1 + \frac{1}{6}x_2 + \frac{1}{4}x_3 = x_2$$

$$\frac{1}{4}x_1 + \frac{1}{2}x_2 + \frac{5}{8}x_3 = x_3$$

Again, since $\mathbf{v} = (x_1 \quad x_2 \quad x_3)$ is a probability vector, we must also have $x_1 + x_2 + x_3 = 1$. Solving these equations for x_1, x_2, and x_3 gives

$$x_1 = \frac{18}{65} \qquad x_2 = \frac{3}{13} \qquad x_3 = \frac{32}{65}$$

Thus,

$$(\tfrac{18}{65} \quad \tfrac{3}{13} \quad \tfrac{32}{65})$$

is the fixed probability vector.

Comment The reader should verify that

$$(\tfrac{18}{65} \quad \tfrac{3}{13} \quad \tfrac{32}{65})$$

is indeed the fixed probability vector by multiplying it by

$$\begin{pmatrix} \frac{1}{2} & \frac{1}{4} & \frac{1}{4} \\ \frac{1}{3} & \frac{1}{6} & \frac{1}{2} \\ \frac{1}{8} & \frac{1}{4} & \frac{5}{8} \end{pmatrix}$$

The answer should be

$$(\tfrac{18}{65} \quad \tfrac{3}{13} \quad \tfrac{32}{65})$$

Often when working with transition matrices, we may find that the transition matrix has some zero elements in it, whereas some powers of the transition matrix contains only positive elements. For example, the transition matrix

$$\mathbf{A} = \begin{pmatrix} \frac{1}{2} & \frac{1}{2} \\ 1 & 0 \end{pmatrix}$$

discussed in Example 3 has some zero elements in it, whereas

$$\mathbf{A}^2 = \begin{pmatrix} \frac{3}{4} & \frac{1}{4} \\ \frac{1}{2} & \frac{1}{2} \end{pmatrix}$$

has only positive elements in it. Such a transition matrix is called regular.

Definition 11.4 A Markov chain is said to be *regular* if some power of its transition matrix contains only positive elements.

Example 9 a) The transition matrix

$$\mathbf{M} = \begin{pmatrix} \frac{1}{5} & \frac{4}{5} \\ 1 & 0 \end{pmatrix}$$

is regular since

$$\mathbf{M}^2 = \begin{pmatrix} \frac{21}{25} & \frac{4}{25} \\ \frac{1}{5} & \frac{4}{5} \end{pmatrix}$$

contains only positive elements.

b) The transition matrix

$$\mathbf{A} = \begin{pmatrix} \frac{1}{4} & \frac{1}{2} & \frac{1}{4} \\ 0 & \frac{1}{2} & \frac{1}{2} \\ \frac{1}{3} & \frac{1}{3} & \frac{1}{3} \end{pmatrix}$$

is regular since

$$\mathbf{A}^2 = \begin{pmatrix} \frac{7}{48} & \frac{11}{24} & \frac{19}{48} \\ \frac{1}{6} & \frac{5}{12} & \frac{5}{12} \\ \frac{7}{36} & \frac{4}{9} & \frac{13}{36} \end{pmatrix}$$

contains only positive elements.

c) The transition matrix

$$\mathbf{M} = \begin{pmatrix} 1 & 0 \\ \frac{1}{9} & \frac{8}{9} \end{pmatrix}$$

is *not* regular since it contains some zero elements and every power of $\mathbf{M}$ will have 0 as the entry in the first row, second column. You should convince yourself of the truth of this statement by raising $\mathbf{M}$ to some powers, say $\mathbf{M}^{10}$ or $\mathbf{M}^{15}$.

d) The identity matrix is an example of a transition matrix that is *not* regular.

Comment If a Markov chain is regular, then sooner or later (that is, after n trials) all the states must occur.

We can now summarize our discussion of transition matrices with the following theorem, which we state without proof.

Theorem 11.1 If $\mathbf{M}$ is a transition matrix of a regular Markov chain, then:

1. $\mathbf{M}$ has a fixed probability vector $\mathbf{v}$ whose elements are all positive.
2. As n gets very large, then $\mathbf{M}^1, \mathbf{M}^2, \mathbf{M}^3, \ldots, \mathbf{M}^n$ approaches a matrix, all the rows of which are identical and equal to the fixed probability vector $\mathbf{v}$.
3. If P is any probability vector, then as n increases, $P\mathbf{M}^n$ approaches the fixed probability vector $\mathbf{v}$.

We will use Theorem 11.1 in the next section when we discuss applications of Markov chains.

There are certain Markov chains in which it is possible to enter some states; however, once we are in these states, it is impossible to leave. To illustrate, consider the following example.

Example 10 *Random walk.* Suppose the six consecutive concrete blocks on a sidewalk along Main Street are arranged in a straight line and are numbered from left to right with the numbers 0, 1, 2, 3, 4, and 5 (of course, there is one number to each block). A child performs a random walk among the six blocks by tossing a coin. If heads comes up, then the child jumps one box to the right. If tails comes up, then the child jumps one box to the left. The possible positions are 0, 1, 2, 3, 4, and 5. When the child reaches either block 0 or block 5, the walk is over. We can set up the following transition matrix to illustrate the child's walk.

		Probability that the child will next jump to					
		Block 0	Block 1	Block 2	Block 3	Block 4	Block 5
Child's position now	Block 0	1	0	0	0	0	0
	Block 1	$\frac{1}{2}$	0	$\frac{1}{2}$	0	0	0
	Block 2	0	$\frac{1}{2}$	0	$\frac{1}{2}$	0	0
	Block 3	0	0	$\frac{1}{2}$	0	$\frac{1}{2}$	0
	Block 4	0	0	0	$\frac{1}{2}$	0	$\frac{1}{2}$
	Block 5	0	0	0	0	0	1

$= \mathbf{M}$

The states 0 and 5 are absorbing states, since the probability of leaving them and passing to some other states is 0. Notice that $P_{11} = 1$. This means that if the child is in block 0, then the probability that she remains in block 0 is 1. All other entries in the first row are 0. The entry P_{21} is $\frac{1}{2}$, since if the child is in block 1, she may go either to block 2 or to block 0, each with a probability of $\frac{1}{2}$. Similar interpretations can be made for the other entries. The entry P_{66} is 1. This means that if the child enters block 5, then she will remain there. She cannot leave.

We can generalize with the following definition.

Definition 11.5 A state i of a Markov chain is called an *absorbing state* if the transition probability of going from state i to state i is 1, that is, if $P_{ii} = 1$. Once an absorbing state is entered, we cannot leave it ever.

Definition 11.6 A Markov chain is said to be an *absorbing Markov chain* if and only if it contains at least one absorbing state and it is possible to go from a nonabsorbent state to an absorbent state in a finite number of steps.

Comment A Markov chain may have one or more absorbing states.

We use the ideas of regular Markov chains and absorbing Markov chains in the next section when we discuss general applications for Markov chains.

Exercises

In Exercises 1–6 determine which are probability vectors.

1. $\begin{pmatrix}\frac{1}{3} & \frac{1}{3} & \frac{1}{3}\end{pmatrix}$

2. $\begin{pmatrix}1 & 0 & 0\end{pmatrix}$

3. $\begin{pmatrix}\frac{1}{2} & 1 & -\frac{1}{2} & 0\end{pmatrix}$

4. $\begin{pmatrix}\frac{1}{2} & 0\end{pmatrix}$

5. $\begin{pmatrix}\frac{1}{4} & \frac{3}{4} & 0\end{pmatrix}$

6. $\begin{pmatrix}\frac{1}{4} & \frac{2}{9} & \frac{3}{7} & \frac{4}{5}\end{pmatrix}$

In Exercises 7–20 determine which transition matrices are regular.

7. $\begin{pmatrix}1 & 0\\ \frac{1}{4} & \frac{3}{4}\end{pmatrix}$

8. $\begin{pmatrix}\frac{3}{5} & \frac{2}{5}\\ 0 & 1\end{pmatrix}$

9. $\begin{pmatrix}\frac{2}{11} & \frac{9}{11}\\ \frac{1}{3} & \frac{2}{3}\end{pmatrix}$

10. $\begin{pmatrix}0 & 1\\ \frac{3}{5} & \frac{2}{5}\end{pmatrix}$

11. $\begin{pmatrix}\frac{2}{9} & \frac{7}{9}\\ 1 & 0\end{pmatrix}$

12. $\begin{pmatrix}1 & 0\\ \frac{3}{5} & \frac{2}{5}\end{pmatrix}$

13. $\begin{pmatrix}1 & 0\\ 0 & 1\end{pmatrix}$

14. $\begin{pmatrix}\frac{2}{3} & \frac{1}{3}\\ 0 & 1\end{pmatrix}$

15. $\begin{pmatrix}\frac{1}{3} & \frac{1}{3} & \frac{1}{3}\\ 0 & 1 & 0\\ \frac{1}{4} & \frac{1}{2} & \frac{1}{4}\end{pmatrix}$

16. $\begin{pmatrix}\frac{1}{9} & \frac{7}{9} & \frac{1}{9}\\ 1 & 0 & 0\\ \frac{1}{2} & \frac{1}{4} & \frac{1}{4}\end{pmatrix}$

17. $\begin{pmatrix}\frac{1}{5} & \frac{3}{5} & \frac{1}{5}\\ \frac{1}{6} & \frac{1}{3} & \frac{1}{2}\\ 1 & 0 & 0\end{pmatrix}$

18. $\begin{pmatrix}\frac{1}{4} & \frac{1}{2} & \frac{1}{4}\\ \frac{1}{3} & \frac{1}{3} & \frac{1}{3}\\ 0 & 1 & 0\end{pmatrix}$

19. $\begin{pmatrix} \frac{1}{2} & \frac{1}{4} & \frac{1}{4} \\ \frac{1}{3} & \frac{1}{3} & \frac{1}{3} \\ \frac{1}{9} & \frac{2}{9} & \frac{2}{3} \end{pmatrix}$

20. $\begin{pmatrix} \frac{1}{4} & \frac{1}{2} & \frac{1}{4} \\ 0 & \frac{1}{2} & \frac{1}{2} \\ \frac{1}{3} & \frac{1}{3} & \frac{1}{3} \end{pmatrix}$

In Exercises 21–26 find the fixed probability vectors (if they exist) for the given transition matrices.

21. $\begin{pmatrix} \frac{1}{3} & \frac{2}{3} \\ 1 & 0 \end{pmatrix}$

22. $\begin{pmatrix} \frac{5}{7} & \frac{2}{7} \\ 1 & 0 \end{pmatrix}$

23. $\begin{pmatrix} \frac{2}{9} & \frac{7}{9} \\ 1 & 0 \end{pmatrix}$

24. $\begin{pmatrix} \frac{1}{4} & \frac{1}{2} & \frac{1}{4} \\ 0 & \frac{1}{2} & \frac{1}{2} \\ \frac{1}{3} & \frac{1}{3} & \frac{1}{3} \end{pmatrix}$

25. $\begin{pmatrix} \frac{1}{2} & \frac{1}{4} & \frac{1}{4} \\ \frac{2}{7} & \frac{3}{7} & \frac{2}{7} \\ \frac{1}{3} & \frac{1}{3} & \frac{1}{3} \end{pmatrix}$

26. $\begin{pmatrix} \frac{1}{4} & \frac{1}{2} & \frac{1}{4} \\ \frac{1}{3} & \frac{1}{3} & \frac{1}{3} \\ 0 & 1 & 0 \end{pmatrix}$

In Exercises 27–30 determine whether the transition matrices are absorbing Markov chains.

27. $\begin{pmatrix} 1 & 0 & 0 & 0 \\ 0 & \frac{1}{3} & \frac{2}{3} & 0 \\ 0 & 0 & 1 & 0 \\ 0 & 0 & 0 & 1 \end{pmatrix}$

28. $\begin{pmatrix} 0 & \frac{1}{2} & \frac{1}{2} & 0 \\ \frac{1}{3} & 0 & \frac{1}{3} & \frac{1}{3} \\ 0 & \frac{1}{4} & \frac{1}{2} & \frac{1}{4} \\ 0 & 0 & 0 & 1 \end{pmatrix}$

29. $\begin{pmatrix} \frac{1}{2} & \frac{1}{2} & 0 & 0 & 0 & 0 \\ \frac{1}{2} & 0 & \frac{1}{2} & 0 & 0 & 0 \\ 0 & \frac{1}{2} & 0 & \frac{1}{2} & 0 & 0 \\ 0 & 0 & \frac{1}{2} & 0 & \frac{1}{2} & 0 \\ 0 & 0 & 0 & \frac{1}{2} & 0 & \frac{1}{2} \\ 0 & 0 & 0 & 0 & \frac{1}{2} & \frac{1}{2} \end{pmatrix}$

30. $\begin{pmatrix} \frac{1}{3} & \frac{2}{3} & 0 & 0 & 0 & 0 & 0 \\ \frac{1}{3} & 0 & \frac{2}{3} & 0 & 0 & 0 & 0 \\ 0 & \frac{1}{3} & 0 & \frac{2}{3} & 0 & 0 & 0 \\ 0 & 0 & \frac{1}{3} & 0 & \frac{2}{3} & 0 & 0 \\ 0 & 0 & 0 & \frac{1}{3} & 0 & \frac{2}{3} & 0 \\ 0 & 0 & 0 & 0 & \frac{1}{3} & 0 & \frac{2}{3} \\ 0 & 0 & 0 & 0 & 0 & \frac{1}{3} & \frac{1}{3} \end{pmatrix}$

31. Suppose a Markov chain has the transition matrix

$$\begin{pmatrix} \frac{1}{5} & \frac{4}{5} \\ 1 & 0 \end{pmatrix}$$

Find P_{12} after:

a) The second trial
b) The third trial
c) The fourth trial

In Exercises 32–35 find the two-step and three-step transition probability matrices for the indicated transition probability matrices.

32. $\begin{pmatrix} \frac{1}{3} & \frac{2}{3} \\ \frac{1}{4} & \frac{3}{4} \end{pmatrix}$

33. $\begin{pmatrix} \frac{1}{2} & \frac{1}{2} \\ \frac{1}{5} & \frac{4}{5} \end{pmatrix}$

34. $\begin{pmatrix} \frac{1}{3} & \frac{2}{3} & 0 \\ 0 & \frac{1}{4} & \frac{3}{4} \\ \frac{1}{3} & \frac{1}{3} & \frac{1}{3} \end{pmatrix}$

35. $\begin{pmatrix} \frac{1}{5} & \frac{2}{5} & \frac{2}{5} \\ \frac{1}{4} & 0 & \frac{3}{4} \\ \frac{1}{3} & \frac{1}{6} & \frac{1}{2} \end{pmatrix}$

36. *Selecting an outfit.* Margaret has three different sweaters that she can wear to school: a red one, a beige one, and a black one. Her policy is never to wear the same sweater to school two days in a row. If she wears the red one on one day, then she is as likely to wear the beige one as the black one on the next day. If she wears the black one on one day, then the probability that she will wear the red one on the next day is .53. If she wears the beige one on one day, then the probability that she will wear the black one on the next day is .65. If Margaret is observed wearing the beige sweater on Tuesday, find the probability that she will wear the black one on Thursday.

★ 37. *Weekend sports.* Jack is an outdoor sportsman. On weekends he either skis or goes snowmobiling. If he skis on one weekend, then the probability is .7 that he will snowmobile on the following weekend. If he snowmobiles on one weekend, then the probability is .6 that he will ski on the next weekend. In the long run, how often does Jack ski?

Photo: DeWys

★ 38. *New toys.* In Thomasville there are three local toy companies A, B, and C that each control $\frac{1}{3}$ of the market. In an effort to increase sales, each company undertakes an extensive television advertisement program. During the year, the following changes occur.

A keeps 45% of its customers but loses 30% of its customers to B and 25% of its customers to C.

B keeps 35% of its customers but loses 40% of its customers to A and 25% of its customers to C.

C keeps 40% of its customers but loses 30% of its customers to A and 30% of its customers to B.

Assuming that this trend will continue, find the share of that market that each company will have after the time indicated.

a) Two years
b) Three years
c) In the long run

11.4 Applications for Markov Chains

In this section we apply the ideas of Markov chains developed in the previous sections. The reader should analyze each of these examples carefully.

Example 11 *Public opinion polls.* A public opinion survey conducted on July 5 indicated that if the real election were held then, 45% of the people would vote for the Democratic candidate, 35% would vote for the Republican candidate, and 20% would vote for the Conservative candidate. After this poll was released, all the candidates launched major television and magazine advertisement campaigns. A similar survey conducted one month later indicated the following: Of the 45% who said they would vote Democratic, 80% said they would still vote Democratic, whereas 12% said they would now vote Republican and 8% said they would now vote Conservative. Of the original 35% who said they would vote Republican, 75% said they would still vote Republican, whereas 20% said they would now vote Democratic and 5% would now vote Conservative. Of the original 20% who said they would vote Conservative, 90% said they would still vote Conservative, whereas 6% said they would now vote Democtratic and 4% would now vote Republican. Assuming that these trends continue, what percentage of the electorate will each party have one month later on September 5?

Solution We first set up the transition matrix. It is

		Vote in August		
		Democrat	Republican	Conservative
Vote in July	Democrat	0.80	0.12	0.08
	Republican	0.20	0.75	0.05
	Conservative	0.06	0.04	0.90

$= \mathbf{M}$

This transition matrix represents the state of the electorate in August, that is, after one transition. To find the percentage of the electorate that each party will have in September, we must find the transition matrix after two trials. To accomplish this, we use Rule 1 with $n = 2$. we simply square matrix $\mathbf{M}$, getting

$$\mathbf{M}^2 = \begin{pmatrix} 0.80 & 0.12 & 0.08 \\ 0.20 & 0.75 & 0.05 \\ 0.06 & 0.04 & 0.90 \end{pmatrix} \cdot \begin{pmatrix} 0.80 & 0.12 & 0.08 \\ 0.20 & 0.75 & 0.05 \\ 0.06 & 0.04 & 0.90 \end{pmatrix}$$

$$= \begin{pmatrix} 0.6688 & 0.1892 & 0.1420 \\ 0.3130 & 0.5885 & 0.0985 \\ 0.1100 & 0.0732 & 0.8168 \end{pmatrix}$$

We interpret the entries in this matrix as follows: The entry $P_{11} = 0.6688$ in the first row, first column gives us the percentage of the people who originally said they would vote Democratic in July and who would still vote Democratic in September. Thus, of the original 45% of the population, 0.6688 would still vote Democratic, or

$0.6688(0.45) = 0.30096$ of the population would still vote Democratic. The entry $P_{21} = 0.3130$ in the second row, first column gives us the percentage of the people who originally said they would vote Republican in July and who would vote Democratic in September. Thus, of the original 35% of the population, 0.3130 would now vote Democratic, or $0.3130(0.35) = 0.10955$ of the population will switch from Republican and vote Democratic. The entry $P_{31} = 0.1100$ in the third row, first column gives the percentage of the people who originally said they would vote Conservative in July and who would vote Democratic in September. Thus, of the original 20% of the population, 0.1100 would now vote Democratic, or $0.1100(0.20) = 0.02200$ of the population will switch from Conservative and vote Democratic. We now add the percentages of the people who will vote Democratic from the three categories:

Percentage originally voting Democratic and now also voting Democratic		percentage originally voting Republican and now voting Democratic		percentage originally voting Conservative and now voting Democratic
0.30096	+	0.10955	+	0.02200
= 0.43251				

Thus, about 43% of the population would vote for the Democratic candidate if the election were held in September.

In a similar manner, we find the percentages of the people who would vote Republican from the three categories:

Percentage originally voting Democratic and now voting Republican		percentage originally voting Republican and still voting Republican		percentage originally voting Conservative and now voting Republican
0.1892(0.45)	+	0.5885(0.35)	+	0.0732(0.20)
0.085140	+	0.205975	+	0.014640
= 0.305755				

Thus, about 31% of the population would vote for the Republican candidate if the election were held in September.

Also the percentages of the people who would vote Conservative from the three categories are:

Percentage originally voting Conservative and still voting Conservative		percentage originally voting Democratic and now voting Conservative		percentage originally voting Republican and now voting Conservative
0.8168(0.20)	+	0.1420(0.45)	+	0.0985(0.35)
0.163360	+	0.063900	+	0.034475
= 0.261735				

Thus, about 26% of the population would vote for the Conservative candidate if the election were held in September.

Comment Assuming the trend continues, we can use a procedure similar to the one given in Example 11 to predict the winner in the elections held on election day in November and also the percentage of the votes that each party will carry. (The reader will be asked to calculate these percentages in Exercise 39 at the end of this section.)

Example 12 *Predicting the educational level of grandchildren.* Refer to the newspaper article at the beginning of this chapter. In that article we were told that "85% of the children of parents in Morganville who had only an elementary school education went on to complete high school and 25% of them even completed college studies. Only 15% of these children dropped out of school after completing elementary school. For parents who had completed high school, 84% of their children also completed high school, with 48% of them completing their college education. Only 16% of them dropped out of school after completing elementary school. On the other hand, if the parents were college graduates, then 98% of their children completed high school, with 81% of them completing college. A mere 2% dropped out after completing only elementary school."

Set up the transition matrix for this information and use it to find the predicted educational level of the grandchildren.

Solution We first arrange the information given in the article in tabular form.

Highest educational level achieved by parents	Highest educational level achieved by children		
	Elementary school	High school	College
Elementary school	15%	60%	25%
High school	16%	36%	48%
College	2%	17%	81%

The transition matrix is

$$\mathbf{M} = \begin{pmatrix} 0.15 & 0.60 & 0.25 \\ 0.16 & 0.36 & 0.48 \\ 0.02 & 0.17 & 0.81 \end{pmatrix}$$

To find the predicted educational level of the grandchildren, we must find $\mathbf{M}^2$:

$$\mathbf{M}^2 = \begin{pmatrix} 0.15 & 0.60 & 0.25 \\ 0.16 & 0.36 & 0.48 \\ 0.02 & 0.17 & 0.81 \end{pmatrix} \cdot \begin{pmatrix} 0.15 & 0.60 & 0.25 \\ 0.16 & 0.36 & 0.48 \\ 0.02 & 0.17 & 0.81 \end{pmatrix}$$

$$= \begin{pmatrix} 0.1235 & 0.3485 & 0.5280 \\ 0.0912 & 0.3072 & 0.6016 \\ 0.0464 & 0.2109 & 0.7427 \end{pmatrix}$$

This transition matrix represents the predicted educational level of the grandchildren if we know the educational level of the people. For the people of Morganville who have only an elementary school education, 12.35% of their grandchildren will also complete only elementary school, whereas 34.85% of them will complete high school only and 52.80% will also complete college. For the people who have a high school education, 9.12% of their grandchildren will drop out of school after elementary school, 30.72% will complete high school and 60.16% will complete college. For the people who have a college education, 4.64% of their grandchildren will drop out of school after elementary school, 21.09% will complete high school, and 74.27% will complete college.

Example 13 *Spy message.* A spy is transmitting a coded message to an enemy country through a communications system involving enemy agents using only the digits 0 and 1. The system is not functioning properly. There is a probability of $\frac{3}{4}$ that a digit that is received by an agent will be transmitted by that agent unchanged. Each digit that enters the system must pass through two agents before it is received by the enemy country. Find the probability that a digit transmitted by the spy as a 0 will be received by the enemy country as a 0.

Solution We first find the transition matrix. It is

Prob(digit received and transmitted by agent 1 as)

$$\begin{array}{ll} & \begin{array}{cc} 0 & 1 \end{array} \\ \text{Digit entered by spy} \begin{array}{c} 0 \\ 1 \end{array} & \begin{pmatrix} \frac{3}{4} & \frac{1}{4} \\ \frac{1}{4} & \frac{3}{4} \end{pmatrix} = \mathbf{M} \end{array}$$

As the message passes through the various agents and on to the enemy country, we can view the process as a series of transitions. Here we are interested in the transition matrix after the third trial (two agents plus one enemy country), so that $n = 3$. Therefore, we must find $\mathbf{M}^3$:

$$\mathbf{M}^3 = \mathbf{M} \cdot \mathbf{M} \cdot \mathbf{M} = \begin{pmatrix} \frac{3}{4} & \frac{1}{4} \\ \frac{1}{4} & \frac{3}{4} \end{pmatrix} \cdot \begin{pmatrix} \frac{3}{4} & \frac{1}{4} \\ \frac{1}{4} & \frac{3}{4} \end{pmatrix} \cdot \begin{pmatrix} \frac{3}{4} & \frac{1}{4} \\ \frac{1}{4} & \frac{3}{4} \end{pmatrix}$$

$$= \begin{pmatrix} \frac{10}{16} & \frac{6}{16} \\ \frac{6}{16} & \frac{10}{16} \end{pmatrix} \cdot \begin{pmatrix} \frac{3}{4} & \frac{1}{4} \\ \frac{1}{4} & \frac{3}{4} \end{pmatrix} = \begin{pmatrix} \frac{36}{64} & \frac{28}{64} \\ \frac{28}{64} & \frac{36}{64} \end{pmatrix}$$

$$= \begin{pmatrix} \frac{9}{16} & \frac{7}{16} \\ \frac{7}{16} & \frac{9}{16} \end{pmatrix}$$

The entry P_{11} is $\frac{9}{16}$. Thus, the probability is $\frac{9}{16}$ that a digit transmitted by the spy as a 0 will be received by the enemy country as a 0. The probability is $\frac{7}{16}$ that it will be received as a 1.

Example 14 *Gambler's ruin.* Jack and Jill each have \$3 and are playing the following game: Jack tosses a coin that has a probability of $\frac{1}{2}$ of coming up heads. On each toss Jack wins \$1 from Jill if heads comes up and loses \$1 to Jill if tails comes up. The game ends when Jack has no more money left or when Jill has no money left, that is, when either Jack or Jill has \$6. Set up the transition matrix as a random walk Markov chain with absorbing states (see Example 10).

Solution At any time Jack can have \$0, \$1, \$2, \$3, \$4, \$5, or \$6. Then for $i = 0$, 1, 2, 3, 4, 5, 6, we define the chain to be in state i if Jack has i dollars left. The transition matrix can be viewed as the random walk that we discussed in Example 10 with seven possible states. The probability of winning or losing \$1 is $\frac{1}{2}$. Compare this with the probability of moving to the right or to the left that we discussed in the earlier example. For each play the procedure is repeated. Thus, the transition matrix with absorbing states for this situation is:

$$\begin{array}{c} \\ i=0 \\ i=1 \\ i=2 \\ i=3 \\ i=4 \\ i=5 \\ i=6 \end{array} \begin{array}{c} \begin{array}{ccccccc} i=0 & i=1 & i=2 & i=3 & i=4 & i=5 & i=6 \end{array} \\ \begin{pmatrix} 1 & 0 & 0 & 0 & 0 & 0 & 0 \\ \frac{1}{2} & 0 & \frac{1}{2} & 0 & 0 & 0 & 0 \\ 0 & \frac{1}{2} & 0 & \frac{1}{2} & 0 & 0 & 0 \\ 0 & 0 & \frac{1}{2} & 0 & \frac{1}{2} & 0 & 0 \\ 0 & 0 & 0 & \frac{1}{2} & 0 & \frac{1}{2} & 0 \\ 0 & 0 & 0 & 0 & \frac{1}{2} & 0 & \frac{1}{2} \\ 0 & 0 & 0 & 0 & 0 & 0 & 1 \end{pmatrix} \end{array}$$

States i_0 and i_6 are the absorbing states, since i_0 means that Jack has \$0 left and state i_6 means that Jill has no money left. In either case the game is over.

Exercises

39. *Predicting election results.* Refer to Example 11. Using the procedure given, find the predicted percentage of the electorate that each party will carry on election day.

40. *Educational level of great-grandchildren.* Refer to Example 12. Find the predicted educational level of the great-grandchildren.

41. *More spying.* Refer to Example 13. Find the probability that a digit transmitted by the spy as a 0 will be received by the enemy country as a 1.

42. *Continued spying.* Refer to Example 13. Suppose the probability is $\frac{8}{9}$ that a digit that is received by an agent will be transmitted unchanged by that agent. Find the probability that a digit transmitted by the spy as a 0 will be received by the enemy country as a 0.

43. *Still more spying.* Refer to Example 13. Suppose a digit that enters the system must pass through three agents before it is received by the enemy country. Find the probability that a digit transmitted by the spy as a 0 will be received by the enemy country as a 0.

★ 44. *Urban renewal.* A sociologist conducted a survey of the living conditions in a city slum area. It was found that 30% of the houses were in need of minor repairs (category A), 63% were in need of major repairs (category B), and only 7% did not need any repairs (category C). After the report was released, some repairs were made. As a result of these repairs,

60% of those in A remained in A

37% of those in A were reclassified into category C

3% of those in A were reclassified into category B

50% of those in B remained in B

40% of those in B were reclassified into category A

10% of those in B were reclassified into category C

95% of those in C remained in C

4% of those in C were reclassified into category B

1% of those in C were reclassified into category A

A year later a similar survey was conducted. Again the report was publicized and some repairs were made. As a result of these repairs, the same reclassifications as indicated previously were made. After the second year, find the percentages of houses in need of minor repairs, major repairs, and no repairs.

45. *Movie attendance.* A survey of several hundred families living on Long Island disclosed that 75% of them were moviegoers and 25% of them were not moviegoers. Over the year, 90% of the moviegoers remained moviegoers, whereas 10% stopped going to movies. As for those who were not moviegoers, 80% remained the same and 20% became moviegoers. Assuming the trend continues, find

a) The percentage of moviegoers after four years

b) The long-range predictions for moviegoers on Long Island

46. *Competition.* In 1977 there were three major brands of disposable diapers on the market in a large city, each capturing a certain percentage of the market. Brand A had 60% of the market, brand B had 25% of the market, and brand C had 15% of the market. Each company launched a major advertising campaign and the percentages soon changed.

Brand A retained 80% of the business.

Brand A lost 12% of the business to brand B.

Brand A lost 8% of the business to brand C.

Brand B retained 65% of the business.

Brand B lost 21% of the business to brand A.

Brand B lost 14% of the business to brand C.

Brand C retained 75% of the business.

Brand C lost 13% of the business to brand A.

Brand C lost 12% of the business to brand B.

Assuming that these companies continued their same advertising strategy in 1978, find the percentage of the market captured by each of these companies in 1979 (two years later).

47. *Loaded coin.* In the gambler's ruin problem in Example 14, assume that Jack flipped a loaded coin where the probability of heads coming up is .6. Find the transition probability matrix and interpret the entries.

48. *Moving from the suburbs to the city.* Of all the accountants in a large city, 75% reside in the suburbs and 25% reside in the city. Furthermore, there is an 80% chance that a city-dwelling accountant can be convinced to move to the suburbs, and a 40% chance that a suburban-dwelling accountant can be convinced to move back to the city. In the long run, what proportion of the group will be

a) City dwellers
b) Suburban dwellers
c) Convinced to move from the city to the suburbs
d) Convinced to move from the suburbs to the city

49. *Tires.* Three tire companies F, G, and U are competing against each other to capture the largest share of the market for steel-belted tires. Various surveys show that the transition matrix of customers switching from one brand to another is

$$\begin{array}{c} \\ F \\ G \\ U \end{array}\begin{array}{c} \begin{array}{ccc} F & G & U \end{array} \\ \begin{pmatrix} 0.75 & 0.15 & 0.10 \\ 0.19 & 0.69 & 0.12 \\ 0.03 & 0.17 & 0.80 \end{pmatrix} \end{array}$$

Assume that all customers stay with these brands of tires only. If the trend indicated previously continues, find the long-range market distribution.

50. *Kicking child.* A child psychologist is observing a baby in a crib kick at a mobile that is suspended from the sides of the crib. It is observed

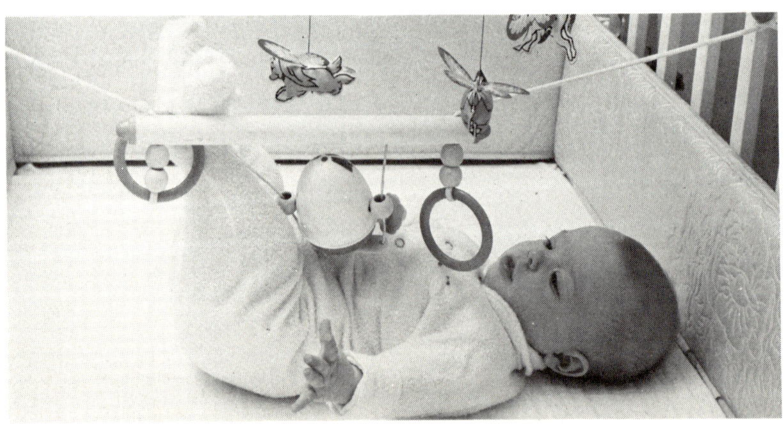

Photo: Lew Merrin, Monkmeyer

that if the baby kicks at it with his left foot, then the probability that the next kick at it will be with his left foot is $\frac{2}{7}$. On the other hand, if he kicks at it with his right foot, then the probability that the next kick at it will be with his right foot is $\frac{4}{7}$.

a) Find the transition matrix.

b) Find the probability that the fifth kick from now will be with his right foot if it is observed that he just kicked at it with his left foot.

51. *Selecting an outfit.* Wendy has three different pairs of shoes, red, blue, and black. She never wears the same pair of shoes two days in a row. If she wears the blue ones on Monday, the probability that she will wear the red ones on Tuesday is 1. On the other hand, if she wears the black ones on Monday, then she is as likely to wear the red ones on Tuesday as she is to wear the blue ones. She never wears the red ones on Monday. Assume that this process continues indefinitely. In the long run, how often does she wear each pair of shoes.

52. *Television watching.* Bill has the following studying habits: If he watches television one evening, then he is 80% sure to do his math homework on the following evening. On the other hand, if he does his math homework on one evening, then he is 40% sure to watch television on the following evening. Assume that this process continues indefinitely. In the long run, how often does Bill do his math homework?

11.5 Summary

In this chapter we introduced the newly developed topic of Markov chains. This topic was investigated by the Russian mathematician A. A. Markov, who was interested in experiments where repeated trials were not independent. Thus, the outcome on any one trial of an experiment depends on (at most) the outcome of the preceding trial. Furthermore, we allow for the possibility that there may be more than two outcomes for each trial of the experiment.

We first studied transition matrices and probability vectors. Each of the entries P_{ij} of a transition matrix gives us the probability that an experiment now in state i will be in state j on the next trial. The rows of the transition matrix are called probability vectors. The sum of the probabilities in any probability row vector is one.

We gave a rule that enables us to find the probability that an experiment now in state i will be in state j after n trials. If some power of a transition matrix contains only positive elements, then we have a regular Markov chain. We also discussed a procedure for finding fixed probability vectors.

Finally we discussed absorbing Markov chains and indicated in several examples how they are used.

As indicated earlier, Markov chains is a newly developed topic. Applications for Markov chains are constantly being found. Our objective in this chapter has been merely to introduce you to this developing idea. Further and more detailed discussions can be found in the suggested readings given at the end of this chapter. In recent years Markov chains have been applied to the field of genetics.

Study Guide

You should be able to demonstrate your knowledge of the following ideas presented in this chapter by giving definitions, descriptions, or specific examples. Page references are given for each term so that you can check your answer.

Markov chain (p. 473)	States (p. 473)
Transition matrix (p. 475)	Probability vectors (p. 475)
P_{ij} (p. 475)	n-Step transition matrix (p. 476)
Fixed probability vector (p. 480)	Regular Markov chain (p. 484)
Random walk (p. 484)	Absorbing state (p. 485)
Absorbing Markov chain (p. 485)	

Formulas to Remember

The following list summarizes all the formulas given in this chapter.

$$\mathbf{M} = \begin{pmatrix} P_{11} & P_{12} & P_{13} & \cdots & P_{1n} \\ P_{21} & P_{22} & P_{23} & \cdots & P_{2n} \\ \vdots & \vdots & \vdots & & \vdots \\ P_{n1} & P_{n2} & P_{n3} & \cdots & P_{nn} \end{pmatrix}$$

is called a transition matrix and consists of probability row vectors each of whose sum is one. P_{ij} is the probability that an experiment in state a_i will move next to state a_j.

The transition matrix $\mathbf{M}$ raised to the nth power gives the probability that an experiment now in state i will be in state j after n trials.

$\mathbf{v} \cdot \mathbf{M} = \mathbf{v}$, where $\mathbf{v}$ is a fixed probability vector.

Mastery Tests

FORM A

1. True or false: The entry P_{ij} in a transition matrix $\mathbf{M}$ of a Markov chain is the probability that the process changes from state a_i to state a_j in one trial.

2. Each entry P_{ij} of a transition matrix must be which of the following?
   a) Nonnegative
   b) Nonpositive
   c) Zero
   d) A fraction

3. Which of the following transition matrices are regular?

a) $\begin{pmatrix} 1 & 0 \\ \frac{1}{2} & \frac{1}{2} \end{pmatrix}$ b) $\begin{pmatrix} 1 & 0 \\ 0 & 1 \end{pmatrix}$

c) $\begin{pmatrix} \frac{1}{3} & \frac{2}{3} \\ \frac{1}{2} & \frac{1}{2} \end{pmatrix}$ d) None of these

4. Which of the following row vectors are probability vectors?

a) $(\frac{1}{3} \quad \frac{2}{3} \quad -1 \quad 1)$ b) $(\frac{1}{4} \quad \frac{1}{8} \quad \frac{3}{7})$

c) $(0.7 \quad 0 \quad 0.15 \quad 0.25)$ d) $(\frac{1}{2} \quad \frac{1}{2} \quad 0)$

5. Which of the following matrices are transition matrices for an absorbing Markov chain?

a) $\begin{pmatrix} \frac{1}{5} & \frac{4}{5} \\ 0 & 1 \end{pmatrix}$ b) $\begin{pmatrix} \frac{1}{3} & 0 & \frac{2}{3} \\ \frac{1}{4} & \frac{3}{4} & 0 \\ 0 & 0 & 1 \end{pmatrix}$

c) $\begin{pmatrix} 1 & 0 & 0 \\ \frac{1}{4} & \frac{1}{2} & \frac{1}{4} \\ 0 & 0 & 1 \end{pmatrix}$ d) $\begin{pmatrix} 1 & 0 & 0 & 0 \\ 0 & 1 & 0 & 0 \\ 0 & 0 & 0 & 1 \\ \frac{1}{3} & \frac{1}{3} & \frac{1}{3} & 0 \end{pmatrix}$

6. Suppose the Markov chain

$$\mathbf{M} = \begin{array}{c} \\ 1 \\ 2 \\ 3 \\ 4 \\ 5 \end{array} \begin{array}{c} \begin{array}{ccccc} 1 & 2 & 3 & 4 & 5 \end{array} \\ \begin{pmatrix} \frac{1}{4} & 0 & \frac{3}{4} & 0 & 0 \\ 0 & 1 & 0 & 0 & 0 \\ \frac{1}{3} & 0 & \frac{1}{2} & \frac{1}{6} & 0 \\ 0 & 1 & 0 & 0 & 0 \\ 0 & 0 & 0 & 0 & 0 \end{pmatrix} \end{array}$$

has five states as labeled. Which of the states are absorbing states?

7. Find the fixed probability vector for this transition matrix

$$\begin{pmatrix} \frac{1}{3} & \frac{2}{3} \\ \frac{1}{4} & \frac{3}{4} \end{pmatrix}$$

8. Find the three-step transition matrix for a Markov chain whose transition matrix is

$$\begin{pmatrix} \frac{1}{3} & \frac{2}{3} \\ \frac{1}{2} & \frac{1}{2} \end{pmatrix}$$

9. A Markov chain is said to be regular if some power of its transition matrix contains only ____________.

10. For a state i of a Markov chain to be called an absorbing state, what must P_{ii} equal?

11. True or false: In an absorbing Markov chain, it is never possible to pass from a nonabsorbent state to an absorbent state.

12. Can a Markov chain have more than one absorbing state?

FORM B

1. *Lion experiments.* A lion is placed in a cell that has doors leading to adjacent cells as shown in the diagram. Every hour both doors from the cell are opened and the lion is able to and does enter *either* adjacent cell with an equal probability. Once the cell is entered, the lion cannot leave that cell until the next hour, at which time the lion again can choose either door. This process is repeated over a period of several hours. The lion is placed in cell 1 at the beginning of the experiment.

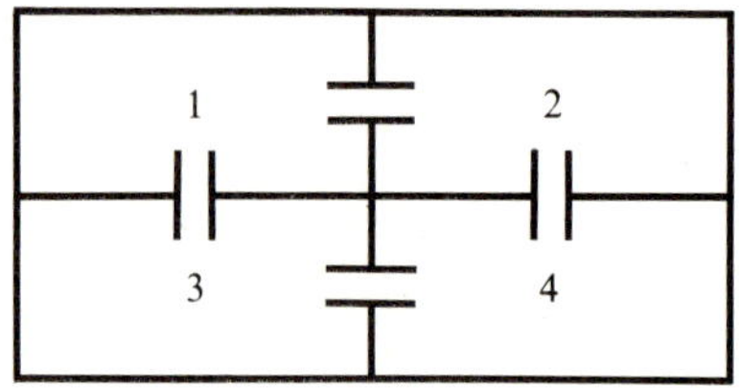

   a) Find the transition matrix.
   b) Find the probability that the lion will be in cell 4 after the fourth hour.

2. *Right-handed or left-handed.* Assume that the probability that a right-handed father has a right-handed son is .9 and the probability that a left-handed father has a left-handed son is .6. Find the probability that a right-handed man will be the great-grandfather of a left-handed grandson.

3. *Changing the stakes.* Refer to the gambler's ruin problem in Example 14. If Jack starts out with \$4 and Jill starts out with \$3, find the transition matrix.

4. *Selecting a dessert.* A waiter has observed the following about Margaret Greene's ordering habits: She either orders a piece of strawberry short-cake, an ice cream sundae, or Jello for dessert. She never orders the same item for dessert two days in a row. However, if she has an ice cream sundae on one day, then the probability is $\frac{1}{3}$ that she will order Jello on the next day. If she has strawberry shortcake on one day, then the probability is $\frac{4}{7}$ that she will order an ice cream sundae on the next day. If she has Jello on one day, then the probability is $\frac{1}{2}$ that she will order an ice cream sundae on the following day.
   a) Determine the transition matrix.
   b) If Margaret orders an ice cream sundae on Monday for dessert, find the probability that she will order an ice cream sundae for dessert on Thursday.

5. *Which airport?* The greater New York metropolitan area has three major airports, Kennedy, LaGuardia, and Newark. Mark is a cab driver who takes passengers every day to one of these three airports.

However, he never goes to the same airport two days in a row. If he goes to Kennedy Airport one day, then the probability that he will go to LaGuardia Airport the following day is $\frac{3}{8}$. If he goes to LaGuardia Airport one day, then the probability that he will go to Newark the following day is $\frac{5}{7}$. On the other hand, if he goes to Newark on one day, then he is equally likely to go to Kennedy or to LaGuardia the following day. In the long run, how often does the cabbie go to each airport?

6. *Baseball.* In baseball the infielders frequently practice by throwing the ball around the infield. In several recent practice sessions, the following pattern was repeatedly observed:

	Probability ball will be thrown to			
Ball presently held by	First baseman	Second baseman	Shortstop	Third baseman
First baseman	0	.8	.05	.15
Second baseman	.6	0	.3	.1
Shortstop	.5	.3	0	.2
Third baseman	.4	.35	.25	0

a) Find the transition matrix.
b) If the shortstop now has the ball, find the probability that he will have the ball again after four throws by the infielders.

7. *Broken telephone.* In the dormitory of State University, there is a pay phone in the lobby that either functions properly or is out of order. If the phone is out of order at the start of a day, then the phone company is called and the probability is .8 that it will be repaired and functioning properly by the start of the next day. If the phone is functioning properly today, however, then the probability is .1 that it will not be functioning properly by the start of the next day.
a) Find the transition matrix.
b) If the phone is functioning properly on Monday morning, find the probability that it will be functioning properly on Friday morning.

8. *Election campaigns.* In the upcoming presidential election, various polls indicate that each candidate has to carry the states of New York, New Jersey, or Connecticut in order to win the election. One of the candidates decides to concentrate his campaign efforts in the closing days as follows: If he campaigns in New York on one day, then he will campaign in New York, New Jersey, or Connecticut the following day with equal probability. If he campaigns in New Jersey on one day, then he will not campaign in New Jersey the following day but is twice as likely to be in New York than in Connecticut the following day. If he campaigns in Connecticut on one day, then he will campaign

in New York or New Jersey the following day with equal probability. Assume that the candidate will campaign in only one of these three states each day.

a) Set up the transition matrix.
b) If the candidate is in New York on Sunday, find the probability that he will be in New York on Thursday.

9. *Car preference.* A survey was taken a year ago of the employees of the Printex Corp. who drive to work. The survey showed that 85% of the employees preferred automatic transmission and only 15% preferred standard transmission. A year later a similar survey revealed that of those initially preferring automatic transmission, 85% still preferred automatic transmission, whereas 15% now preferred standard transmission. Of those initially preferring standard transmission, 70% still preferred standard transmission, whereas 30% now preferred automatic transmission. If it is assumed that the probability of yearly shifts in preference will remain the same, find the percentage of employees preferring automatic transmission:

a) Four years after the initial survey
b) Ten years after the initial survey

10. *The work force.* According to figures released by the labor bureau of a certain city, the breakdown of the male population between the ages of 20 and 25 on February 1, 1977 was students 27%, working 51%, and unemployed 22%. On February 1, 1978, a similar survey revealed the following:

Concerning the students: 80% remained students, 17% began working, and 3% became unemployed.

Concerning the working people: 5% became students, 75% continued working, and 20% became unemployed.

Concerning the unemployed people: 20% became students, 68% began working, and 12% remained unemployed.

Assuming these trends continue, find the long-range prediction for the breakdown of the male population in each category.

11. *Teacher preference.* Professor Gertrude Hoffman teaches calculus, algebra, statistics, and finite math at Richmond Hill College on a selective basis as shown here. This semester Professor Hoffman is

If she is currently teaching	Probability that next semester she will teach			
	Calculus	Algebra	Statistics	Finite math
Calculus	.10	.40	.30	.20
Algebra	.25	.05	.44	.26
Statistics	.20	.30	.10	.40
Finite math	.15	.25	.35	.25

teaching a course in finite math. Find the probability that she will be teaching a finite math course three semesters later.

Suggested Further Reading

Alling, D. W., "The After History of Pulmonary TB: A Stochastic Model," *Biometrics*, vol. 4, pp. 527–547, 1958.

Cox, D. R., and H. D. Miller, *The Theory of Stochastic Processes*, New York: Wiley, 1965.

Kemeny, John, and J. Laurie Snell, *Finite Markov Chains*, New York: Van Nostrand, 1959.

Parzen, Emanuel, *Stochastic Processes*, San Francisco: Holden-Day, 1962.

CHAPTER 12 Computers

Objectives to be Achieved in this Chapter

- *To learn* what a computer is, what it can do, and what are its limitations.
- *To discuss* the various parts of a computer and what each part does.
- *To indicate* how the computer has evolved to its present state. This is accomplished by studying its history.
- *To analyze* what flow charts are and how they are used.
- *To study* programming and the computer language of BASIC.
- *To learn* how to actually write computer programs using BASIC.
- *To apply* the ideas of computer programs to solve several simple problems.

DAILY PRESS MAY 25, 1979

Suspect Arrested

New York City (May 25)—The Federal Bureau of Investigation announced today the arrest of a suspect in the $3 million robbery of a Wells Fargo armored car in Staten Island, New York, during December 1978. Extensive computer traces of fingerprints and criminal records by the National Crime and Information Center of the FBI in Washington led to the arrest. The name of the suspect was not disclosed pending further investigation.

BERGENVILLE HERALD AUGUST 20, 1978

Computer Goof

Bergenville (Aug. 20)—Bank officials were at a loss to explain how a $100,000 mistake went unnoticed for several days. According to Al Carlton, president of County Savings Bank, an unidentified depositor closed his savings account last week. The computer inadvertently made out a check in the amount of $100,000 instead of $1,000. The mistake went unnoticed until a routine audit disclosed the mistake. In the meantime, the check had been cashed. The unidentified depositor has since moved out of town without leaving a forwarding address.

Computers are becoming more a part of our lives with each passing day. When used properly, they can be extremely helpful in such areas as criminal investigations. This is indicated by the first newspaper article. When used improperly, they can produce great monetary losses as illustrated by the second newspaper article. Computers can also be used in a variety of other ways both in the scientific field and in the general business field.

Since a computer is only a huge piece of metal, how does the computer do it? Can a computer think? Will computers replace people?

12.1 Introduction

It is often said that we live in a computer age and that our lives are influenced to some degree and in some way by computers. Computers have been on the market for only a little more than 30 years; nevertheless, in this short time they have become an integral part of our society and our lives. Computers are not electronic beasts as some people claim but are rather very efficient servants of mankind (when used properly).

Today the computer is not merely a luxury. It is no longer confined to the laboratory to be used by research scientists. Because of its wide-ranging capabilities, it has become a necessity for many businesses (both large and small), schools, hospitals, government agencies, and so forth. The following are some specific situations in which the computer has been applied.

1. *Crime prevention and detection.* When a call for police assistance is received, computers are used to locate the nearest police car and dispatch it to the scene of the crime. Often computers are used in tracking stolen vehicles or in recovering stolen stock securities. Law enforcement officials use computers to search millions of fingerprints in a few seconds and come up with a suspect.
2. *Traffic control.* The flow of traffic on the arterial highways of many large cities is often controlled by means of computers that monitor the traffic patterns and control the traffic signals accordingly.
3. *Business use.* Most businesses use the computer not only for such simple clerical functions as preparing paychecks and punch-in cards but also for inventory control and market forecasting. A bank processes your monthly credit card and checking account with a computer. Our monthly utility bills (telephone, electricity, gas, water, etc.) are prepared by a computer.
4. *Education.* Computers are used by colleges to store student records, student programs, faculty schedules, and other information. Computers are also used to provide a student with individualized *computer-assisted instruction.* With this method of instruction, the computer asks a student a question. If the student answers the question correctly, another question is asked. If the student answers the question incorrectly, another problem of the same type is posed so that the student can try to solve it again.
5. *Medicine.* In a hospital computers are used not only in research but also in different units of the hospital where patients are hooked up to various machines that monitor their vital signs. Nurses can remain at their stations and use a computer terminal to monitor patient progress. Doctors can also recall this information from the computer if needed later. Computers can tell the nurse how much medicine to administer to a particular patient.
6. *Industry.* The railroad industry uses the computer to keep track of its rolling stock. The airplane industry uses computers to determine the location of its planes, to determine when a particular plane needs a major overhaul, and to schedule departures.
7. *Spacecraft launchings.* Computers are used by scientists to check that all systems are functioning properly before a spaceship is launched. In addition, computers are used extensively to *simulate* different conditions that could arise.
8. *Urban analysis.* Many cities use computers to analyze:
    a. Where urban decay is greatest
    b. Which areas are in need of new police stations, fire houses, hospitals, schools, and other facilities

Figure 12.1
Photo: Bell Labs

c. Population trends
d. Crime statistics
e. Traffic scofflaws

9. *Ecology.* Computers are used by ecologists to keep track of any change in the air we breathe, the water we drink, how we dispose of our garbage and nuclear wastes, and other conditions.
10. *The arts.* Computers can be programmed to play chess, generate music, and create pieces of art. (See Figure 12.1.) Computers can even simulate certain games. A classic example of such simulation was done by Edward Thorp in his book *Beat the Dealer*. In this book Thorp simulated the blackjack card game and discovered from these results how to devise a scheme for winning. This scheme was so successful that the gambling casinos in Nevada had to modify their rules of the game pretty quickly.
11. *National defense.* The U.S. government uses computers extensively not only to keep track of its arsenal of weapons and strategic capabilities but also to simulate "war games" and to test new equipment.
12. *Computer matches.* Computers are often used to arrange dates by matching two people with similar qualities. Computers are also used to match prospective job seekers with any job openings that exist.
13. *Computer projections.* Politicians use computers to analyze election returns on a district basis and to make projections about upcoming elections. Thus, their campaign strategy is often dictated by computer projections.
14. *Computer games.* In the past few years there has been a steady increase in the number of electronic computer games that are available on the market.

These examples are only some ways in which the computer is used. Today the computer is playing an ever-increasing role in our lives as new ways are found in which it can be used to make life more pleasant.

While we are all affected by computers, many people are unaware of what a computer really is, how it works, or what it can and cannot do. In this chapter we discuss some of these issues.

12.2 What Is a Computer?

One of the earliest computers, an astrolable made by the Parisian astronomer-mathematician Jean Fusoris (1365–1436) in 1390. (*Photo:* Adler Planetarium, Chicago)

What is a computer? To some, it is a hunk of metal with flashing lights that stores all sorts of information about us and also sends us incorrect bills. (Actually computer errors occur only when there is a mechanical failure or when the human operator makes a mistake. This, of course, explains the mistake described in the second newspaper article at the beginning of this chapter.) To others, the computer is our salvation since it relieves people of immensely complicated and time-consuming computations that they might encounter.

Actually a *computer* is simply any device that helps us compute. Nevertheless, we must realize that computers cannot think for us, nor can they do anything a human could not do if given enough time. Their usefulness lies in their tremendous speed and accuracy, since they are able to perform calculations accurately at lightning speed. Thus, computers relieve us of the often time-consuming and boring calculations associated with mathematical, scientific, and business problems.

Perhaps you are wondering if the modern-day computer is the first human excursion into the computational field? The answer is no. Ever since people learned how to count, they have been looking for ways to simplify computations. At first they used pebbles and sticks to help them add and subtract. Then other devices were invented.

One of the earliest such computational devices was the *abacus* or *counting board* (see Figure 12.2). To use this device one had to move

Figure 12.2 An abacus. (*Photo:* Elinor S. Beckwith)

beads back and forth along the rods. The abacus was popular in the Far East. It did not enjoy great popularity in Europe, however, as people mistrusted it or simply did not know how to operate it.

In 1617 the Scottish mathematician John Napier (1550–1617) invented a device that today is referred to as *Napier's rods* or *Napier's bones* and that could be used to perform multiplication and division. Napier's rods became the forerunner of the slide rule, which was used until the 1970s, when inexpensive electronic hand-held calculators became very popular.

Figure 12.3 Babbage's multiplier. (*Photo:* The Granger Collection)

In 1642 Blaise Pascal invented the first mechanical adding machine, which enabled the user to perform the operations of addition and subtraction much like our modern-day adding machines. The device utilized six wheels, each of which could register any digit from 0 to 9. The user had to enter the appropriate number on each dial and then pull a handle to register it. Unfortunately, it took a great deal of time to perform calculations with it.

In 1812 the Englishman Charles Babbage (1791–1871) began building a "difference engine," which would be capable of performing calculations with numbers up to 26 decimal places and would also print mathematical tables (see Figure 12.3). His attempt failed. Undaunted by his first failure, Babbage attempted in 1833 to build an "analytical machine." Both of Babbage's brilliant ideas were doomed to failure since the technology needed to build them was not yet developed.

Figure 12.4 Jacquard's weaving loom. (*Photo:* The Granger Collection)

In 1804 Joseph Marie Jacquard made a loom for weaving patterns. The loom was able to produce different designs by using punched cards. "Instructions" were punched on a card, which caused the loom to weave the appropriate design (see Figure 12.4). This use of punched cards for storing information was similar to our present-day use of them.

The first major breakthrough in computers occurred during the 1890s and was a direct consequence of the 1890 U.S. census. As a result of the 1880 census, the U.S. Census Bureau estimated that it would take considerably more time to collect and tabulate the data for the 1890 census. At the time Herman Hollerith (1860–1927) was working in the U.S. Patent Office. Being familiar with Jacquard's use of punched cards to store information on the loom, Hollerith invented machines that could do exactly the same thing, read cards and compile the information. Under Hollerith's direction, the U.S. Census Bureau took three years to tabulate the results of the 1890 census as opposed to the nine years needed for the 1880 census. This was done in spite of the fact that the population of the United States had increased considerably over the ten-year period. In 1896 Hollerith founded the Tabulating Machine Corporation, which was later to become the International Business Machines Corporation, commonly known as IBM.

The twentieth century has really seen the computer come into being. In 1944 Howard Aiken and IBM developed the first completely automatic electromechanical computer called MARK I. Instead of performing computations by mechanical means (such as rotating wheels), the MARK I computer performed them by means of electronic impulses.

At about the same time, the army developed the ENIAC (*E*lectronic *N*umerical *I*ntegration *A*nd *C*alculator) computer for use in World War II to help with its ballistic tables. The ENIAC computer contained more than 19,000 vacuum tubes (to be used in place of electronic relays), needed many technicians to service it, and weighed 30 tons. It was capable of performing 5000 additions or subtractions a second (rather slow by today's standards) but was not capable of storing any instructions internally. All instructions had to be read into the computer one at a time by means of punched cards, paper tapes, or some other means.

At that time John Von Neumann (1903–1957), who was working on the EDVAC (*E*lectronic *D*iscrete *V*ariable *A*utomatic *C*omputer), visited the ENIAC project at Aberdeen. He believed that a computer should be capable of storing instructions (programs) *as well as* numbers, so he began working on developing such a computer. This is exactly where our present-day computer comes in. In 1952 the EDVAC computer was completed. This computer was capable of doing exactly what Von Neumann had envisioned. It could store its programs internally. This development was a major breakthrough in the computer field, since it freed the programmer to work on new projects while the computer worked through a problem. Supervising the computer and telling the computer what to do next were no longer necessary.

After this breakthrough, improvements began to appear very rapidly. Since World War II computer technology has gone through three and perhaps four distinct phases, often called *generations* by computer professionals.

The early computers, that is, those built prior to 1958, are referred to as *first-generation computers*. They used vacuum tubes and were capable of performing calculations in milliseconds (thousandths of a second). The UNIVAC computer, which is the business-oriented version of the ENIAC computer, is an example of a first-generation computer.

The vacuum tube soon gave way to the transistor and computers built with transistors instead of vacuum tubes began appearing on the market in 1958. Such *second-generation computers* were capable of performing calculations in a few microseconds (millionths of a second). These computers lasted until the early 1960s, when integrated circuits began replacing transistors.

These *third-generation computers* that use integrated circuits are capable of performing more than 500,000 calculations in a second. This immense speed enables them to work on several programs at the same time through time-sharing plans. Computer progress was further advanced with the development of Monolithic Systems Technology, which made it possible for computers to be built with microscopic logic circuits capable of performing well over a million computations a second.

In 1971 a miniature computer (as opposed to the original 30-ton ENIAC computer of 1946) utilizing a microprocessor was invented. This invention has opened the way for much smaller but just as efficient computers.

Who knows what lies ahead with the computer? Have we come to the end of the line? All indications point to the fact that computers are here to stay and that they will be an integral part of our life. In the not too distant future we may find computers in every household in America helping us keep track of our household accounts. Furthermore, the level of sophistication needed to operate these computers will be greatly diminished, so that we have definitely not come to the end of the line but are rather just beginning to recognize the usefulness of the computer.

12.3 How a Computer Works

As indicated earlier, a computer is not capable of doing its own thinking.[1] It can merely do what it has been programmed to do and it does this at lightning speed. How does a computer work? There are several main parts to any computer; these are pictured in Figure 12.5.

In order for the computer to do anything, we must have some way of entering data and instructions into the machine. This is done in the

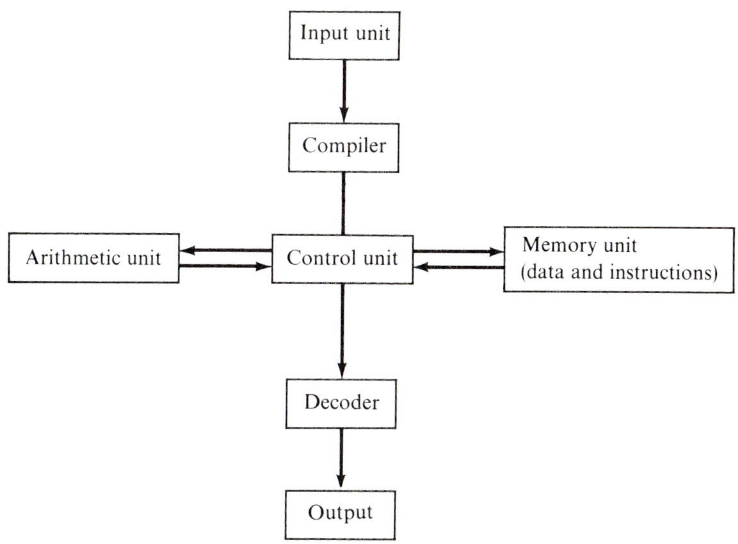

Figure 12.5 The parts of a modern computer.

[1] We restrict our discussion to the most popular types of computers in use today. These are the *digital computers* as opposed to the *analog computers*, which are used primarily in science.

input unit. Input into the machine can be accomplished by using punched cards, which have been prepared on machines called *key-punches.* The computer reads data from punched cards by noting the presence or absence of a hole in a specified location. Input into the machine can also be accomplished by using paper tapes, magnetic tapes, discs, or console typewriters. Information obtained from any input device is transferred by means of electric impulses to a compiler.

Since almost all computers do their work internally in base 2 as opposed to the decimal system, any information that the computer receives must be in a language it understands. In *base 2* there are only two digits, 0 and 1. Thus, all instructions and data have to be converted into base 2. This is the job of the *compiler.* Before the advent of compilers in the early 1950s, much of this conversion had to be done by the programmer. Thus, the compiler enables the programmer to talk to the computer with familiar words and algebraic symbols. The computer does the rest.

After the input information passes through the compiler, it is transmitted to the *control unit,* which both receives and interprets all the instructions it gets. In essence, the control unit is the "brains" of the computer. The control unit then decides which of the messages received should go to the arithmetic unit, which to the memory unit, and which to the output unit, as well as the sequence in which these operations should be done. All the computations needed in a problem are performed in the *arithmetic unit* of the computer. Of course, these computations are done in base 2. Any information that has to be used at a later time, whether it be a number or an instruction, is stored in the *memory unit.*

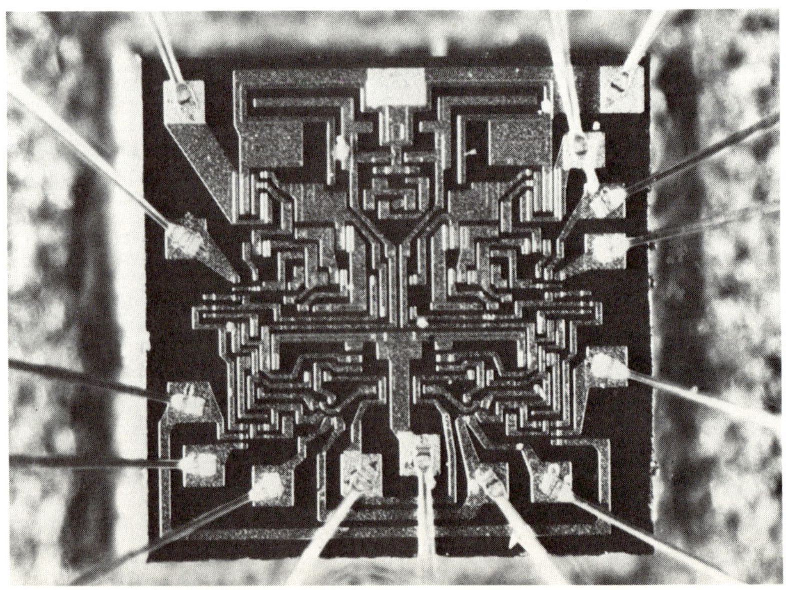

Figure 12.6 An enlarged memory chip. This one contains 46 transistors, 52 resistors, and 12 diodes; it is the size of two pinheads. (*Photo:* Honeywell Electronic Data Processing)

Many computers store information by using magnetic cores, which are tiny doughnut-shaped pieces of metal called *bits*. These are no larger than the size of a pinhead. When magnetized in one direction they represent the number 1, and when magnetized in the other direction they represent the number 0. (Remember that a computer works in base 2, in which the only digits are 0 and 1.) The size of these magnetic cores, which are usually grouped into small *chips*, has been constantly decreasing, thereby increasing the memory capabilities of the machine (see Figure 12.6).

After the computer solves the problem, the control unit sends the answer to the *decoder*, where it is translated from binary notation into the language being used by the programmer, and then to the *output unit*, which presents the answer in a predetermined form, such as on a display console, on punched cards, or on a teletype.

The physical parts of the computer are referred to as the *hardware* as opposed to the *software*, which includes such items as the actual programs and the programming techniques.

12.4 Programming

As indicated earlier, the computer can execute instructions only when they are written clearly. This is the job of the programmer, who writes a program (a list of instructions) for the computer to follow.

Programmers can use one of the many possible *programming languages* that are available, each designed with a particular need in mind. FORTRAN, BASIC, ALGOL, COBOL, PL-1, LISP, and APL are some of the more popular computer languages. COBOL (an abbreviation for *CO*mmon *B*usiness-*O*riented *L*anguage was designed to be used primarily in business applications. FORTRAN (*FOR*mula *TRAN*slation) is a general-purpose language used mostly in scientific applications.

In this chapter we discuss BASIC (*B*eginner's *A*ll-purpose *S*ymbolic *I*nstruction *C*ode). This language was designed to help people learn how to use a computer without having to spend a considerable amount of time learning one of the more complicated languages.

12.5 Flowcharts

Often the steps involved in writing a program are lengthy and quite complicated. In such cases it is usually advisable to draw a *flowchart* first. This is a diagram indicating the logical sequence of steps to be followed by the computer; that is, flowcharts show us what is to be done and in what order. They also indicate when decisions have to be made. Flowcharts can help us analyze programs since they provide visual insight into what is happening.

When drawing a flowchart, it is standard to use circles (or ovals) to indicate the beginning or end of a program and rectangles to indicate the processing of a program. Input/output is usually indicated by the

Table 12.1 Symbols Commonly Used in Flowcharts

Symbol	Use
	To indicate start or end of program
	To indicate input or output of data
	To indicate the processing of the program
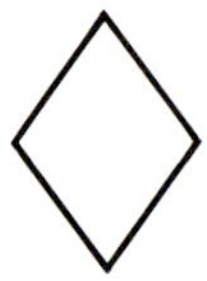	To indicate that a decision has to be made and the branching to different alternatives depending on the decision made
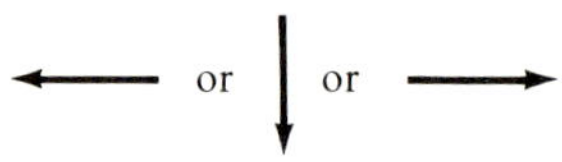	To indicate the direction to go in the flowchart to find the next step

parallelogram symbol. We also use the diamond symbol to indicate that a decision must be made. These *symbols* are illustrated in Table 12.1.

Let us construct flowcharts for several different situations.

Example 1 Construct a flowchart to compute the area, A, of a rectangle that has a length, L, of 16 centimeters and a width, W, of 24 centimeters.

Solution This is a rather simple program, since no decisions have to be made. The flowchart is

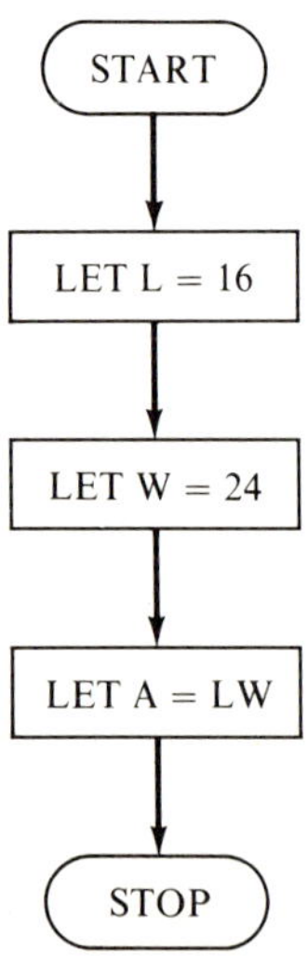

A program written from this flowchart will tell the computer to find the area by multiplying the length, 16, by the width, 24, obtaining 384. Since the computer cannot talk to us, however, how are we to be told the answer? Obviously, we must tell the computer to print out the answer it obtains. The flowchart for this problem would now be

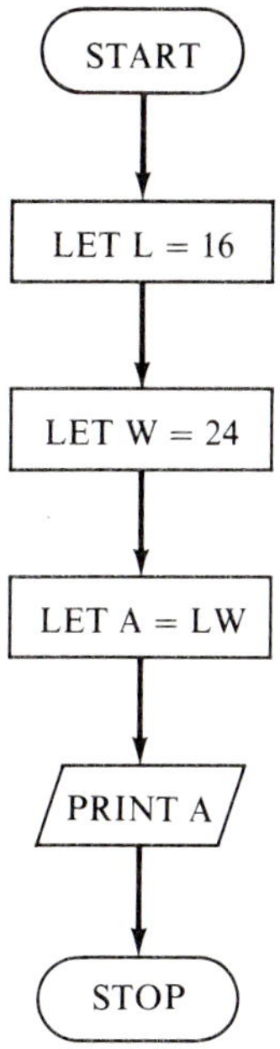

Comment When using flowcharts, we follow the arrows from step to step.

Example 2 Professor Kim is teaching a computer class. Twenty students took the midterm. Write a flowchart to help the professor find the average grade on the midterm.

Solution To calculate the mean, we must first tell the computer what the grades are. This is accomplished by a Read statement, such as Read A_1, A_2, $A_3, \ldots, A_{20}$, where the A_i values represent the grades of the students in the class. This Read instruction gets the numbers into the computer. Once the grades have been entered, we have the computer add them and then divide the sum by 20. Finally, we want the computer to print out the result. The flowchart that accomplishes these things is

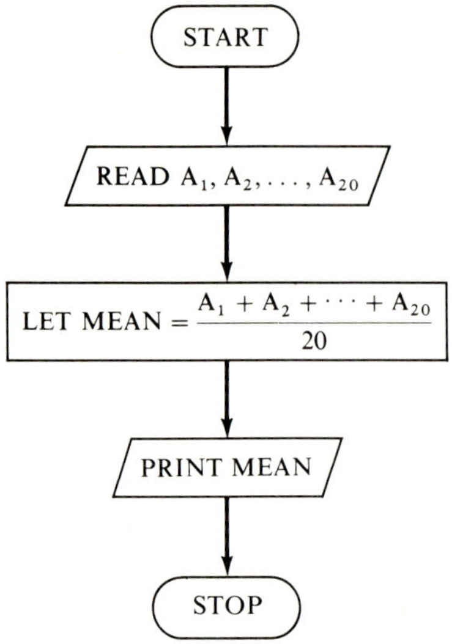

Example 3 In Example 2 suppose the professor is teaching several classes, each of which has 20 students. Write a flowchart to help the professor find the average grade in the midterm for each class.

Solution In this case the same program can be used over and over again for each class. The only thing that will be different is the data for the different classes. We indicate that the computer is to do the same thing by *looping* it back to the Read statement to start again. The flowchart that accomplishes this is

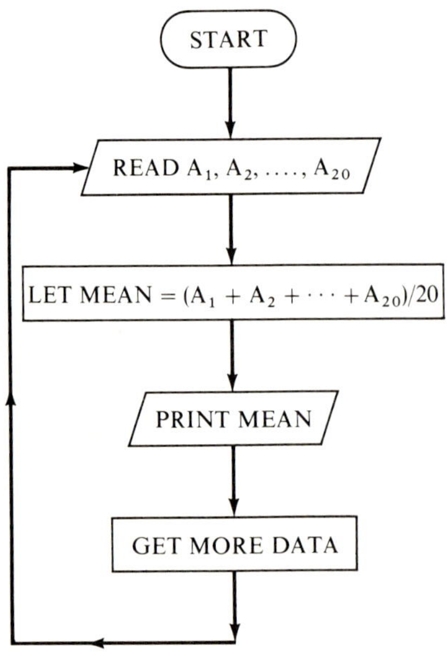

When the computer runs out of data, it will stop.

Let us now construct a flowchart for a situation where a decision is involved.

Example 4 A large company rents photocopying machines to its customers and also arranges to service these machines. The monthly costs for these services are as follows:

Cost for each photocopy made: 5¢

Rental fee including service charge for month for the first 15 service calls: $50

Charge for each service call over 15 calls: $25

Write a flowchart to compute a customer's monthly bill.

Solution In this case we must first tell the machine how many photocopies were made and how many service calls were made. The computer then has to decide whether or not the number of service calls is greater than 15 since there is a different charge in each case. Let N represent the number of photocopies made, let S be the number of service calls made, let C be the charge for making N photocopies, and let P be the charge for making S service calls. Then the total charge, T, can be found by using the following flowchart:

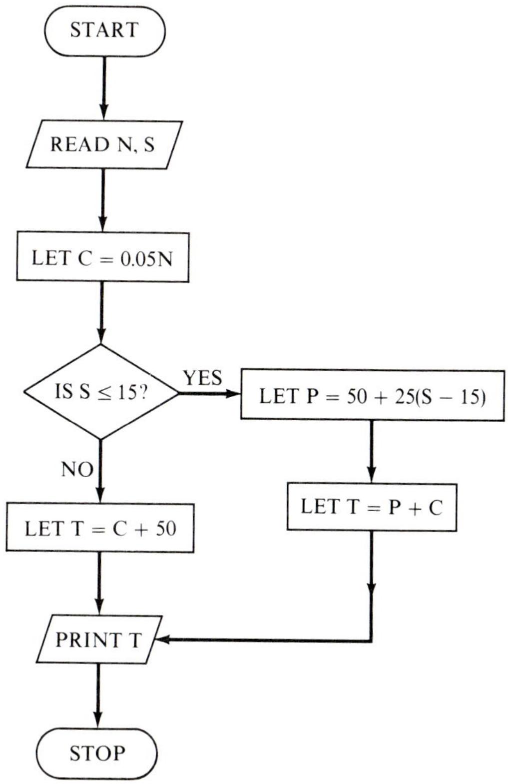

Comment When using the diamond symbol ◊ to indicate that a decision is to be made, we must tell the computer what to do in each alternative, that is, how to proceed if the answer to the question is yes or if the answer is no.

Example 5 Write a flowchart to print the square root of the first 25 (natural) numbers.

Solution In this case the flowchart will involve a loop and a decision symbol. Let us analyze the following flowchart. The first step tells the computer that $N = 1$ and then that $A = \sqrt{N} = \sqrt{1} = 1$. The computer prints this information. The computer must now decide whether $N > 25$ or not. If the answer is yes, then the program is done and the computer stops. If the answer is no, then the machine increases the value of N by 1. In BASIC computer language, we write $N = N + 1$. This is not to be interpreted as an algebraic equation. Instead, it is an instruction to the computer that the new value of N, wherever it appears from now on, is to be the old value with 1 added to it. With this change in mind, the computer now loops back to find the value of A. Since N is now 2, the value of A is $\sqrt{2}$, or 1.414. The same procedure is followed until the value of N reaches 25, when the program ends and the computer stops.

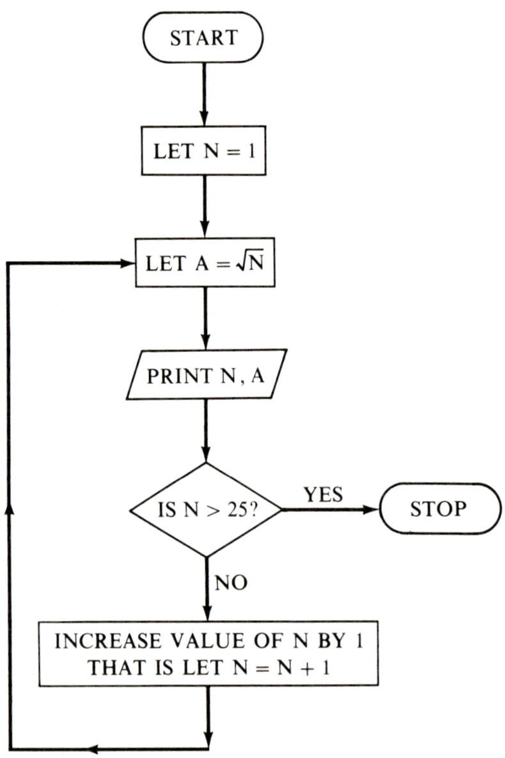

Example 6 What does the flowchart on the next page describe?

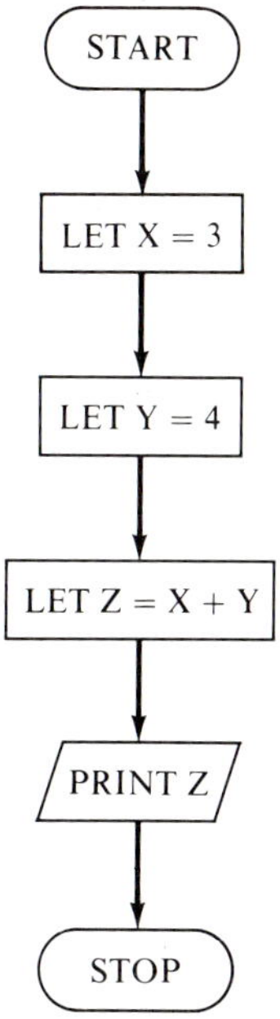

Solution The flowchart tells the computer to add 3 and 4 and to print the sum.

In the next section, when we actually write computer programs in BASIC, we suggest that you always draw a flowchart appropriate for the problem so that you can actually see how a logical thought or process becomes an instruction to be executed by the computer.

Exercises

1. Write a flowchart to tell the computer to find the average of the numbers 71, 63, 21, and 16.
2. Write a flowchart to determine the amount of interest earned on \$1000 deposited in a bank that pays 5% simple interest annually. (Remember, interest $= P \cdot R$, where $P = 1000$ and $R = 0.05$.)
3. Write a flowchart to determine the cost of sending a telegram of 30 words if the cost for the first 10 words is a flat 50¢ and the cost for each additional word over 10 is 4¢.
4. Write a flowchart to determine the cost of admission to an amusement park for a group of 8 adults and 12 children if the charge is \$4.00 per adult and \$2.75 per child.
5. Write a flowchart to determine the amount of money accumulated in two years if \$1000 is deposited in a bank that pays 5% interest compounded quarterly. (*Hint:* Use the compound interest formula discussed in Chapter 10.)
6. Write a flowchart to find the squares of the numbers from 1 to 50.
7. Write a flowchart to find the cubes of the numbers from 17 to 35 inclusive.

8. Write a flowchart to find the range of the numbers 16 and 48.

9. Write a flowchart to find the range of the numbers 16, 24, 81, 32, 71, and 12.

10. Write a flowchart to find the standard deviation of the numbers 16, 24, 81, 32, 71, and 12.

11. Write a flowchart to determine the total charge for a car carrying seven people across a river by a ferryboat if the charge is \$2 for the car and driver and 50¢ for each additional person.

12. Write a flowchart to determine your telephone bill if the monthly charge is \$6 for the first 75 calls and 10¢ for each additional call.

13. Write a flowchart to convert a Fahrenheit temperature reading to a centigrade reading.

14. Write a flowchart to determine a salesman's commission if it is dependent on the sale price of the item sold as shown in the accompanying table.

Sale price	Commission
Less than \$100	3%
From \$100 to \$300	7%
Over \$300	9%

15. Write a flowchart to help a bank determine the total charge to its checking account customers if it charges the following fees: \$2 monthly service charge, 5¢ for each check written, and \$5 for each overdraft check that is presented for payment. (An overdraft agreement allows the bank to honor your checks even though you do not have sufficient funds in your account to cover the check.)

★ 16. Write a flowchart to determine the maximum finance charge that can be charged by banks and oil companies on any outstanding indebtedness as shown in the table.

State	Average daily balance	Monthly rate of interest
Ohio	Less than \$400	$1\frac{1}{2}\%$
	\$400 and over	1%
Minnesota	All	1%
New York	Less than \$500	$1\frac{1}{2}\%$
	\$500 and over	1%
Arkansas	All	$\frac{5}{6}\%$

★ 17. Write a flowchart to solve a quadratic equation using the quadratic formula

$$x = \frac{-b \pm \sqrt{b^2 - 4ac}}{2a}$$

18. Write a flowchart to indicate how Professor Giorelli computes each student's term grade if she gives three tests plus a midterm and a final,

where the midterm counts twice as much as a test and the final counts three times as much as a test.

19. Write a flowchart to determine the sales tax on an item purchased in New York City if the tax rate is 4% for New York State residents plus an *additional* 4% for residents of New York City.

20. Write a flowchart to compute a worker's salary when he works 70 hours one week and his salary is determined as shown in the table. The hourly rate of pay is $6.

Hours worked	Rate of pay
0–40 hours	Regular hourly rate
Between 40 and 60 hours	Double the hourly rate for each hour over 40
Over 60 hours	Triple the hourly rate for each hour over 60

In Exercises 21–25 tell what each flowchart does.

21.

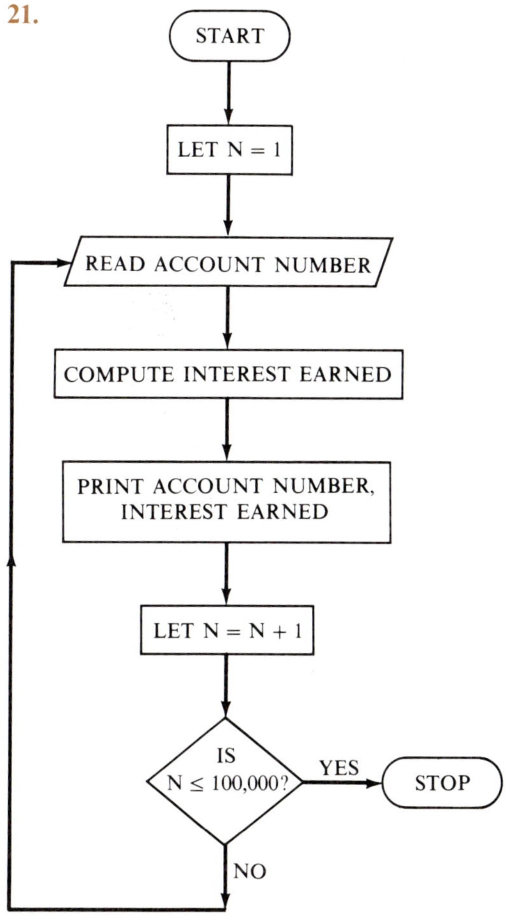

22.

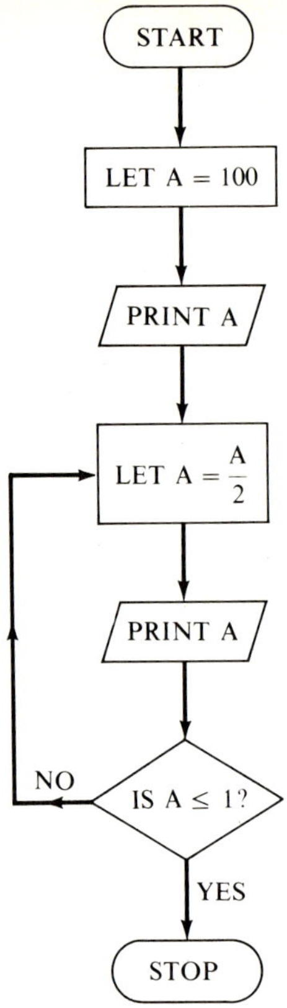
START
LET A = 100
PRINT A
LET A = A/2
PRINT A
IS A ≤ 1?
NO
YES
STOP

23.

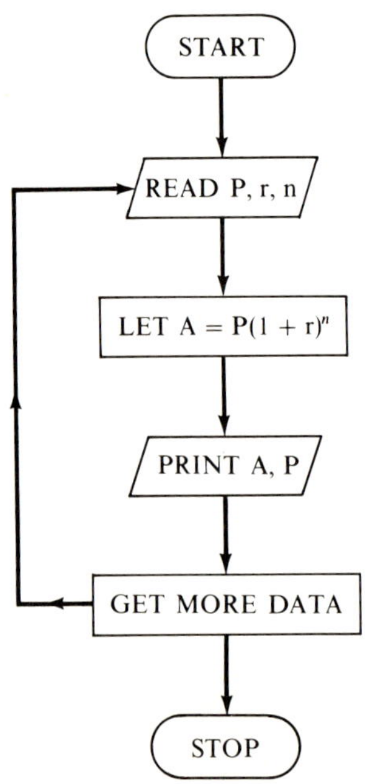
START
READ P, r, n
LET A = P(1 + r)^n
PRINT A, P
GET MORE DATA
STOP

24.

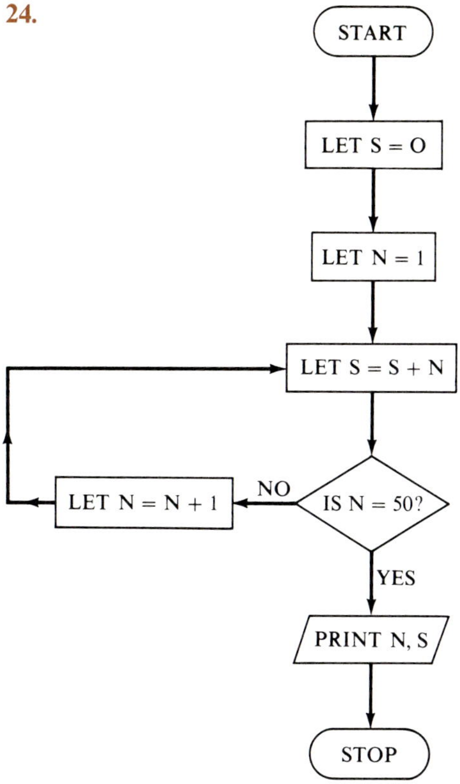
START
LET S = O
LET N = 1
LET S = S + N
IS N = 50?
NO
LET N = N + 1
YES
PRINT N, S
STOP

25.

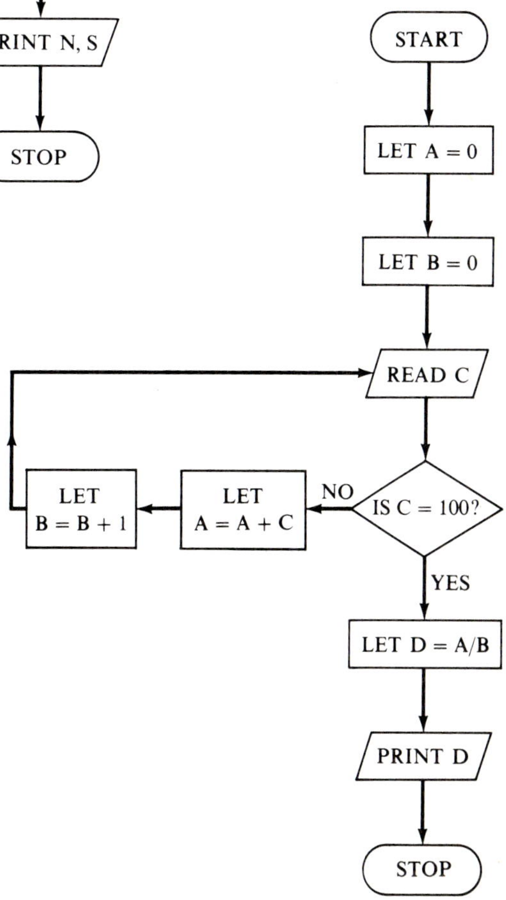
START
LET A = 0
LET B = 0
READ C
IS C = 100?
NO
LET A = A + C
LET B = B + 1
YES
LET D = A/B
PRINT D
STOP

12.6 BASIC

As indicated earlier, a computer is capable of doing only what it has been programmed to do. Thus, when computers were first developed, programmers had to spend a considerable amount of their time writing programs in a language that the computer could understand. This job was very time-consuming and very prone to errors. In due time the language of FORTRAN was developed to serve as a translation language. Soon other languages were developed. Today there are more than 250 different computer languages, each serving a particular purpose in science or in industry.

In this section we discuss BASIC, one of the more popular computer languages in use today. This language was developed by John Kemeny at Dartmouth College during the 1960s. The language BASIC is easy to learn because it is similar in many respects to ordinary English and mathematics. As indicated earlier, however, a computer is merely a heap of metal and is not capable of thinking on its own. Thus, a computer program must be written with great care, no matter what computer language is involved. Often when writing a sentence or composition in English, the meaning is clear even if a comma, quotation marks, or the like is omitted. The same is not true for a computer; the computer will not overlook minor mistakes.

It should be pointed out that our discussion is not intended to be a complete and detailed analysis of the BASIC language. Our discussion is merely intended to give you the feeling of the kind of detail that is necessary to get a program to run on a machine. A more complete discussion can be found in any manual on BASIC.

To get us started, let us consider the Abel Office Equipment Company, which rents photocopying machines for a flat rental fee of \$100 a month plus a charge of 4¢ for each copy made. Since each customer pays the same rental fee of \$100 and the same 4¢ for each copy made, the company is interested in writing a BASIC program that will determine the amount of money to bill each customer depending on the number of copies made. How can such a program be written?

As we mentioned in the last section, it is usually advisable to first draw a flowchart that indicates the logical sequence to be followed by the computer. To be specific, let us solve the following example.

Example 7 Suppose one customer of the Abel Office Equipment Company made 1505 copies during a month on the rented photocopier. Write a BASIC program to determine the amount of money that the company should bill the customer.

Solution Let N be the number of copies made during the month and let C be the charge for these N copies. Then a flowchart and a BASIC program that can be used to determine the amount of money, A, that the customer should be billed are shown here.

Flow Chart

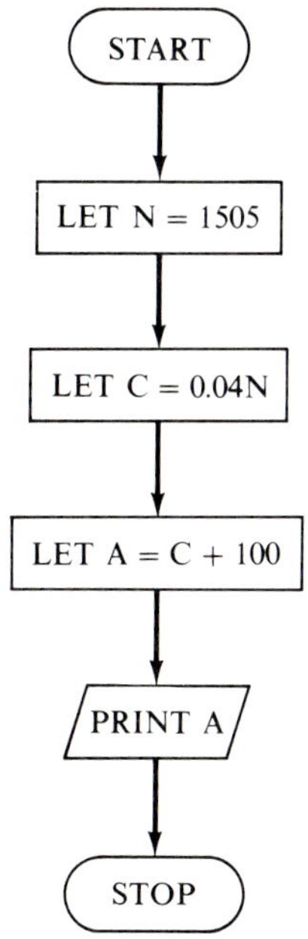

BASIC Program

```
     COPIER
100  LET N = 1505
200  LET C = N * 0.04N
300  LET A = C + 100
400  PRINT A
500  END
     RUN
     160.20
```

Although this BASIC program will accomplish the job, there are many steps that should be added so that the program is really useful. As we proceed with our discussion of this program, we will add extra steps, so that by the time we are done, we will have a "useful" program. In each case we will indicate why the extra step (instruction to the computer) was added.

Let us now analyze our simple program. We notice that this program has been named COPIER. Each program run on the computer must have a name that can be no longer than six characters. Although we are free to choose any name we want for any program, it is best to select a name that will identify the program being run. In this case we have selected the six-letter word COPIER as the name of the program for obvious reasons.

Notice that each line in the program has a number. We can use any number from 1 to 99999 as the line number provided the numbers are chosen in numerical order with respect to the order of the steps and each line number is different. Most programmers skip numbers between lines to allow for the insertion of other lines between two existing lines without having to change all the line numbers in the program. Skipping numbers also makes correction a rather simple process. For example, in this case, if the company decides to charge 6¢ a copy instead of 4¢ we can simply insert the new line 250.

```
     COPIER
100  LET N = 1505
200  LET C = N * 0.04
250  LET C = N * 0.06
300  LET A = C + 100
400  PRINT A
500  END
     RUN
     190.30
```

The computer will now use this new line 250 to replace the previous value of C that was obtained when line 200 was used.

Line 100 contains a LET *statement*. This is a very important instruction, since it tells the computer to replace N by 1505. Similarly, line 200 tells the computer to replace C by the value obtained when N is multiplied by 0.04. Line 300 tells the computer that A is to be replaced by the value obtained for C + 100. In BASIC an instruction like N = N + 1 is also perfectly meaningful, since it tells the computer that N now has the value N + 1.

Line 300, 300 LET A = C + 100, has two types of characters, a variable and a constant. A *constant* is a number that never changes, whereas a *variable* may change in value as the problem proceeds. In BASIC a variable name can consist of either a single letter or a single letter followed by a single digit. In our example N, C, and A are one-character variable names. Examples of two-character BASIC names are X6, A7, and B3.

In BASIC all mathematical calculations are performed much like ordinary algebraic or arithmetic calculations. The following symbols are used in BASIC.

BASIC symbol	Meaning	BASIC expression	Algebraic expression
+	Addition	A + B	$a + b$
−	Subtraction	A − B	$a - b$
=	Equal	A = B	$a = b$
*	Multiplication	A * B	ab
/	Division	A/B	$\frac{a}{b}$, or $a \div b$
↑	Powers	A ↑ 3	a^3
( )	Grouping symbol	(A + B)/3	$\frac{a+b}{3}$

In BASIC each symbol must be indicated; nothing is taken for granted. Thus, while we may write *ab* in ordinary algebra to stand for *a* times *b*, in BASIC we must write A * B. AB does not mean A times B.

Generally speaking, when the computer has to perform several calculations, powers are evaluated first, then multiplications and divisions (from left to right), and finally additions and subtractions (again from left to right). Parentheses are used to group or to change the order of operations in a problem. Thus, the statement A + B/3 tells the computer to first divide B by 3 and then to add the results to A as opposed to the statement (A + B)/3, which tells the computer to first add A and B together and then to divide the sum by 3.

Example 8 Which of the following are correct BASIC statements? If incorrect, explain why.

a) 100 LET A = B + 400
b) 300 LET X34 = B + 75
c) LET T = 5
d) 400 LET A + B = 31
e) 500 LET N = N + 1

Solution

a) Correct.
b) Incorrect, the variable has three characters. In BASIC at most two-character variables are permitted.
c) Incorrect, there is no line number.
d) Incorrect, we cannot have two variables on the left of the equal sign.
e) Correct.

Example 9 Translate each of the following mathematical equations or formulas into BASIC statements.

a) $y = 3x + 2$

b) $a = \frac{2b + c}{3}$

c) $a = \frac{b^2}{3}$

d) $\mu = \frac{x_1 + x_2 + x_3 + x_4}{4}$

e) $y = \frac{(x + 2)^3}{7x}$

f) $y = \sqrt{x + 1}$

g) $x = \frac{-b + \sqrt{b^2 - 4ac}}{2a}$

h) $v = \frac{4}{3}\pi r^3$ (Use $\pi = 3.14159$)

Solution The BASIC statements corresponding to each of the mathematical equations or formulas are:

```
a) 100  LET Y = 3 * X + 2
b) 120  LET A = (2 * B + C)/3
c) 130  LET A = (B ↑ 2)/3
d) 140  LET U = (X1 + X2 + X3 + X4)/4
e) 150  LET Y = (X + 2) ↑ 3/(7 * X)
f) 160  LET Y = (X + 1) ↑ (1/2) or
   161  LET Y = (X + 1) ↑ 0.5
g) 170  LET X = (- B + (B ↑ 2 - 4 * A * C) ↑ (1/2))/(2 * A)
h) 180  LET V = (4 * 3.14159 * R ↑ 3)/3
```

Example 10 What is wrong with each of the following BASIC statements? Can you suggest a correction?

a) 100 LET A = 2(B + 7)
b) 110 A = 2
c) 400 LET A = (X + 5) ↑ 1/2

Solution a) There is no multiplication symbol. Possible correction:

```
101   LET A = 2 * (B + 7)
```

b) There is no LET statement. Correction:

```
111   LET A = 2
```

c) If square root is intended, then parentheses must be used. Possible correction:

```
401   LET A = (X + 5) ↑ (1/2)
```

As it stands, statement 400 tells the computer to raise X + 5 to the first power and then divide the result by 2; in other words, evaluate

$$\frac{X+5}{2}$$

Let us now return to line 400 of the original program in Example 7. It is

```
400   PRINT A
```

When the computer determines that the cost is $160.20, it will simply print $160.20. We could change line 400 to read as follows:

```
401   PRINT N, C, A
```

This would cause the computer to print the following:

```
1505      60.20      160.20
```

The commas in the PRINT *statement* tell the computer to skip 15 spaces called *fields* before printing the next numbers or words. By using this type of PRINT statement, we are able to have the original data printed out along with the result.

It is also possible to use a PRINT statement to have the computer print out words. We simply enclose the message to be printed within quotation marks followed by a semicolon. The semicolon tells the computer what to print next and to disregard the predetermined *fields*. Thus, in our case, we could replace line 400 by

```
402   PRINT "TOTAL COST TO CUSTOMER IS"; C
```

The computer output for this would then be

```
TOTAL COST TO CUSTOMER IS 160.20
```

Suppose the Abel Office Equipment Corporation wants to use this program to calculate the total cost for several of its customers. With the program presently available to us, a new program would have to be reentered with different data for each customer. How can we use this same program and change only the data? This can be accomplished by using READ and DATA statements as shown in Example 11.

Example 11 Write a BASIC program that will determine the total cost to a customer of the Abel Office Equipment Corporation who made 1505 photocopies during the month. The program will also be used to compute the costs to be charged to other customers.

Solution The BASIC program accomplishing this is as follows:

```
    COPIER
100 PRINT "CUSTOMER", "COPIES USED", "TOTAL COST"
200 PRINT
300 READ M, N
400 LET C = N * 0.04
500 LET A = C + 100
600 PRINT M, N, A
700 DATA 1, 1505
800 END
    RUN
```

The READ *statement* tells the computer to find the values of M and N in the DATA *statement*. We always list the variables to be read after the READ statement and the values of these variables in the DATA statement. The numbers in the DATA statement must be separated by commas, and the number of data terms must be a multiple of the number of variables used. DATA statements are usually placed right before the END statement.

This BASIC program will result in the following printout:

```
CUSTOMER      COPIES USED      TOTAL COST

1             1505             160.20
```

Notice that line 100 tells the computer to print headings, whereas line 200, which contains only the word PRINT, causes the computer to skip a line. Line 800 is the END *statement*. Every program written in BASIC must have an END statement to tell the computer that the program has been completed.

You will also notice the use of the RUN *statement*. We never assign a line number to a RUN statement. A RUN statement is always placed after an END statement and tells the computer to begin executing the program.

Exercises

26. Which of the following are not acceptable variable names in BASIC?

a) J2 b) 3X c) J32
d) DC10 e) JIP f) M17

27. Which of the following are correct BASIC statements? If incorrect, explain why.

a) A = 100
b) 100 LET X ↑ 2 = −4 * Y + 3
c) 105 READ A, B, C
   110 DATA 70, 81, 46, 32, 99
d) 108 LET X = Y + −3
e) 150 PRINT THE ANSWER IS X
f) 110 PRINT "THE AREA IS
g) 130 PRINT "MEAN, STANDARD DEVIATION"

28. Translate each of the following mathematical equations or formulas into a BASIC statement.

a) $x = 2y + 3z$ b) $y = \dfrac{4x + 3}{7}$

c) $z = \sqrt{3x + 7}$ d) $20x^2 = 5y^3 + 7$

e) $z = \dfrac{x^2 + 3y^3}{7xy}$ f) $a = (b + c^3 + dx^4)^3$

29. The following are BASIC statements. Rewrite them as ordinary algebraic expressions.

a) 100 LET A = 5 * 3 ↑ 2
b) 101 LET B = X + 4 * (5 + 8)
c) 102 LET C = X − 3 ↑ 0.5
d) 103 LET D = (2 + 3 ↑ 2)/(4 * 5)
e) 104 LET E = A * B/(C + D)
f) 105 LET F = A * B/C + D

30. What is the computer printout at the end of the following BASIC programs?

a)
```
    EXMPLE
100 READ A, B, C, D
150 LET X = (A + B)/(C − D)
200 PRINT X
250 DATA 10, 70, 90, 40
300 END
    RUN
```

```
b)      EXMPLE
   200  READ A
   250  READ B
   275  LET C = A * B
   300  PRINT A
   350  PRINT B
   400  PRINT C
   425  DATA 750
   450  DATA 21
   500  END
        RUN
```

12.7 Some Programming Examples

All the programs we discussed in the previous section were relatively simple. None of them required the computer to make any decisions or to repeat the same type of calculation several times. The computer can be programmed to perform the same type of operation many times by using a GO TO *statement* as shown in Example 12.

Example 12 Write a BASIC program that would allow the Abel Office Equipment Corporation to determine the total cost to customer 1, who made 1505 photocopies, customer 2, who made 1227 copies, and customer 3, who made 1807 copies during the month.

Solution The BASIC program accomplishing this is shown here. Notice that in

```
     COPIER
100  READ M, N
200  LET C = N * 0.04
300  LET A = C + 100
350  PRINT "CUSTOMER NUMBER"; M
375  PRINT "NUMBER OF COPIES USED"; N
400  PRINT "TOTAL COST $"; A
425  PRINT
450  PRINT
500  PRINT
600  PRINT
700  GO TO 100
800  DATA 1, 1505, 2, 1227, 3, 1807
900  END
     RUN
```

line 700 we have added a GO TO statement. This causes the computer to go to the line number indicated. After the computer determines the total charge for customer 1, we want the computer to go to the READ statement, read new values for M and N, and determine the total cost for these values. Thus, we insert the GO TO 100 statement right after the PRINT statement and before the DATA statement. The computer goes to line 100, computes the total cost for customer 2, and then goes through the same procedure for customer 3. The computer is now in a loop.

When all the data in line 800 are used up, the computer will stop and tell us OUT OF DATA IN LINE 100. This means that there are no more numbers in the data list to read.

The previous program will cause the computer to print out the following results:

```
CUSTOMER NUMBER 1
NUMBER OF COPIES USED 1505
TOTAL COST $160.20



CUSTOMER NUMBER 2
NUMBER OF COPIES USED 1227
TOTAL COST $149.08



CUSTOMER NUMBER 3
NUMBER OF COPIES USED 1807
TOTAL COST $172.28



OUT OF DATA IN LINE 100
```

Example 13 Write a BASIC program to compute the area of three different circles, one with a radius of 7, one with radius 9, and one with radius 10.6. (Remember that the area of a circle is given by $A = \pi r^2$ or $A = 3.14159r^2$ where r is the radius of the circle and where we have used 3.14159 as the value of π.)

Solution The BASIC program that will accomplish this is given here.

```
     AREA
100  READ R
125  LET A = 3.14159 * R ↑ 2
150  PRINT "IF RADIUS OF CIRCLE IS"; R
175  PRINT "AREA OF CIRCLE IS"; A
200  PRINT
300  PRINT
400  PRINT
500  PRINT
600  GO TO 100
700  DATA, 7, 9, 10.6
800  END
     RUN
```

When the computer executes this program, it will print out:

```
IF RADIUS OF CIRCLE IS 7
AREA OF CIRCLE IS 153.93791




IF RADIUS OF CIRCLE IS 9
AREA OF CIRCLE IS 254.46879




IF RADIUS OF CIRCLE IS 10.6
AREA OF CIRCLE IS 352.9890524




OUT OF DATA IN LINE 100
```

Example 14 Write a BASIC program to compute the squares and cubes of the first five numbers.

Solution As indicated earlier, it is advisable to first draw a flowchart to picture the logical steps that are necessary to execute the program. Using the flowchart symbols of Section 12.5, we have

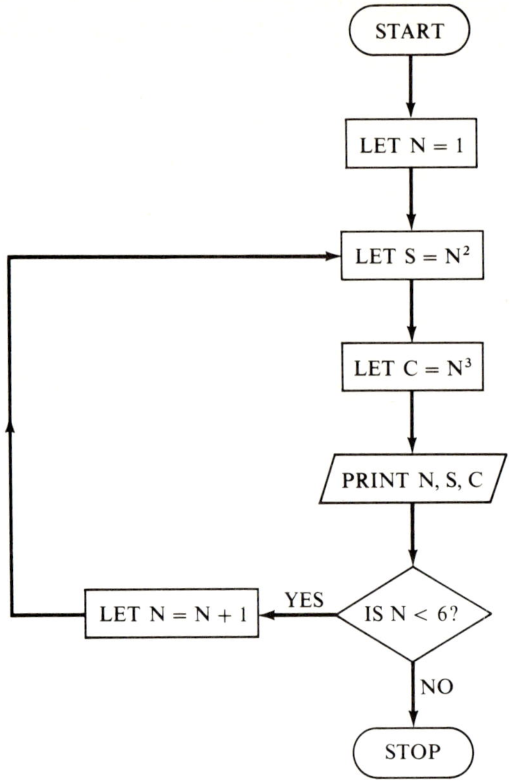

A BASIC program based on this flowchart is as follows:

```
      SQUARE
100   PRINT
200   PRINT
300   LET N = 1
400   LET S = N ↑ 2
500   LET C = N ↑ 3
525   PRINT "IF NUMBER IS"; N
550   PRINT "SQUARE IS"; S
575   PRINT "CUBE IS"; C
600   PRINT
625   PRINT
650   PRINT
700   LET N = N + 1
800   IF N < 6 THEN 400
900   END
      RUN
```

Notice that in line 800 we have used an IF THEN *statement*. Such a statement tells the computer to decide whether the relation given is true or false. If the relation given is true, the computer will jump to the line number given in the statement. If the relation is false, the computer proceeds to the next BASIC statement in sequence.

When the computer executes this program, the following will be printed out:

```
IF NUMBER IS 1
SQUARE IS 1
CUBE IS 1



IF NUMBER IS 2
SQUARE IS 4
CUBE IS 8



IF NUMBER IS 3
SQUARE IS 9
CUBE IS 27



IF NUMBER IS 4
SQUARE IS 16
CUBE IS 64



IF NUMBER IS 5
SQUARE IS 25
CUBE IS 125
```

For Example 14, a BASIC program can also be written which will have the computer print out the results in the following format:

```
NUMBER      SQUARE OF NUMBER      CUBE OF NUMBER
   1                1                    1
   2                4                    8
   3                9                   27
   4               16                   64
   5               25                  125
```

Table 12.2 summarizes the various relational *symbols* that can be used in BASIC IF THEN statements.

Table 12.2 Some BASIC Symbols and Their Meanings

BASIC Symbol	Meaning
=	Is equal to
<	Is less than
>	Is greater than
< =	Is less than or equal to
> =	Is greater than or equal to
< >	Is not equal to

Example 15 The Cal Finance Company charges different interest rates on outstanding monthly balances depending on the amount of money involved as shown in the table. Write a BASIC program to determine the finance charge for five accounts with outstanding monthly balances of \$58.17, \$423.17, \$312.98, \$1400.16, and \$817.64.

Outstanding monthly balance	Annual interest rate
Up to \$200.00	18%
From \$200.01 to \$500.00	15%
Above \$500.00	12%

Solution It is best to draw a flowchart for the problem first. Let B be the outstanding balance and let I be the finance charge. We then have the following flowchart for this problem:

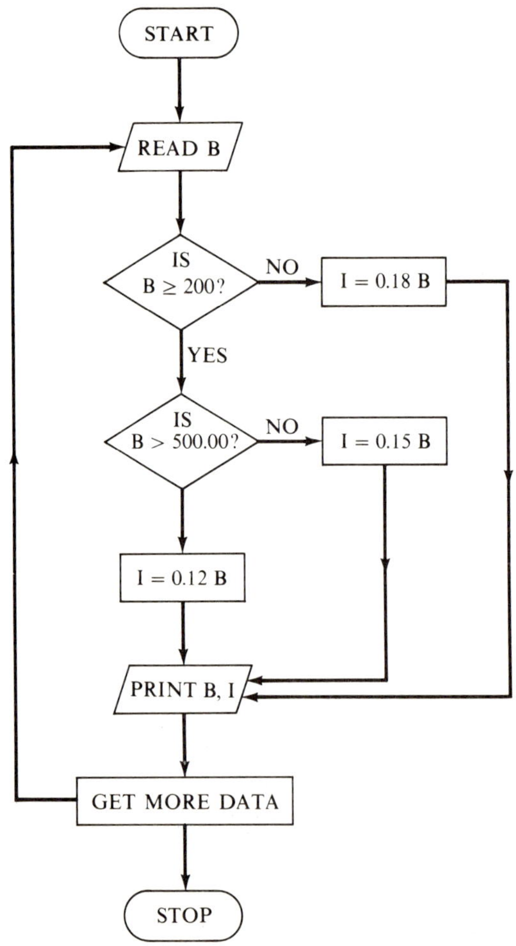

Using this flowchart, we set up the following BASIC program:

```
      CHARGE
 100  PRINT
 200  PRINT
 300  READ B
 400  IF B > 200.00 THEN 700
 500  LET I = B * 0.18
 600  GO TO 1100
 700  IF B > 500.00 THEN 1000
 800  LET I = B * 0.15
 900  GO TO 1100
1000  LET I = B * 0.12
1100  PRINT "IF OUTSTANDING BALANCE IS $"; B
1125  PRINT "FINANCE CHARGE IS $"; I
1130  PRINT
1150  PRINT
1175  PRINT
1180  PRINT
1200  GO TO 300
1300  DATA 58.17, 423.17, 312.98, 1400.16, 817.64
1400  END
      RUN
```

When this program is executed, the computer printout will be as follows:

```
IF OUTSTANDING BALANCE IS $58.17
FINANCE CHARGE IS $10.47




IF OUTSTANDING BALANCE IS $423.17
FINANCE CHARGE IS $63.48




IF OUTSTANDING BALANCE IS $312.98
FINANCE CHARGE IS $46.95
```

```
IF OUTSTANDING BALANCE IS $1400.16
FINANCE CHARGE IS $168.02



IF OUTSTANDING BALANCE IS $817.64
FINANCE CHARGE IS $98.12



OUT OF DATA IN LINE 300
```

Exercises

Length	Width
16	28
41	18.3
12	842.78
17	19
44	1235

31. Write a BASIC program for computing the area of several rectangles with the lengths and widths shown in the table. What will the computer print out?

32. Write a BASIC program for converting the following Fahrenheit temperature readings to equivalent centigrade temperature readings: 10°, 25°, 32°, 40°, 88°, and 98.6°. Indicate the computer printout for your program.

33. Write a BASIC program to find the mean and the standard deviation of the following numbers: 12, 17, 42, 19, 28, and 20. Indicate the computer printout for your program.

34. Write a BASIC program to compute the value of 10!.

35. Write a BASIC program to compute the square root of the integers from 1 to 15. Indicate the computer printout for your program.

36. Write a BASIC program to compute the cube root of the integers from 1 to 15. Indicate the computer printout for your program.

★ **37.** Write a BASIC program to compute the amount of money that will accumulate in a bank in n years when P dollars are deposited and the bank pays $r\%$ interest compounded quarterly.

38. Write a BASIC program to compute a salesman's commission if it is dependent on the sale price of the item as shown in the table.

Sale price	Commission
Up to $100	3%
From $100 to $300	7%
Over $300	9%

★ **39.** Write a BASIC program for the problem described in Exercise 16 earlier in this chapter.

★ **40.** Write a BASIC program for the problem described in Exercise 17.

41. Write a BASIC program for the problem described in Exercise 18.

42. Write a BASIC program for the problem described in Exercise 19.

43. Write a BASIC program for the problem described in Exercise 20.

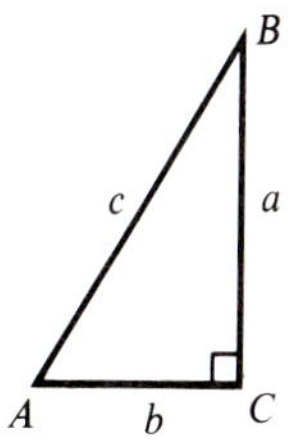

44. Write a BASIC program to compute the value of the hypotenuse c in the accompanying diagram. (*Hint:* Remember, in the right triangle pictured, $c = \sqrt{a^2 + b^2}$.)

45. Write a BASIC program to find the sum of the integers from 1 to 20. (*Hint:* Use the formula, sum of the integers from 1 to $n = n(n + 1)/2$.)

46. Write a BASIC program to find the sum of the squares of the integers from 1 to 10 inclusive.

47. Write a BASIC program to determine which of the following numbers is the largest: 17, 23, 14, 18, or 20.63.

48. Explain what the computer will print out when it executes the following programs.

a)

```
     EXMPLE
100  LET R = 7
200  LET PI = 3.14159
300  LET A = PI * R ↑ 2
400  PRINT A
500  END
     RUN
```

b)

```
     EXMPLE
100  READ X
200  READ Y
300  IF X < Y THEN 500
400  IF X > = Y THEN 700
500  LET C = X + 200
600  PRINT C
700  LET D = X − Y
800  PRINT D
900  DATA 700, 900
1000 END
     RUN
```

c)
```
     EXMPLE
100  READ A, B
200  IF A > B THEN 500
300  PRINT B; "IS BIGGER THAN"; A
400  GO TO 100
500  PRINT A; "IS BIGGER THAN"; B
600  GO TO 100
700  DATA 480, 360, 999, 999.01, -3, -4
800  END
     RUN
```

d)
```
     EXMPLE
100  LET A = 0
200  LET N = 1
300  LET A = A + N
400  IF A = 25 THEN 700
500  LET A = A + 1
600  GO TO 200
700  PRINT A
800  END
     RUN
```

e)
```
     EXMPLE
100  LET A = 1000
200  LET D = A/2
300  IF D < 1 THEN 600
400  LET D = D/2
500  GO TO 300
600  PRINT D
700  END
     RUN
```

49. Each of the following BASIC programs contains one or more mistakes. Locate the mistake and indicate how it can be corrected.

a)

```
    SAMPLE
100 READ A, B
200 LET N = A + B
300 PRINT A, B, N
400 DATA 5, 9, 8, 6, 7
```

b)

```
    AREA
100 LET R = 4
200 LET P = 3.14159
300 LET A = P(R ↑ 2)
400 PRINT “RADIUS, “AREA IS”
500 PRINT
600 PRINT R, A
700 END
    RUN
```

c)

```
100 LET N = 5
200 LET M = 25
400 IF M < N THEN 300
300 PRINT M
500 IF N = <M THEN 600
600 PRINT N
700 END
    RUN
```

d)

```
    EXMPLE
100 READ A, B
200 IF A > B, THEN 300
300 PRINT A “IS BIGGEST”
400 IF B < A THEN 500
500 PRINT B, “IS BIGGEST”
600 GO TO 100
700 DATA 40, 70, 60, 80, 100.7, 10.70
800 END
    RUN
```

50. What is the printout of the following BASIC programs?

a)
```
    EXMPLE
100 READ A
200 LET C = 2.54 * A
250 PRINT A; "INCHES EQUAL"; C; "CENTIMETERS"
300 PRINT
350 GO TO 100
375 DATA 7, 15, 36, 50, 80
400 END
    RUN
```

b)
```
    EXMPLE
300 READ A, B
400 LET C = A/B
500 PRINT "A/B = "; C
600 PRINT
700 GO TO 300
780 DATA 68, 4, 250, 4, 400, 3, 79, 7
800 END
    RUN
```

c)
```
    EXMPLE
100 READ A
150 IF A = 50 THEN 350
175 LET B = A ↑ 3
200 PRINT A, B
225 PRINT
250 GO TO 100
300 DATA 10, 20, 30, 40, 50
350 END
    RUN
```

d)

```
    EXMPLE
100 PRINT "NUMBER", "SQUARE", "SQUARE ROOT"
150 READ A
200 LET S = A ↑ 2
225 LET T = A ↑ (1/2)
250 PRINT
300 PRINT A, S, T
325 GO TO 150
350 DATA 1, 7, 19.6, 25.4, 48.32
375 END
    RUN
```

12.8 Summary

In this chapter we traced the development of the computer from its infancy to its present stage. We discussed numerous ways in which the computer is used today in a variety of fields. We also discussed how a computer works from both a hardware and a software point of view.

Among the many different computing languages available, we chose to discuss BASIC. This is the simplest language to learn. We indicated how a computer program is written, first making a flowchart and then translating it into the appropriate BASIC form. We actually analyzed several programs that you might find helpful in your studies.

Since computers are fast becoming part of our normal life, it is important to know how a computer works, its capabilities and its limitations. Our discussion was not intended to make expert programmers out of anybody. Rather, it was intended to give you an appreciation of what a computer is. Hopefully you should now be able to write flowcharts and some simple programs in BASIC and also be able to recognize what a program that has already been written in BASIC actually accomplishes.

Study Guide

You should be able to demonstrate your knowledge of the following ideas presented in this chapter by giving definitions, descriptions, or specific examples, Page references are given for each term so that you can check your answer.

Computer-assisted instruction (p. 504)
Computer (p. 506)
Napier's rods or Napier's bones (p. 507)
First-generation computers (p. 508)
Analog computers (p. 509)
Computer simulation (p. 504)
Abacus or counting board (p. 506)
Second-generation computers (p. 508)
Third-generation computers (p. 508)
Digital computers (p. 509)

Arithmetic unit (p. 510)
Compiler (p. 510)
Input unit (p. 510)
Memory unit (p. 510)
Chips (p. 511)
Decoder (p. 511)
FORTRAN (p. 511)
Output unit (p. 511)
Software (p. 511)
Flowchart symbols (p. 512)
BASIC (p. 522)
BASIC variable (p. 524)
PRINT statements (p. 526)
END and RUN statements (p. 527)
IF THEN statements (p. 533)

Base 2 (p. 510)
Control unit (p. 510)
Keypunches (p. 510)
Bits (p. 511)
COBOL (p. 511)
Flowchart (p. 511)
Hardware (p. 511)
Programming languages (p. 511)
Looping (p. 514)
LET statements (p. 524)
BASIC constant (p. 524)
READ statements (p. 527)
DATA statements (p. 527)
GO TO statements (p. 529)
IF THEN symbols (p. 533)

Manual for Quick Reference

The following list of symbols and key words gives the BASIC statements and symbols or flowchart symbols discussed in this chapter.

Flowchart symbols

Start or end of program

Input or output of data

Processing of program

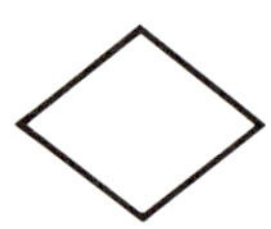
Decision symbol

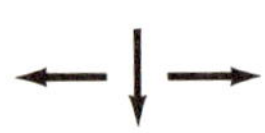
Direction to go in flowchart

BASIC statements

LET tells the computer that a variable has a particular value.

PRINT tells the computer to print out the results that it obtained.

READ tells the computer to obtain data from the statement in sequential order.

DATA stores the data to be used as input information.

GO TO tells the computer to go out of sequence and go to the line number indicated.

IF THEN tells the computer to make a decision and to proceed according to the answer to the decision.

END tells the computer that the program is done. It is always the last numbered statement in the program.

RUN tells the computer to start executing the program.

BASIC operations	
+	Addition
−	Subtraction
*	Multiplication
/	Division
↑	Exponents (powers)
( )	Grouping symbol
=	Equality

BASIC decision symbols	
=	Is equal to
<	Is less than
>	Is greater than
< =	Is less than or equal to
> =	Is greater than or equal to
< >	Is not equal to

Mastery Tests

FORM A

1. True or false: A DATA statement in a BASIC program does not have to be placed in any special sequence in the program.
2. True or false: In a BASIC program, a GO TO statement must be placed between a PRINT statement and a DATA statement.
3. True or false: In a BASIC program, the last statement must always be a RUN statement.
4. Can a program name be longer than six characters?
5. The word BASIC stands for several words. What are these words?
6. In what base do all computers do their mathematical calculations?
7. What do we call the part of the computer that receives and interprets all the instructions received?
8. What are the physical parts of the computer called?
9. Computers store information on tiny doughnut-shaped pieces of magnetic cores called ____________.
10. What do we call a diagram indicating the logical sequence of steps to be followed by the computer?
11. AB is the correct BASIC expression for A times B. True or false?
12. Is the statement

    $$100 \quad \text{LET X} = \frac{1}{2}$$

    a meaningful statement in BASIC? Explain your answer.
13. What is the value of the following calculation performed in BASIC?

    (2 ↑ 4) * 5
14. Rewrite the following algebraic calculation as a BASIC statement:

    $$\left(\frac{4 + 5}{9 \div 3}\right)^3$$

15. Which of the following is a valid BASIC variable name?
  a) B-10   b) BVD   c) U2   d) 10X

FORM B

1. Consult an encyclopedia or a detailed computer textbook to find the difference between digital and analog computers.

2. A first-degree equation of the form $ax + b = c$ has as its solution

$$x = \frac{c - b}{a}$$

Write a BASIC program that can be used to solve each of the following first-degree equations.

  a) $2x + 5 = 11$ b) $2x - 7 = 3$
  c) $0.2x + 3 = 5$ d) $\frac{7x}{4} - 8 = 6$
  e) $\frac{2x}{5} + 3 = 5$ f) $4x - 2 = 0.8$

3. A quadratic equation of the form $ax^2 + bx + c = 0$ has as one of its solutions

$$x = \frac{-b + \sqrt{b^2 - 4ac}}{2a}$$

Using the BASIC program written for Exercise 40, solve the following quadratic equations.

  a) $x^2 - 5x + 6 = 0$ b) $2x^2 + x - 21 = 0$
  c) $x^2 - 4x - 5 = 0$ d) $x^2 - 4x - 3 = 0$
  e) $x^2 + 5x + 1 = 0$ f) $3x^2 - 5x - 2 = 0$

4. Write a BASIC program for calculating the number of permutations of n things n at a time, that is, for evaluating $n!$.

5. Write a BASIC program to find the product of the integers from 13 to 23.

6. Write a BASIC program to compute the take-home pay of several workers of the Acme Metal Company, which pays its workers according to the following rules.
   a) The number of hours worked is multiplied by $5, which is the hourly rate.
   b) The number of exemptions claimed is multiplied by 8.
   c) The result obtained in step b is subtracted from the result in step a.
   d) The result obtained in step c is multiplied by 0.10 to get the federal tax.
   e) The take-home pay is obtained by subtracting the result of step d from the result of step c.

Honor's Project[2]

1. a) Write a BASIC program for solving a linear programming problem of the type

$$Ax + By \leq E$$
$$Cx + Dy \leq F$$
$$\max z = Gx + Hy$$

by the simplex method.

b) Using the results obtained in part a, solve the following linear programming problem.

$$0.2x + 0.4y \leq 30$$
$$0.2x + 0.2y \leq 20$$
$$x \geq 0, y \geq 0$$
$$\max z = 10x + 12y$$

2. Read the article "The Role of the Computer" by Louis N. Ridenour in *Mathematics—An Introduction to Its Spirit and Use*, Scientific American, edited by Morris Kline, 1979. In this article the author indicates how large analog computers are being used to simulate the performance of guided missiles, aircraft, ships, submarines, and other machines.

3. Read the article "Computer Logic and Memory" by David C. Evans in *Mathematics—An Introduction to Its Spirit and Use*, Scientific American, edited by Morris Kline, 1979. In this article the author indicates how the computer actually converts decimal digits to binary digits and also traces how computer logic and memory have evolved.

4. Suppose we wish to add the fractions (a/b) and (c/d). (Assume b and d are not 0.) In algebra we are told that

$$\frac{a}{b} + \frac{c}{d} = \frac{ad + bc}{bd}$$

Write a BASIC program that will add the fractions $\frac{2}{3}$ and $\frac{5}{7}$ and express the answer in fraction form, not as decimals.

5. Write a BASIC program to arrange the following numbers in order from lowest to highest: 61, 42, 83, 79, 58, 21, and 114.

6. The following are the term averages for seven students in Professor Chinn's class: 49, 82, 68, 72, 98, 59, and 42. The professor intends to give A's to those students whose average is between 90 and 100, B's to those whose average is between 80 and 90, and so on, and F's to those whose average is less than 60. Write a BASIC program that will tell the professor how many A's, B's, C's, D's, and F's he has assigned.

[2] This section is slightly more difficult and may require some additional time and thought; it can be omitted without affecting the continuity of the text.

Suggested Further Reading

Adler, I., *Thinking Machines*, New York: John Day, 1961.

Bernstein, J., *The Analytical Engine*, New York: Random House, 1964.

Computers and Computation (readings from *Scientific American*), San Francisco: Freeman, 1971.

Corliss, W. R., *Computers*, Oak Ridge, Tenn.: U.S. Atomic Energy Commission, 1967.

Eames, C., and R. Eames, *A Computer Perspective*, Cambridge; Harvard University Press, 1973.

Halacy, D. S., Jr., *Computers, The Machines We Think With*, New York: Dell, 1964.

Hawkes, N., *The Computer Revolution*, World of Science Library, New York: Dutton, 1972.

Kemeny, John G., *Man and the Computer*, New York: Scribner, 1972.

Nolan, Richard I., *Introduction to Computing the BASIC Language*, New York: Holt, Rinehart and Winston, 1969.

Reichardt, J., *The Computer in the Art*, New York: Van Nostrand, 1971.

Rosen, S., "Electronic Computers: A Historical Survey," *Computing Surveys*, Vol. 1, no. 1, 1969.

Springer, C. H., et al., *The Mathematics for Management Series*, Homewood, Ill.: Irwin, 1966.

Appendix

Statistical Tables

Table 1 Factorials

n	$n!$
0	1
1	1
2	2
3	6
4	24
5	120
6	720
7	5,040
8	40,320
9	362,880
10	3,628,800
11	39,916,800
12	479,001,600
13	6,227,020,800
14	87,178,291,200
15	1,307,674,368,000
16	20,922,789,888,000
17	355,687,428,096,000
18	6,402,373,705,728,000
19	121,645,100,408,832,000
20	2,432,902,008,176,640,000

Table 2 Binomial Probabilities

		p										
n	*x*	0.05	0.1	0.2	0.3	0.4	0.5	0.6	0.7	0.8	0.9	0.95
2	0	0.902	0.810	0.640	0.490	0.360	0.250	0.160	0.090	0.040	0.010	0.002
	1	0.095	0.180	0.320	0.420	0.480	0.500	0.480	0.420	0.320	0.180	0.095
	2	0.002	0.010	0.040	0.090	0.160	0.250	0.360	0.490	0.640	0.810	0.902
3	0	0.857	0.729	0.512	0.343	0.216	0.125	0.064	0.027	0.008	0.001	
	1	0.135	0.243	0.384	0.441	0.432	0.375	0.288	0.189	0.096	0.027	0.007
	2	0.007	0.027	0.096	0.189	0.288	0.375	0.432	0.441	0.384	0.243	0.135
	3		0.001	0.008	0.027	0.064	0.125	0.216	0.343	0.512	0.729	0.857
4	0	0.815	0.656	0.410	0.240	0.130	0.062	0.026	0.008	0.002		
	1	0.171	0.292	0.410	0.412	0.346	0.250	0.154	0.076	0.026	0.004	
	2	0.014	0.049	0.154	0.265	0.346	0.375	0.346	0.265	0.154	0.049	0.014
	3		0.004	0.026	0.076	0.154	0.250	0.346	0.412	0.410	0.292	0.171
	4			0.002	0.008	0.026	0.062	0.130	0.240	0.410	0.656	0.815
5	0	0.774	0.590	0.328	0.168	0.078	0.031	0.010	0.002			
	1	0.204	0.328	0.410	0.360	0.259	0.156	0.077	0.028	0.006		
	2	0.021	0.073	0.205	0.309	0.346	0.312	0.230	0.132	0.051	0.008	0.001
	3	0.001	0.008	0.051	0.132	0.230	0.312	0.346	0.309	0.205	0.073	0.021
	4			0.006	0.028	0.077	0.156	0.259	0.360	0.410	0.328	0.204
	5				0.002	0.010	0.031	0.078	0.168	0.328	0.590	0.774
6	0	0.735	0.531	0.262	0.118	0.047	0.016	0.004	0.001			
	1	0.232	0.354	0.393	0.303	0.187	0.094	0.037	0.010	0.002		
	2	0.031	0.098	0.246	0.324	0.311	0.234	0.138	0.060	0.015	0.001	
	3	0.002	0.015	0.082	0.185	0.276	0.312	0.276	0.185	0.082	0.015	0.002
	4		0.001	0.015	0.060	0.138	0.234	0.311	0.324	0.246	0.098	0.031
	5			0.002	0.010	0.037	0.094	0.187	0.303	0.393	0.354	0.232
	6				0.001	0.004	0.016	0.047	0.118	0.262	0.531	0.735
7	0	0.698	0.478	0.210	0.032	0.028	0.008	0.002				
	1	0.257	0.372	0.367	0.247	0.131	0.055	0.017	0.004			
	2	0.041	0.124	0.275	0.318	0.261	0.164	0.077	0.025	0.004		
	3	0.004	0.023	0.115	0.227	0.290	0.273	0.194	0.097	0.029	0.003	
	4		0.003	0.029	0.097	0.194	0.273	0.290	0.227	0.115	0.023	0.004
	5			0.004	0.025	0.077	0.164	0.261	0.318	0.275	0.124	0.041
	6				0.004	0.017	0.055	0.131	0.247	0.367	0.372	0.257
	7					0.002	0.008	0.028	0.082	0.210	0.478	0.698
8	0	0.663	0.430	0.168	0.058	0.017	0.004	0.001				
	1	0.279	0.383	0.336	0.198	0.090	0.031	0.008	0.001			
	2	0.051	0.149	0.294	0.296	0.209	0.109	0.041	0.010	0.001		
	3	0.005	0.033	0.147	0.254	0.279	0.219	0.124	0.047	0.009		
	4		0.005	0.046	0.136	0.232	0.273	0.232	0.136	0.046	0.005	
	5			0.009	0.047	0.124	0.219	0.279	0.254	0.147	0.033	0.005
	6			0.001	0.010	0.041	0.109	0.209	0.296	0.294	0.149	0.051
	7				0.001	0.008	0.031	0.090	0.198	0.336	0.383	0.279
	8					0.001	0.004	0.017	0.058	0.168	0.430	0.663

Table 2 Binomial Probabilities (*Continued*)

						p						
n	x	0.05	0.1	0.2	0.3	0.4	0.5	0.6	0.7	0.8	0.9	0.95
9	0	0.630	0.387	0.134	0.040	0.010	0.002					
	1	0.299	0.387	0.302	0.156	0.060	0.018	0.004				
	2	0.063	0.172	0.302	0.267	0.161	0.070	0.021	0.004			
	3	0.008	0.045	0.176	0.267	0.251	0.164	0.074	0.021	0.003		
	4	0.001	0.007	0.066	0.172	0.251	0.246	0.167	0.074	0.017	0.001	
	5		0.001	0.017	0.074	0.167	0.246	0.251	0.172	0.066	0.007	0.001
	6			0.003	0.021	0.074	0.164	0.251	0.267	0.176	0.045	0.008
	7				0.004	0.021	0.070	0.161	0.267	0.302	0.172	0.063
	8					0.004	0.018	0.060	0.156	0.302	0.387	0.299
	9						0.002	0.010	0.040	0.134	0.387	0.630
10	0	0.599	0.349	0.107	0.028	0.006	0.001					
	1	0.315	0.387	0.268	0.121	0.040	0.010	0.002				
	2	0.075	0.194	0.302	0.233	0.121	0.044	0.011	0.001			
	3	0.010	0.057	0.201	0.267	0.215	0.117	0.042	0.009	0.001		
	4	0.001	0.011	0.088	0.200	0.251	0.205	0.111	0.037	0.006		
	5		0.001	0.026	0.103	0.201	0.246	0.201	0.103	0.026	0.001	
	6			0.006	0.037	0.111	0.205	0.251	0.200	0.088	0.011	0.001
	7			0.001	0.009	0.042	0.117	0.215	0.267	0.201	0.057	0.010
	8				0.001	0.011	0.044	0.121	0.233	0.302	0.194	0.075
	9					0.002	0.010	0.040	0.121	0.268	0.387	0.315
	10						0.001	0.006	0.028	0.107	0.349	0.599
11	0	0.569	0.314	0.086	0.020	0.004						
	1	0.329	0.384	0.236	0.093	0.027	0.005	0.001				
	2	0.087	0.213	0.295	0.200	0.089	0.027	0.005	0.001			
	3	0.014	0.071	0.221	0.257	0.177	0.081	0.023	0.004			
	4	0.001	0.016	0.111	0.220	0.236	0.161	0.070	0.017	0.002		
	5		0.002	0.039	0.132	0.221	0.226	0.147	0.057	0.010		
	6			0.010	0.057	0.147	0.226	0.221	0.132	0.039	0.002	
	7			0.002	0.017	0.070	0.161	0.236	0.220	0.111	0.016	0.001
	8				0.004	0.023	0.081	0.177	0.257	0.221	0.071	0.014
	9				0.001	0.005	0.027	0.089	0.200	0.295	0.213	0.087
	10					0.001	0.005	0.027	0.093	0.236	0.384	0.329
	11							0.004	0.020	0.086	0.314	0.569
12	0	0.540	0.282	0.069	0.014	0.002						
	1	0.341	0.377	0.206	0.071	0.017	0.003					
	2	0.099	0.230	0.283	0.168	0.064	0.016	0.002				
	3	0.017	0.085	0.236	0.240	0.142	0.054	0.012	0.001			
	4	0.002	0.021	0.133	0.231	0.213	0.121	0.042	0.008	0.001		
	5		0.004	0.053	0.158	0.227	0.193	0.101	0.029	0.003		
	6			0.016	0.079	0.177	0.226	0.177	0.079	0.016		
	7			0.003	0.029	0.101	0.193	0.227	0.158	0.053	0.004	
	8			0.001	0.008	0.042	0.121	0.213	0.231	0.133	0.021	0.002
	9				0.001	0.012	0.054	0.142	0.240	0.236	0.085	0.017
	10					0.002	0.016	0.064	0.168	0.283	0.230	0.099
	11						0.003	0.017	0.071	0.206	0.377	0.341
	12							0.002	0.014	0.069	0.282	0.540

Table 2 Binomial Probabilities (*Continued*)

						p						
n	*x*	0.05	0.1	0.2	0.3	0.4	0.5	0.6	0.7	0.8	0.9	0.95
13	0	0.513	0.254	0.055	0.010	0.001						
	1	0.351	0.367	0.179	0.054	0.011	0.002					
	2	0.111	0.245	0.268	0.139	0.045	0.010	0.001				
	3	0.021	0.100	0.246	0.218	0.111	0.035	0.006	0.001			
	4	0.003	0.028	0.154	0.234	0.184	0.087	0.024	0.003			
	5		0.006	0.069	0.180	0.221	0.157	0.066	0.014	0.001		
	6		0.001	0.023	0.103	0.197	0.209	0.131	0.044	0.006		
	7			0.006	0.044	0.131	0.209	0.197	0.103	0.023	0.001	
	8			0.001	0.014	0.066	0.157	0.221	0.180	0.069	0.006	
	9				0.003	0.024	0.087	0.184	0.234	0.154	0.028	0.003
	10				0.001	0.006	0.035	0.111	0.218	0.246	0.100	0.021
	11					0.001	0.010	0.045	0.139	0.268	0.245	0.111
	12						0.002	0.011	0.054	0.179	0.367	0.351
	13							0.001	0.010	0.055	0.254	0.513
14	0	0.488	0.229	0.044	0.007	0.001						
	1	0.359	0.356	0.154	0.041	0.007	0.001					
	2	0.123	0.257	0.250	0.113	0.032	0.006	0.001				
	3	0.026	0.114	0.250	0.194	0.085	0.022	0.003				
	4	0.004	0.035	0.172	0.229	0.155	0.061	0.014	0.001			
	5		0.008	0.086	0.196	0.207	0.122	0.041	0.007			
	6		0.001	0.032	0.126	0.207	0.183	0.092	0.023	0.002		
	7			0.009	0.062	0.157	0.209	0.157	0.062	0.009		
	8			0.002	0.023	0.092	0.183	0.207	0.126	0.032	0.001	
	9				0.007	0.041	0.122	0.207	0.196	0.086	0.008	
	10				0.001	0.014	0.061	0.155	0.229	0.172	0.035	0.004
	11					0.003	0.022	0.085	0.194	0.250	0.114	0.026
	12					0.001	0.006	0.032	0.113	0.250	0.257	0.123
	13						0.001	0.007	0.041	0.154	0.356	0.359
	14							0.001	0.007	0.044	0.229	0.488
15	0	0.463	0.206	0.035	0.005							
	1	0.366	0.343	0.132	0.031	0.005						
	2	0.135	0.267	0.231	0.092	0.022	0.003					
	3	0.031	0.129	0.250	0.170	0.063	0.014	0.002				
	4	0.005	0.043	0.188	0.219	0.127	0.042	0.007	0.001			
	5	0.001	0.010	0.103	0.206	0.186	0.092	0.024	0.003			
	6		0.002	0.043	0.147	0.207	0.153	0.061	0.012	0.001		
	7			0.014	0.081	0.177	0.196	0.118	0.035	0.003		
	8			0.003	0.035	0.118	0.196	0.177	0.081	0.014		
	9			0.001	0.012	0.061	0.153	0.207	0.147	0.043	0.002	
	10				0.003	0.024	0.092	0.186	0.206	0.103	0.010	0.001
	11				0.001	0.007	0.042	0.127	0.219	0.188	0.043	0.005
	12					0.002	0.014	0.063	0.170	0.250	0.129	0.031
	13						0.003	0.022	0.092	0.231	0.267	0.135
	14							0.005	0.031	0.132	0.343	0.366
	15								0.005	0.035	0.206	0.463

Table 3 Areas of a Standard Normal Distribution

An entry in the table is the proportion under the entire curve which is between $z = 0$ and a positive value of z. Areas for negative values of z are obtained by symmetry.

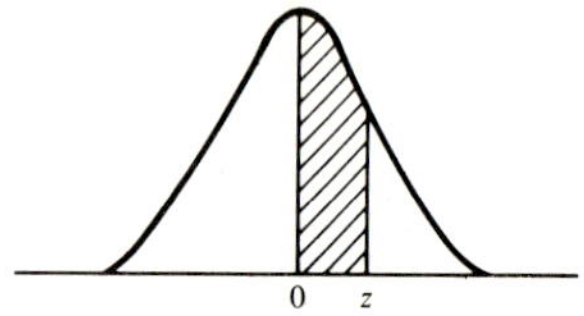

z	0.00	0.01	0.02	0.03	0.04	0.05	0.06	0.07	0.08	0.09
0.0	0.0000	0.0040	0.0080	0.0120	0.0160	0.0199	0.0239	0.0279	0.0319	0.0359
0.1	0.0398	0.0438	0.0478	0.0517	0.0557	0.0596	0.0636	0.0675	0.0714	0.0753
0.2	0.0793	0.0832	0.0871	0.0910	0.0948	0.0987	0.1026	0.1064	0.1103	0.1141
0.3	0.1179	0.1217	0.1255	0.1293	0.1331	0.1368	0.1406	0.1443	0.1480	0.1517
0.4	0.1554	0.1591	0.1628	0.1664	0.1700	0.1736	0.1772	0.1808	0.1844	0.1879
0.5	0.1915	0.1950	0.1985	0.2019	0.2054	0.2088	0.2123	0.2157	0.2190	0.2224
0.6	0.2257	0.2291	0.2324	0.2357	0.2389	0.2422	0.2454	0.2486	0.2517	0.2549
0.7	0.2580	0.2611	0.2642	0.2673	0.2703	0.2734	0.2764	0.2794	0.2823	0.2852
0.8	0.2881	0.2910	0.2939	0.2967	0.2995	0.3023	0.3051	0.3078	0.3106	0.3133
0.9	0.3159	0.3186	0.3212	0.3238	0.3264	0.3289	0.3315	0.3340	0.3365	0.3389
1.0	0.3413	0.3438	0.3461	0.3485	0.3508	0.3531	0.3554	0.3577	0.3599	0.3621
1.1	0.3643	0.3665	0.3686	0.3708	0.3729	0.3749	0.3770	0.3790	0.3810	0.3830
1.2	0.3849	0.3869	0.3888	0.3907	0.3925	0.3944	0.3962	0.3980	0.3997	0.4015
1.3	0.4032	0.4049	0.4066	0.4082	0.4099	0.4115	0.4131	0.4147	0.4162	0.4177
1.4	0.4192	0.4207	0.4222	0.4236	0.4251	0.4265	0.4279	0.4292	0.4306	0.4319
1.5	0.4332	0.4345	0.4357	0.4370	0.4382	0.4394	0.4406	0.4418	0.4429	0.4441
1.6	0.4452	0.4463	0.4474	0.4484	0.4495	0.4505	0.4515	0.4525	0.4535	0.4545
1.7	0.4554	0.4564	0.4573	0.4582	0.4591	0.4599	0.4608	0.4616	0.4625	0.4633
1.8	0.4641	0.4649	0.4656	0.4664	0.4671	0.4678	0.4686	0.4693	0.4699	0.4706
1.9	0.4713	0.4719	0.4726	0.4732	0.4738	0.4744	0.4750	0.4756	0.4761	0.4767
2.0	0.4772	0.4778	0.4783	0.4788	0.4793	0.4798	0.4803	0.4808	0.4812	0.4817
2.1	0.4821	0.4826	0.4830	0.4834	0.4838	0.4842	0.4846	0.4850	0.4854	0.4857
2.2	0.4861	0.4864	0.4868	0.4871	0.4875	0.4878	0.4881	0.4884	0.4887	0.4890
2.3	0.4893	0.4896	0.4898	0.4901	0.4904	0.4906	0.4909	0.4911	0.4913	0.4916
2.4	0.4918	0.4920	0.4922	0.4925	0.4927	0.4929	0.4931	0.4932	0.4934	0.4936
2.5	0.4938	0.4940	0.4941	0.4943	0.4945	0.4946	0.4948	0.4949	0.4951	0.4952
2.6	0.4953	0.4955	0.4956	0.4957	0.4959	0.4960	0.4961	0.4962	0.4963	0.4964
2.7	0.4965	0.4966	0.4967	0.4968	0.4969	0.4970	0.4971	0.4972	0.4973	0.4974
2.8	0.4974	0.4975	0.4976	0.4977	0.4977	0.4978	0.4979	0.4979	0.4980	0.4981
2.9	0.4981	0.4982	0.4982	0.4983	0.4984	0.4984	0.4985	0.4985	0.4986	0.4986
3.0	0.4987	0.4987	0.4987	0.4988	0.4988	0.4989	0.4989	0.4989	0.4990	0.4990

Table 4 The Exponential Function

x	e^x	x	e^x	x	e^x	x	e^x	x	e^x
0.00	1.00000	0.40	1.49182	0.80	2.22554	1.20	3.32012	1.60	4.95303
0.01	1.01005	0.41	1.50682	0.81	2.24791	1.21	3.35348	1.61	5.00281
0.02	1.02020	0.42	1.52196	0.82	2.27050	1.22	3.38719	1.62	5.05309
0.03	1.03045	0.43	1.53726	0.83	2.29332	1.23	3.42123	1.63	5.10387
0.04	1.04081	0.44	1.55271	0.84	2.31637	1.24	3.45561	1.64	5.15517
0.05	1.05127	0.45	1.56831	0.85	2.33965	1.25	3.49034	1.65	5.20698
0.06	1.06184	0.46	1.58407	0.86	2.36316	1.26	3.52542	1.66	5.25931
0.07	1.07251	0.47	1.59999	0.87	2.38691	1.27	3.56085	1.67	5.31217
0.08	1.08329	0.48	1.61607	0.88	2.41090	1.28	3.59664	1.68	5.36556
0.09	1.09417	0.49	1.63232	0.89	2.43513	1.29	3.63279	1.69	5.41948
0.10	1.10517	0.50	1.64872	0.90	2.45960	1.30	3.66930	1.70	5.47395
0.11	1.11628	0.51	1.66529	0.91	2.48432	1.31	3.70617	1.71	5.52896
0.12	1.12750	0.52	1.68203	0.92	2.50929	1.32	3.74342	1.72	5.58453
0.13	1.13883	0.53	1.69893	0.93	2.53451	1.33	3.78104	1.73	5.64065
0.14	1.15027	0.54	1.71601	0.94	2.55998	1.34	3.81904	1.74	5.69734
0.15	1.16183	0.55	1.73325	0.95	2.58571	1.35	3.85743	1.75	5.75460
0.16	1.17351	0.56	1.75067	0.96	2.61170	1.36	3.89619	1.76	5.81244
0.17	1.18530	0.57	1.76827	0.97	2.63794	1.37	3.93535	1.77	5.87085
0.18	1.19722	0.58	1.78604	0.98	2.66446	1.38	3.97490	1.78	5.92986
0.19	1.20925	0.59	1.80399	0.99	2.69123	1.39	4.01485	1.79	5.98945
0.20	1.22140	0.60	1.82212	1.00	2.71828	1.40	4.05520	1.80	6.04965
0.21	1.23368	0.61	1.84043	1.01	2.74560	1.41	4.09596	1.81	6.11045
0.22	1.24608	0.62	1.85893	1.02	2.77319	1.42	4.13712	1.82	6.17186
0.23	1.25860	0.63	1.87761	1.03	2.80107	1.43	4.17870	1.83	6.23389
0.24	1.27125	0.64	1.89648	1.04	2.82922	1.44	4.22070	1.84	6.29654
0.25	1.28403	0.65	1.91554	1.05	2.85765	1.45	4.26311	1.85	6.35982
0.26	1.29693	0.66	1.93479	1.06	2.88637	1.46	4.30596	1.86	6.42374
0.27	1.30996	0.67	1.95424	1.07	2.91538	1.47	4.34924	1.87	6.48830
0.28	1.32313	0.68	1.97388	1.08	2.94468	1.48	4.39295	1.88	6.55350
0.29	1.33643	0.69	1.99372	1.09	2.97427	1.49	4.43710	1.89	6.61937
0.30	1.34986	0.70	2.01375	1.10	3.00417	1.50	4.48169	1.90	6.68589
0.31	1.36343	0.71	2.03399	1.11	3.03436	1.51	4.52673	1.91	6.75309
0.32	1.37713	0.72	2.05443	1.12	3.06485	1.52	4.57223	1.92	6.82096
0.33	1.39097	0.73	2.07508	1.13	3.09566	1.53	4.61818	1.93	6.88951
0.34	1.40495	0.74	2.09594	1.14	3.12677	1.54	4.66459	1.94	6.95875
0.35	1.41907	0.75	2.11700	1.15	3.15819	1.55	4.71147	1.95	7.02869
0.36	1.43333	0.76	2.13828	1.16	3.18993	1.56	4.75882	1.96	7.09933
0.37	1.44773	0.77	2.15977	1.17	3.22199	1.57	4.80665	1.97	7.17068
0.38	1.46228	0.78	2.18147	1.18	3.25437	1.58	4.85496	1.98	7.24274
0.39	1.47698	0.79	2.20340	1.19	3.28708	1.59	4.90375	1.99	7.31553

Table 4 The Exponential Function (*Continued*)

x	e^x	x	e^x	x	e^x	x	e^x	x	e^x
2.00	7.38906	2.40	11.02318	2.80	16.44465	3.20	24.53253	3.60	36.59823
2.01	7.46332	2.41	11.13396	2.81	16.60992	3.21	24.77909	3.61	36.96605
2.02	7.53832	2.42	11.24586	2.82	16.77685	3.22	25.02812	3.62	37.33757
2.03	7.61409	2.43	11.35888	2.83	16.94546	3.23	25.27966	3.63	37.71282
2.04	7.69061	2.44	11.47304	2.84	17.11577	3.24	25.53372	3.64	38.09184
2.05	7.76790	2.45	11.58835	2.85	17.28778	3.25	25.79034	3.65	38.47467
2.06	7.84597	2.46	11.70481	2.86	17.46153	3.26	26.04954	3.66	38.86134
2.07	7.92482	2.47	11.82245	2.87	17.63702	3.27	26.31134	3.67	39.25191
2.08	8.00447	2.48	11.94126	2.88	17.81427	3.28	26.57577	3.68	39.64639
2.09	8.08491	2.49	12.06128	2.89	17.99331	3.29	26.84286	3.69	40.04485
2.10	8.16617	2.50	12.18249	2.90	18.17414	3.30	27.11264	3.70	40.44730
2.11	8.24824	2.51	12.30493	2.91	18.35680	3.31	27.38512	3.71	40.85381
2.12	8.33114	2.52	12.42860	2.92	18.54129	3.32	27.66035	3.72	41.26439
2.13	8.41487	2.53	12.55351	2.93	18.72763	3.33	27.93834	3.73	41.67911
2.14	8.49944	2.54	12.67967	2.94	18.91585	3.34	28.21913	3.74	42.09799
2.15	8.58486	2.55	12.80710	2.95	19.10595	3.35	28.50273	3.75	42.52108
2.16	8.67114	2.56	12.93582	2.96	19.29797	3.36	28.78919	3.76	42.94843
2.17	8.75828	2.57	13.06582	2.97	19.49192	3.37	29.07853	3.77	43.38006
2.18	8.84631	2.58	13.19714	2.98	19.68782	3.38	29.37077	3.78	43.81604
2.19	8.93521	2.59	13.32977	2.99	19.88568	3.39	29.66595	3.79	44.25640
2.20	9.02501	2.60	13.46374	3.00	20.08554	3.40	29.96410	3.80	44.70118
2.21	9.11572	2.61	13.59905	3.01	20.28740	3.41	30.26524	3.81	45.15044
2.22	9.20733	2.62	13.73572	3.02	20.49129	3.42	30.56941	3.82	45.60421
2.23	9.29987	2.63	13.87377	3.03	20.69723	3.43	30.87664	3.83	46.06254
2.24	9.39333	2.64	14.01320	3.04	20.90524	3.44	31.18696	3.84	46.52547
2.25	9.48774	2.65	14.15404	3.05	21.11534	3.45	31.50039	3.85	46.99306
2.26	9.58309	2.66	14.29629	3.06	21.32756	3.46	31.81698	3.86	47.46535
2.27	9.67940	2.67	14.43997	3.07	21.54190	3.47	32.13674	3.87	47.94238
2.28	9.77668	2.68	14.58509	3.08	21.75840	3.48	32.45972	3.88	48.42421
2.29	9.87494	2.69	14.73168	3.09	21.97708	3.49	32.78595	3.89	48.91089
2.30	9.97418	2.70	14.87973	3.10	22.19795	3.50	33.11545	3.90	49.40245
2.31	10.07442	2.71	15.02928	3.11	22.42104	3.51	33.44827	3.91	49.89895
2.32	10.17567	2.72	15.18032	3.12	22.64638	3.52	33.78443	3.92	50.40044
2.33	10.27794	2.73	15.33289	3.12	22.87398	3.53	34.12397	3.93	50.90698
2.34	10.38124	2.74	15.48698	3.14	23.10387	3.54	34.46692	3.94	51.41860
2.35	10.48557	2.75	15.64263	3.15	23.33606	3.55	34.81332	3.95	51.93537
2.36	10.59095	2.76	15.79984	3.16	23.57060	3.56	35.16320	3.96	52.45732
2.37	10.69739	2.77	15.95863	3.17	23.80748	3.57	35.51659	3.97	52.98453
2.38	10.80490	2.78	16.11902	3.18	24.04675	3.58	35.87354	3.98	53.51703
2.39	10.91349	2.79	16.28102	3.19	24.28843	3.59	36.23408	3.99	54.05489

Table 4 The Exponential Function (*Continued*)

x	e^x	x	e^x	x	e^x	x	e^x	x	e^x
4.00	54.59815	4.40	81.45087	4.80	121.51041	7.00	1,096.63309	11.00	59,874.13477
4.01	55.14687	4.41	82.26946	4.81	122.73161	7.10	1,211.96703	11.10	66,171.15430
4.02	55.70110	4.42	83.09628	4.82	123.96509	7.20	1,339.43076	11.20	73,130.43652
4.03	56.26091	4.43	83.93141	4.83	125.21096	7.30	1,480.29985	11.30	80,821.63379
4.04	56.82634	4.44	84.77494	4.84	126.46935	7.40	1,635.98439	11.40	89,321.72168
4.05	57.39745	4.45	85.62694	4.85	127.74039	7.50	1,808.04231	11.50	98,715.75879
4.06	57.97431	4.46	86.48751	4.86	129.02420	7.60	1,998.19582	11.60	109,097.78906
4.07	58.55696	4.47	87.35672	4.87	130.32091	7.70	2,208.34796	11.70	120,571.70605
4.08	59.14547	4.48	88.23467	4.88	131.63066	7.80	2,440.60187	11.80	133,252.34570
4.09	59.73989	4.49	89.12144	4.89	132.95357	7.90	2,697.28226	11.90	147,266.62109
4.10	60.34029	4.50	90.01713	4.90	134.28978	8.00	2,980.95779		
4.11	60.94671	4.51	90.92182	4.91	135.63941	8.10	3,294.46777		
4.12	61.55924	4.52	91.83560	4.92	137.00261	8.20	3,640.95004		
4.13	62.17792	4.53	92.75856	4.93	138.37951	8.30	4,023.87219		
4.14	62.80282	4.54	93.69080	4.94	139.77024	8.40	4,447.06665		
4.15	63.43400	4.55	94.63240	4.95	141.17496	8.50	4,914.76886		
4.16	64.07152	4.56	95.58347	4.96	142.59379	8.60	5,431.65906		
4.17	64.71545	4.57	96.54411	4.97	144.02688	8.70	6,002.91180		
4.18	65.36585	4.58	97.51439	4.98	145.47438	8.80	6,634.24371		
4.19	66.02279	4.59	98.49443	4.99	146.93642	8.90	7,331.97339		
4.20	66.68633	4.60	99.48431	5.00	148.41316	9.00	8,103.08295		
4.21	67.35654	4.61	100.48415	5.10	164.02190	9.10	8,955.29187		
4.22	68.03348	4.62	101.49403	5.20	181.27224	9.20	9,897.12830		
4.23	68.71723	4.63	102.51406	5.30	200.33680	9.30	10,938.01868		
4.24	69.40785	4.64	103.54435	5.40	221.40641	9.40	12,088.38049		
4.25	70.10541	4.65	104.58498	5.50	244.69192	9.50	13,359.72522		
4.26	70.80998	4.66	105.63608	5.60	270.42640	9.60	14,764.78015		
4.27	71.52163	4.67	106.69774	5.70	298.86740	9.70	16,317.60608		
4.28	72.24044	4.68	107.77007	5.80	330.29955	9.80	18,033.74414		
4.29	72.96647	4.69	108.85318	5.90	365.03746	9.90	19,930.36987		
4.30	73.69979	4.70	109.94717	6.00	403.42877	10.00	22,026.46313		
4.31	74.44049	4.71	111.05216	6.10	445.85775	10.10	24,343.00708		
4.32	75.18863	4.72	112.16825	6.20	492.74903	10.20	26,903.18408		
4.33	75.94429	4.73	113.29556	6.30	544.57188	10.30	29,732.61743		
4.34	76.70754	4.74	114.43420	6.40	601.84502	10.40	32,859.62500		
4.35	77.47846	4.75	115.58428	6.50	665.14159	10.50	36,315.49854		
4.36	78.25713	4.76	116.74592	6.60	735.09516	10.60	40,134.83350		
4.37	79.04363	4.77	117.91924	6.70	812.40582	10.70	44,355.85205		
4.38	79.83803	4.78	119.10435	6.80	897.84725	10.80	49,020.79883		
4.39	80.64042	4.79	120.30136	6.90	992.27469	10.90	54,176.36230		

Table 5 Amount of an Annuity (*Continued*)

Period	$1\frac{1}{2}\%$	$1\frac{3}{4}\%$	2%	$2\frac{1}{4}\%$	$2\frac{1}{2}\%$	$2\frac{3}{4}\%$	3%
1	1.000 000	1.000 000	1.000 000	1.000 000	1.000 000	1.000 000	1.000 000
2	2.002 500	2.003 333	2.004 166	2.005 000	2.007 500	2.010 000	2.012 500
3	3.007 506	3.010 011	3.012 517	3.015 025	3.022 556	3.030 100	3.037 656
4	4.015 025	4.020 044	4.025 069	4.030 100	4.045 225	4.060 401	4.075 626
5	5.025 062	5.033 444	5.041 840	5.050 250	5.075 564	5.101 005	5.126 572
6	6.037 625	6.050 222	6.062 848	6.075 501	6.113 631	6.152 015	6.190 654
7	7.052 719	7.070 390	7.088 110	7.105 879	7.159 483	7.213 535	7.268 037
8	8.070 351	8.093 958	8.117 643	8.141 408	8.213 179	8.285 670	8.358 888
9	9.090 526	9.120 938	9.151 467	9.182 115	9.274 778	9.368 527	9.463 374
10	10.113 253	10.151 341	10.189 598	10.228 026	10.344 339	10.462 212	10.581 666
11	11.138 536	11.185 178	11.232 055	11.279 166	11.421 921	11.566 834	11.713 937
12	12.166 382	12.222 462	12.278 855	12.335 562	12.507 586	12.682 503	12.860 361
13	13.196 798	13.263 204	13.330 017	13.397 240	13.601 393	13.809 328	14.021 115
14	14.229 790	14.307 415	14.385 559	14.464 226	14.703 403	14.947 421	15.196 379
15	15.265 365	15.355 106	15.445 498	15.536 547	15.813 679	16.096 895	16.386 334
16	16.303 528	16.406 290	16.509 855	16.614 230	16.932 281	17.257 864	17.591 163
17	17.344 287	17.460 977	17.578 646	17.697 301	18.059 273	18.430 443	18.811 053
18	18.387 648	18.519 181	18.651 890	18.785 787	19.194 718	19.614 747	20.046 191
19	19.433 617	19.580 911	19.729 606	19.879 716	20.338 678	20.810 895	21.296 768
20	20.482 201	20.646 181	20.811 813	20.979 115	21.491 218	22.019 003	22.562 978
21	21.533 406	21.715 001	21.898 529	22.084 011	22.652 403	23.239 194	23.845 015
22	22.587 240	22.787 385	22.989 773	23.194 431	23.822 296	24.471 585	25.143 078
23	23.643 708	23.863 343	24.085 564	24.310 403	25.000 963	25.716 301	26.457 366
24	24.702 817	24.942 887	25.185 920	25.431 955	26.188 470	26.973 464	27.788 084
25	25.764 574	26.026 030	26.290 861	26.559 115	27.384 884	28.243 199	29.135 435
26	26.828 986	27.112 784	27.400 407	27.691 910	28.590 270	29.525 631	30.499 628
27	27.896 058	28.203 160	28.514 575	28.830 370	29.804 697	30.820 887	31.880 873
28	28.965 798	29.297 170	29.633 386	29.974 521	31.028 233	32.129 096	33.279 384
29	30.038 213	30.394 827	30.756 858	31.124 394	32.260 944	33.450 387	34.695 376
30	31.113 308	31.496 143	31.885 012	32.280 016	33.502 901	34.784 891	36.129 068
31	32.191 092	32.601 131	33.017 866	33.441 416	34.754 173	36.132 740	37.580 682
32	33.271 569	33.709 801	34.155 440	34.608 623	36.014 829	37.494 067	39.050 440
33	34.354 748	34.822 167	35.297 755	35.781 666	37.284 941	38.869 008	40.538 571
34	35.440 635	35.938 241	36.444 829	36.960 575	38.564 578	40.257 698	42.045 303
35	36.529 237	37.058 035	37.596 682	38.145 378	39.853 812	41.660 275	43.570 869
36	37.620 560	38.181 562	38.753 335	39.336 104	41.152 716	43.076 878	45.115 505
37	38.714 611	39.308 834	39.914 807	40.532 785	42.461 361	44.507 647	46.679 449
38	39.811 398	40.439 863	41.081 119	41.735 449	43.779 821	45.952 723	48.262 942
39	40.910 926	41.574 663	42.252 290	42.944 126	45.108 170	47.412 250	49.866 229
40	42.013 204	42.713 245	43.428 341	44.158 847	46.446 481	48.886 373	51.489 557
41	43.118 237	43.855 622	44.609 293	45.379 641	47.794 830	50.375 237	53.133 176
42	44.226 032	45.001 808	45.795 165	46.606 539	49.153 291	51.878 989	54.797 341
43	45.336 597	46.151 814	46.985 978	47.839 572	50.521 941	53.397 779	56.482 308
44	46.449 939	47.305 653	48.181 753	49.078 770	51.900 855	54.931 757	58.188 336
45	47.566 064	48.463 339	49.382 510	50.324 164	53.290 112	56.481 074	59.915 691
46	48.684 979	49.624 883	50.588 271	51.575 784	54.689 787	58.045 885	61.664 637
47	49.806 691	50.790 299	51.799 055	52.833 663	56.099 961	59.626 344	63.435 445
48	50.931 208	51.959 600	53.014 885	54.097 832	57.520 711	61.222 607	65.228 388
49	52.058 536	53.132 799	54.235 780	55.368 321	58.952 116	62.834 833	67.043 743
50	53.188 682	54.309 908	55.461 762	56.645 162	60.394 257	64.463 182	68.881 789

Table 5 Amount of an Annuity (*Continued*)

Period	$1\frac{1}{2}\%$	$1\frac{3}{4}\%$	2%	$2\frac{1}{4}\%$	$2\frac{1}{2}\%$	$2\frac{3}{4}\%$	3%
51	54.321 654	55.490 942	56.692 853	57.928 388	61.847 214	66.107 814	70.742 812
52	55.457 458	56.675 911	57.929 073	59.218 030	63.311 068	67.768 892	72.627 097
53	56.596 102	57.864 831	59.170 445	60.514 120	64.785 901	69.446 581	74.534 936
54	57.737 592	59.057 714	60.416 988	61.816 691	66.271 795	71.141 046	76.466 622
55	58.881 936	60.254 573	61.668 725	63.125 774	67.768 834	72.852 457	78.422 455
56	60.029 141	61.455 421	62.925 679	64.441 403	69.277 100	74.580 981	80.402 736
57	61.179 214	62.660 273	64.187 869	65.763 610	70.796 678	76.326 791	82.407 770
58	62.332 162	63.869 140	65.455 318	67.092 428	72.327 653	78.090 059	84.437 867
59	63.487 992	65.082 038	66.728 049	68.427 891	73.870 111	79.870 960	86.493 340
60	64.646 712	66.298 978	68.006 082	69.770 030	75.424 136	81.669 669	88.574 507
61	65.808 329	67.519 974	69.289 441	71.118 880	76.989 817	83.486 366	90.681 689
62	66.972 850	68.745 041	70.578 147	72.474 475	78.567 241	85.321 230	92.815 210
63	68.140 282	69.974 191	71.872 223	73.836 847	80.156 495	87.174 442	94.975 400
64	69.310 633	71.207 438	73.171 690	75.206 031	81.757 669	89.046 186	97.162 592
65	70.483 909	72.444 796	74.476 572	76.582 061	83.370 852	90.936 648	99.377 125
66	71.660 119	73.686 279	75.786 891	77.964 972	84.996 133	92.846 015	101.619 339
67	72.839 269	74.931 900	77.102 670	79.354 797	86.633 604	94.774 475	103.889 581
68	74.021 367	76.181 673	78.423 931	80.751 570	88.283 356	96.722 220	106.188 200
69	75.206 421	77.435 612	79.750 698	82.155 328	89.945 481	98.689 442	108.515 553
70	76.394 437	78.693 731	81.082 992	83.566 105	91.620 072	100.676 336	110.871 997
71	77.585 423	79.956 043	82.420 838	84.983 936	93.307 223	102.683 100	113.257 897
72	78.779 387	81.222 563	83.764 258	86.408 855	95.007 027	104.709 931	115.673 621
73	79.976 335	82.493 305	85.113 276	87.840 899	96.719 580	106.757 030	118.119 541
74	81.176 276	83.768 283	86.467 914	89.280 104	98.444 977	108.824 600	120.596 035
75	82.379 217	85.047 510	87.828 197	90.726 504	100.183 314	110.912 846	123.103 486
76	83.585 165	86.331 002	89.194 148	92.180 137	101.934 689	113.021 975	125.642 280
77	84.794 127	87.618 772	90.565 791	93.641 038	103.699 199	115.152 195	128.212 808
78	86.006 113	88.910 835	91.943 148	95.109 243	105.476 943	117.303 717	130.815 468
79	87.221 128	90.207 204	93.326 244	96.584 789	107.268 020	119.476 754	133.450 661
80	88.439 181	91.507 895	94.715 104	98.067 713	109.072 530	121.671 521	136.118 795
81	89.660 279	92.812 921	96.109 750	99.558 052	110.890 574	123.888 236	138.820 280
82	90.884 430	94.122 298	97.510 207	101.055 842	112.722 254	126.127 119	141.555 533
83	92.111 641	95.436 039	98.916 500	102.561 121	114.567 670	128.388 390	144.324 977
84	93.341 920	96.754 159	100.328 652	104.073 927	116.426 928	130.672 274	147.129 040
85	94.575 275	98.076 673	101.746 688	105.594 296	118.300 130	132.978 997	149.968 153
86	95.811 713	99.403 595	103.170 633	107.122 268	120.187 381	135.308 787	152.842 755
87	97.051 242	100.734 940	104.600 510	108.657 879	122.088 786	137.661 874	155.753 289
88	98.293 870	102.070 723	106.036 346	110.201 169	124.004 452	140.038 493	158.700 205
89	99.539 605	103.410 959	107.478 164	111.752 174	125.934 486	142.438 878	161.683 958
90	100.788 454	104.755 662	108.925 990	113.310 935	127.878 994	144.863 267	164.705 007
91	102.040 425	106.104 848	110.379 848	114.877 490	129.838 087	147.311 900	167.763 820
92	103.295 526	107.458 531	111.839 764	116.451 877	131.811 872	149.785 019	170.860 867
93	104.553 765	108.816 726	113.305 763	118.034 137	133.800 461	152.282 869	173.996 628
94	105.815 149	110.179 448	114.777 870	119.624 308	135.803 965	154.805 698	177.171 586
95	107.079 687	111.546 713	116.256 111	121.222 429	137.822 495	157.353 755	180.386 231
96	108.347 386	112.918 535	117.740 512	122.828 541	139.856 163	159.927 292	183.641 059
97	109.618 255	114.294 930	119.231 097	124.442 684	141.905 084	162.526 565	186.936 572
98	110.892 300	115.675 913	120.727 894	126.064 897	143.969 373	165.151 831	190.273 279
99	112.169 531	117.061 500	122.230 926	127.695 222	146.049 143	167.803 349	193.651 695
100	113.449 955	118.451 705	123.740 222	129.333 698	148.144 512	170.481 382	197.072 341

Table 5 Amount of an Annuity (*Continued*)

Period	$1\frac{1}{2}\%$	$1\frac{3}{4}\%$	2%	$2\frac{1}{4}\%$	$2\frac{1}{2}\%$	$2\frac{3}{4}\%$	3%
1	1.000 000	1.000 000	1.000 000	1.000 000	1.000 000	1.000 000	1.000 000
2	2.015 000	2.017 500	2.020 000	2.022 500	2.025 000	2.027 500	2.030 000
3	3.045 225	3.052 806	3.060 400	3.068 006	3.075 625	3.083 256	3.090 900
4	4.090 903	4.106 230	4.121 608	4.137 036	4.152 515	4.168 045	4.183 627
5	5.152 266	5.178 089	5.204 040	5.230 119	5.256 328	5.282 667	5.309 135
6	6.229 550	6.268 705	6.308 120	6.347 797	6.387 736	6.427 940	6.468 409
7	7.322 994	7.378 408	7.434 283	7.490 622	7.547 430	7.604 708	7.662 462
8	8.432 839	8.507 530	8.582 969	8.659 161	8.736 115	8.813 838	8.892 336
9	9.559 331	9.656 412	9.754 628	9.853 993	9.954 518	10.056 218	10.159 106
10	10.702 721	10.825 399	10.949 720	11.075 707	11.203 381	11.332 764	11.463 879
11	11.863 262	12.014 843	12.168 715	12.324 911	12.483 466	12.644 415	12.807 795
12	13.041 211	13.225 103	13.412 089	13.602 221	13.795 552	13.992 137	14.192 029
13	14.236 829	14.456 543	14.680 331	14.908 271	15.140 441	15.376 921	15.617 790
14	15.450 382	15.709 532	15.973 938	16.243 707	16.518 952	16.799 786	17.086 324
15	16.682 137	16.984 449	17.293 416	17.609 191	17.931 926	18.261 780	18.598 913
16	17.932 369	18.281 677	18.639 285	19.005 398	19.380 224	19.763 979	20.156 881
17	19.201 355	19.601 606	20.012 070	20.433 019	20.864 730	21.307 488	21.761 587
18	20.489 375	20.944 634	21.412 312	21.892 762	22.386 348	22.893 444	23.414 435
19	21.796 716	22.311 165	22.840 558	23.385 349	23.946 007	24.523 014	25.116 868
20	23.123 667	23.701 611	24.297 369	24.911 520	25.544 657	26.197 397	26.870 374
21	24.470 522	25.116 389	25.783 317	26.472 029	27.183 274	27.917 825	28.676 485
22	25.837 579	26.555 926	27.298 983	28.067 649	28.862 855	29.685 566	30.536 780
23	27.225 143	28.020 654	28.844 963	29.699 172	30.584 427	31.501 919	32.452 883
24	28.633 520	29.511 016	30.421 862	31.367 403	32.349 037	33.368 221	34.426 470
25	30.063 023	31.027 459	32.030 299	33.073 169	34.157 763	35.285 848	36.459 264
26	31.513 968	32.570 439	33.670 905	34.817 316	36.011 708	37.256 208	38.553 042
27	32.986 678	34.140 422	35.344 323	36.600 705	37.912 000	39.280 754	40.709 633
28	34.481 478	35.737 879	37.051 210	38.424 221	39.859 800	41.360 975	42.930 922
29	35.998 700	37.363 292	38.792 234	40.288 766	41.856 295	43.498 402	45.218 850
30	37.538 681	39.017 150	40.568 079	42.195 264	43.902 703	45.694 608	47.575 415
31	39.101 761	40.699 950	42.379 440	44.144 657	46.000 270	47.951 210	50.002 678
32	40.688 288	42.412 199	44.227 029	46.137 912	48.150 277	50.269 868	52.502 758
33	42.298 612	44.154 413	46.111 570	48.176 015	50.354 034	52.652 289	55.077 841
34	43.933 091	45.927 115	48.033 801	50.259 975	52.612 885	55.100 227	57.730 176
35	45.592 087	47.730 839	49.994 477	52.390 825	54.928 207	57.615 483	60.462 081
36	47.275 969	49.566 129	51.994 367	54.569 618	57.301 412	60.199 909	63.275 944
37	48.985 108	51.433 536	54.034 254	56.797 435	59.733 947	62.855 407	66.174 222
38	50.719 885	53.333 623	56.114 939	59.075 377	62.227 296	65.583 930	69.159 449
39	52.480 683	55.266 962	58.237 238	61.404 573	64.782 979	68.387 489	72.234 232
40	54.267 893	57.234 133	60.401 983	63.786 176	67.402 553	71.268 144	75.401 259
41	56.081 912	59.235 731	62.610 022	66.221 365	70.087 617	74.228 018	78.663 297
42	57.923 141	61.272 356	64.862 223	68.711 345	72.839 807	77.269 289	82.023 196
43	59.791 988	63.344 622	67.159 467	71.257 351	75.660 803	80.394 194	85.483 892
44	61.688 867	65.453 153	69.502 657	73.860 641	78.552 323	83.605 035	89.048 409
45	63.614 200	67.598 583	71.892 710	76.522 506	81.516 131	86.904 173	92.719 861
46	65.568 413	69.781 559	74.330 564	79.244 262	84.554 034	90.294 038	96.501 457
47	67.551 940	72.002 736	76.817 175	82.027 258	87.667 885	93.777 124	100.396 500
48	69.565 219	74.262 784	79.353 519	84.872 871	90.859 582	97.355 995	104.408 395
49	71.608 697	76.562 382	81.940 589	87.782 511	94.131 071	101.033 285	108.540 647
50	73.682 828	78.902 224	84.579 401	90.757 617	97.484 348	104.811 700	112.796 867

Table 5 Amount of an Annuity (*Continued*)

Period	$1\frac{1}{2}\%$	$1\frac{3}{4}\%$	2%	$2\frac{1}{4}\%$	$2\frac{1}{2}\%$	$2\frac{3}{4}\%$	3%
51	75.788 070	81.283 013	87.270 989	93.799 664	100.921 457	108.694 022	117.180 173
52	77.924 891	83.705 466	90.016 409	96.910 156	104.444 493	112.683 108	121.696 196
53	80.093 764	86.170 312	92.816 737	100.090 635	108.055 606	116.781 893	126.347 082
54	82.295 171	88.678 292	95.673 072	103.342 674	111.756 996	120.993 395	131.137 494
55	84.529 598	91.230 162	98.586 533	106.667 884	115.550 921	125.320 714	136.071 619
56	86.797 542	93.826 690	101.558 264	110.067 911	119.439 694	129.767 033	141.153 768
57	89.099 506	96.468 657	104.589 429	113.544 440	123.425 686	134.335 627	146.388 381
58	91.435 998	99.156 859	107.681 218	117.099 189	127.511 328	139.029 856	151.780 032
59	93.807 538	101.892 104	110.834 842	120.733 921	131.699 112	143.853 177	157.333 433
60	96.214 651	104.675 215	114.051 539	124.450 434	135.991 589	148.809 140	163.053 436
61	98.657 871	107.507 032	117.332 570	128.250 569	140.391 379	153.901 391	168.945 039
62	101.137 739	110.388 405	120.679 221	132.136 207	144.901 164	159.133 680	175.013 391
63	103.654 805	113.320 202	124.092 806	136.109 272	149.523 693	164.509 856	181.263 792
64	106.209 627	116.303 305	127.574 662	140.171 730	154.261 785	170.033 877	187.701 706
65	108.802 772	119.338 613	131.126 155	144.325 594	159.118 330	175.709 808	194.332 757
66	111.434 813	122.427 039	134.748 678	148.572 920	164.096 288	181.541 828	201.162 740
67	114.106 335	125.569 512	138.443 652	152.915 811	169.198 695	187.534 228	208.197 622
68	116.817 930	128.766 979	142.212 525	157.356 417	174.428 663	193.691 420	215.443 551
69	119.570 199	132.020 401	146.056 775	161.896 936	179.789 379	200.017 934	222.906 857
70	122.363 752	135.330 758	149.977 911	166.539 617	185.284 114	206.518 427	230.594 063
71	125.199 209	138.699 046	153.977 469	171.286 758	190.916 217	213.197 684	238.511 885
72	128.077 197	142.126 279	158.057 018	176.140 711	196.689 122	220.060 620	246.667 242
73	130.998 355	145.613 489	162.218 159	181.103 877	202.606 350	227.112 287	255.067 259
74	133.963 330	149.161 725	166.462 522	186.178 714	208.671 509	234.357 875	263.719 277
75	136.972 780	152.772 056	170.791 772	191.367 735	214.888 297	241.802 717	272.630 855
76	140.027 372	156.445 566	175.207 608	196.673 509	221.260 504	249.452 291	281.809 781
77	143.127 782	160.183 364	179.711 760	202.098 663	227.792 017	257.312 229	291.264 074
78	146.274 699	163.986 573	184.305 995	207.645 883	234.486 817	265.388 316	301.001 996
79	149.468 820	167.856 338	188.992 115	213.317 915	241.348 987	273.686 494	311.032 056
80	152.710 852	171.793 824	193.771 957	219.117 568	248.382 712	282.212 873	321.363 018
81	156.001 515	175.800 216	198.647 396	225.047 714	255.592 280	290.973 727	332.003 909
82	159.341 537	179.876 719	203.620 344	231.111 287	262.982 087	299.975 504	342.964 026
83	162.731 661	184.024 562	208.692 751	237.311 291	270.556 639	309.224 831	354.252 947
84	166.172 635	188.244 992	213.866 606	243.650 795	278.320 555	318.728 514	365.880 535
85	169.665 225	192.539 279	219.143 938	250.132 938	286.278 569	328.493 548	377.856 951
86	173.210 203	196.908 717	224.526 817	256.760 929	294.435 533	338.527 120	390.192 660
87	176.808 356	201.354 619	230.017 354	263.538 050	302.796 422	348.836 616	402.898 440
88	180.460 482	205.878 325	235.617 701	270.467 656	311.366 332	359.429 623	415.985 393
89	184.167 389	210.481 196	241.330 055	277.553 179	320.150 491	370.313 938	429.464 955
90	187.929 900	215.164 617	247.156 656	284.798 125	329.154 253	381.497 571	443.348 903
91	191.748 848	219.929 997	253.099 789	292.206 083	338.383 109	392.988 754	457.649 370
92	195.625 081	224.778 772	259.161 785	299.780 720	347.842 687	404.795 945	472.378 851
93	199.559 457	229.712 401	265.345 020	307.525 786	357.538 754	416.927 834	487.550 217
94	203.552 849	234.732 368	271.651 921	315.445 116	367.477 223	429.393 349	503.176 723
95	207.606 142	239.840 184	278.084 959	323.542 631	377.664 153	442.201 666	519.272 025
96	211.720 234	245.037 388	284.646 658	331.822 340	388.105 757	455.362 212	535.850 186
97	215.896 038	250.325 542	291.339 592	340.288 343	398.808 401	468.884 673	552.925 692
98	220.134 478	255.706 239	298.166 383	348.944 831	409.778 611	482.779 001	570.513 462
99	224.436 495	261.181 098	305.129 711	357.796 090	421.023 077	497.055 424	588.628 866
100	228.803 043	266.751 767	312.232 305	366.846 502	432.548 654	511.724 448	607.287 732

Table 5 Amount of an Annuity (*Continued*)

Period	$3\frac{1}{2}\%$	4%	5%	6%	7%	8%
1	1.000 000	1.000 000	1.000 000	1.000 000	1.000 000	1.000 000
2	2.035 000	2.040 000	2.050 000	2.060 000	2.070 000	2.080 000
3	3.106 225	3.121 600	3.152 500	3.183 600	3.214 900	3.246 400
4	4.214 942	4.246 464	4.310 125	4.374 616	4.439 943	4.506 112
5	5.362 465	5.416 322	5.525 631	5.637 092	5.750 739	5.866 600
6	6.550 152	6.632 975	6.801 912	6.975 318	7.153 290	7.335 929
7	7.779 407	7.898 294	8.142 008	8.393 837	8.654 021	8.922 803
8	9.051 686	9.214 226	9.549 108	9.897 467	10.259 802	10.636 627
9	10.368 495	10.582 795	11.026 564	11.491 315	11.977 988	12.487 557
10	11.731 393	12.006 107	12.577 892	13.180 794	13.816 447	14.486 562
11	13.141 991	13.486 351	14.206 787	14.971 642	15.783 599	16.645 487
12	14.601 961	15.025 805	15.917 126	16.869 941	17.888 451	18.977 126
13	16.113 030	16.626 837	17.712 982	18.882 137	20.140 642	21.495 296
14	17.676 986	18.291 911	19.598 631	21.015 065	22.550 487	24.214 920
15	19.295 680	20.023 587	21.578 563	23.275 969	25.129 022	27.152 113
16	20.971 029	21.824 531	23.657 491	25.672 528	27.888 053	30.324 283
17	22.705 015	23.697 512	25.840 366	28.212 879	30.840 217	33.750 225
18	24.499 691	25.645 412	28.132 384	30.905 652	33.999 032	37.450 243
19	26.357 180	27.671 229	30.539 003	33.759 991	37.378 964	41.446 263
20	28.279 681	29.778 078	33.065 954	36.785 591	40.995 492	45.761 964
21	30.269 470	31.969 201	35.719 251	39.992 726	44.865 176	50.422 921
22	32.328 902	34.247 969	38.505 214	43.392 290	49.005 739	55.456 755
23	34.460 413	36.617 888	41.430 475	46.995 827	53.436 140	60.893 295
24	36.666 528	39.082 604	44.501 998	50.815 577	58.176 670	66.764 759
25	38.949 856	41.645 908	47.727 098	54.864 511	63.249 037	73.105 939
26	41.313 101	44.311 744	51.113 453	59.156 382	68.676 470	79.954 415
27	43.759 060	47.084 214	54.669 126	63.705 765	74.483 823	87.350 768
28	46.290 627	49.967 582	58.402 582	68.528 111	80.697 690	95.338 829
29	48.910 799	52.966 286	62.322 711	73.639 798	87.346 529	103.965 936
30	51.622 677	56.084 937	66.438 847	79.058 186	94.460 786	113.283 211
31	54.429 470	59.328 335	70.760 789	84.801 677	102.073 041	123.345 868
32	57.334 502	62.701 468	75.298 829	90.889 778	110.218 154	134.213 537
33	60.341 210	66.209 527	80.063 770	97.343 164	118.933 425	145.950 620
34	63.453 152	69.857 908	85.066 959	104.183 754	128.258 764	158.626 670
35	66.674 012	73.652 224	90.320 307	111.434 779	138.236 878	172.316 803
36	70.007 603	77.598 313	95.836 322	119.120 866	148.913 459	187.102 147
37	73.457 869	81.702 246	101.628 138	127.268 118	160.337 402	203.070 319
38	77.028 894	85.970 336	107.709 545	135.904 205	172.561 020	220.315 945
39	80.724 906	90.409 149	114.095 023	145.058 458	185.640 291	238.941 221
40	84.550 277	95.025 515	120.799 774	154.761 965	199.635 111	259.056 518
41	88.509 537	99.826 536	127.839 762	165.047 683	214.609 569	280.781 040
42	92.607 371	104.819 597	135.231 751	175.950 544	230.632 239	304.243 523
43	96.848 629	110.012 381	142.993 338	187.507 577	247.776 496	329.583 005
44	101.238 331	115.412 876	151.143 005	199.758 031	266.120 851	356.949 645
45	105.781 672	121.029 392	159.700 155	212.743 513	285.749 310	386.505 617
46	110.484 031	126.870 567	168.685 163	226.508 124	306.751 762	418.426 066
47	115.350 972	132.945 390	178.119 421	241.098 612	329.224 385	452.900 152
48	120.388 256	139.263 206	188.025 392	256.564 528	353.270 092	490.132 164
49	125.601 845	145.833 734	198.426 662	272.958 400	378.998 999	530.342 737
50	130.997 910	152.667 083	209.347 995	290.335 904	406.528 929	573.770 156

Table 5 Amount of an Annuity (*Continued*)

Period	$3\frac{1}{2}\%$	4%	5%	6%	7%	8%
51	136.582 837	159.773 767	220.815 395	308.756 058	435.985 954	620.671 768
52	142.363 236	167.164 717	232.856 165	328.281 422	467.504 971	671.325 510
53	148.345 949	174.851 306	245.498 973	348.978 307	501.230 319	726.031 551
54	154.538 057	182.845 358	258.773 922	370.917 006	537.316 441	785.114 075
55	160.946 889	191.159 172	272.712 618	394.172 026	575.928 592	848.923 201
56	167.580 030	199.805 539	287.348 249	418.822 348	617.243 594	917.837 057
57	174.445 332	208.797 761	302.715 661	444.951 689	661.450 645	992.264 022
58	181.550 918	218.149 671	318.851 444	472.648 790	708.752 190	1,072.645 143
59	188.905 200	227.875 658	335.794 017	502.007 717	759.364 844	1,159.456 755
60	196.516 882	237.990 685	353.583 717	533.128 180	813.520 383	1,253.213 295
61	204.394 973	248.510 312	372.262 903	566.115 871	871.466 810	1,354.470 359
62	212.548 797	259.450 725	391.876 048	601.082 824	933.469 486	1,463.827 988
63	220.988 005	270.828 754	412.469 851	638.147 793	999.812 350	1,581.934 227
64	229.722 585	282.661 904	434.093 343	677.436 661	1,070.799 215	1,709.488 965
65	238.762 876	294.968 380	456.798 011	719.082 860	1,146.755 160	1,847.248 082
66	248.119 577	307.767 115	480.637 911	763.227 832	1,228.028 021	1,996.027 929
67	257.803 762	321.077 800	505.669 807	810.021 502	1,314.989 983	2,156.710 163
68	267.826 894	334.920 912	531.953 297	859.622 792	1,408.039 282	2,330.246 976
69	278.200 835	349.317 748	559.550 962	912.200 160	1,507.602 032	2,517.666 734
70	288.937 864	364.290 458	588.528 510	967.932 169	1,614.134 174	2,720.080 073
71	300.050 689	379.862 077	618.954 936	1,027.008 099	1,728.123 566	2,938.686 479
72	311.552 463	396.056 560	650.902 683	1,089.628 585	1,850.092 216	3,174.781 398
73	323.456 800	412.898 822	684.447 817	1,156.006 300	1,980.598 671	3,429.763 909
74	335.777 788	430.414 775	719.670 208	1,226.366 679	2,120.240 578	3,705.145 022
75	348.530 010	448.631 366	756.653 718	1,300.948 679	2,269.657 418	4,002.556 624
76	361.728 561	467.576 621	795.486 404	1,380.005 600	2,429.533 437	4,323.761 154
77	375.389 060	487.279 686	836.260 724	1,463.805 936	2,600.600 778	4,670.662 046
78	389.527 677	507.770 873	879.073 760	1,552.634 292	2,783.642 833	5,045.315 010
79	404.161 146	529.081 708	924.027 448	1,646.792 350	2,979.497 831	5,449.940 211
80	419.306 786	551.244 976	971.228 821	1,746.599 891	3,189.062 679	5,886.935 428
81	434.982 524	574.294 775	1,020.790 262	1,852.395 884	3,413.297 067	6,358.890 262
82	451.206 912	598.266 566	1,072.829 775	1,964.539 637	3,653.227 861	6,868.601 483
83	467.999 154	623,197 229	1,127.471 264	2,083.412 016	3,909.953 812	7,419.089 602
84	485.379 125	649.125 118	1,184.844 827	2,209.416 737	4,184.650 579	8,013.616 770
85	503.367 394	676.090 123	1,245.087 068	2,342.981 741	4,478.576 119	8,655.706 112
86	521.985 253	704.133 728	1,308.341 422	2,484.560 645	4,793.076 448	9,349.162 601
87	541.254 737	733.299 077	1,374.758 493	2,634.634 284	5,129.591 799	10,098.095 609
88	561.198 652	763.631 040	1,444.496 418	2,793.712 341	5,489.663 225	10,906.943 257
89	581.840 605	795.176 282	1,517.721 239	2,962.335 082	5,874.939 651	11,780.498 718
90	603.205 027	827.983 333	1,594.607 300	3,141.075 187	6,287.185 426	12,723.938 615
91	625.317 202	862.102 666	1,675.337 666	3,330.539 698	6,728.288 406	13,742.853 705
92	648.203 305	897.586 773	1,760.104 549	3,531.372 080	7,200.268 595	14,843.282 001
93	671.890 420	934,490 244	1,849.109 776	3,744.254 405	7,705.287 396	16,031.744 561
94	696.406 585	972.869 854	1,942.565 265	3,969.909 669	8,245.657 514	17,315.284 126
95	721.780 815	1,012.784 648	2,040.693 528	4,209.104 249	8,823.853 540	18,701.506 856
96	748.043 144	1,054.296 034	2,143.728 205	4,462.650 504	9,442.523 288	20,198.627 405
97	775.224 654	1,097.467 875	2,251.914 615	4,731.409 534	10,104.499 918	21,815.517 597
98	803.357 517	1,142.366 590	2,365.510 346	5,016.294 106	10,812.814 912	23,561.759 005
99	832.475 030	1,189.061 254	2,484.785 863	5,318.271 753	11,570.711 956	25,447.699 726
100	862.611 656	1,237.623 704	2,610.025 156	5,638.368 058	12,381.661 793	27,484.515 704

Table 6 Present Value of an Annuity

Period	0.0025($\frac{1}{4}$%)	0.004167($\frac{5}{12}$%)	0.005($\frac{1}{2}$%)	0.005833($\frac{7}{12}$%)	0.0075($\frac{3}{4}$%)	0.01(1%)	0.01125($1\frac{1}{8}$%)	0.0125($1\frac{1}{4}$%)	0.015($1\frac{1}{2}$%)
1	0.997 506	0.995 850	0.995 024	0.994 200	0.992 555	0.990 099	0.988 875	0.987 654	0.985 221
2	1.992 524	1.987 569	1.985 099	1.982 635	1.977 722	1.970 395	1.966 749	1.963 115	1.955 883
3	2.985 062	2.975 172	2.970 248	2.965 337	2.955 556	2.940 985	2.933 744	2.926 533	2.912 200
4	3.975 124	3.958 678	3.950 495	3.942 340	3.926 110	3.901 965	3.889 982	3.878 057	3.854 384
5	4.962 717	4.938 102	4.925 866	4.913 677	4.889 439	4.853 431	4.835 582	4.817 835	4.782 644
6	5.947 848	5.913 463	5.896 384	5.879 380	5.845 597	5.795 476	5.770 662	5.746 009	5.697 187
7	6.930 521	6.884 776	6.862 074	6.839 483	6.794 637	6.728 194	6.695 339	6.662 725	6.598 213
8	7.910 744	7.852 059	7.822 959	7.7 018	7.736 613	7.651 677	7.609 730	7.568 124	7.485 925
9	8.888 523	8.815 329	8.779 063	8.743 017	8.671 576	8.566 017	8.513 948	8.462 344	8.360 517
10	9.863 863	9.774 601	9.730 411	9.686 513	9.599 579	9.471 304	9.408 106	9.345 525	9.222 184
11	10.836 771	10.729 893	10.677 026	10.624 536	10.520 674	10.367 628	10.292 318	10.217 803	10.071 117
12	11.807 253	11.681 222	11.618 932	11.557 120	11.434 912	11.255 077	11.166 693	11.079 311	10.907 505
13	12.775 315	12.628 602	12.556 151	12.484 295	12.342 345	12.133 740	12.031 340	11.930 184	11.731 532
14	13.740 963	13.572 052	13.488 707	13.406 092	13.243 022	13.003 703	12.886 368	12.770 552	12.543 381
15	14.704 202	14.511 587	14.416 624	14.322 544	14.136 994	13.865 052	13.731 885	13.600 545	13.343 233
16	15.665 040	15.447 224	15.339 925	15.233 681	15.024 312	14.717 873	14.567 995	14.420 292	14.131 264
17	16.623 481	16.378 978	16.258 631	16.139 534	15.905 024	15.562 251	15.394 803	15.229 918	14.907 649
18	17.579 532	17.306 866	17.172 768	17.040 133	16.779 181	16.398 268	16.212 413	16.029 548	15.672 560
19	18.533 199	18.230 904	18.082 356	17.935 509	17.646 829	17.226 008	17.020 928	16.819 307	16.426 168
20	19.484 488	19.151 108	18.987 419	18.825 693	18.508 019	18.045 552	17.820 448	17.599 316	17.168 638
21	20.433 404	20.067 493	19.887 979	19.710 713	19.362 798	18.856 983	18.611 073	18'369 694	17.900 136
22	21.379 954	20.980 076	20.784 058	20.590 602	20.211 214	19.660 379	19.392 903	19.130 562	18.620 824
23	22.324 144	21.888 872	21.675 680	21.465 387	21.053 314	20.455 821	20.166 035	19.882 037	19.330 861
24	23.265 979	22.793 898	22.562 866	22.335 099	21.889 146	21.243 387	20.930 566	20.624 234	20.030 405
25	24.205 465	23.695 168	23.445 638	23.199 767	22.718 755	22.023 155	21.686 592	21.357 268	20.719 611
26	25.142 609	24.592 698	24.324 017	24.059 420	23.542 189	22.795 203	22.434 207	22.081 252	21.398 631
27	26.077 415	25.486 505	25.198 027	24.914 088	24.359 492	23.559 607	23.173 505	22.796 299	22.067 617
28	27.009 891	26.376 602	26.067 689	25.763 799	25.170 712	24.316 443	23.904 579	23.502 517	22.726 716
29	27.940 041	27.263 006	26.933 024	26.608 582	25.975 893	25.965 785	24.627 519	24.200 017	23.376 075
30	28.867 871	28.145 732	27.794 053	27.448 466	26.775 080	25.807 708	25.342 417	24.888 906	24.015 838
31	29.793 387	29.024 796	28.650 799	28.283 479	27.568 317	26.542 285	26.049 362	25.569 290	24.646 145
32	30.716 596	29.900 212	29.503 283	29.113 650	28.355 650	27.269 589	26.748 442	26.241 274	25.267 138
33	31.637 502	30.771 995	30.351 525	29.939 006	29.137 122	27.989 692	27.439 745	26.904 962	25.878 954
34	32.556 112	31.640 161	31.195 548	30.759 575	29.912 776	28.702 665	28.123 357	27.560 456	26.481 728
35	33.472 431	32.504 725	32.035 371	31.575 385	30.682 656	29.408 580	28.799 364	28.207 858	27.075 594
36	34.386 465	33.365 701	32.871 016	32.386 464	31.446 805	30.107 505	29.467 851	28.847 267	27.660 684
37	35.298 219	34.223 105	33.702 503	33.192 839	32.205 265	30.799 509	30.128 901	29.478 782	28.237 127
38	36.207 700	35.076 951	34.529 854	33.994 538	32.958 080	31.484 663	30.782 596	30.102 501	28.805 051
39	37.114 913	35.927 254	35.353 089	34.791 587	33.705 290	32.163 032	31.429 020	30.718 519	29.364 582
40	38.019 863	36.774 029	36.172 227	35.584 013	34.446 938	32.834 686	32.068 252	31.326 933	29.915 845
41	38.922 556	37.617 290	36.987 291	36.371 844	35.183 065	33.499 689	32.700 373	31.927 835	30.458 960
42	39.822 999	38.457 052	37.798 299	37.155 106	35.913 712	34.158 108	33.325 461	32.521 318	30.994 050
43	40.721 196	39.293 330	38.605 273	37.933 825	36.638 920	34.810 008	33.943 596	33.107 475	31.521 231
44	41.617 153	40.126 138	39.408 232	38.708 029	37.358 730	35.455 453	34.554 854	33.686 395	32.040 622
45	42.510 876	40.955 490	40.207 196	39.477 742	38.073 181	36.094 508	35.159 312	34.258 168	32.552 337
46	43.402 370	41.781 401	41.002 185	40.242 991	38.782 314	36.727 236	35.757 045	34.822 882	33.056 489
47	44.291 641	42.603 884	41.793 219	41.003 802	39.486 167	37.353 699	36.348 128	35.380 624	33.553 191
48	45.178 694	43.422 955	42.580 317	41.760 201	40.184 781	37.973 959	36.932 636	35.931 480	34.042 553
49	46.063 535	44.238 628	43.363 500	42.512 213	40.878 195	38.588 078	37.510 642	36.475 536	34.524 683
50	46.946 170	45.050 916	44.142 786	43.259 864	41.566 447	39.196 117	38.082 217	37.012 875	34.999 688

Table 6 Present Value of an Annuity (*Continued*)

Period	0.0025($\frac{1}{4}$%)	0.004167($\frac{5}{12}$%)	0.005($\frac{1}{2}$%)	0.005833($\frac{7}{12}$%)	0.0075($\frac{3}{4}$%)	0.01(1%)	0.01125($1\frac{1}{8}$%)	0.0125($1\frac{1}{4}$%)	0.015($1\frac{1}{2}$%)
51	47.826 603	45.859 833	44.918 195	44.003 179	42.249 575	39.798 136	38.647 433	37.543 580	35.467 672
52	48.704 841	46.665 394	45.689 746	44.742 183	42.927 618	40.394 194	39.206 361	38.067 734	35.928 741
53	49.580 889	47.467 612	46.457 459	45.476 901	43.600 613	40.984 350	39.759 072	38.585 416	36.382 996
54	50.454 752	48.266 502	47.221 352	46.207 358	44.268 599	41.568 664	40.305 633	39.096 707	36.830 538
55	51.326 436	49.062 076	47.981 445	46.933 578	44.931 611	42.147 192	40.846 115	39.601 686	37.271 466
56	52.195 946	49.854 350	48.737 756	47.655 588	45.589 689	42.719 992	41.380 583	40.100 431	37.705 878
57	53.063 288	50.643 336	49.490 305	48.373 409	46.242 867	43.287 121	41.909 106	40.593 018	38.133 870
58	53.928 467	51.429 048	50.239 109	49.087 068	46.891 183	43.848 634	42.431 748	41.079 524	38.555 537
59	54.791 488	52.211 500	50.984 188	49.796 588	47.534 673	44.404 588	42.948 577	41.560 024	38.970 972
60	55.652 357	52.990 706	51.725 560	50.501 993	48.173 373	44.955 038	43.459 656	42.034 591	39.380 268
61	56.511 079	53.766 678	52.463 244	51.203 307	48.807 318	45.500 038	43.965 049	42.503 300	39.783 516
62	57.367 660	54.539 430	53.197 258	51.900 554	49.436 544	46.039 641	44.464 820	42.966 222	40.180 804
63	58.222 105	55.308 976	53.927 620	52.593 757	50.061 086	46.573 902	44.959 031	43.423 429	40.572 220
64	59.074 419	56.075 329	54.654 348	53.282 940	50.680 979	47.102 873	45.447 744	43.874 992	40.957 852
65	59.924 608	56.838 502	55.377 461	53.968 126	51.296 257	47.626 607	45.931 020	44.320 980	41.337 786
66	60.772 676	57.598 508	56.096 976	54.649 338	51.906 954	48.145 156	46.408 919	44.761 461	41.712 104
67	61.618 629	58.355 361	56.812 911	55.326 599	52.513 106	48.658 570	46.881 502	45.196 505	42.080 891
68	62.462 473	59.109 073	57.525 285	55.999 933	53.114 746	49.166 901	47.348 828	45.626 178	42.444 227
69	63.304 213	59.859 658	58.234 114	56.669 362	53.711 906	49.670 199	47.810 955	46.050 546	42.802 194
70	64.143 853	60.607 128	58.939 417	57.334 908	54.304 622	50.168 514	48.267 940	46.469 675	43.154 871
71	64.981 399	61.351 497	59.641 211	57.996 595	54.892 925	50.661 895	48.719 842	46.883 630	43.502 336
72	65.816 857	62.092 777	60.339 513	58.654 444	55.476 848	51.150 391	49.166 717	47.292 474	43.844 666
73	66.650 232	62.830 981	61.034 342	59.308 478	56.056 425	51.634 050	49.608 620	47.696 270	44.181 937
74	67.481 528	63.566 122	61.725 713	59.958 718	56.631 687	52.112 921	50.045 607	48.095 082	44.514 224
75	68.310 751	64.298 213	62.413 645	60.605 188	57.202 667	52.587 051	50.477 732	48.488 970	44.841 600
76	69.137 906	65.027 266	63.098 154	61.247 909	57.769 397	53.056 486	50.905 050	48.877 995	45.164 138
77	69.962 999	65.753 294	63.779 258	61.886 902	58.331 908	53.521 273	51.327 615	49.262 217	45.481 909
78	70.786 034	66.476 310	64.456 973	62.522 189	58.890 231	53.981 459	51.745 478	49.641 696	45.794 984
79	71.607 016	67.196 325	65.131 316	63.153 792	59.444 398	54.437 088	52.158 693	50.016 490	46.103 433
80	72.425 951	67.913 353	65.802 305	63.781 732	59.994 440	54.888 206	52.567 310	50.386 657	46.407 323
81	73.242 844	68.627 405	66.469 955	64.406 030	60.540 387	55.334 857	52.971 382	50.752 253	46.706 722
82	74.057 700	69.338 495	67.134 284	65.026 707	61.082 270	55.777 086	53.370 959	51.113 337	47.001 697
83	74.870 524	70.046 634	67.795 307	65.643 785	61.620 119	56.214 937	53.766 091	51.469 962	47.292 312
84	75.681 320	70.751 834	68.453 042	66.257 285	62.153 964	56.648 452	54.156 826	51.822 185	47.578 633
85	76.490 095	71.454 109	69.107 504	66.867 226	62.683 835	57.077 676	54.543 215	52.170 059	47.860 722
86	77.296 853	72.153 469	69.758 711	67.473 630	63.209 762	57.502 649	54.925 305	52.513 639	48.138 642
87	78.101 599	72.849 928	70.406 677	68.076 517	63.731 774	57.923 415	55.303 145	52.852 976	48.412 455
88	78.904 338	73.543 497	71.051 420	68.675 907	64.249 900	58.340 015	55.676 781	53.188 125	48.682 222
89	79.705 075	74.234 188	71.692 956	69.271 821	64.764 168	58.752 490	56.046 261	53.519 136	48.948 002
90	80.503 816	74.922 013	72.331 299	69.864 280	65.274 609	59.160 881	56.411 630	53.846 060	49.209 854
91	81.300 564	75.606 984	72.966 467	70.453 302	65.781 249	59.565 229	56.772 934	54.168 948	49.467 836
92	82.095 326	76.289 112	73.598 474	71.038 909	66.284 118	59.965 573	57.130 219	54.487 850	49.722 006
93	82.888 106	76.968 411	74.227 338	71.621 119	66.783 244	60.361 953	57.483 530	54.802 815	49.972 420
94	83.678 909	77.644 890	74.853 072	72.199 952	67.278 654	60.754 409	57.832 909	55.113 891	50.219 133
95	84.467 739	78.318 563	75.475 694	72.775 429	67.770 376	61.142 980	58.178 402	55.421 127	50.462 200
96	85.254 603	78.989 440	76.095 218	73.347 568	66.258 438	61.527 702	58.520 052	55.724 570	50.701 675
97	86.039 504	79.657 534	76.711 659	73.916 389	68.742 867	61.908 616	58.857 900	56.024 266	50.937 611
98	86.822 448	80.322 855	77.325 034	74.481 911	69.223 689	62.285 759	59.191 991	56.320 263	51.170 060
99	87.603 439	80.985 416	77.935 357	75.044 154	69.700 932	62.659 167	59.522 364	56.612 606	51.399 074
100	88.382 483	81.645 227	78.542 644	75.603 136	70.174 622	63.028 878	59.849 062	56.901 339	51.624 703

Table 6 Present Value of an Annuity (*Continued*)

Period	0.0175($1\frac{3}{4}\%$)	0.02(2%)	0.0225 ($2\frac{1}{4}\%$)	0.025 ($2\frac{1}{2}\%$)	0.0275 ($2\frac{3}{4}\%$)	0.03 (3%)	0.035 ($3\frac{1}{2}\%$)	0.04 (4%)	0.045 ($4\frac{1}{2}\%$)
1	0.982 800	0.980 392	0.977 995	0.975 609	0.973 236	0.970 873	0.966 183	0.961 538	0.956 937
2	1.948 698	1.941 560	1.934 469	1.927 424	1.920 424	1.913 469	1.899 694	1.886 094	1.872 667
3	2.897 984	2.883 883	2.869 896	2.856 023	2.842 262	2.828 611	2.801 636	2.775 091	2.748 964
4	3.830 942	3.807 728	3.784 740	3.761 974	3.739 427	3.717 098	3.673 079	3.629 895	3.587 525
5	4.747 855	4.713 459	4.679 452	4.645 828	4.612 581	4.579 707	4.515 052	4.451 822	4.389 976
6	5.648 997	5.601 430	5.554 476	5.508 125	5.462 366	5.417 191	5.328 553	5.242 136	5.157 872
7	6.534 641	6.471 991	6.410 246	6.349 390	6.289 408	6.230 282	6.114 543	6.002 054	5.892 700
8	7.405 052	7.325 481	7.247 184	7.170 137	7.094 314	7.019 692	6.873 955	6.732 744	6.595 886
9	8.260 494	8.162 236	8.065 706	7.970 865	7.877 678	7.786 108	7.607 686	7.435 331	7.268 790
10	9.101 222	8.982 585	8.866 216	8.752 063	8.640 076	8.530 202	8.316 605	8.110 895	7.912 718
11	9.927 491	9.786 848	9.649 111	9.514 208	9.382 069	9.252 624	9.001 551	8.760 476	8.528 916
12	10.739 549	10.575 341	10.414 778	10.257 764	10.104 203	9.954 003	9.663 334	9.385 073	9.118 580
13	11.537 640	11.348 373	11.163 597	10.983 184	10.807 010	10.634 955	10.302 738	9.985 647	9.682 852
14	12.322 005	12.106 248	11.895 939	11.690 912	11.491 008	11.296 073	10.920 520	10.563 122	10.222 825
15	13.092 880	12.849 263	12.612 165	12.381 377	12.156 698	11.937 935	11.517 410	11.118 387	10.739 545
16	13.850 496	13.577 709	13.312 631	13.055 002	12.804 573	12.561 102	12.094 116	11.652 295	11.234 015
17	14.595 082	14.291 871	13.997 683	13.712 197	13.435 107	13.166 118	12.651 320	12.165 668	11.707 191
18	15.326 862	14.992 031	14.667 661	14.353 363	14.048 766	13.753 513	13.189 681	12.659 296	12.159 991
19	16.046 056	15.678 462	15.322 895	14.978 891	14.646 001	14.323 799	13.709 837	13.133 939	12.593 293
20	16.752 881	16.351 433	15.963 712	15.589 162	15.227 252	14.877 474	14.212 403	13.590 326	13.007 936
21	17.447 549	17.011 209	16.590 427	16.184 548	15.792 946	15.415 024	14.697 974	14.029 159	13.404 723
22	18.130 269	17.658 048	17.203 352	16.765 413	16.343 499	15.936 916	15.167 124	14.451 115	13.784 424
23	18.801 247	18.292 204	17.802 789	17.332 110	16.879 318	16.443 608	15.620 410	14.856 841	14.147 774
24	19.460 685	18.913 925	18.389 036	17.884 985	17.400 796	16.935 542	16.058 367	15.246 963	14.495 478
25	20.108 781	19.523 456	18.962 382	18.424 376	17.908 317	17.413 147	16.481 514	15.622 079	14.828 208
26	20.745 731	20.121 035	19.523 112	18.950 611	18.402 255	17.876 842	16.890 352	15.982 769	15.146 611
27	21.371 726	20.706 897	20.071 503	19.464 010	18.882 974	18.327 031	17.285 364	16.329 585	15.451 302
28	21.986 954	21.281 272	20.607 827	19.964 888	19.350 826	18.764 108	17.667 018	16.663 063	15.742 873
29	22.591 601	21.844 384	21.132 349	20.453 549	19.806 157	19.188 454	18.035 767	16.983 714	16.021 888
30	23.185 849	22.396 455	21.645 329	20.930 292	20.249 301	19.600 441	18.392 045	17.292 033	16.288 888
31	23.769 876	22.937 701	22.147 021	21.395 407	20.680 585	20.000 428	18.736 275	17.588 493	16.544 390
33	24.343 858	23.468 334	22.637 674	21.849 177	21.100 326	20.388 765	19.068 865	17.873 551	16.788 890
33	24.907 969	23.988 563	23.117 529	22.291 880	21.508 833	20.765 791	19.390 208	18.147 645	17.022 862
34	25.462 377	24.498 591	23.586 826	22.723 786	21.906 407	21.131 836	19.700 684	18.411 197	17.246 757
35	26.007 251	24.998 619	24.045 795	23.145 157	22.293 340	21.487 220	20.000 661	18.664 613	17.461 012
36	26.542 752	25.488 842	24.494 665	23.556 251	22.669 917	21.832 252	20.290 493	18.908 281	17.666 040
37	27.069 044	25.969 453	24.933 658	23.957 318	23.036 416	22.167 235	20.570 525	19.142 578	17.862 239
38	27.586 284	26.440 640	25.362 991	24.348 603	23.393 105	22.492 461	20.841 087	19.367 864	18.049 990
39	28.094 628	26.902 588	25.782 876	24.730 344	23.740 248	22.808 215	21.102 499	19.584 484	18.229 655
40	28.594 229	27.355 479	26.193 522	25.102 775	24.078 101	23.114 771	21.355 072	19.792 773	18.401 584
41	29.085 237	27.799 489	26.595 131	25.466 122	24.406 911	23.412 399	21.599 103	19.993 051	18.566 109
42	29.567 801	28.234 793	26.987 903	25.820 606	24.726 920	23.701 359	21.834 882	20.185 626	18.723 549
43	30.042 065	28.661 562	27.372 033	26.166 445	25.038 365	23.981 902	22.062 688	20.370 794	18.874 210
44	30.508 172	29.079 963	27.747 709	26.503 849	25.341 475	24.254 273	22.282 791	20.548 841	19.018 383
45	30.966 262	29.490 159	28.115 119	26.833 023	25.636 472	24.518 712	22.495 450	20.720 039	19.156 347
46	31.416 474	29.892 313	28.474 444	27.154 169	25.923 573	24.775 449	22.700 918	20.884 653	19.288 370
47	31.858 942	30.286 581	28.825 862	27.467 482	26.202 991	25.024 707	22.899 437	21.042 936	19.414 708
48	32.293 801	30.673 119	29.169 547	27.773 153	26.474 930	25.266 706	23.091 244	21.195 130	19.535 606
49	32.721 180	31.052 078	29.505 670	28.071 369	26.739 592	25.501 656	23.276 564	21.341 472	19.651 298
50	33.141 209	31.423 605	29.834 396	28.362 311	26.997 169	25.729 764	23.455 617	21.482 184	19.762 007

Table 6 Present Value of an Annuity (*Continued*)

Period	0.0175 ($1\frac{3}{4}\%$)	0.02 (2%)	0.0225 ($2\frac{1}{4}\%$)	0.025 ($2\frac{1}{2}\%$)	0.0275 ($2\frac{3}{4}\%$)	0.03 (3%)	0.035 ($3\frac{1}{2}\%$)	0.04 (4%)	0.045 ($4\frac{1}{2}\%$)
51	33.554 014	31.787 848	30.155 888	28.646 157	27.247 853	25.951 227	23.628 616	21.617 485	19.867 950
52	33.959 719	32.144 949	30.470 306	28.923 080	27.491 828	26.166 239	23.795 764	21.747 581	19.969 330
53	34.358 446	32.495 048	30.777 806	29.193 249	27.729 273	26.374 990	23.957 260	21.872 674	20.066 344
54	34.750 315	32.838 283	31.078 539	29.456 828	27.960 363	26.577 660	24.113 295	21.992 956	20.159 181
55	35.135 445	33.174 787	31.372 654	29.713 979	28.185 268	26.774 427	24.264 053	22.108 612	20.248 020
56	35.513 951	33.504 693	31.660 297	29.964 857	28.404 154	26.965 463	24.409 713	22.219 819	20.333 034
57	35.885 947	33.828 131	31.941 611	30.209 617	28.617 182	27.150 935	24.550 447	22.326 749	20.414 386
58	36.251 545	34.145 226	32.216 734	30.448 407	28.824 508	27.331 005	24.686 422	22.429 566	20.492 236
59	36.610 855	34.456 104	32.485 804	30.681 372	29.026 285	27.505 830	24.817 799	22.528 429	20.566 733
60	36.963 985	34.760 886	32.748 952	30.908 656	29.222 662	27.675 563	24.944 734	22.623 489	20.638 022
61	37.311 042	35.059 692	33.006 310	31.130 396	29.413 782	27.840 353	25.067 375	22.714 894	20.706 241
62	37.652 130	35.352 640	33.258 005	31.346 728	29.599 788	28.000 342	25.185 870	22.802 782	20.771 522
63	37.987 351	35.639 843	33.504 162	31.557 783	29.780 816	28.155 672	25.300 357	22.887 291	20.833 992
64	38.316 807	35.921 414	33.744 901	31.763 691	29.956 998	28.306 478	25.410 973	22.968 549	20.893 773
65	38.640 596	36.197 465	33.980 344	31.964 577	30.128 466	28.452 891	25.517 849	23.046 681	20.950 979
66	38.958 817	36.468 103	34.210 605	32.160 562	30.295 344	28.595 040	25.621 110	23.121 809	21.005 721
67	39.271 565	36.733 434	34.435 799	32.351 768	30.457 755	28.733 048	25.720 879	23.194 047	21.058 106
68	39.578 933	36.993 563	34.656 039	32.538 310	30.615 820	28.867 037	25.817 274	23.263 507	21.108 236
69	39.881 015	37.248 591	34.871 431	32.720 303	30.769 655	28.997 123	25.910 410	23.330 295	21.156 206
70	40.177 902	37.498 619	35.082 084	32.897 856	30.919 372	29.123 421	26.000 396	23.394 514	21.202 111
71	40.469 683	37.743 744	35.288 102	33.071 079	31.065 082	29.246 040	26.087 339	23.456 264	21.246 040
72	40.756 445	37.984 063	35.489 586	33.240 078	31.206 893	29.365 087	26.171 342	23.515 638	21.288 076
73	41.038 275	38.219 669	35.686 637	33.404 954	31.344 908	29.480 667	26.252 505	23.572 729	21.328 302
74	41.315 258	38.450 656	35.879 352	33.565 808	31.479 229	29.592 881	26.330 922	23.627 624	21.366 797
75	41.587 477	38.677 114	36.067 826	33.722 740	31.609 955	29.701 826	26.406 688	23.680 408	21.403 633
76	41.855 014	38.899 131	36.252 152	33.875 844	31.737 183	29.807 598	26.479 892	23.731 161	21.438 883
77	42.117 950	39.116 795	36.432 423	34.025 213	31.861 005	29.910 289	26.550 620	23.779 963	21.472 616
78	42.376 364	39.330 191	36.608 726	34.170 940	31.981 513	30.009 989	26.618 957	23.826 887	21.504 895
79	42.630 333	39.539 403	36.781 150	34.313 112	32.098 796	30.106 786	26.684 982	23.872 007	21.535 785
80	42.879 934	39.744 513	36.949 780	34.451 817	32.212 940	30.200 763	26.748 775	23.915 391	21.565 344
81	43.125 242	39.945 601	37.114 700	34.587 138	32.324 030	30.292 003	26.810 411	23.957 107	21.593 631
82	43.366 332	40.142 746	37.275 990	34.719 159	32.432 146	30.380 585	26.869 962	23.997 218	21.620 700
83	43.603 274	40.336 026	37.433 731	34.847 960	32.537 368	30.466 588	26.927 500	24.035 787	21.646 602
84	43.836 142	40.525 515	37.588 001	34.973 620	32.639 774	30.550 085	26.983 091	24.072 872	21.671 390
85	44.065 004	40.711 289	37.738 876	35.096 214	32.739 440	30.631 151	27.036 803	24.108 531	21.695 110
86	44.289 930	40.893 421	37.886 431	35.215 819	32.836 438	30.709 855	27.088 699	24.142 818	21.717 808
87	44.510 988	41.071 981	38.030 740	35.332 506	32.930 839	30.786 267	27.138 839	24.175 786	21.739 530
88	44.728 244	41.247 041	38.171 873	35.446 348	33.022 715	30.860 453	27.187 284	24.207 487	21.760 315
89	44.941 763	41.418 667	38.309 900	35.557 412	33.112 131	30.932 479	27.234 091	24.237 968	21.780 206
90	45.151 610	41.586 929	38.444 890	35.665 768	33.199 154	31.002 407	27.279 315	24.267 277	21.799 240
91	45.357 848	41.751 891	38.576 909	35.771 481	33.283 849	31.070 298	27.323 010	24.295 459	21.817 455
92	45.560 538	41.913 618	38.706 024	35.874 616	33.366 276	31.136 211	27.365 227	24.322 556	21.834 885
93	45.759 743	42.072 175	38.832 297	35.975 235	33.446 497	31.200 205	27.406 016	24.348 612	21.851 564
94	45.955 521	42.227 622	38.955 792	36.073 400	33.524 572	31.262 335	27.445 426	24.373 665	21.867 526
95	46.147 932	42.380 022	39.076 569	36.169 170	33.600 556	31.322 655	27.483 504	24.397 755	21.882 800
96	46.337 034	42.529 433	39.194 688	36.262 605	33.674 507	31.381 219	27.520 293	24.420 916	21.897 416
97	46.522 884	42.675 915	39.310 209	36.353 761	33.746 479	31.438 077	27.555 839	24.443 191	21.911 403
98	46.705 537	42.819 525	39.423 187	36.442 694	33.816 525	31.493 278	27.590 183	24.464 606	21.924 787
99	46.885 048	42.960 318	39.533 679	36.529 457	33.884 695	31.546 872	27.623 365	24.485 198	21.937 596
100	47.061 473	43.098 351	39.641 740	36.614 105	33.951 042	31.598 905	27.655 425	24.504 998	21.949 852

Table 6 Present Value of an Annuity (*Continued*)

Period	0.05 (5%)	0.055 ($5\frac{1}{2}$%)	0.06 (6%)	0.065 ($6\frac{1}{2}$%)	0.07 (7%)	0.07 ($7\frac{1}{2}$%)	0.08 (8%)
1	0.952 380	0.947 867	0.943 396	0.938 967	0.934 579	0.930 232	0.925 925
2	1.859 410	1.846 319	1.833 392	1.820 626	1.808 018	1.795 565	1.783 264
3	2.723 248	2.697 933	2.673 011	2.648 475	2.624 316	2.600 525	2.577 096
4	3.545 950	3.505 150	3.465 105	3.425 798	3.387 211	3.349 326	3.312 126
5	4.329 476	4.270 284	4.212 363	4.155 679	4.100 197	4.045 884	3.992 710
6	5.075 692	4.995 530	4.917 324	4.841 013	4.766 539	4.693 846	4.622 879
7	5.786 373	5.682 967	5.582 381	5.484 519	5.389 289	5.296 601	5.206 370
8	6.463 212	6.334 565	6.209 793	6.088 750	5.971 298	5.857 303	5.746 638
9	7.107 821	6.952 195	6.801 692	6.656 104	6.515 232	6.378 887	6.246 887
10	7.721 734	7.537 625	7.360 087	7.188 830	7.023 581	6.864 080	6.710 081
11	8.306 414	8.092 536	7.886 874	7.689 942	7.498 674	7.315 424	7.138 964
12	8.863 251	8.618 517	8.383 843	8.158 725	7.942 686	7.735 278	7.536 078
13	9.393 572	9.117 078	8.852 682	8.599 742	8.357 650	8.125 840	7.903 775
14	9.898 640	9.589 647	9.294 983	9.013 842	8.745 467	8.489 153	8.244 236
15	10.379 658	10.037 580	9.712 248	9.402 668	9.107 914	8.827 119	8.559 478
16	10.837 769	10.462 162	10.105 895	9.767 764	9.446 648	9.141 506	8.851 369
17	11.274 066	10.864 608	10.477 259	10.110 576	9.763 222	9.433 959	9.121 638
18	11.689 586	11.246 074	10.827 603	10.432 466	10.059 086	9.706 009	9.371 887
19	12.085 320	11.607 653	11.158 116	10.734 710	10.335 595	9.595 078	9.603 599
20	12.462 210	11.950 382	11.469 921	11.018 507	10.594 014	10.194 491	9.818 147
21	12.821 152	12.275 244	11.764 076	11.284 983	10.835 527	10.413 480	10.016 803
22	13.163 002	12.583 169	12.041 581	11.535 195	11.061 240	10.617 191	10.200 743
23	13.488 573	12.875 042	12.303 378	11.770 136	11.272 187	10.806 689	10.371 058
24	13.798 641	13.151 698	12.550 357	11.990 738	11.469 334	10.982 966	10.528 758
25	14.093 944	13.413 932	12.783 356	12.197 876	11.653 583	11.146 945	10.674 776
26	14.375 185	13.662 495	13.003 166	12.392 372	11.825 778	11.299 484	10.809 977
27	14.643 033	13.898 099	13.210 534	12.574 997	11.986 709	11.441 380	10.935 164
28	14.898 127	14.121 421	13.406 164	12.746 476	12.137 111	11.573 377	11.051 078
29	15.141 073	14.333 101	13.590 721	12.907 489	12.277 674	11.696 165	11.158 406
30	15.372 451	14.533 745	13.764 831	13.058 675	12.409 041	11.810 386	11.257 783
31	15.592 810	14.723 929	13.929 085	13.200 634	12.531 814	11.916 638	11.349 799
32	15.802 676	14.904 198	14.084 043	13.333 929	12.646 555	12.015 477	11.434 999
33	16.002 549	15.075 069	14.230 229	13.459 088	12.753 790	12.107 420	11.513 888
34	16.192 904	15.237 032	14.368 141	13.576 608	12.854 009	12.192 949	11.586 933
35	16.374 194	15.390 552	14.498 246	13.686 956	12.947 672	12.272 511	11.654 568
36	16.546 851	15.536 068	14.620 987	13.790 569	13.035 207	12.346 522	11.717 192
37	16.711 287	15.673 998	14.736 780	13.887 858	13.117 016	12.415 369	11.775 178
38	16.867 892	15.804 737	14.846 019	13.979 210	13.193 473	12.479 413	11.828 868
39	17.017 040	15.928 661	14.949 074	14.064 986	13.264 928	12.538 989	11.878 582
40	17.159 086	16.046 124	15.046 296	14.145 526	13.331 708	12.594 408	11.924 613
41	17.294 367	16.157 464	15.138 015	14.221 151	13.392 120	12.645 961	11.967 234
42	17.423 207	16.262 999	15.224 543	14.292 161	13.452 448	12.693 917	12.006 698
43	17.545 911	16.363 032	15.306 172	14.358 837	13.506 961	12.738 528	12.043 239
44	17.662 773	16.457 850	15.383 182	14.421 443	13.557 908	12.780 026	12.077 073
45	17.774 069	16.547 725	15.455 832	14.480 228	13.605 521	12.818 628	12.108 401
46	17.880 066	16.632 915	15.524 369	14.535 425	13.650 020	12.854 538	12.137 408
47	17.981 015	16.713 663	15.589 028	14.587 254	13.691 607	12.887 942	12.164 267
48	18.077 157	16.790 202	15.650 026	14.635 919	13.730 474	12.919 016	12.189 136
49	18.168 721	16.862 751	15.707 572	14.681 614	13.766 798	12.947 922	12.212 163
50	18.255 925	16.931 517	15.761 860	14.724 520	13.800 746	12.974 811	12.233 484

Table 6 Present Value of an Annuity (*Continued*)

Period	0.05 (5%)	0.055 ($5\frac{1}{2}$%)	0.06 (6%)	0.065 ($6\frac{1}{2}$%)	0.07 (7%)	0.07 ($7\frac{1}{2}$%)	0.08 (8%)
51	18.338 976	16.996 699	15.813 076	14.764 808	13.832 473	12.999 824	12.253 226
52	18.418 072	17.058 482	15.861 392	14.802 636	13.862 124	13.023 092	12.271 506
53	18.493 402	17.117 045	15.906 974	14.838 156	13.889 835	13.044 737	12.288 431
54	18.565 145	17.172 554	15.949 975	14.871 508	13.915 734	13.064 872	12.304 103
55	18.633 471	17.225 170	15.990 542	14.902 824	13.939 938	13.083 601	12.318 614
56	18.698 544	17.275 043	16.028 814	14.932 229	13.962 559	13.101 025	12.332 050
57	18.760 518	17.322 315	16.064 918	14.959 840	13.983 700	13.117 232	12.344 490
58	18.819 541	17.367 123	16.098 980	14.985 765	14.003 458	13.132 309	12.356 010
59	18.875 754	17.409 596	16.131 113	15.010 108	14.021 923	13.146 334	12.366 675
60	18.929 289	17.449 854	16.161 427	15.032 965	14.039 181	13.159 380	12.376 551
61	18.980 275	17.488 013	16.190 026	15.054 427	14.055 309	13.171 516	12.385 696
62	19.028 834	17.524 183	16.217 005	15.074 580	14.070 382	13.182 806	12.394 163
63	19.075 080	17.558 467	16.242 458	15.093 502	14.084 469	13.193 308	12.402 002
64	19.119 123	17.590 964	16.266 470	15.111 269	14.097 635	13.203 077	12.409 261
65	19.161 070	17.621 767	16.289 122	15.127 853	14.109 939	13.212 165	12.415 983
66	19.201 019	17.650 964	16.310 493	15.143 617	14.121 438	13.220 618	12.422 206
67	19.239 066	17.678 639	16.330 653	15.158 326	14.132 185	13.228 482	12.427 969
68	19.275 301	17.704 871	16.349 673	15.172 137	14.142 229	13.235 797	12.433 304
69	19.309 810	17.729 735	16.367 616	15.185 105	14.151 616	13.242 602	12.438 245
70	19.342 676	17.753 304	16.384 543	15.197 282	14.160 389	13.248 932	12.442 819
71	19.373 977	17.775 643	16.400 513	15.208 715	14.168 588	13.254 821	12.447 055
72	19.403 788	17.796 818	16.415 578	15.219 451	14.176 250	13.260 298	12.450 977
73	19.432 179	17.816 889	16.429 790	15.229 531	14.183 411	13.265 394	12.454 608
74	19.459 218	17.835 914	16.443 198	15.238 997	14.190 104	13.270 134	12.457 970
75	19.484 969	17.853 947	16.455 848	15.247 884	14.196 359	13.274 543	12.461 083
76	19.509 495	17.871 040	16.467 781	15.256 229	14.202 204	13.278 644	12.463 966
77	19.532 852	17.887 241	16.479 038	15.264 065	14.207 668	13.282 460	12.466 635
78	19.555 097	17.902 598	16.489 659	15.271 422	14.212 774	13.286 009	12.469 107
79	19.576 283	17.917 155	16.499 678	15.278 331	14.217 545	13.289 311	12.471 395
80	19.596 460	17.930 952	16.509 130	15.284 818	14.222 005	13.292 382	12.473 514
81	19.615 676	17.944 031	16.518 047	15.290 909	14.226 173	13.295 239	12.475 476
82	19.633 977	17.956 427	16.526 460	15.296 628	14.230 068	13.297 897	12.477 292
83	19.651 407	17.968 177	16.534 396	15.301 998	14.233 708	13.300 369	12.478 974
84	19.668 007	17.979 315	16.541 883	15.307 040	14.237 111	13.302 669	12.480 532
85	19.683 816	17.989 872	16.548 946	15.311 775	14.240 290	13.304 808	12.481 974
86	19.698 872	17.999 879	16.555 610	15.316 221	14.243 262	13.306 798	12.483 309
87	19.713 211	18.009 364	16.561 896	15.320 395	14.246 039	13.308 650	12.484 545
88	19.726 868	18.018 354	16.567 826	15.324 314	14.248 635	13.310 372	12.485 690
89	19.739 874	18.026 876	16.573 421	15.327 995	14.251 060	13.311 974	12.486 750
90	19.752 261	18.034 953	16.578 699	15.331 450	14.253 327	13.313 464	12.487 732
91	19.764 058	18.042 610	16.583 678	15.334 695	14.255 446	13.314 850	12.488 640
92	19.775 294	18.049 867	16.588 376	15.337 742	14.257 426	13.316 140	12.489 482
93	19.785 994	18.056 746	16.592 807	15.340 603	14.259 277	13.317 339	12.490 261
94	19.796 185	18.063 266	16.596 988	15.343 289	14.261 006	13.318 455	12.490 982
95	19.805 890	18.069 447	16.600 932	15.345 811	14.262 623	13.319 493	12.491 650
96	19.815 133	18.075 305	16.604 653	15.348 179	14.264 133	13.320 458	12.492 269
97	19.823 937	18.080 858	16.608 163	15.350 403	14.265 545	13.321 357	12.492 841
98	19.832 320	18.086 121	16.611 474	15.352 491	14.266 865	13.322 192	12.493 372
99	19.840 305	18.091 110	16.614 599	15.354 452	14.268 098	13.322 969	12.493 862
100	19.847 910	18.095 839	16.617 546	15.356 293	14.269 250	13.323 692	12.494 317

Table 7 Amortization Payments Table

n	$\frac{1}{4}\%$	$\frac{1}{2}\%$	$\frac{3}{4}\%$	1%	$1\frac{1}{4}\%$	$1\frac{1}{2}\%$	$1\frac{3}{4}\%$	2%	$2\frac{1}{2}\%$	3%
1	1.002500	1.005000	1.007500	1.010000	1.012500	1.015000	1.017500	1.020000	1.025000	1.030000
2	0.501876	0.503753	0.505632	0.507512	0.509394	0.511278	0.513163	0.515050	0.518827	0.522611
3	0.335002	0.336672	0.338346	0.340022	0.341701	0.343383	0.345067	0.346755	0.350137	0.353530
4	0.251565	0.253133	0.254705	0.256281	0.257861	0.259445	0.261032	0.262624	0.265818	0.260927
5	0.201503	0.203010	0.204522	0.206040	0.207562	0.209089	0.210621	0.212158	0.215247	0.218355
6	0.168128	0.169595	0.171069	0.172548	0.174034	0.175525	0.177023	0.178526	0.181550	0.184598
7	0.144289	0.145729	0.147175	0.148628	0.150089	0.151556	0.153031	0.154512	0.157495	0.160506
8	0.126410	0.127829	0.129256	0.130690	0.132133	0.133584	0.135043	0.136510	0.139467	0.142456
9	0.112505	0.113907	0.115319	0.116740	0.118171	0.119610	0.121058	0.122515	0.125457	0.128434
10	0.101380	0.102771	0.104171	0.105582	0.107003	0.108434	0.109875	0.111327	0.114259	0.117231
11	0.092278	0.093659	0.095051	0.096454	0.097868	0.099294	0.100730	0.102178	0.105106	0.108077
12	0.084694	0.086066	0.087451	0.088849	0.090258	0.091680	0.093114	0.094560	0.097487	0.100462
13	0.078276	0.079642	0.081022	0.082415	0.083821	0.085240	0.086673	0.088118	0.091048	0.094030
14	0.072775	0.074136	0.075511	0.076901	0.078305	0.079723	0.081156	0.082602	0.085537	0.088526
15	0.068008	0.069364	0.070736	0.072124	0.073526	0.074944	0.076377	0.077825	0.080766	0.083767
16	0.063836	0.065189	0.066559	0.067945	0.069347	0.070765	0.072200	0.073650	0.076599	0.079611
17	0.060156	0.061506	0.062873	0.064258	0.065660	0.067080	0.068516	0.069970	0.072928	0.075953
18	0.056884	0.058232	0.059598	0.060982	0.062385	0.063806	0.065245	0.066702	0.069670	0.072709
19	0.053957	0.055303	0.056667	0.058052	0.059455	0.060878	0.062321	0.063782	0.066761	0.069814
20	0.051323	0.052666	0.054031	0.055415	0.056820	0.058246	0.059691	0.061157	0.064147	0.067216
21	0.048939	0.050282	0.051645	0.053031	0.054437	0.055865	0.057315	0.058785	0.061787	0.064872
22	0.046773	0.048114	0.049477	0.050864	0.052272	0.053703	0.055156	0.056631	0.059647	0.062747
23	0.044795	0.046135	0.047498	0.048886	0.050297	0.051731	0.053188	0.054668	0.057696	0.060814
24	0.042981	0.044321	0.045685	0.047073	0.048487	0.049924	0.051386	0.052871	0.055913	0.059047
25	0.041313	0.042652	0.044016	0.045407	0.046822	0.048263	0.049730	0.051220	0.054276	0.057428
26	0.039773	0.041112	0.042477	0.043869	0.045287	0.046732	0.048203	0.049699	0.052769	0.055938
27	0.038347	0.039686	0.041052	0.042446	0.043867	0.045315	0.046791	0.048293	0.051377	0.054564
28	0.037023	0.038362	0.039729	0.041124	0.042549	0.044001	0.045482	0.046990	0.050088	0.053293
29	0.035791	0.037129	0.038497	0.039895	0.041322	0.042779	0.044264	0.045778	0.048891	0.052115
30	0.034641	0.035979	0.037348	0.038748	0.040179	0.041639	0.043130	0.044650	0.047778	0.051019
31	0.033565	0.034903	0.036274	0.037676	0.039109	0.040574	0.042070	0.043596	0.046739	0.049999
32	0.032556	0.033895	0.035266	0.036671	0.038108	0.039577	0.041078	0.042611	0.045768	0.049047
33	0.031608	0.032947	0.034320	0.035727	0.037168	0.038641	0.040148	0.041687	0.044859	0.048156
34	0.030716	0.032056	0.033431	0.034840	0.036284	0.037762	0.039274	0.040819	0.044007	0.047322
35	0.029875	0.031215	0.032592	0.034004	0.035451	0.036934	0.038451	0.040002	0.043206	0.046539
36	0.029081	0.030422	0.031800	0.033214	0.034665	0.036152	0.037675	0.039233	0.042452	0.045804
37	0.028330	0.029671	0.031051	0.032468	0.033923	0.035414	0.036943	0.038507	0.041741	0.045112
38	0.027618	0.028960	0.030342	0.031761	0.033220	0.034716	0.036250	0.037821	0.041070	0.044459
39	0.026943	0.028286	0.029669	0.031092	0.032554	0.034055	0.035594	0.037171	0.040436	0.043844
40	0.026302	0.027646	0.029030	0.030456	0.031921	0.033427	0.034972	0.036556	0.039836	0.043262
41	0.025692	0.027036	0.028423	0.029851	0.031321	0.032831	0.034382	0.035972	0.039268	0.042712
42	0.025111	0.026456	0.027845	0.029276	0.030749	0.032264	0.033821	0.035417	0.038729	0.042192
43	0.024557	0.025903	0.027293	0.028727	0.030205	0.031725	0.033287	0.034890	0.038217	0.041698
44	0.024029	0.025375	0.026768	0.028204	0.029686	0.031210	0.032778	0.034388	0.037730	0.041230
45	0.023523	0.024871	0.026265	0.027705	0.029190	0.030720	0.032293	0.033910	0.037268	0.040785
46	0.023040	0.024389	0.025785	0.027228	0.028717	0.030251	0.031830	0.033453	0.036827	0.040363
47	0.022578	0.023927	0.025325	0.126771	0.028264	0.029803	0.031388	0.033018	0.036407	0.039961
48	0.022134	0.023485	0.024885	0.026334	0.027831	0.029375	0.030966	0.032602	0.036006	0.039578
49	0.021709	0.023601	0.024463	0.025915	0.027416	0.028965	0.030561	0.032204	0.035623	0.039213
50	0.021301	0.022654	0.024058	0.025513	0.027018	0.028572	0.030174	0.031823	0.035258	0.038865

Table 7 Amortization Payments Table (*Continued*)

n	$\frac{1}{4}\%$	$\frac{1}{2}\%$	$\frac{3}{4}\%$	1%	$1\frac{1}{4}\%$	$1\frac{1}{2}\%$	$1\frac{3}{4}\%$	2%	$2\frac{1}{2}\%$	3%
51	0.020909	0.022263	0.023669	0.025127	0.026636	0.028195	0.029803	0.031459	0.034909	0.038534
52	0.020532	0.021887	0.023295	0.024756	0.026269	0.027833	0.029447	0.031109	0.034574	0.038217
53	0.020169	0.021525	0.022935	0.024400	0.025917	0.027485	0.029105	0.030774	0.034254	0.037915
54	0.019820	0.021177	0.022589	0.024057	0.025578	0.027151	0.028777	0.030452	0.033948	0.037626
55	0.019483	0.020841	0.022256	0.023726	0.025251	0.026830	0.028461	0.030143	0.033654	0.037349
56	0.019159	0.020518	0.021935	0.023408	0.024937	0.026521	0.028158	0.029847	0.033372	0.037084
57	0.018845	0.020206	0.021625	0.023102	0.024635	0.026223	0.027866	0.029561	0.033102	0.036831
58	0.018543	0.019905	0.021326	0.022806	0.024343	0.025937	0.027585	0.029287	0.032842	0.036588
59	0.018251	0.019614	0.021037	0.022520	0.024062	0.025660	0.027314	0.029022	0.032593	0.036356
60	0.017969	0.019333	0.020758	0.022244	0.023790	0.025393	0.027053	0.028768	0.032353	0.036133
61	0.017696	0.019061	0.020489	0.021978	0.023528	0.025136	0.026802	0.028523	0.032123	0.035919
62	0.017431	0.018798	0.020228	0.021720	0.023274	0.024888	0.026559	0.028286	0.021901	0.035714
63	0.017176	0.018543	0.019976	0.021471	0.023029	0.024647	0.026325	0.028058	0.031688	0.035517
64	0.016928	0.018297	0.019731	0.021230	0.022792	0.024415	0.026098	0.027839	0.031482	0.035328
65	0.016688	0.018058	0.019495	0.020997	0.022563	0.024191	0.025880	0.027626	0.031285	0.035146
66	0.016455	0.017826	0.019265	0.020771	0.022341	0.023974	0.025668	0.027421	0.031094	0.034971
67	0.016229	0.017602	0.019043	0.020551	0.022126	0.023764	0.025464	0.027223	0.030910	0.034803
68	0.016010	0.017384	0.018827	0.020339	0.021917	0.023560	0.025266	0.027032	0.030733	0.034642
69	0.015797	0.017172	0.018618	0.020133	0.021715	0.023363	0.025075	0.026847	0.030562	0.034486
70	0.015590	0.016967	0.018415	0.019933	0.021519	0.023172	0.024889	0.026668	0.030397	0.034337
71	0.015389	0.016767	0.018217	0.019739	0.021329	0.022987	0.024710	0.026494	0.030238	0.034193
72	0.015194	0.016573	0.018026	0.019550	0.021145	0.022808	0.024536	0.026327	0.030084	0.030084
73	0.015004	0.016384	0.017839	0.019367	0.020966	0.022634	0.024367	0.026165	0.029936	0.033921
74	0.014819	0.016201	0.017658	0.019189	0.020792	0.022465	0.024204	0.026007	0.029792	0.033792
75	0.014639	0.016022	0.017482	0.019016	0.020623	0.022301	0.024046	0.025855	0.029654	0.033668
76	0.014464	0.015848	0.017310	0.018848	0.020459	0.022141	0.023892	0.025708	0.029520	0.033548
77	0.014293	0.015679	0.017143	0.018684	0.020300	0.021987	0.023743	0.025564	0.029390	0.033433
78	0.014127	0.015514	0.016981	0.018525	0.020144	0.021836	0.023598	0.025426	0.029265	0.033322
79	0.013965	0.015354	0.016822	0.018370	0.019993	0.021690	0.023457	0.025291	0.029143	0.033215
80	0.013807	0.015197	0.016668	0.018219	0.019847	0.021548	0.023321	0.025161	0.029026	0.033112
81	0.013653	0.015044	0.016518	0.018072	0.019704	0.021410	0.023188	0.025034	0.028912	0.033012
82	0.013503	0.014896	0.016371	0.017929	0.019564	0.021276	0.023059	0.024911	0.028803	0.032916
83	0.013356	0.014750	0.016228	0.017789	0.019429	0.021145	0.022934	0.024792	0.028696	0.032823
84	0.013213	0.014609	0.016089	0.017653	0.019297	0.021018	0.022812	0.024676	0.028593	0.032733
85	0.013074	0.014470	0.015953	0.017520	0.019168	0.020894	0.022694	0.024563	0.028493	0.032647
86	0.012937	0.014335	0.015820	0.017391	0.019043	0.020773	0.022578	0.024454	0.028396	0.032563
87	0.012804	0.014203	0.015691	0.017264	0.018920	0.020656	0.022466	0.024347	0.028303	0.032482
88	0.012674	0.014074	0.015564	0.017141	0.018801	0.020541	0.022357	0.024244	0.028212	0.032404
89	0.012546	0.013948	0.015441	0.017021	0.018685	0.020430	0.022251	0.024144	0.028124	0.032328
90	0.012422	0.013825	0.015320	0.016903	0.018571	0.020321	0.022148	0.024046	0.028038	0.032256
91	0.012300	0.013705	0.015202	0.016788	0.018461	0.020215	0.022047	0.023951	0.027955	0.032185
92	0.012181	0.013587	0.015087	0.016676	0.018353	0.020112	0.021949	0.023859	0.027875	0.032117
93	0.012064	0.013472	0.014974	0.016567	0.018247	0.020011	0.021853	0.023769	0.027797	0.032051
94	0.011950	0.013360	0.014864	0.016460	0.018144	0.019913	0.021760	0.023681	0.027721	0.031987
95	0.011839	0.013249	0.014756	0.016355	0.018044	0.019817	0.021669	0.023596	0.027648	0.031926
96	0.011730	0.013141	0.014650	0.016253	0.017945	0.019723	0.021581	0.023513	0.027577	0.031866
97	0.011623	0.013036	0.014547	0.016153	0.017849	0.019632	0.021495	0.023432	0.027507	0.031809
98	0.011518	0.012932	0.014446	0.016055	0.017756	0.019543	0.021411	0.023354	0.027440	0.031753
99	0.011415	0.012831	0.014347	0.015959	0.017664	0.019456	0.021329	0.023277	0.027375	0.031699
100	0.011314	0.012732	0.014250	0.015866	0.017574	0.019371	0.021249	0.023203	0.927312	0.931647

Table 7 Amortization Payments Table (*Continued*)

n	$3\frac{1}{2}\%$	4%	$4\frac{1}{2}\%$	5%	$5\frac{1}{2}\%$	6%	$6\frac{1}{2}\%$	7%	$7\frac{1}{2}\%$	8%
1	1.035000	1.040000	1.045000	1.050000	1.055000	1.060000	1.065000	1.070000	1.075000	1.080000
2	0.526401	0.530196	0.533998	0.537805	0.541618	0.545437	0.549262	0.553092	0.556928	0.560769
3	0.356934	0.360349	0.363773	0.367209	0.370654	0.374110	0.377576	0.381052	0.384538	0.388034
4	0.272251	0.275490	0.278744	0.282012	0.285294	0.288591	0.291903	0.295228	0.298568	0.301921
5	0.221481	0.224627	0.227792	0.230975	0.234176	0.237196	0.240635	0.243891	0.247165	0.250456
6	0.187668	0.190762	0.193878	0.197017	0.200179	0.203363	0.206568	0.209796	0.213045	0.216315
7	0.163544	0.166610	0.169701	0.172820	0.175964	0.179135	0.182331	0.185553	0.188800	0.192072
8	0.145477	0.148528	0.151610	0.154722	0.157864	0.161036	0.164237	0.167468	0.170727	0.174015
9	0.131446	0.134493	0.137574	0.140690	0.143839	0.147022	0.150238	0.153486	0.156767	0.160080
10	0.120241	0.123291	0.126379	0.129505	0.132668	0.135868	0.139105	0.142378	0.145686	0.149029
11	0.111092	0.114149	0.117248	0.120389	0.123571	0.126793	0.130055	0.133357	0.136697	0.140076
12	0.103484	0.106552	0.109666	0.112825	0.116029	0.119277	0.122568	0.125902	0.129278	0.132695
13	0.097062	0.100144	0.103275	0.106456	0.109684	0.112960	0.116283	0.119651	0.123064	0.126522
14	0.091571	0.094669	0.097820	0.101024	0.104279	0.107585	0.110940	0.114345	0.117797	0.121297
15	0:086825	0.089941	0.903114	0.096342	0.099626	0.102963	0.106353	0.109795	0.113287	0.116830
16	0.082685	0.085820	0.089015	0.092270	0.095583	0.098952	0.102378	0.105858	0.109391	0.112977
17	0.079043	0.082199	0.085418	0.088699	0.092042	0.095445	0.098906	0.102425	0.106000	0.109629
18	0.075817	0.078993	0.082237	0.085546	0.088920	0.092357	0.095855	0.099413	0.103029	0.106702
19	0.072940	0.076139	0.079407	0.082745	0.086150	0.089621	0.093156	0.096753	0.100411	0.104128
20	0.070361	0.073582	0.076876	0.080243	0.083679	0.087185	0.090756	0.094393	0.098092	0.101852
21	0.068037	0.071280	0.074601	0.077996	0.081465	0.085005	0.088613	0.092289	0.096029	0.099832
22	0.065932	0.069199	0.072546	0.075971	0.079471	0.083046	0.086691	0.090406	0.094187	0.098032
23	0.064019	0.067309	0.070682	0.074137	0.077670	0.081278	0.084961	0.088714	0.092535	0.096422
24	0.062273	0.065587	0.068987	0.072471	0.076036	0.079679	0.083398	0.087189	0.091050	0.094978
25	0.060674	0.064012	0.067439	0.070952	0.074549	0.078227	0.081981	0.085811	0.089711	0.093679
26	0.059205	0.062567	0.066021	0.069564	0.073193	0.076904	0.080695	0.084561	0.088500	0.092507
27	0.057852	0.061239	0.064719	0.068292	0.071952	0.075697	0.079523	0.083426	0.087402	0.091448
28	0.056603	0.060013	0.063521	0.067123	0.070814	0.074593	0.078453	0.082392	0.086405	0.090489
29	0.055445	0.058880	0.062415	0.066046	0.069769	0.073580	0.077474	0.081449	0.085498	0.089619
30	0.054371	0.057830	0.061392	0.065051	0.068805	0.072649	0.076577	0.080586	0.084671	0.088827
31	0.053372	0.056855	0.060443	0.064132	0.067917	0.071792	0.075754	0.079797	0.083916	0.088107
32	0.052442	0.055949	0.059563	0.063280	0.067095	0.071002	0.074997	0.079073	0.083226	0.087451
33	0.051572	0.055104	0.058745	0.062490	0.066335	0.070273	0.074299	0.078408	0.082594	0.086852
34	0.050760	0.054315	0.057982	0.061755	0.065630	0.069598	0.073656	0.077797	0.082015	0.086304
35	0.049998	0.053577	0.057270	0.061072	0.064975	0.068974	0.073062	0.077234	0.081483	0.085803
36	0.049284	0.052887	0.056606	0.060434	0.064366	0.068395	0.072513	0.076715	0.080994	0.085345
37	0.048613	0.052240	0.055984	0.059840	0.063800	0.067857	0.072005	0.076237	0.080545	0.084924
38	0.047982	0.051632	0.055402	0.059284	0.063272	0.067358	0.071535	0.075795	0.080132	0.084539
39	0.047388	0.051061	0.054856	0.058765	0.062780	0.066894	0.071099	0.075387	0.079751	0.084185
40	0.046827	0.050523	0.054343	0.058278	0.062320	0.066462	0.070694	0.075009	0.079400	0.083860
41	0.046298	0.050017	0.053862	0.057822	0.061891	0.066059	0.070318	0.074660	0.079077	0.083561
42	0.045798	0.049540	0.053409	0.057395	0.061489	0.065683	0.069968	0.074336	0.078778	0.083287
43	0.045325	0.049090	0.052982	0.056993	0.061113	0.065333	0.069644	0.074036	0.078502	0.083034
44	0.044878	0.048665	0.052581	0.056616	0.060761	0.065006	0.069341	0.073758	0.078247	0.082802
45	0.044453	0.048262	0.052202	0.056262	0.060431	0.064700	0.069060	0.073500	0.078011	0.082587
46	0.044051	0.047882	0.051845	0.055928	0.060122	0.064415	0.068797	0.073260	0.077794	0.082390
47	0.043669	0.047522	0.051507	0.055614	0.059831	0.064148	0.068553	0.073037	0.077592	0.082208
48	0.043306	0.047181	0.051189	0.055318	0.059559	0.063898	0.068325	0.072831	0.077405	0.082040
49	0.042962	0.046857	0.050887	0.055040	0.059302	0.063664	0.068112	0.072639	0.077232	0.081886
50	0.042634	0.046550	0.050602	0.054777	0.059061	0.063444	0.067914	0.072460	0.077072	0.081743

Table 7 Amortization Payments Table (*Continued*)

n	$3\frac{1}{2}\%$	4%	$4\frac{1}{2}\%$	5%	$5\frac{1}{2}\%$	6%	$6\frac{1}{2}\%$	7%	$7\frac{1}{2}\%$	8%
51	0.042322	0.046259	0.050332	0.054529	0.058835	0.063239	0.067729	0.072294	0.076924	0.081611
52	0.042024	0.045982	0.050077	0.054294	0.058622	0.063046	0.067556	0.072139	0.076787	0.081490
53	0.041741	0.045719	0.049835	0.054073	0.058421	0.062866	0.067394	0.071995	0.076659	0.081377
54	0.041471	0.045469	0.049605	0.053864	0.058232	0.062696	0.067243	0.071861	0.076541	0.081274
55	0.041213	0.045231	0.049388	0.053667	0.058055	0.062537	0.067101	0.071736	0.076432	0.081178
56	0.040967	0.045005	0.049181	0.053480	0.057887	0.062388	0.066969	0.071620	0.076330	0.081090
57	0.040732	0.044789	0.048985	0.053303	0.057729	0.062247	0.066846	0.071512	0.076236	0.081008
58	0.040508	0.044584	0.048799	0.053136	0.057580	0.062116	0.066730	0.071411	0.076148	0.080932
59	0.040294	0.044388	0.048622	0.052978	0.057440	0.061992	0.066622	0.071317	0.076067	0.080862
60	0.040089	0.044202	0.048454	0.052828	0.057307	0.061876	0.066520	0.071229	0.075991	0.080798
61	0.039892	0.044024	0.048295	0.052686	0.057182	0.061766	0.066426	0.071147	0.075921	0.080738
62	0.039705	0.043854	0.048143	0.052552	0.057064	0.061664	0.066337	0.071071	0.075856	0.080683
63	0.039525	0.043692	0.047998	0.052424	0.056953	0.061567	0.066254	0.071000	0.075796	0.080632
64	0.039353	0.043538	0.047861	0.052304	0.056847	0.061476	0.066176	0.070934	0.075740	0.080585
65	0.039188	0.043390	0.047730	0.052189	0.056748	0.061391	0.066103	0.070872	0.075688	0.080541
66	0.039030	0.043249	0.047606	0.052081	0.056654	0.061310	0.066034	0.070814	0.075639	0.080501
67	0.038879	0.043115	0.047488	0.051978	0.056565	0.061235	0.065970	0.070760	0.075594	0.080464
68	0.038734	0.042986	0.047375	0.051880	0.056482	0.061163	0.065910	0.070710	0.075553	0.080429
69	0.038595	0.042863	0.047267	0.051787	0.056402	0.061096	0.065854	0.070663	0.075514	0.080397
70	0.038461	0.042745	0.047165	0.051699	0.056328	0.061033	0.065801	0.070620	0.075478	0.080368
71	0.038333	0.042633	0.047068	0.051616	0.056257	0.060974	0.065752	0.070579	0.075444	0.080340
72	0.038210	0.042525	0.046975	0.051536	0.056190	0.060918	0.065705	0.070541	0.075413	0.080315
73	0.038092	0.042422	0.046886	0.051461	0.056127	0.060865	0.065662	0.070505	0.075384	0.080292
74	0.037978	0.042323	0.046802	0.051390	0.056067	0.060815	0.065621	0.070472	0.075357	0.080270
75	0.037869	0.042229	0.046721	0.051322	0.056010	0.060769	0.065583	0.070441	0.075332	0.080250
76	0.037765	0.042139	0.046644	0.051257	0.055956	0.060725	0.065547	0.070412	0.075304	0.080231
77	0.037664	0.042052	0.046571	0.051196	0.055906	0.060683	0.065513	0.070385	0.075287	0.082214
78	0.037567	0.041969	0.046501	0.051138	0.055858	0.060644	0.065482	0.070359	0.075267	0.080198
79	0.037474	0.041890	0.046434	0.051082	0.055812	0.060607	0.065452	0.070336	0.075248	0.080183
80	0.037385	0.041814	0.046371	0.051030	0.055769	0.060573	0.065424	0.070314	0.075231	0.080170
81	0.037299	0.041741	0.046310	0.050980	0.055729	0.060540	0.065398	0.070293	0.075215	0.080157
82	0.037216	0.041671	0.046252	0.050932	0.055690	0.060509	0.065374	0.070274	0.075200	0.080146
83	0.037137	0.041605	0.046197	0.050887	0.055654	0.060480	0.065351	0.070256	0.075186	0.080135
84	0.037060	0.041541	0.046144	0.050844	0.055619	0.060453	0.065329	0.070239	0.075173	0.080125
85	0.036987	0.041479	0.046093	0.050803	0.055587	0.060427	0.065309	0.070223	0.075161	0.080116
86	0.036916	0.041420	0.046045	0.050764	0.055556	0.060402	0.065290	0.070209	0.075150	0.080107
87	0.036848	0.041364	0.045999	0.050727	0.055527	0.060380	0.065272	0.070195	0.075139	0.080099
88	0.036782	0.041310	0.045955	0.050692	0.055499	0.060358	0.065256	0.070182	0.075129	0.080092
89	0.036719	0.041258	0.045913	0.050659	0.055473	0.060338	0.065240	0.070170	0.075120	0.080085
90	0.036658	0.041208	0.045873	0.050627	0.055448	0.060318	0.065225	0.070159	0.075112	0.080079
91	0.036599	0.041160	0.045835	0.050597	0.055424	0.060300	0.065212	0.070149	0.075104	0.080073
92	0.036543	0.041114	0.045798	0.050568	0.055402	0.060283	0.065199	0.070139	0.075097	0.080067
93	0.036488	0.041070	0.045763	0.050541	0.055381	0.060267	0.065186	0.070130	0.075090	0.080062
94	0.036436	0.041028	0.045730	0.050515	0.055361	0.060252	0.065175	0.070121	0.075084	0.080058
95	0.036385	0.040987	0.045698	0.050490	0.055342	0.060238	0.065164	0.070113	0.075078	0.080053
96	0.036337	0.040949	0.045667	0.050466	0.055324	0.060224	0.065154	0.070106	0.075072	0.080050
97	0.036290	0.040911	0.045638	0.050444	0.055307	0.060211	0.065145	0.070097	0.075067	0.080046
98	0.036245	0.040875	0.045610	0.050423	0.055291	0.060199	0.065136	0.070092	0.075063	0.080042
99	0.036201	0.040841	0.045584	0.050402	0.055276	0.060188	0.065128	0.070086	0.075058	0.080039
100	0.036159	0.040808	0.045558	0.050383	0.055261	0.060177	0.065120	0.070081	0.075054	0.080036

Table 8 Sinking Fund Payments Table

n	$\frac{1}{4}\%$	$\frac{1}{2}\%$	$\frac{3}{4}\%$	1%	$1\frac{1}{4}\%$	$1\frac{1}{2}\%$	$1\frac{3}{4}\%$	2%	$2\frac{1}{2}\%$	3%
1	1.000000	1.000000	1.000000	1.000000	1.000000	1.000000	1.000000	1.000000	1.000000	1.000000
2	0.499376	0.498753	0.498132	0.497512	0.496894	0.496278	0.495663	0.495050	0.493827	0.492611
3	0.332502	0.331672	0.330846	0.330022	0.329201	0.328383	0.327567	0.326755	0.325137	0.323530
4	0.249065	0.248133	0.247205	0.246281	0.245361	0.244445	0.243532	0.242624	0.240818	0.239027
5	0.199003	0.198010	0.197022	0.196040	0.195062	0.194089	0.193121	0.192158	0.190247	0.188355
6	0.165628	0.164595	0.163569	0.162548	0.161534	0.160525	0.159523	0.158526	0.156550	0.154598
7	0.141789	0.140729	0.139675	0.138628	0.137589	0.136556	0.135531	0.134512	0.132495	0.130506
8	0.123910	0.122829	0.121756	0.120690	0.119633	0.118584	0.117543	0.116510	0.114467	0.112456
9	0.110005	0.108907	0.107819	0.106740	0.105671	0.104610	0.103558	0.102515	0.100457	0.098434
10	0.098880	0.097771	0.096671	0.095582	0.094503	0.093434	0.092375	0.091327	0.089259	0.087231
11	0.089778	0.088659	0.087551	0.086454	0.085368	0.084294	0.083230	0.082178	0.080106	0.078077
12	0.082194	0.081066	0.079951	0.078849	0.077758	0.076680	0.075614	0.074560	0.072487	0.070462
13	0.075776	0.074642	0.073522	0.072415	0.071321	0.070240	0.069173	0.068118	0.066048	0.064030
14	0.070275	0.069136	0.068011	0.066901	0.065805	0.064723	0.063656	0.062602	0.060537	0.058526
15	0.065508	0.064364	0.063236	0.062124	0.061026	0.059944	0.058877	0.057825	0.055766	0.053767
16	0.061336	0.060189	0.058059	0.057945	0.056847	0.055765	0.054700	0.053650	0.051599	0.049611
17	0.057656	0.056506	0.055373	0.054258	0.053160	0.052080	0.051016	0.049970	0.047928	0.045953
18	0.054384	0.053232	0.052098	0.050982	0.049885	0.048806	0.047745	0.046702	0.044670	0.042709
19	0.051457	0.050303	0.049167	0.048052	0.046955	0.045878	0.044821	0.043782	0.041761	0.039814
20	0.048823	0.047666	0.046531	0.045415	0.044320	0.043246	0.042191	0.041157	0.039147	0.037216
21	0.046439	0.045282	0.044145	0.043031	0.041937	0.040865	0.039815	0.038785	0.036787	0.034872
22	0.044273	0.043114	0.041977	0.040864	0.039772	0.038703	0.037656	0.036631	0.034647	0.032747
23	0.042295	0.041135	0.039998	0.038886	0.037797	0.036731	0.035688	0.034668	0.032696	0.030814
24	0.040481	0.039321	0.038185	0.037073	0.035987	0.034924	0.033886	0.032871	0.030913	0.029047
25	0.038813	0.037652	0.036516	0.035407	0.034322	0.033263	0.032230	0.031220	0.029276	0.027428
26	0.037273	0.036112	0.034977	0.033869	0.032787	0.031732	0.030703	0.029699	0.027769	0.025938
27	0.035847	0.034686	0.035552	0.032446	0.031367	0.030315	0.029291	0.028293	0.026377	0.024564
28	0.034523	0.033362	0.032229	0.031124	0.030049	0.029001	0.027982	0.026990	0.025088	0.023293
29	0.033291	0.032129	0.030997	0.029895	0.028822	0.027779	0.026764	0.025778	0.023891	0.022115
30	0.032141	0.030979	0.029848	0.028748	0.027679	0.026639	0.025630	0.024650	0.022788	0.021019
31	0.031065	0.029903	0.028774	0.027676	0.026609	0.025574	0.024570	0.023596	0.021739	0.019999
32	0.030056	0.028895	0.027766	0.026671	0.025608	0.024577	0.023578	0.022611	0.020768	0.019047
33	0.029108	0.027947	0.026820	0.025727	0.024668	0.023641	0.022648	0.021687	0.019859	0.018156
34	0.028216	0.027056	0.025931	0.024840	0.023784	0.022762	0.021774	0.020819	0.019007	0.017322
35	0.027375	0.026215	0.025092	0.024004	0.022951	0.021934	0.020951	0.020002	0.018206	0.016539
36	0.026581	0.025422	0.024300	0.023214	0.022165	0.021152	0.020175	0.019233	0.017452	0.015804
37	0.025830	0.024671	0.023551	0.022468	0.021423	0.020414	0.019443	0 018507	0.016741	0.015112
38	0.025118	0.023960	0.022842	0.021761	0.020720	0.019716	0.018750	0.017821	0.016070	0.014459
39	0.024443	0.023286	0.022169	0.021092	0.020054	0.019055	0.018094	0.017171	0.015436	0.013844
40	0.023802	0.022646	0.021530	0.020456	0.019421	0.018427	0.017472	0.016556	0.014836	0.013262
41	0.023192	0.022036	0.020923	0.019851	0.018821	0.017831	0.016882	0.015972	0.014268	0.012712
42	0.022611	0.021456	0.020345	0.019276	0.018249	0.017264	0.016321	0.015417	0.013729	0.012192
43	0.022057	0.020903	0.019793	0.018727	0.017705	0.016725	0.015787	0.014890	0.013217	0.011698
44	0.021529	0.020375	0.019268	0.018204	0.017186	0.016210	0.015278	0.014388	0.012730	0.011230
45	0.021023	0.019871	0.018765	0.017705	0.016690	0.015720	0.014793	0.013910	0.012268	0.010785
46	0.020540	0.019389	0.018285	0.017228	0.016217	0.015251	0.014330	0.013453	0.011827	0.010363
47	0.020078	0.018927	0.017825	0.016771	0.015764	0.014803	0.013888	0.013018	0.011407	0.009961
48	0.019634	0.018485	0.017385	0.016334	0.015331	0.014375	0.013466	0.012602	0.011006	0.009578
49	0.019209	0.018061	0.016963	0.015915	0.014916	0.013965	0.013061	0.012204	0.010623	0.009213
50	0.018801	0.017654	0.016558	0.015513	0.014518	0.013572	0.012674	0.011823	0.010258	0.008865

Table 8 Sinking Fund Payments Table (*Continued*)

n	$\frac{1}{4}\%$	$\frac{1}{2}\%$	$\frac{3}{4}\%$	1%	$1\frac{1}{4}\%$	$1\frac{1}{2}\%$	$1\frac{3}{4}\%$	2%	$2\frac{1}{2}\%$	3%
51	0.018409	0.017263	0.016169	0.015127	0.014136	0.013195	0.012303	0.011459	0.009909	0.008534
52	0.018032	0.016887	0.015795	0.014756	0.013769	0.012833	0.011947	0.011109	0.009574	0.008217
53	0.017669	0.016525	0.015435	0.014400	0.013417	0.012485	0.011605	0.010774	0.009254	0.007915
54	0.017320	0.016177	0.015089	0.014057	0.013078	0.012151	0.011277	0.010452	0.008948	0.007626
55	0.016983	0.015841	0.014756	0.013726	0.012751	0.011830	0.010961	0.010143	0.008654	0.007349
56	0.016659	0.015518	0.014435	0.013408	0.012437	0.011521	0.010658	0.009847	0.008372	0.007084
57	0.016345	0.015206	0.014125	0.013102	0.012135	0.011223	0.010366	0.009561	0.008102	0.006831
58	0.016043	0.014905	0.013826	0.012806	0.011843	0.010937	0.010085	0.009287	0.007842	0.006588
59	0.015751	0.014614	0.013537	0.012520	0.011562	0.010660	0.009814	0.009022	0.007593	0.006356
60	0.015469	0.014333	0.013258	0.012244	0.011290	0.010393	0.009553	0.008768	0.007353	0.006133
61	0.015196	0.014061	0.012989	0.011978	0.011028	0.010136	0.009302	0.008523	0.007123	0.005919
62	0.014931	0.013798	0.012728	0.011720	0.010774	0.009888	0.009059	0.008286	0.006901	0.005714
63	0.014676	0.013543	0.012476	0.011471	0.010529	0.009647	0.008825	0.008058	0.006688	0.005517
64	0.014428	0.013297	0.012231	0.011230	0.010292	0.009415	0.008598	0.007839	0.006482	0.005328
65	0.014188	0.013058	0.011995	0.010997	0.010063	0.009191	0.008380	0.007626	0.006285	0.005146
66	0.013955	0.012826	0.011765	0.010771	0.009841	0.008974	0.008168	0.007421	0.006094	0.004971
67	0.013729	0.012602	0.011543	0.010551	0.009626	0.008764	0.007964	0.007223	0.005910	0.004803
68	0.013510	0.012384	0.011327	0.010339	0.009417	0.008560	0.007766	0.007032	0.005733	0.004642
69	0.013297	0.012172	0.011118	0.010133	0.009215	0.008363	0.007575	0.006847	0.005562	0.004486
70	0.013090	0.011967	0.010915	0.009933	0.009019	0.008172	0.007389	0.006668	0.005397	0.004337
71	0.012889	0.011767	0.010717	0.009739	0.008829	0.007987	0.007210	0.006494	0.005238	0.004193
72	0.012694	0.011573	0.010526	0.009550	0.008645	0.007808	0.007036	0.006327	0.005084	0.004054
73	0.012504	0.011384	0.010339	0.009367	0.008466	0.007634	0.006867	0.006165	0.004936	0.003921
74	0.012319	0.011201	0.010158	0.009189	0.008292	0.007465	0.006704	0.006007	0.004792	0.003792
75	0.012139	0.011022	0.009982	0.009016	0.008123	0.007301	0.006546	0.005855	0.004654	0.003668
76	0.011964	0.010848	0.009810	0.008848	0.007959	0.007141	0.006392	0.005708	0.004520	0.003548
77	0.011793	0.010679	0.009643	0.008684	0.007800	0.006987	0.006243	0.005564	0.004390	0.003433
78	0.011627	0.010514	0.009481	0.008525	0.007644	0.006836	0.006098	0.005426	0.004265	0.003322
79	0.011465	0.010354	0.009322	0.008370	0.007493	0.006690	0.005957	0.005291	0.004143	0.003215
80	0.011307	0.010197	0.009168	0.008219	0.007347	0.006548	0.005821	0.005161	0.004026	0.003112
81	0.011153	0.010044	0.009018	0.008072	0.007204	0.006410	0.005688	0.005034	0.003912	0.003012
82	0.011003	0.009896	0.008871	0.007929	0.007064	0.006276	0.005559	0.004911	0.003803	0.002916
83	0.010856	0.009750	0.008728	0.007789	0.006929	0.006145	0.005434	0.004792	0.003696	0.002823
84	0.010713	0.009609	0.008589	0.007653	0.006797	0.006018	0.005312	0.004676	0.003593	0.002733
85	0.010574	0.009470	0.008453	0.007520	0.006668	0.005894	0.005194	0.004563	0.003493	0.002647
86	0.010437	0.009335	0.008320	0.007391	0.006543	0.005773	0.005078	0.004454	0.003396	0.002563
87	0.010304	0.009203	0.008191	0.007264	0.006420	0.005656	0.004966	0.004347	0.003303	0.002482
88	0.010174	0.009074	0.008064	0.007141	0.006301	0.005541	0.004857	0.004244	0.003212	0.002404
89	0.010046	0.008948	0.007941	0.007021	0.006185	0.005430	0.004751	0.004144	0.003124	0.002328
90	0.009922	0.008825	0.007820	0.006903	0.006071	0.005321	0.004648	0.004046	0.003038	0.002256
91	0.009800	0.008705	0.007702	0.006788	0.005961	0.005215	0.004547	0.003951	0.002955	0.002185
92	0.009681	0.008587	0.007587	0.006676	0.005853	0.005112	0.004449	0.003859	0.002875	0.002117
93	0.009564	0.008472	0.007474	0.006567	0.005747	0.005011	0.004353	0.003769	0.002797	0.002051
94	0.009450	0.008360	0.007364	0.006460	0.005644	0.004913	0.004260	0.003681	0.002721	0.001987
95	0.009339	0.008249	0.007256	0.006355	0.005544	0.004817	0.004169	0.003596	0.002648	0.001926
96	0.009230	0.008141	0.007150	0.006253	0.005445	0.004723	0.004081	0.003513	0.002577	0.001866
97	0.009123	0.008036	0.007047	0.006153	0.005349	0.004632	0.003995	0.003432	0.002507	0.001809
98	0.009018	0.007932	0.006946	0.006055	0.005256	0.004543	0.003911	0.003354	0.002440	0.001753
99	0.008915	0.007831	0.006847	0.005959	0.005164	0.004456	0.003829	0.003277	0.002375	0.001699
100	0.008814	0.007732	0.006750	0.005866	0.005074	0.004371	0.003749	0.003203	0.002312	0.001647

Table 8 Sinking Fund Payments Table (*Continued*)

n	$3\frac{1}{2}\%$	4%	$4\frac{1}{2}\%$	5%	$5\frac{1}{2}\%$	6%	$6\frac{1}{2}\%$	7%	$7\frac{1}{2}\%$	8%
1	1.000000	1.000000	1.000000	1.000000	1.000000	1.000000	1.000000	1.000000	1.000000	1.000000
2	0.491401	0.490196	0.488998	0.487805	0.486618	0.485437	0.484262	0.483092	0.481928	0.480769
3	0.321934	0.320349	0.318773	0.317209	0.315654	0.314110	0.312576	0.311052	0.309538	0.308034
4	0.237251	0.235490	0.233744	0.232012	0.230294	0.228591	0.226903	0.225228	0.223568	0.221921
5	0.186481	0.184627	0.182792	0.180975	0.179176	0.177396	0.175635	0.173891	0.173165	0.170456
6	0.152668	0.150762	0.148878	0.147017	0.145179	0.143363	0.141568	0.139796	0.138045	0.136315
7	0.128544	0.126610	0.124701	0.122820	0.120964	0.119135	0.117331	0.115553	0.113800	0.112072
8	0.110477	0.108528	0.106610	0.104722	0.102864	0.101036	0.099237	0.097468	0.095727	0.094015
9	0.096446	0.094493	0.092574	0.090690	0.088839	0.087022	0.085238	0.083486	0.081767	0.080080
10	0.085241	0.038291	0.081379	0.079505	0.077668	0.075868	0.074105	0.072378	0.070686	0.069029
11	0.076092	0.074149	0.072248	0.070389	0.068571	0.066793	0.065055	0.063357	0.061697	0.060076
12	0.068484	0.066552	0.064666	0.062825	0.061029	0.059277	0.057568	0.055902	0.054278	0.052695
13	0.062062	0.060144	0.058275	0.056456	0.054684	0.052960	1.051283	0.049651	0.048064	0.046522
14	0.056571	0.054669	0.052820	0.051024	0.049279	0.047585	0.045940	0.044345	0.042797	0.041297
15	0.051825	0.049941	0.048114	0.046342	0.044626	0.042963	0.041353	0.039795	0.038287	0.036830
16	0.047685	0.045820	0.044015	0.042270	0.040583	0.038952	0.037378	0.035858	0.034391	0.032977
17	0.044043	0.042199	0.040418	0.038699	0.037042	0.035445	0.033906	0.032425	0.031000	0.029629
18	0.040817	0.038993	0.037237	0.035546	0.033920	0.032357	0.030855	0.029413	0.028029	0.026702
19	0.037940	0.036139	0.034407	0.032745	0.031150	0.029621	0.028156	0.026753	0.025411	0.024128
20	0.035361	0.033582	0.031876	0.030243	0.028679	0.027185	0.025756	0.024393	0.023092	0.021852
21	0.033037	0.031280	0.029601	0.027996	0.026465	0.025005	0.023613	0.022289	0.021029	0.019832
22	0.030932	0.029199	0.027546	0.025971	0.024471	0.023046	0.021691	0.020406	0.019187	0.018032
23	0.029019	0.027309	0.025682	0.024137	0.022670	0.021278	0.019961	0.018714	0.017535	0.016422
24	0.027273	0.025587	0.023987	0.022471	0.021036	0.019679	0.018398	0.017189	0.016050	0.014978
25	0.025674	0.024012	0.022439	0.020952	0.019549	0.018227	0.016981	0.015811	0.014711	0.013679
26	0.024205	0.022567	0.021021	0.019564	0.018193	0.016904	0.015695	0.014561	0.013500	0.012507
27	0.022852	0.021239	0.019719	0.018293	0.016952	0.015697	0.014523	0.013426	0.012402	0.011448
28	0.021603	0.020013	0.018521	0.017123	0.015814	0.014593	0.013453	0.012392	0.011405	0.010489
29	0.020445	0.018880	0.017415	0.016046	0.014769	0.013580	0.012474	0.011449	0.010498	0.009619
30	0.019371	0.017830	0.016392	0.015051	0.013805	0.012649	0.011577	0.010586	0.009671	0.008827
31	0.018372	0.016855	0.015443	0.014132	0.012917	0.011792	0.010754	0.009797	0.008916	0.008107
32	0.017442	0.015949	0.014563	0.013280	0.012095	0.011002	0.009997	0.009073	0.008226	0.007451
33	0.016572	0.015104	0.013745	0.012490	0.011335	0.010273	0.009299	0.008408	0.007594	0.006852
34	0.015760	0.014315	0.012982	0.011755	0.010630	0.009598	0.008656	0.007797	0.007015	0.006304
35	0.014998	0.013577	0.012270	0.011072	0.009975	0.008974	0.008062	0.007234	0.006483	0.005803
36	0.014284	0.012887	0.011606	0.010434	0.009366	0.008395	0.007513	0.006715	0.005994	0.005345
37	0.013613	0.012240	0.010984	0.009840	0.008800	0.007857	0.007005	0.006237	0.005545	0.004924
38	0.012982	0.011632	0.010402	0.009284	0.008272	0.007358	0.006535	0.005795	0.005132	0.004539
39	0.012388	0.011061	0.009856	0.008765	0.007780	0.006894	0.006099	0.005387	0.004751	0.004185
40	0.011827	0.010523	0.009343	0.008278	0.007320	0.006462	0.005694	0.005009	0.004400	0.003860
41	0.011298	0.010017	0.008862	0.007822	0.006891	0.006059	0.005318	0.004660	0.004077	0.003561
42	0.010798	0.009540	0.008409	0.007395	0.006489	0.005683	0.004968	0.004336	0.003778	0.003287
43	0.010325	0.009090	0.007982	0.006993	0.006113	0.005333	0.004644	0.004036	0.003502	0.003034
44	0.009878	0.008665	0.007581	0.006616	0.005761	0.005006	0.004341	0.003758	0.003247	0.002802
45	0.009453	0.008262	0.007202	0.006262	0.005431	0.004700	0.004060	0.003500	0.003011	0.002587
46	0.009051	0.007882	0.006845	0.005928	0.005122	0.004415	0.003797	0.003260	0.002794	0.002390
47	0.008669	0.007522	0.006507	0.005614	0.004831	0.004148	0.003553	0.003037	0.002592	0.002208
48	0.008306	0.007181	0.006189	0.005318	0.004559	0.003898	0.003325	0.002831	0.002405	0.002040
49	0 007962	0.006857	0.005887	0.005040	0.004302	0.003664	0.003112	0.002639	0.002232	0.001886
50	0.007634	0.006550	0.005602	0.004777	0.004061	0.003444	0.002914	0.002460	0.002072	0.001743

Table 8 Sinking Fund Payments Table (*Continued*)

n	$3\frac{1}{2}\%$	4%	$4\frac{1}{2}\%$	5%	$5\frac{1}{2}\%$	6%	$6\frac{1}{2}\%$	7%	$7\frac{1}{2}\%$	8%
51	0.007322	0.006259	0.005332	0.004529	0.003835	0.003239	0.002729	0.002294	0.001924	0.001611
52	0.007024	0.005982	0.005077	0.004294	0.003622	0.003046	0.002556	0.002139	0.001787	0.001490
53	0.006741	0.005719	0.004835	0.004073	0.003421	0.002866	0.002394	0.001995	0.001659	0.001377
54	0.006471	0.005469	0.004605	0.003864	0.003232	0.002696	0.002243	0.001861	0.001541	0.001274
55	0.006213	0.005231	0.004388	0.003667	0.003055	0.002537	0.002101	0.001736	0.001432	0.001178
56	0.005967	0.005005	0.004181	0.003480	0.002887	0.002388	0.001969	0.001620	0.001330	0.001090
57	0.005732	0.004789	0.003985	0.003303	0.002729	0.002247	0.001846	0.001512	0.001236	0.001008
58	0.005508	0.004584	0.003799	0.003136	0.002580	0.002116	0.001730	0.001411	0.001148	0.000932
59	0.005294	0.004388	0.003622	0.002978	0.002440	0.001992	0.001622	0.001317	0.001067	0.000862
60	0.005089	0.004202	0.003454	0.002828	0.002307	0.001876	0.001520	0.001229	0.000991	0.000798
61	0.004892	0.004024	0.003295	0.002686	0.002182	0.001766	0.001426	0.001147	0.000921	0.000738
62	0.004705	0.003854	0.003143	0.002552	0.002064	0.001664	0.001337	0.001071	0.000856	0.000683
63	0.004525	0.003692	0.002998	0.002424	0.001953	0.001567	0.001254	0.001000	0.000796	0.000632
64	0.004353	0.003538	0.002861	0.002304	0.001847	0.001476	0.001176	0.000934	0.000740	0.000585
65	0.004188	0.003390	0.002730	0.002189	0.001748	0.001391	0.001103	0.000872	0.000688	0.000541
66	0.004030	0.003249	0.002606	0.002081	0.001654	0.001310	0.001034	0.000814	0.000639	0.000501
67	0.003879	0.003115	0.002488	0.001978	0.001565	0.001235	0.000970	0.000760	0.000594	0.000464
68	0.003734	0.002986	0.002375	0.001880	0.001482	0.001163	0.000910	0.000710	0.000553	0.000429
69	0.003595	0.002863	0.002267	0.001787	0.001402	0.001096	0.000854	0.000663	0.000514	0.000397
70	0.003461	0.002745	0.002165	0.001699	0.001328	0.001033	0.000801	0.000620	0.000478	0.000368
71	0.003333	0.002633	0.002068	0.001616	0.001257	0.000974	0.000752	0.000579	0.000444	0.000340
72	0.003210	0.002525	0.001975	0.001536	0.001190	0.000918	0.000705	0.000541	0.000413	0.000315
73	0.003092	0.002422	0.001886	0.001461	0.001127	0.000865	0.000662	0.000505	0.000384	0.000292
74	0.002978	0.002323	0.001802	0.001390	0.001067	0.000815	0.000621	0.000472	0.000357	0.000270
75	0.002869	0.002229	0.001721	0.001322	0.001010	0.000769	0.000583	0.000441	0.000332	0.000250
76	0.002765	0.002139	0.001644	0.001257	0.000956	0.000725	0.000547	0.000412	0.000309	0.000231
77	0.002664	0.002052	0.001571	0.001196	0.000906	0.000683	0.000513	0.000385	0.000287	0.000214
78	0.002567	0.001969	0.001501	0.001138	0.000858	0.000644	0.000482	0.000359	0.000267	0.000198
79	0.002474	0.001890	0.001434	0.001082	0.000812	0.000607	0.000452	0.000336	0.000248	0.000183
80	0.002385	0.001814	0.001371	0.001030	0.000769	0.000573	0.000424	0.000314	0.000231	0.000170
81	0.002299	0.001741	0.001310	0.000980	0.000729	0.000540	0.000398	0.000293	0.000215	0.000157
82	0.002216	0.001671	0.001252	0.000832	0.000690	0.000509	0.000374	0.000274	0.000200	0.000146
83	0.002137	0.001605	0.001197	0.000887	0.000654	0.000480	0.000351	0.000256	0.000186	0.000135
84	0.002060	0.001541	0.001144	0.000844	0.000619	0.000453	0.000329	0.000239	0.000173	0.000125
85	0.001987	0.001479	0.001093	0.000803	0.000587	0.000427	0.000309	0.000223	0.000161	0.000116
86	0.001916	0.001420	0.001045	0.000764	0.000556	0.000402	0.000290	0.000209	0.000150	0.000107
87	0.001848	0.001364	0.000999	0.000727	0.000527	0.000380	0.000272	0.000195	0.000139	0.000099
88	0.001782	0.001310	0.000955	0.000692	0.000499	0.000358	0.000256	0.000182	0.000129	0.000092
89	0.001719	0.001258	0.000913	0.000659	0.000473	0.000338	0.000240	0.000170	0.000120	0.000085
90	0.001658	0.001208	0.000873	0.000627	0.000448	0.000318	0.000225	0.000159	0.000112	0.000079
91	0.001599	0.001160	0.000835	0.000597	0.000424	0.000300	0.000212	0.000149	0.000104	0.000073
92	0.001543	0.001114	0.000798	0.000568	0.000402	0.000283	0.000199	0.000139	0.000097	0.000067
93	0.001488	0.001070	0.000763	0.000541	0.000381	0.000267	0.000186	0.000130	0.000090	0.000062
94	0.001436	0.001028	0.000730	0.000515	0.000361	0.000252	0.000175	0.000121	0.000084	0.000058
95	0.001385	0.000987	0.000698	0.000490	0.000342	0.000238	0.000164	0.000113	0.000078	0.000053
96	0.001337	0.000949	0.000667	0.000466	0.000324	0.000224	0.000154	0.000106	0.000072	0.000050
97	0.001290	0.000911	0.000638	0.000444	0.000307	0.000211	0.000145	0.000099	0.000067	0.000046
98	0.001245	0.000875	0.000610	0.000423	0.000291	0.000199	0.000136	0.000092	0.000063	0.000042
99	0.001201	0.000841	0.000584	0.000402	0.000276	0.000188	0.000128	0.000086	0.000058	0.000039
100	0.001159	0.000808	0.000558	0.000383	0.000261	0.000177	0.000120	0.000081	0.000054	0.000036

Answers to Selected Exercises and Tests

Chapter 1

Section 1.2

1. a) No b) No c) No d) Yes e) No f) Yes g) No h) Yes

3. a) $\{x|x$ is a car car manufacturer that is based in the United States$\}$

b) $\{x|x$ is a vowel in the English alphabet$\}$

c) $\{x|x$ is a prime number$\}$

d) $\{x|x$ is an airline carrier in the United States (or the world)$\}$

e) $\{x|x$ is a state in the United States that begins with the letter N$\}$

f) $\{x|x$ is a popular sport in the United States$\}$

g) $\{x|x$ is a position on a baseball team$\}$

h) $\{x|x$ is the national bird of the United States$\}$

i) $\{x|x$ is the first letter of the integers One, Two, Three, Four, ... $\}$

5. False **7.** False **9.** False **11.** True

13. False **15.** False **17.** False

19. a) 1 or 6 b) 3 or 6 c) 4 or 6 d) 2 or 6 e) 6 f) 5 or 1 or 6

21. $A' = \{2, 3, 5\}$ b) $B' = \{1, 3, 4, 6\}$ c) ϕ d) U

23.

Number of elements in set	1	2	3	4	5	6	7	8	10	25	50	n
Total number of possible subsets	2	4	8	16	32	64	128	256	1024	2^{25}	2^{50}	2^n

25. 16

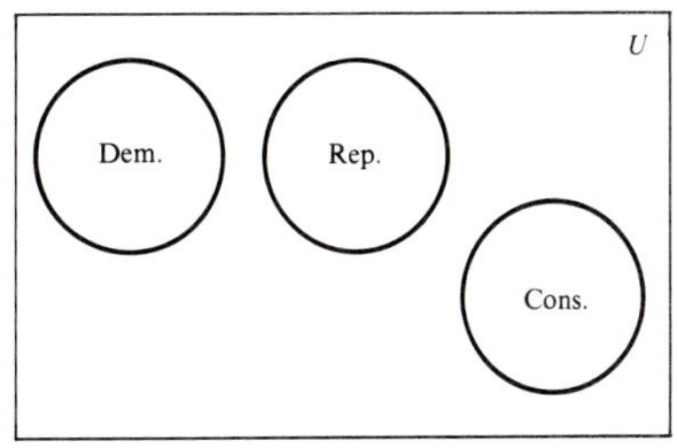

27. This assumes that a Democrat cannot be a Conservative and that a Republican cannot be a Conservative and that "Conservatives" is a full-fledged political party.

29. 10

Section 1.3

31. a) $\{c, d, e\}$ b) $\{e, g\}$ c) $\{e\}$ d) $\{a, b, c, d, e, f, g\}$ e) $\{a, b, c, d, e, g\}$ f) $\{a, e, g\}$ g) $\{a, e\}$ h) ϕ i) $\{c, d, f\}$ j) ϕ

33. a)

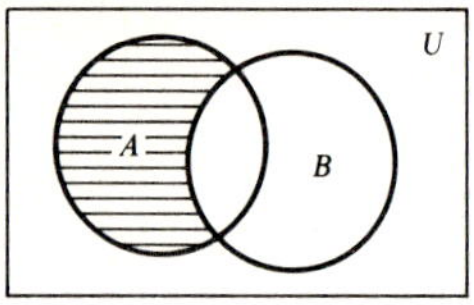

b)

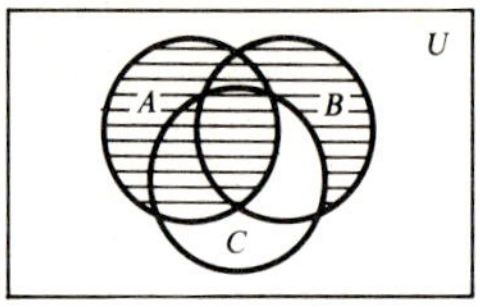

c)

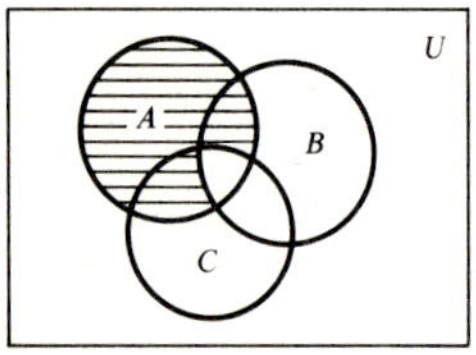

d)

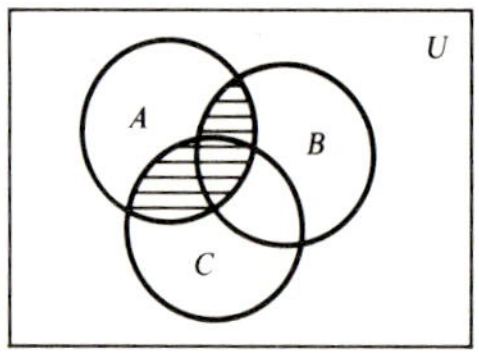

e)

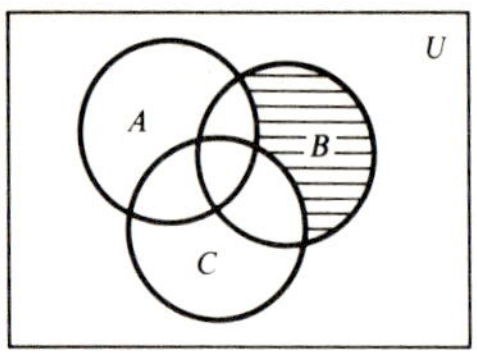

f)

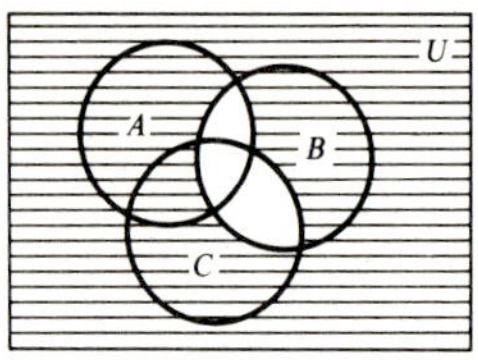

35. a) The set of all male patients who write with their right hand
b) The set of all tall male patients who write with their right hand
c) The set of all male patients or short patients who don't write with their right hand
d) The set of all male or female patients
e) The set of all tall female patients who write with their left hand

37. The set of all male applicants who earn above
a) $20,000 and who are applying for a home improvement loan
b) The set of all male or female applicants for car purchase loans
c) The set of all female applicants whose salary is $20,000 or below and who are not applying for a personal finance loan
d) The set of all male applicants whose salary is above $20,000 and who are applying for a home improvement and car purchase loan
e) The set of all female applicants whose salary is above $20,000 and who are applying for a home improvement loan
f) The set of all male applicants whose salary is above $20,000 and who are not applying for a home improvement loan

39.

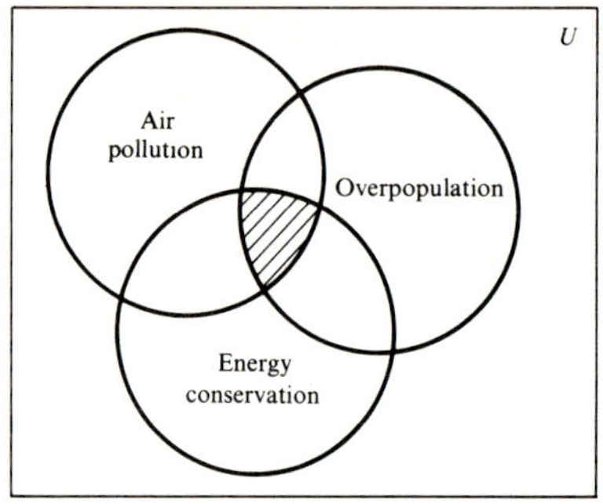

41.

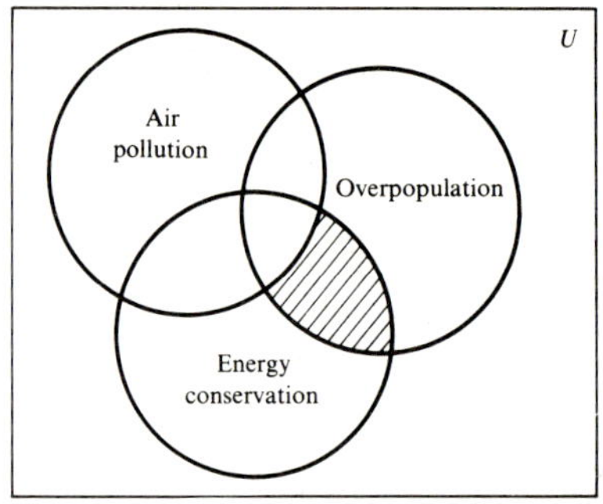

43.

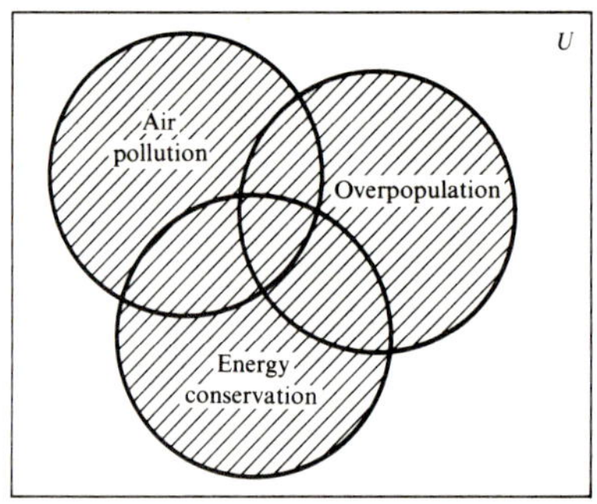

45.

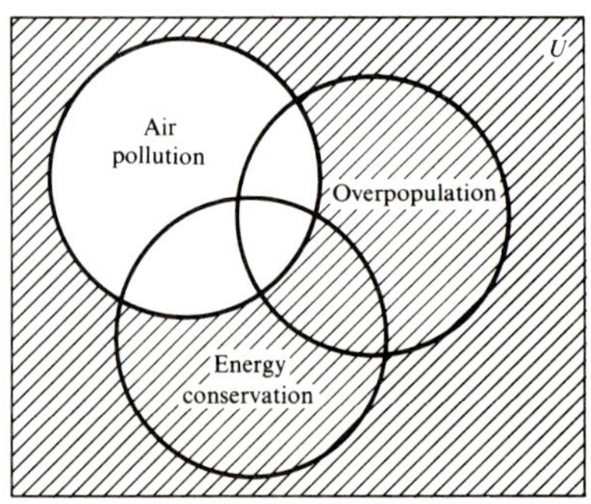

Section 1.4

47. a) 4 b) 172
49. a) 30 b) 10 c) 8
51. The numbers add up to 55 and not 50.
53. a) 8 b) 18 c) 28 d) 6 e) Type O positive f) Type A positive
55. a) 193 b) 130 c) 48 d) 7

Mastery Tests

FORM A

1. (d) **2.** (c) **3.** (d) **4.** (a) **5.** (d)
6. (c) **7.** (c) **8.** (c) **9.** (d)
10. (a) **11.** (b) **12.** (d) **13.** (b) **14.** (a)
15. (b) **16.** (d) **17.** (a) **18.** (c)

FORM B

1. The numbers are inconsistent.
2. Yes, Paul ate them.
3. $B = \phi$
4. $C = \{b, c, d, f\}$. Other answers are possible
5. a) 4 b) 3 c) 53 d) 77 e) 92 f) 129
6. 89
7. a) 1748 b) 7783 c) 8238 d) 0 e) 8238 f) 8175
8. An infinite number

Chapter 2

Section 2.2

1. a) Function b) Function c) Not a function d) Function
(Actually there are 6 functions depending upon whether each male or female selects the annual, semi-annual or quarterly payment plan.)
e) Function f) Function

3. a) Yes b) \$1400 c) $y = 150 + 50x$

5. (see table below)
7. a) No b) Approximately 9.15

5. $f(3)$	$f(-5)$	$f(0)$	$f(\frac{1}{3})$	$f(c)$	$f(x+h)$
a) 15	-1	9	$9\frac{2}{3}$	$2c + 9$	$2x + 2h + 9$
b) 26	74	-1	$-\frac{2}{3}$	$3c^2 - 1$	$3x^2 + 6xh + 3h^2 - 1$
c) $\frac{18}{5}$	$+2$	-3	-14	$\frac{5c + 3}{2c - 1}$	$\frac{5x + 5h + 3}{2x + 2h - 1}$
d) 3	$\sqrt{-7}$ is not a real number	$\sqrt{3}$	$\sqrt{\frac{11}{3}}$	$\sqrt{2c + 3}$	$\sqrt{2x + 2h + 3}$
e) $\sqrt{\frac{9}{7}} = \frac{3}{\sqrt{7}}$	$\sqrt{7}$	$\frac{\sqrt{3}}{2}$	$\sqrt{\frac{11}{13}}$	$\sqrt{\frac{2c + 3}{c + 4}}$	$\sqrt{\frac{2x + 2h + 3}{x + h + 4}}$
f) 27	-125	0	$\frac{1}{27}$	c^3	$(x + h)^3$ or $x^3 + 3x^2h + 3xh^2 + h^3$
g) 76	76	1	$\frac{4}{27}$	$c^3 + 7c^2 - 5c + 1$	$x^3 + 3x^2h + 3xh^2 + h^3 + 7x^2 + 14xh + 7h^2 - 5x - 5h + 1$
h) 3	13	3	$\frac{7}{3}$	$\lvert 2c - 3 \rvert$	$\lvert 2x + 2h - 3 \rvert$

Section 2.3

9.

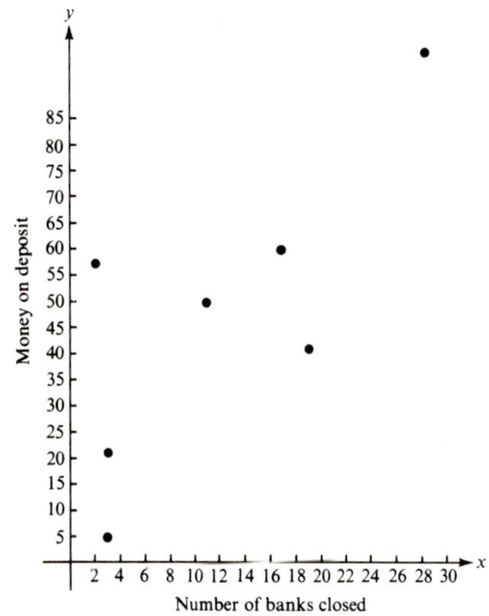

11.

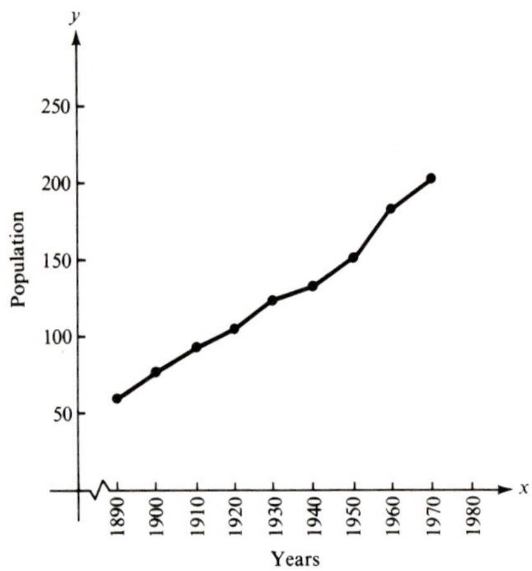

About 210 million

13. b) Yes

Section 2.4

15.

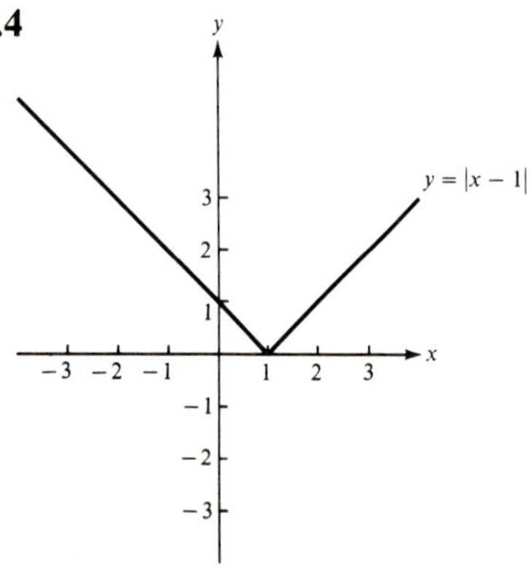

17.

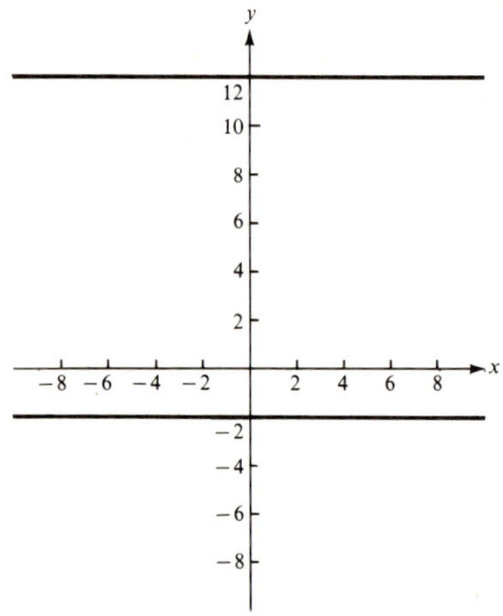

Graph of $|y - 5| = 7$

19.

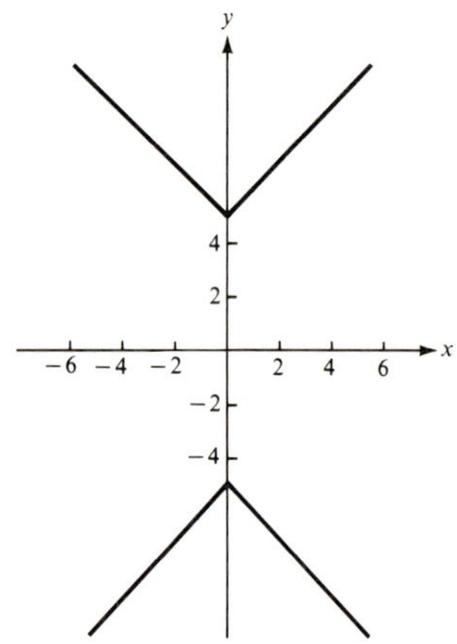

Graph of $|y| - |x| = 5$

Section 2.5

21. a) $\frac{7}{5}$ b) $\frac{7}{8}$ c) 1 d) $\frac{3}{5}$ e) $\frac{5}{2}$ f) 1 g) -6 h) $\frac{1}{2}$ i) 0 j) Undefined slope

23. No

25. a) $y = -4x - 9$ b) $2y = 3x + 1$ c) $y = -7x + 28$ d) $y = -2x - 15$ e) $2y + 3x = 0$

27. $y = 2x + 5$

29. a) Yes b) Yes c) Yes

Section 2.6

31.

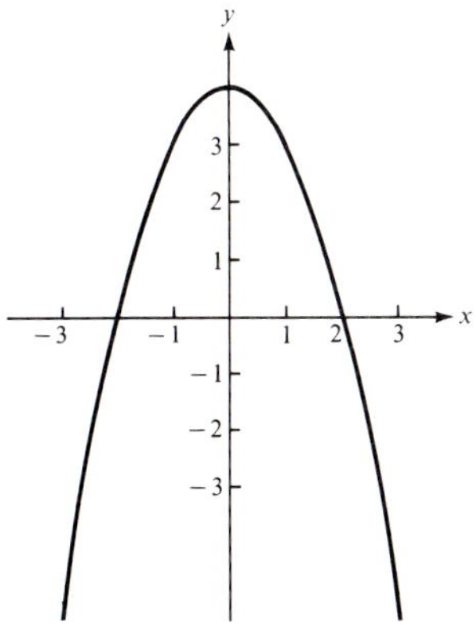

Graph of $y + x^2 = 4$

33.

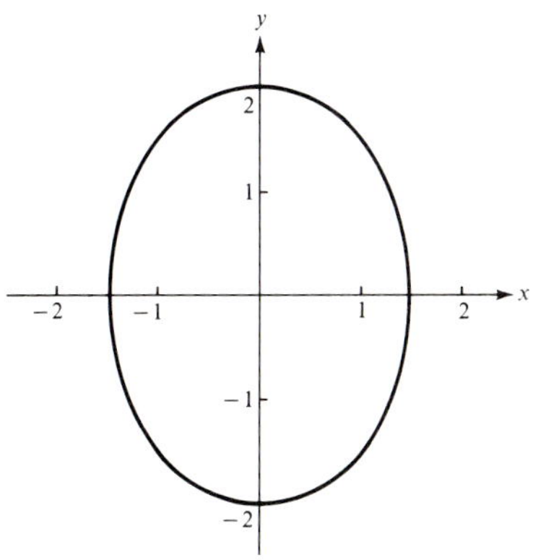

Graph of $2x^2 + y^2 = 4$

35.

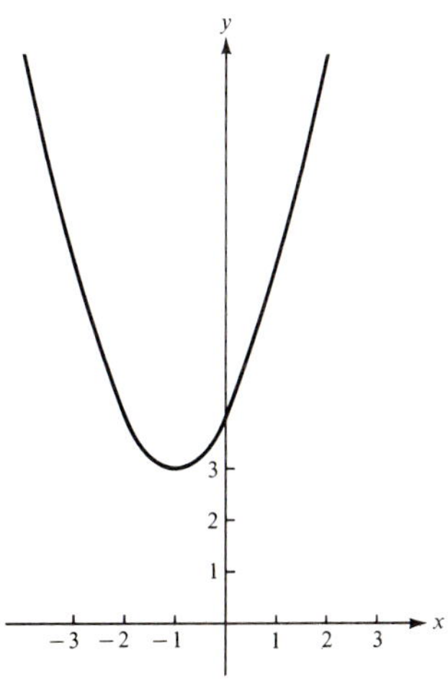

Graph of $y = x^2 + 2x + 4$

37.

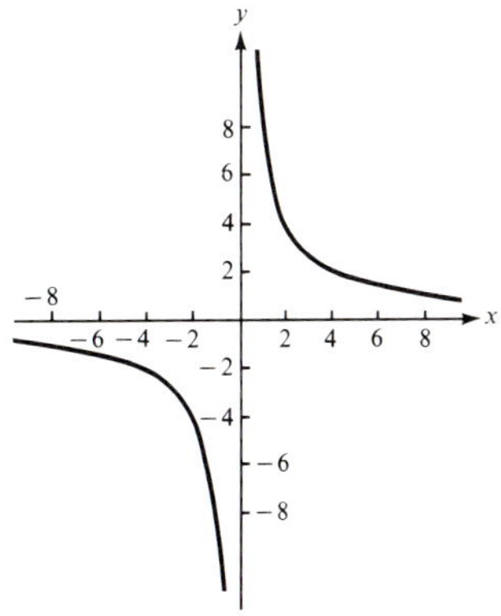

Graph of $xy = 8$

39. b) At 8:00 A.M.

41. b) $-32°$F c) 8 quarts

Section 2.7

43.

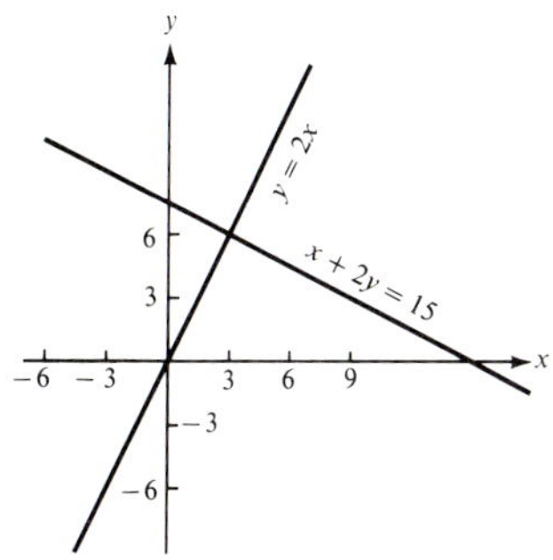

$x = 3$, $y = 6$

45.

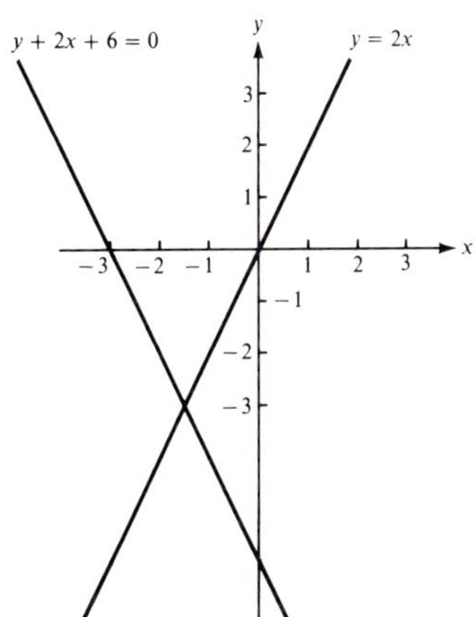

$x = -\frac{3}{2}$, $y = -3$

47.

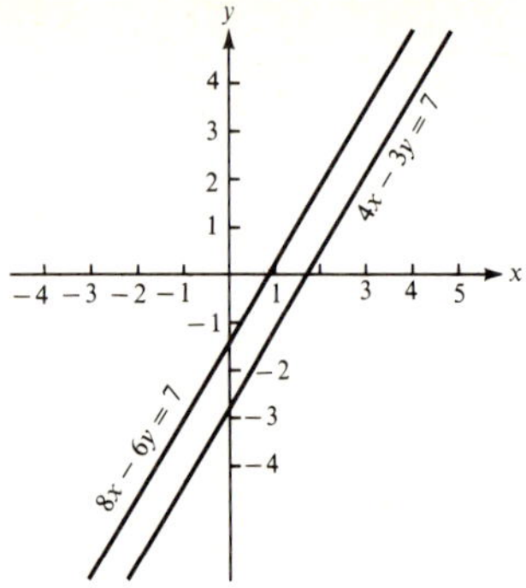

No point of intersection; the lines are parallel.

49.

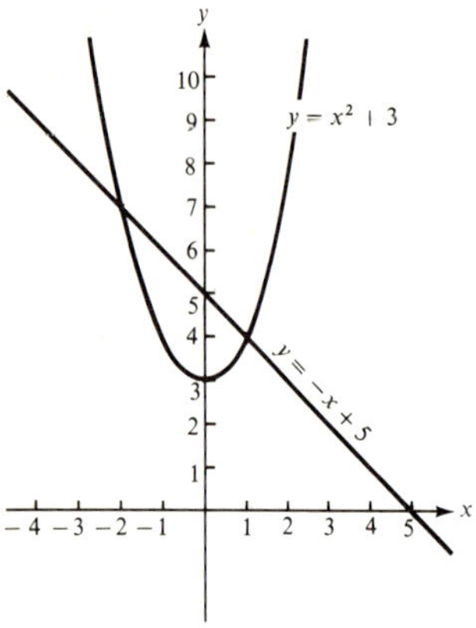

$x = -2, y = 7$ and $x = 1, y = 4$

51.

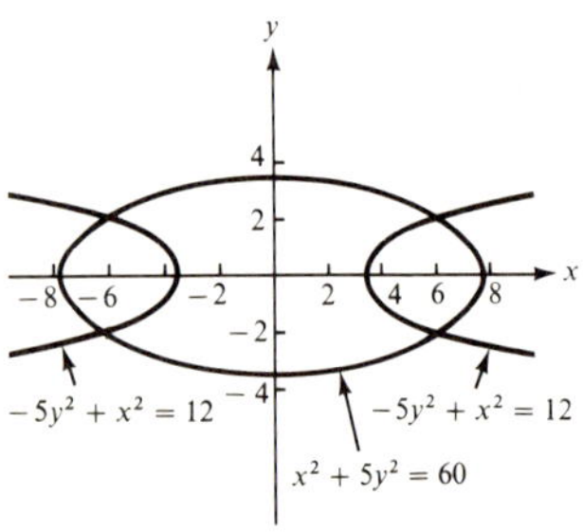

$x = +6, y = 2.19$
$x = +6, y = -2.19$
$x = -6, y = 2.19$
$x = -6, y = -2.19$

53. a) 480 b) 200 c) 333.33

55. a) 0; 200 b) 100; 180; 180

Section 2.8

57. 95 milligrams per 5000 liters of water

59. a)

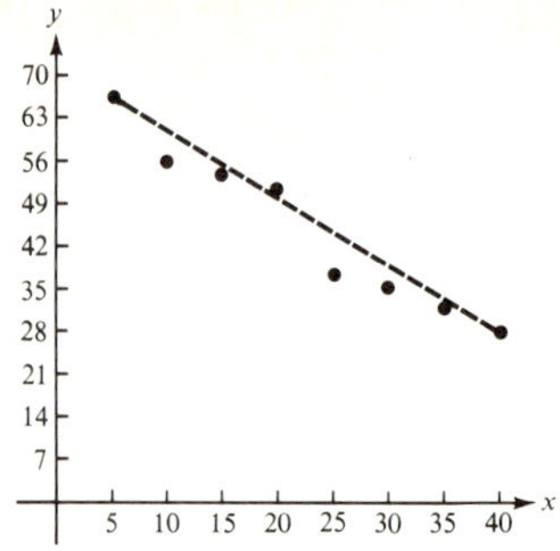

b) $y = -1.09x + 71.6$
c) 77.05 fires

61. a) \$2500 b) \$20,000 c) \$30,000

63. 16

Section 2.9

65. a) 64 b) 32 c) 1 d) $\frac{1}{2}$ e) 8 f) 1024
g) $\frac{1}{343}$ h) 1024 i) 16 j) 15,625 k) 18
l) $2^6 \cdot 3^9$ or $(108)^3$ or 1259712

67. a) $\frac{1}{4^2}$ b) $\frac{1}{16^{1/4}}$ c) $\frac{1}{10}$ d) $\frac{7}{4}$ e) $\left(\frac{5}{2}\right)^3$

f) $\left(\frac{7}{5}\right)^{2/3}$ g) $\frac{1}{xy^2}$ h) $\frac{2}{x^2}$ i) $\frac{4}{x^{1/3}}$ j) $\frac{1}{x^{1/2}y^{1/3}}$

k) $\frac{2}{x^{1/2}y^{3/7}}$ l) $25x^3y^4$

Mastery Tests

FORM A

1. Choice (b) **2.** 7, 5, or 6 **3.** 0 **4.** No
5. 5 **6.** −18 **7.** 175 **8.** Choice (a)
9. Choice (d) **10.** $y = .65x + 125$
11. a) No; Total cost also depends on whether payment is enclosed
b) \$8.94 without COD charges and \$10.44 with COD charges c) \$8.94

FORM B

1. b) $F = \frac{9}{5}C + 32$ or $C = \frac{5}{9}(F - 32)$ c) 20°C
2. a) \$2000 b) \$8000
3. a) Year 1 = \$142.22; Year 2 = \$124.44;
Year 3 = \$106.67; Year 4 = \$88.89;
Year 5 = \$71.11; Year 6 = \$53.33;
Year 7 = \$35.56; Year 8 = \$17.78
b) \$80
4. By the sum of the year's digit method the depreciation is: year 1 = \$8333.33; year 2 = \$6666.67;

year 3 = \$5000; year 4 = \$3333.33; year 5 = \$1666.67 By the straight line method, the annual depreciation is \$5000.

5. Year 1 = \$500; year 2 = \$375; year 3 = \$281.25; year 4 = \$210.94; year 5 = \$158.20; year 6 = \$118.65; year 7 = \$88.99; year 8 = \$66.74

6. a) The depreciation allowable under each method is

Year	200% declining balance	Sum of year's digit	Straight line
1	\$2360.00	\$1966.67	\$1180
2	1416.00	1573.33	1180
3	849.60	1180.00	1180
4	509.76	786.67	1180
5	305.86	393.33	1180
	\$5441.22	\$5900.00	\$5900

b) It depends upon the situation.

7. It must sell 75 calculators.

8. \$30,000 in bonds and \$50,000 in time deposit certificates

9. $x = 15$

11. a)

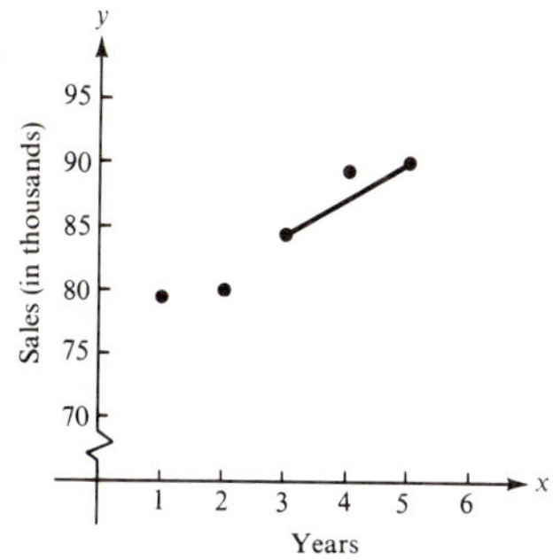

c) $y = 3x + 75$ d) \$93,000

12.

Year	Predicted sales (in thousands)	Actual sales (in thousands)
0	\$75	\$75
1	78	79
2	81	80
3	84	84
4	87	89
5	90	90

The prediction equation is a fairly good estimator of the actual sales.

Chapter 3

Section 3.2

1. a) 3×4 b) 4×3 c) 3×3 d) 1×3 e) 2×3 f) 3×1 g) 1×1

3. No

5. $x = 5, y = 8$ 7. Not equal

9. $\begin{pmatrix} 7 & 4 & 8 & 2 \\ 6 & 5 & 9 & 0 \\ 2 & 3 & 0 & 6 \\ 5 & 1 & 5 & 3 \\ 9 & 2 & 0 & 2 \\ 3 & 5 & 8 & 0 \end{pmatrix}$

11. a) $\begin{pmatrix} 21{,}282 & 8454 & 25{,}468 & 72{,}900 & 80{,}600 \\ 19{,}158 & 8412 & 21{,}926 & 91{,}600 & 84{,}400 \\ 18{,}537 & 7680 & 24{,}125 & 60{,}300 & 58{,}600 \\ 18{,}502 & 7785 & 49{,}222 & 78{,}700 & 66{,}200 \\ 17{,}546 & 6755 & 15{,}857 & 52{,}100 & 42{,}400 \\ 17{,}186 & 6889 & 8450 & 67{,}800 & 54{,}700 \\ 17{,}112 & 7096 & 44{,}340 & 64{,}000 & 57{,}800 \\ 16{,}909 & 6716 & 17{,}891 & 61{,}400 & 52{,}000 \\ 15{,}166 & 7568 & 12{,}530 & 78{,}400 & 66{,}100 \\ 14{,}961 & 7529 & 39{,}322 & 85{,}800 & 81{,}600 \end{pmatrix}$

b) St. Louis

Section 3.3

13. a) $\begin{pmatrix} +5 & +18 & +3 & +13 & +62 \\ +12 & +6 & +2 & +7 & +46 \\ -1 & -2 & +1 & +5 & +47 \end{pmatrix}$

b) A pound of meat in city A

c) A dozen eggs in city C

15. 370 calories, 14 grams protein, 57 grams carbohydrates, 10 grams fat

17. In both cases our answer is $\begin{pmatrix} 13 & 22 & 0 \\ 6 & 8 & 2 \\ 18 & 11 & 17 \end{pmatrix}$

Section 3.4

19. $\begin{pmatrix} -1 & 16 \\ 26 & 48 \end{pmatrix}$ 21. $\begin{pmatrix} 71 & 64 \\ 20 & -41 \\ 37 & 49 \end{pmatrix}$

23. $\begin{pmatrix} 94 & 25 & 53 \\ 84 & 39 & 27 \\ -52 & -19 & -23 \end{pmatrix}$

25. $\begin{pmatrix} 35 & 30 & 15 & 40 \\ 7 & 6 & 3 & 8 \\ 56 & 48 & 24 & 64 \\ 21 & 18 & 9 & 24 \end{pmatrix}$

27. Can't be done

29. $\begin{pmatrix} 6 & -48 & 12 \\ 60 & -72 & -18 \\ 36 & 30 & 0 \end{pmatrix}$ **31.** $\begin{pmatrix} 69 & -27 \\ 12 & 38 \end{pmatrix}$

33. $\begin{pmatrix} -80 & -30 & -46 \\ 28 & -102 & 0 \\ 2 & -10 & 112 \end{pmatrix}$ **35.** $\begin{pmatrix} 0 & 1 \\ 1 & 0 \end{pmatrix}$

37. $b = -c, a = d$

39. $A = \begin{pmatrix} 1 & 0 \\ 0 & 0 \end{pmatrix}$. Other answers are possible.

41. a) $(12 \quad 4 \quad \frac{1}{3} \quad 2 \quad 3 \quad 2)$ b) $\begin{pmatrix} 102 \\ 110 \\ 66 \\ 100 \\ 325 \\ 90 \end{pmatrix}$

c) 3041 calories

43. $\begin{pmatrix} 125{,}000 \\ 166{,}200 \\ 135{,}100 \\ 188{,}800 \end{pmatrix}$

45. a) Bill, \$12,340; Mike \$21,675
b) Bill, \$25,410; Mike \$44,725

55. $\left(\begin{array}{cc|c} 4 & 3 & 7 \\ 3 & 5 & 11 \end{array}\right)$ **57.** $\left(\begin{array}{cc|c} 3 & -2 & 5 \\ 5 & 3 & -8 \end{array}\right)$

59. $\left(\begin{array}{ccc|c} 7 & -5 & 3 & 8 \\ 4 & 7 & -9 & -3 \end{array}\right)$ **61.** $\left(\begin{array}{cc|c} 1 & 2 & 7 \\ 3 & 4 & 8 \\ 7 & -8 & 1 \end{array}\right)$

63. $\left(\begin{array}{ccc|c} 1 & 2 & 0 & 8 \\ 0 & 3 & -4 & 10 \\ 3 & 0 & 5 & 11 \end{array}\right)$

65. a) $7x + 2y = 12$
$3x - 4y = 14$
b) $2x + 5y = 15$
$-3x + 6y = 1$
c) $7x - 2y - 9z + 6w = 1$
$-3x + 2y + 5z + 8w = 10$
d) $7x + 6y + 2z = 6$
$4x - 5y - 9z = 6$
$5x + 3y + 2z = 4$
e) $x + 3y + 2z + 5w = 5$
$3x + 4y + z = -2$
$3x - 2y + 7z + 6w = 4$

67. $x = 3$, $y = 2$ **69.** $x = 3$, $y = -2$ **71.** $x = \frac{1}{3}$, $y = \frac{1}{2}$

73. $x = 3$, $y = 2$, $z = 4$ **75.** $x = 4$, $y = -5$, $z = 2$ **77.** $x = 4$, $y = 3$, $z = 5$

79. $x = 1$, $y = -2$, $z = 3$, $w = -4$

81. Bat \$4.00 and ball \$2.50

83. A white mouse needs three units and a black mouse needs two units.

85. 1200 males and 1000 females

87. Motorboat, \$7; rowboat, \$3

Section 3.6

47. Multiply row 1 by 5.

49. Multiply row 1 by $\frac{1}{5}$. Multiply row 2 by $\frac{1}{3}$. Finally, multiply row 3 by $\frac{1}{4}$.

51. Interchange rows 1 and 2, then multiply row 1 by $\frac{1}{3}$, then multiply row 1 by 5 and add to row 2, then interchange rows 2 and 3. Finally, multiply row 1 by -8 and add to row 2.

53. Multiply row 1 by $\frac{1}{4}$. Multiply row 1 by -6 and add to row 2. Interchange rows 2 and 3. Multiply row 1 by -5 and add to row 2. Multiply row 2 by $\frac{1}{2}$.

Section 3.7

89. $x = 3$, $y = 2$ **91.** $x = 2$, $y = -3$ **93.** $x = 5$, $y = -3$

95. $x = 7$, $y = -2$ **97.** $x = \frac{1}{2}$, $y = \frac{1}{3}$ **99.** $x = 3$, $y = -1$, $z = -2$

101. $x = 2$, $y = 4$, $z = -5$ **103.** $x = 3$, $y = 1$, $z = 7$

105. 2.78 cups of skim milk and 0.48 cups of orange juice

107. 36 members at a cost of \$100 to each

Section 3.8

109. $\begin{pmatrix} 1 & 0 & 1 \\ 1 & 1 & 0 \\ 0 & 1 & 1 \end{pmatrix}$

111.

	Calvin	Doug	Felice	Andrea	Mike	Trudy
Calvin	0	1	0	0	0	0
Doug	0	0	0	1	0	1
Felice	0	0	0	0	0	0
Andrea	0	0	0	0	0	0
Mike	0	0	1	0	0	0
Trudy	0	0	0	0	1	0

In this matrix, "1" stands for "has the number of."

113. SAEYXVP-YBA-HXNSJA-TNXEY-OVZ
115. "HAVE THE FILMS"

Mastery Tests

FORM A

1. False **2.** No **3.** Yes **4.** False
5. Square matrix **6.** No **7.** True

8. $\begin{pmatrix} 1 & 0 & 0 & 0 \\ 0 & 1 & 0 & 0 \\ 0 & 0 & 1 & 0 \\ 0 & 0 & 0 & 1 \end{pmatrix}$ **9.** No

10. $\left(\begin{array}{cc|c} 5 & -3 & 7 \\ 3 & 8 & 26 \end{array}\right)$

FORM B

1. a) $\begin{pmatrix} 16 & 20 \\ 0 & 2 \\ 0 & 3 \\ 7 & 11 \\ 8 & 8 \\ 466 & 702 \\ 566 & 735 \\ 456 & 660 \\ 5365 & 6961 \end{pmatrix}$

b) In 1974, 6884 elective jobs were held by women.
In 1978, 9102 elective jobs were held by women.

2. a) $\begin{pmatrix} 1 & 2 & 3 & 4 & 5 & 6 & 7 & 8 & 9 & 10 \\ 2 & 4 & 6 & 8 & 10 & 12 & 14 & 16 & 18 & 20 \\ 3 & 6 & 9 & 12 & 15 & 18 & 21 & 24 & 27 & 30 \\ 4 & 8 & 12 & 16 & 20 & 24 & 28 & 32 & 36 & 40 \\ 5 & 10 & 15 & 20 & 25 & 30 & 35 & 40 & 45 & 50 \\ 6 & 12 & 18 & 24 & 30 & 36 & 42 & 48 & 54 & 60 \\ 7 & 14 & 21 & 28 & 35 & 42 & 49 & 56 & 63 & 70 \\ 8 & 16 & 24 & 32 & 40 & 48 & 56 & 64 & 72 & 80 \\ 9 & 18 & 27 & 36 & 45 & 54 & 63 & 72 & 91 & 90 \\ 10 & 20 & 30 & 40 & 50 & 60 & 70 & 80 & 90 & 100 \end{pmatrix}$

b) This is the familiar multiplication table.

3. $A = \begin{pmatrix} 3 & 7 & 1 \\ 0 & 2 & 4 \\ 1 & 7 & 6 \end{pmatrix}, \quad B = \begin{pmatrix} 5 & -3 & 2 \\ 7 & 6 & 5 \\ 2 & 1 & 0 \end{pmatrix}$

4. In both cases our answer is $\begin{pmatrix} 129 & -321 \\ 238 & -494 \end{pmatrix}$

5. a) 68 b) 92

6. a) 8.1 units of vitamin A
2.7 units of vitamin B
2.9 units of vitamin C
4.3 units of vitamin D

b) Food I costs 19¢
Food II costs 24.2¢
Food III costs 18.2¢

7. Produce 12 type I packages, 6 type II packages, and 3 type III packages.

8. Type I costs 87¢, type II costs 82¢, and type III costs 92¢.

9. $\begin{pmatrix} 55.3 \\ 71.00 \\ 74.55 \\ 77.55 \\ 89.35 \end{pmatrix}$

10. a) $\begin{pmatrix} 76 & 81 & 83 & 93 & 97 & 108 & 120 & 134 & 139 & 144 & 149 & 157 & 163 & 169 \\ 13 & 12 & 7 & 7 & 7 & 8 & 9 & 9 & 9 & 11 & 12 & 13 & 14 & 16 \\ 15 & 17 & 16 & 17 & 19 & 22 & 26 & 27 & 17 & 16 & 17 & 18 & 18 & 19 \\ 22 & 16 & 15 & 8 & 5 & 3 & 1 & 1 & 0 & 0 & 0 & 0 & 0 & 0 \end{pmatrix}$

b) 1713 c) 71

11. In both cases the answer is $\begin{pmatrix} 70 & 45 \\ 28 & 11 \end{pmatrix}$.

12. $A = \begin{pmatrix} 5 & -6 \\ 2 & 1 \end{pmatrix} \quad A^{-1} = \begin{pmatrix} \frac{1}{17} & \frac{6}{17} \\ -\frac{2}{17} & \frac{5}{17} \end{pmatrix}$

$$(A^{-1})^{-1} = \begin{pmatrix} 5 & -6 \\ 2 & 1 \end{pmatrix}$$

13. Any matrix which has a row of zeros has no inverse.

Chapter 4

Section 4.2

1.

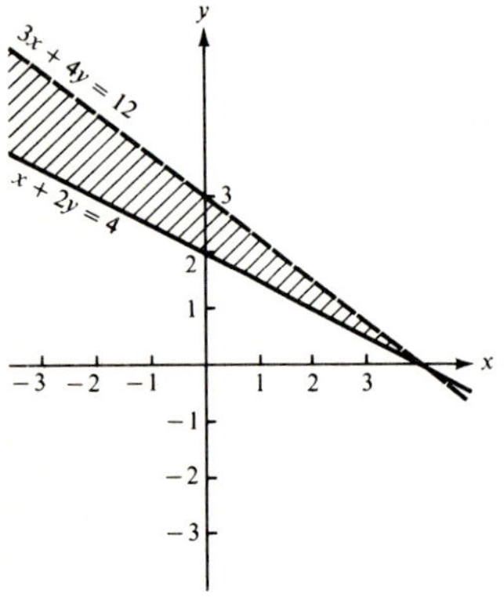

3.

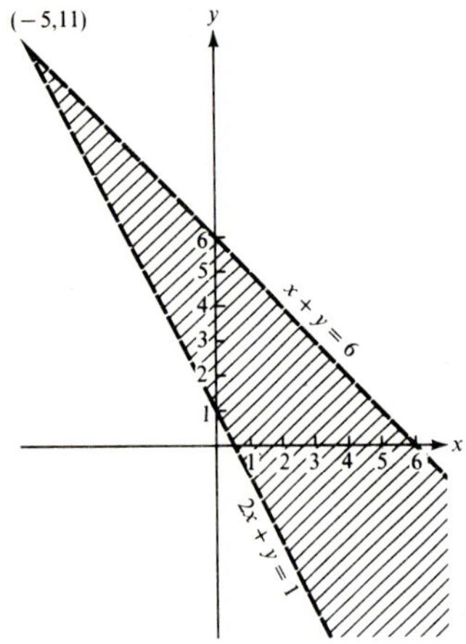

5.

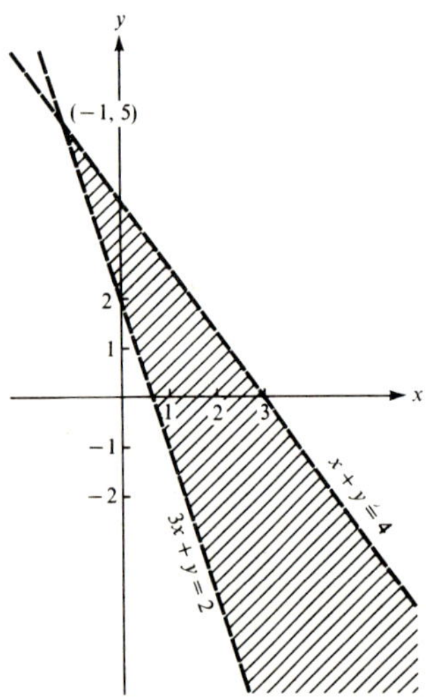

7.

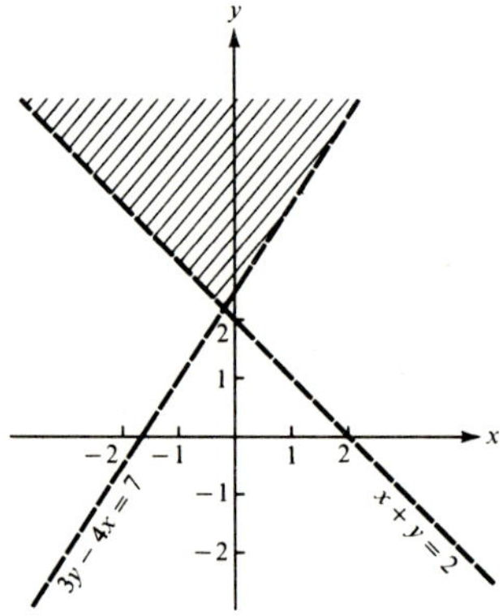

9.

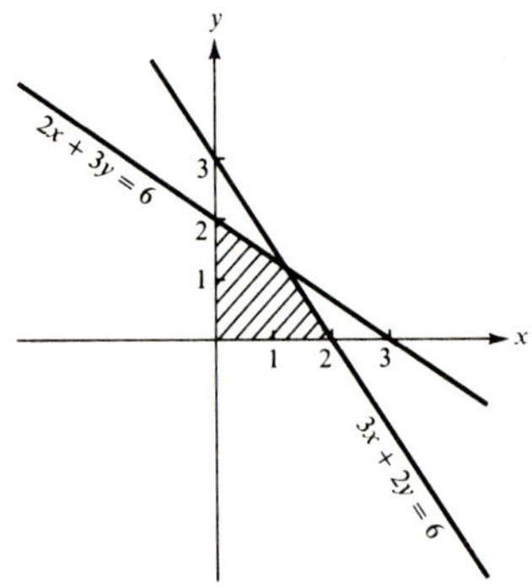

11.

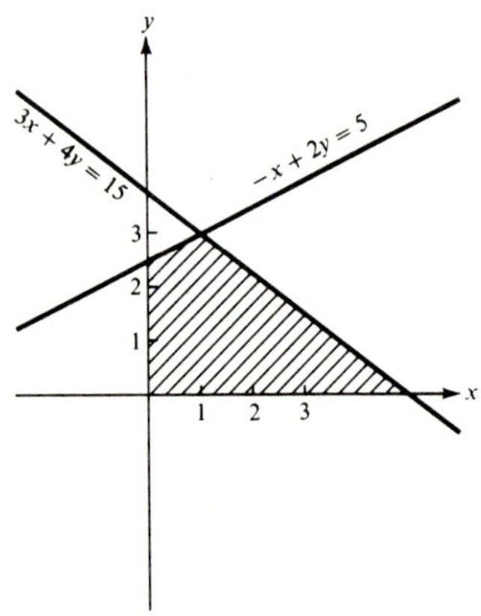

13.

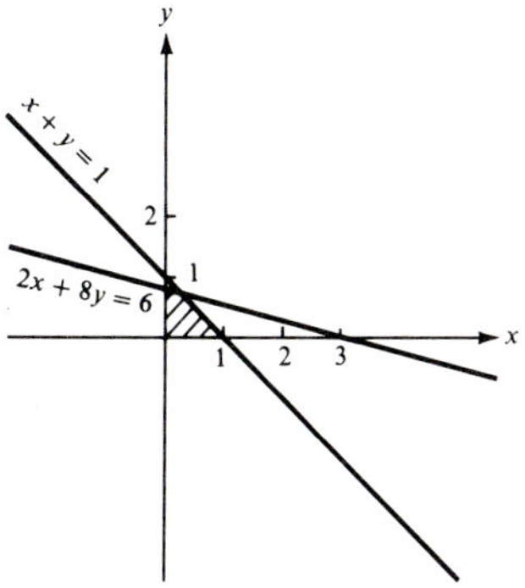

15.

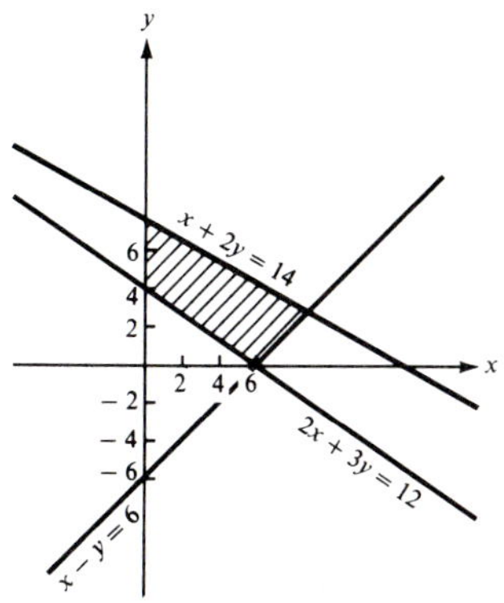

Section 4.3

17. $x = 4$, $y = 0$: Maximum z–value is 20
19. $x = 5$, $y = 0$: Maximum z–value is 25
21. $x = 0$, $y = 2$: Maximum z–value is 8
23. $x = 0$, $y = 5$: Maximum z–value is 35
25. $x = 0$, $y = 4$: Minimum z–value is -12
27. $x = 0$, $y = 4$: Minimum z–value is -8
29. $x = 0$, $y = 6$: Minimum z–value is -12
31. $x = 4$, $y = 0$: Minimum z–value is 12
33. 100 high school students and 0 elementary school students
35. $\frac{200}{3}$ office audits and $\frac{100}{3}$ "on-site" audits
37. 38.4 pages of newspaper advertisements and 1 page of magazine advertisements
39. 28.57 acres of alfafa and 128.57 acres of spinach

Mastery Tests

FORM A

1. A corner, assuming there is only one.

2.

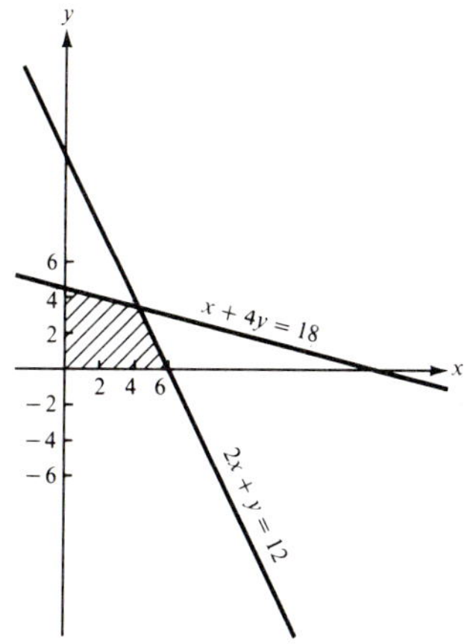

3.

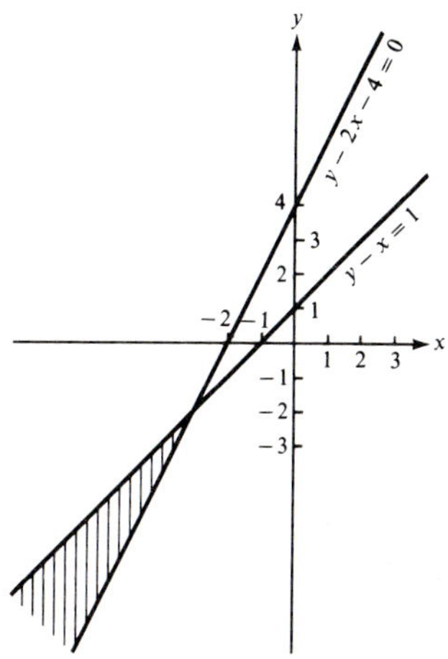

4. Is less than bc
5. The region above the line or the region to the right of the line (assuming that the number in front of y is positive)
6. Smaller 7. Constraints 8. Corner 9. True
10. False; it depends upon the inequalities.

FORM B

1. Maximum value of 1120 occurs when $x = 0$ and $y = 160$.
2. Minimum value of 130 occurs when $x = 3$ and $y = 1$.
3. Minimum value of 41 occurs when $x = 4$ and $y = 5$.
4. Minimum value of 9 occurs when $x = 2$ and $y = 3$.
6. Produce 30 gallons of safe and 15 gallons of secure.
7. Produce two cheap radios and two expensive radios.
8. Invest \$6666.67 at 8% and \$3333.33 at 10%.

Chapter 5

Section 5.2

1. $$\begin{aligned} 8x + 5y + r \quad &= 40 \\ 2x + 3y \quad + s &= 18 \\ x \geq 0, y \geq 0, r \geq 0, s \geq 0 & \\ \max z = 5x + 3y + 0r + 0s & \end{aligned}$$

3. $$\begin{aligned} 2x + 3y - r \quad &= 6 \\ 4x + 24y \quad - s &= 24 \\ x \geq 0, y \geq 0, r \geq 0, s \geq 0 & \\ \min z = 6x + 4y + 0r + 0s & \end{aligned}$$

5. $$\begin{aligned} x + y + r \quad &= 5 \\ 8x + 2y \quad + s &= 16 \\ x \geq 0, y \geq 0, r \geq 0, s \geq 0 & \\ \max z = 5x + 7y + 0r + 0s & \end{aligned}$$

7. $$\begin{aligned} 4x + y - r \quad &= 4 \\ 3x + 5y \quad - s &= 15 \\ x \geq 0, y \geq 0, r \geq 0, s \geq 0 & \\ \min z = 2x - 3y + 0r + 0s & \end{aligned}$$

9. $$\begin{aligned} x + y + r \quad &= 4 \\ x + 2y \quad - s &= 2 \\ x \geq 0, y \geq 0, r \geq 0, s \geq 0 & \\ \min z = 3x - 2y + 0r + 0s & \end{aligned}$$

11. $$\begin{aligned} x + 3y + r \quad &= 18 \\ 2x + 3y \quad - s &= 12 \\ 2x + y \quad + t &= 14 \\ x \geq 0, y \geq 0, r \geq 0, s \geq 0, t \geq 0 & \\ \min z = 9x - 2y + 0r + 0s + 0t & \end{aligned}$$

Section 5.3

13. $$\begin{array}{ccccc|c} a_1 & a_2 & a_3 & a_4 & b & \text{Vectors in basis} \\ 4 & 2 & 1 & 0 & 6 & a_3 \\ 2 & 4 & 0 & 1 & 6 & a_4 \\ \hline 2 & -4 & 0 & 0 & 0 & \end{array}$$

15. $$\begin{array}{ccccc|c} a_1 & a_2 & a_3 & a_4 & b & \text{Vectors in basis} \\ 1 & 1 & 1 & 0 & 3 & a_3 \\ 1 & 2 & 0 & 1 & 4 & a_4 \\ \hline -3 & -2 & 0 & 0 & 0 & \end{array}$$

17. $$\begin{array}{ccccc|c} a_1 & a_2 & a_3 & a_4 & b & \text{Vectors in basis} \\ 1 & 1 & 1 & 0 & 5 & a_3 \\ 8 & 2 & 0 & 1 & 16 & a_4 \\ \hline -5 & -7 & 0 & 0 & 0 & \end{array}$$

19. $$\begin{array}{ccccc|c} a_1 & a_2 & a_3 & a_4 & b & \text{Vectors in basis} \\ 1 & 1 & 1 & 0 & 4 & a_3 \\ 1 & -1 & 0 & 1 & 2 & a_4 \\ \hline -5 & -3 & 0 & 0 & 0 & \end{array}$$

21. $$\begin{array}{ccccccc|c} a_1 & a_2 & a_3 & a_4 & a_5 & a_6 & b & \text{Vectors in basis} \\ 1 & 2 & 1 & 1 & 0 & 0 & 1 & a_4 \\ 3 & 4 & -1 & 0 & 1 & 0 & 8 & a_5 \\ 2 & 3 & 1 & 0 & 0 & 1 & 6 & a_6 \\ \hline -3 & -4 & -5 & 0 & 0 & 0 & 0 & \end{array}$$

Section 5.4

23. $x = 10.5$, $y = 4.5$: Maximum z-value is 87
25. $x = 5$, $y = 0$: Maximum z-value is 45
27. $x = \frac{12}{5}$, $y = \frac{16}{5}$: Maximum z-value is 25.6
29. $x = 0$, $y = 0$, $w = 8$: Maximum z-value is 32
31. $x = \frac{78}{25}$, $y = \frac{98}{25}$, $w = 0$: Maximum z-value is $\frac{626}{25}$
33. $x = 15$, $y = \frac{65}{7}$: Maximum z-value is $\frac{4575}{7}$
35. Produce $\frac{6}{5}$ cassette-tape recorders and $\frac{21}{10}$ eight-track stereo sets.
37. Produce 22 tricycles, 0 10-speed bicycles, and 13 mopeds.
39. Plant 0 acres of sunflower seeds and $\frac{100}{9}$ acres of soy beans.
41. Produce 0 Type A packages, 0 Type B packages, and 1000 Type C packages.
43. Operate mine I $\frac{50}{3}$ days and mine II, 0 days.

Section 5.6

59. Use 0 half crews and 12 full crews.

Mastery Tests

FORM A

1. Slack or surplus variables
2. False
3. a) $x + 4y \leq 10, x \geq 0, y \geq 0$
   b) $z = 7x + 2y + 3$
   c) Yes, $x = 10$, $y = 0$ are solutions. Maximum z is 73
4. No, the same slack variable r was used twice in different equations.
5. Objective
6. Pivotal column
7. We can read our solution from the tableau.

8. Maximize $z = 25y_1 + 175y_2$ subject to

$$\begin{cases} y_1 + 2y_2 \le 2 \\ y_1 + y_2 \le 3 \\ y_1 + 3y_2 \le 4 \\ y_1 + y_2 \le 2 \\ y_1 \ge 0, y_2 \ge 0 \end{cases}$$

9. $\begin{pmatrix} 2 & 3 & 5 \\ 1 & 2 & -2 \\ 7 & 8 & 0 \\ 9 & 6 & 1 \end{pmatrix}$

10. When we cannot find a pivotal row because the pivotal column contains zeros or negative numbers only

FORM B

1. $x = 0, y = 0, w = 16$: Maximum z–value is 112
2. $x = 0, y = 0, w = 3$: Maximum z–value is 12
3. $x = \frac{20}{3}, y = 4, w = 0$: Minimum z–value is $\frac{32}{3}$
5. Make 0 packages of type I and $\frac{1600}{11}$ packages of type II.
7. Use $\frac{10}{9}$ units of food A and $\frac{14}{3}$ units of food B
8. Do 0 theory problems, 0 applications problems, and 60 combination problems.
9. Invest $75,000 in medium-grade bonds and $25,000 in speculative stocks.

Chapter 6

Section 6.2

1. $\frac{1}{5}$ **3.** $\frac{1}{8}$ **5.** a) $\frac{29}{300}$ b) $\frac{37}{75}$ c) $\frac{152}{300} = \frac{38}{75}$ d) $\frac{61}{300}$ **7.** $\frac{1}{28}$ assuming no leap year **9.** $\frac{1}{10}$

11. a) 4, 5, 6, 7, 8, 9, 10, 11, 12 b) 0 c) 1 d) $\frac{5}{9}$
13. $\frac{1}{24}$
15. $\frac{7}{8}$
17. a) $\frac{1}{50}$ b) 0 c) $\frac{21}{50}$
19. a) $\frac{36}{42} = \frac{6}{7}$ b) $\frac{28}{42} = \frac{2}{3}$

35.

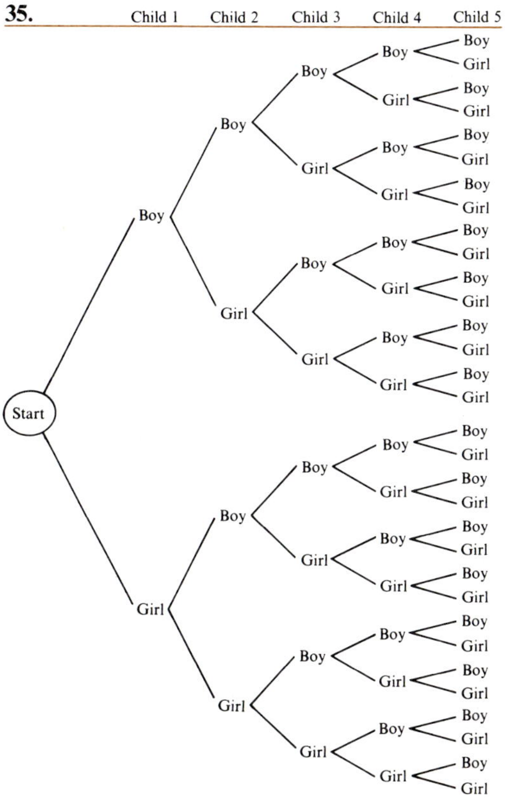

Section 6.3

21. 28 **23.** a) 1,757,600 b) 1,404,000
25. a) 999,999,999 assuming that nobody's S.S. numbers consist of only zeros
b) 999,999,999 × 26 with the same assumptions as in part (a)
27. 251,563 **29.** 6 **31.** 1024 **33.** $\frac{36}{720} = \frac{1}{20}$
35. Art in lst column
37. 5040

Section 6.4

39. 40,320 **41.** 720 **43.** 680
45. 12! = 479,001,600 **47.** 720 **49.** 210
51. a) 24 b) $\frac{1}{24}$ **53.** 116,280
55. a) $\frac{9}{38}$ b) $\frac{6}{95}$ c) $\frac{3}{95}$

Section 6.5

57. a) $\frac{4}{13}$ b) $\frac{4}{13}$ **59.** $\frac{19}{24}$ **61.** $\frac{183}{1080} = \frac{61}{360}$
63. $\frac{25}{56}$ **65.** 1

Section 6.6

67. $\frac{11}{26}$ **69.** .50 **71.** .8178
73. a) $\frac{56}{125}$ b) $\frac{69}{200}$ c) $\frac{69}{448}$ **75.** $\frac{28}{45}$ **77.** $\frac{43}{79}$
79. .6141 **81.** $\frac{3}{4}$

Section 6.7

83. $\frac{36}{121}$ **85.** $\frac{3}{7}$ **87.** $\frac{15}{42} = \frac{5}{14}$ **89.** $\frac{2}{15}$
91. $\frac{2}{11}$

Section 6.8

93. .2613 **95.** .2787 **97.** .7969 **99.** .9916
101. .7351 **103.** .9736 **105.** .9326
107. $(.25)^{10}$ or .00000095

Section 6.9

109. \$−2,777.78 **111.** $-\$\frac{1}{2}$ or −\$0.50
113. 3:4 **115.** 1:19
117. Go to stores D and E. **119.** 2.89

Section 6.10

121. a) $\frac{1}{2}$ b) $\frac{1}{4}$ c) $\frac{1}{4}$ **123.** a) $\frac{1}{2}$ b) $\frac{1}{2}$ c) 0
125. Probabilities are now $\frac{1}{4}$ and $\frac{3}{4}$, respectively.

Mastery Tests

FORM A

1. (a) **2.** (a) **3.** (b) **4.** (b) **5.** (c)
6. (d) **7.** (d) **8.** (b) **9.** (d) **10.** (a)
11. (c) **12.** (b) **13.** (c) **14.** (e) **15.** (d)

FORM B

2. No **3.** $\frac{1}{2^{48}}$ **4.** No **5.** Unlikely

6. a) .125 b) .25 c) .3125 d) .3125
These answers give the probabilities that the world series will end when the indicated number of games are played, no matter which team wins.
7. Sunday, $\frac{687}{4800}$; Monday, $\frac{685}{4800}$; Tuesday, $\frac{685}{4800}$; Wednesday, $\frac{687}{4800}$; Thursday, $\frac{684}{4800}$; Friday, $\frac{688}{4800}$; Saturday, $\frac{684}{4800}$ b) Friday
8. 0 **9.** No

Chapter 7

Section 7.3

1. a) Saddle point is 2.
b) Optimal strategies:
player R, play row 1.
player C, play column 1. value is 2
3. a) Saddle point is 1.
b) Optimal strategies:
player R, play row 1.
player C, play column 2. value is 1
5. a) Saddle point is 1.
b) Optimal strategies:
player R, play row 4.
player C, play column 3. value is 1
7. a) Saddle point is 3.
b) Optimal strategies:
player R, play row 3.
player C, play column 2. value is 3
9. a) Saddle point is 0.
b) Optimal strategies:
player R, play row 1 or row 3.
player C, play column 3. value is 0.

11.

		C		
		1 finger	2 fingers	3 fingers
	1 finger	0	−3	−4
R	2 fingers	+3	0	−5
	3 fingers	+4	+5	0

a) Yes b) Each player should show three fingers.
13. If $a < b$ and $b < d$, then saddle point is a; then player R should play row 1 or row 2 and player C should play column 1. If $a < b$ and $b > d$, then player R should play row 1 if $a > d$ also and either row 1 or row 2 if $a < d$. Player C should always play column 1.
If $a = b$ and $b < d$, then player R can play row 1 or row 2 and player C can play column 1.
If $a > b$ and $b < d$, then player R should play row 2 and player C should play column 2 if $d < a$ and player R should play row 2 and player C should play column 1 if $a < d$.

If $a > b$ and $b = d$, then player R should play row 1 or row 2 and player C should play column 2.
If $a > b$ and $b > d$, then player R should play row 1 and player C should play column 2.
If $a = b = d$ then players R and C can play either row or column.

Section 7.4

15. a) $R_1 = \frac{2}{3}, R_2 = \frac{1}{3}, C_1 = \frac{1}{3}, C_2 = \frac{2}{3}$
b) Player R: play row 1 $\frac{2}{3}$ of time and row 2 $\frac{1}{3}$ of time.
Player C: play column 1 $\frac{1}{3}$ of time and column 2 $\frac{2}{3}$ of time.
c) value of game is $\frac{11}{3}$. Game is not fair.

17. a) $R_1 = \frac{2}{15}, R_2 = \frac{13}{15}, C_1 = \frac{8}{15}, C_2 = \frac{7}{15}$, and $v = \frac{16}{15}$
b) Player R: play row 1 $\frac{2}{15}$ of time and row 2 $\frac{13}{15}$ of time.
Player C: play column 1 $\frac{8}{15}$ of time and column 2 $\frac{7}{15}$ of time.
c) Game is not fair.

19. a) $R_1 = \frac{3}{5}, R_2 = \frac{2}{5}, C_1 = \frac{2}{5}, C_2 = \frac{3}{5}$, and $v = -\frac{6}{5}$
b) Player R: play row 1 $\frac{3}{5}$ of time and row 2 $\frac{2}{5}$ of time.
Player C: play column 1 $\frac{2}{5}$ of time and column 2 $\frac{3}{5}$ of time.
c) Game is not fair.

21. a) $R_1 = \frac{1}{2}, R_2 = \frac{1}{2}, C_1 = \frac{1}{2}, C_2 = \frac{1}{2}$, and $v = \frac{7}{2}$
b) Player R: play row 1 $\frac{1}{2}$ of time and row 2 $\frac{1}{2}$ of time.
Player C: play column 1 $\frac{1}{2}$ of time and column 2 $\frac{1}{2}$ of time.
c) Game is not fair.

23. a)

		Mary guesses	
		black 2	red 5
Bill hides	Black 2	−2	+3
	Red 5	+3	−5

b) No
c) Bill: Hide black 2 $\frac{8}{13}$ of time and red 5 $\frac{5}{13}$ of time.
Mary: Guess black 2 $\frac{8}{13}$ of time and red 5 $\frac{5}{13}$ of time.
d) $-\frac{1}{13}$ e) No

25. a)

		Government agent	
		Through town	Deserted road
Spy	Through town	−50	100
	Deserted road	100	−1000

b) No
c) Spy: Select "through town" route $\frac{22}{25}$ of time and "deserted road" route $\frac{3}{25}$ of time.
Agent: Select "through town" route $\frac{22}{25}$ of time and "deserted road" route $\frac{3}{25}$ of time.

Section 7.5

27. Optimal strategies: Player R: play row 1 $\frac{13}{21}$ of time and row 2 $\frac{8}{21}$ of time.
Player C: play column 1 $\frac{4}{7}$ of time and column 2 $\frac{3}{7}$ of time.

29. Optimal strategies: Player R: play row 1 $\frac{17}{18}$ of time and row 2 $\frac{1}{18}$ of time.
Player C: play column 1 $\frac{7}{9}$ of time and column 2 $\frac{2}{9}$ of time.

31. Optimal strategies: Player R: play row 1 $\frac{1}{8}$ of time and row 2 $\frac{7}{8}$ of time.
Player C: play column 1 $\frac{1}{4}$ of time and column 2 $\frac{3}{4}$ of time.

33. Optimal strategies: Player R: play row 1 $\frac{4}{5}$ of time and row 2 $\frac{1}{5}$ of time.
Player C: play column 1 never, column 2 $\frac{7}{10}$ of time, and column 3 $\frac{3}{10}$ of time.

35. Optimal strategies: Player R: play row 1 $\frac{8}{15}$ of time and row 2 $\frac{7}{15}$ of time.
Player C: play column 1 never, column 2 $\frac{1}{15}$ of time, column 3 $\frac{14}{15}$ of time, and column 4 never.

37. Optimal strategies: Player R: play row 1 never, row 2 $\frac{5}{8}$ of time, and row 3 $\frac{3}{8}$ of time.
Player C: play column 1 $\frac{1}{8}$ of time, column 2 $\frac{7}{8}$ of time, and column 3 never.

39. Optimal strategies: Player R: play row 1 $\frac{35}{692}$ of time, row 2 $\frac{172}{692}$ of time, row 3 $\frac{424}{692}$ of time, and row 4 $\frac{61}{692}$ of time.
Player C: play column 1 $\frac{25}{173}$ of time, column 2 $\frac{79}{173}$ of time, column 3 $\frac{50}{173}$ of time, and column 4 $\frac{19}{173}$ of time.

41. Louis should build the sports stadium.

43. a)

		Phil guesses		
		$1 bill	$5 bill	$10 bill
Dealer hides	$1 bill	+2	+3	+3
	$5 bill	+3	−2	+3
	$10 bill	+3	+3	−7

b) $v = \frac{120}{13}$, no

c) Dealer: hide $1 bill $\frac{10}{13}$ of time, $5 bill $\frac{2}{13}$ of time, and $10 bill $\frac{1}{13}$ of time.
Bill: Guess $1 bill $\frac{10}{13}$ of time, $5 bill $\frac{2}{13}$ of time, and guess $10 bill $\frac{1}{13}$ of time.

45. a)

		Sharon 1	2	3	4	5	6
Rochelle	1	−2	0	4	−5	0	7
	2	0	4	−5	0	7	−8
	3	4	−5	0	7	−8	0
	4	−5	0	7	−8	0	10
	5	0	7	−8	0	10	−11
	6	7	−8	0	10	−11	0

c) No

47. Invest in real estate $\frac{17}{30}$ of time, do not invest in high-grade municipal bonds, and invest in stocks $\frac{13}{30}$ of time.

Section 7.6

49. a)

		Company B Dolly-town	Sky-ville	Morgan-ville
Company A	Dollytown	$\frac{1}{2}$	$\frac{1}{3}$	$\frac{1}{3}$
	Skyville	$\frac{2}{3}$	$\frac{1}{2}$	$\frac{2}{3}$
	Morganville	$\frac{2}{3}$	$\frac{1}{3}$	$\frac{1}{2}$

b) Each should open in Skyville.

51. a)

		Joe: Confess	Don't confess
Al	Confess	2	35
	Don't confess	−35	10

b) Each should confess.

53. a) It depends upon where the first move is made.
b) No.
c) It depends upon what the first move is. Every perfectly played game should end in a tie.

55. a)

		Roosevelt Smith gets off at: Albany	Bingham-ton	Canada
Clayton Jones gets off at	Albany	400	−50	−200
	Binghamton	−50	400	−200
	Canada	−50	−50	400

b) Value is $\frac{800}{11}$. No

c) Clayton should get off at Albany $\frac{3}{11}$ of time, at Binghamton $\frac{3}{11}$ of time, and go on to Canada $\frac{5}{11}$ of time.
Roosevelt should get off at Albany $\frac{4}{11}$ of time, at Binghamton $\frac{4}{11}$ of time, and go on to Canada $\frac{3}{11}$ of time.
(This assumes that we interpret the probabilities as if they both can be sent back to New York.)

Mastery Tests

FORM A

1. Payoff matrix.
2. An optimal strategy if the game is strictly determined.
3. Yes. **4.** Strictly determined games.
5. It favors player C.

6. Nonstrictly determined games. **7.** 2.

8. Player R: play row 1 $\frac{1}{2}$ of time and row 2 $\frac{1}{2}$ of time.
Player C: play column 1 $\frac{1}{2}$ of time and column 2 $\frac{1}{2}$ of time.

9. (b) and (c) **10.** (a)

FORM B

1. a)

John guesses even and sum of dots is:

Sum of dots	Roland guesses even or odd
2	2
3	−3
4	4
5	−5
6	6
7	−7
8	8
9	−9
10	10
11	−11
12	12

John guesses odd and sum of dots is:

Sum of dots	Roland guesses even or odd
2	−2
3	3
4	−4
5	5
6	−6
7	7
8	−8
9	9
10	−10
11	11
12	−12

b) Not fair.

2. a)

		Carolyn 3	Carolyn 4
Robin	3	6	−7
	4	−7	8

b) $v = -\frac{1}{28}$, no

c) Robin: show 3 fingers $\frac{15}{28}$ of time and 4 fingers $\frac{13}{28}$ of time.
Carolyn: show 3 fingers $\frac{15}{28}$ of time and 4 fingers $\frac{13}{28}$ of time.

3. a)

		Enemy attacks Radar	Enemy attacks Airfield
Colonel Smith defends	Radar	+200	−150
	Airfield	−100	+300

b) 60

c) Colonel Smith: defend radar installations $\frac{8}{15}$ of time and airfield $\frac{7}{15}$ of time.
Enemy: attack radar installations $\frac{3}{5}$ of time and airfield $\frac{2}{5}$ of time.

4. a) Play row 1 $\frac{2}{3}$ of time and row 2 $\frac{1}{3}$ of time.
Play column 1 $\frac{11}{15}$ of time and column 2 $\frac{4}{15}$ of time.

5. a) Simplified payoff matrix is $\begin{pmatrix} 1 & 2 \\ 3 & 1 \end{pmatrix}$

Optimal strategies are: player R play row 1 $\frac{2}{3}$ of time and row 2 $\frac{1}{3}$ of time.
player C play column 1 $\frac{1}{3}$ of time and column 2 $\frac{2}{3}$ of time.

b) Simplified payoff matrix is $\begin{pmatrix} 1 & 0 \\ -2 & 3 \end{pmatrix}$

Optimal strategies are: player R play row 1 $\frac{5}{6}$ of time and row 2 $\frac{1}{6}$ of time.
player C play column 1 $\frac{1}{2}$ of time and column 2 $\frac{1}{2}$ of time.

Chapter 8

Section 8.2

1. $316,000 3. 2.5 girls 5. $3028.57
7. Location A
9. a) Tract A $70.6 million; Tract B $66.13 million; Tract C $57.44 million; Tract D $48.46 million
   b) Tract A
11. $17.25 13. $150

Section 8.3

15.

Number of cases of milk sold	Probability
29	$\frac{3}{50}$
30	$\frac{6}{50}$
31	$\frac{8}{50}$
32	$\frac{9}{50}$
33	$\frac{10}{50}$
35	$\frac{7}{50}$
38	$\frac{4}{50}$
39	$\frac{2}{50}$
40	$\frac{1}{50}$

17.

If 40 cases are ordered and number of cases sold is	Profit is
29	$ − 19
30	$ − 10
31	$ − 1
32	$ 8
33	$ 17
35	$ 35
38	$ 62
39	$ 71
40	$ 80

19.

If 40 cases are ordered and number of cases sold is	Profit is
29	$ − 4.50
30	$ 5.00
31	$ 14.50
32	$ 24.00
33	$ 33.50
35	$ 52.50
38	$ 81.00
39	$ 90.50
40	$100.00

21.

Number of loaves of bread sold	Probability
89	$\frac{10}{90}$
92	$\frac{14}{90}$
93	$\frac{16}{90}$
96	$\frac{18}{90}$
97	$\frac{11}{90}$
98	$\frac{10}{90}$
99	$\frac{7}{90}$
100	$\frac{4}{90}$

23.

If 100 loaves are ordered and number of loaves of bread sold is	Profit is
89	$2.71
92	$3.88
93	$4.27
96	$5.44
97	$5.83
98	$6.22
99	$6.61
100	$7.00

25.

If 100 loaves are ordered and number of loaves of bread sold is	Profit is
89	$4.71
92	$5.88
93	$6.27
96	$7.44
97	$7.83
98	$8.22
99	$8.61
100	$9.00

27. Martha makes less money when she raises her prices than when the price that she pays for a loaf of bread is lowered.

Section 8.4

29. 0.9224 31. 0.6177
33. At least $125.50
35. At least $173.75 This is the premium for a one year $25,000 policy at age 48. (In reality, the company

would most likely charge a higher premium if the annual premium is to be the same for each of the five years.)

37. a) 71.43 years b) 71.16 years c) 68.30 years
d) $78.75 assuming the premium will be the same for each of the 18 years
e) $72.80 assuming the premium will be the same for each of the 18 years

Mastery Tests

FORM A

1. 0.5717 **2.** 0.7315 **3.** 0.9714 **4.** False

5. Opportunity loss is defined as the difference between the greatest profit that could be made had there been sufficient quantity of stock and the profit that was actually made.

6. 4 **7.** 12 **8.** Life expectancy

9. Greatest expected profit **10.** −$5.14

FORM B

1. Between $9000 and $11,000

2. $760

4.

Number of lunches sold	Probability
185	$\frac{1}{31}$
188	$\frac{3}{31}$
192	$\frac{2}{31}$
196	$\frac{4}{31}$
197	$\frac{3}{31}$
198	$\frac{4}{31}$
199	$\frac{8}{31}$
200	$\frac{6}{31}$

5. 196.52

6.

Number of lunches sold	Expected profit (in cents)
185	650
188	920
192	1280
196	1640
197	1730
198	1820
199	1910
200	2000

7. 200 lunches

8.

Number of lunches sold	Expected profit (in cents)
185	1575
188	1860
192	2240
196	2620
197	2715
198	2810
199	2905
200	3000

9.

Number of lunches sold	Expected profit (in cents)
185	1695
188	1956
192	2304
196	2652
197	2739
198	2826
199	2913
200	3000

10. 200 lunches

11. We cannot calculate the company's expected profit since we do not know how much money can be earned if oil is found.

Chapter 9

Section 9.2

1. Mean = $975, median = $1057.50, mode = none

3. Mean = $16,100, median = $19,000
mode = $19,000

5. Probably the mode

7. Management is referring to the mean salary and the union is referring to the modal salary.

9. $12,336

11. No

13. (a) $\Sigma x^2 = 196$ b) $(\Sigma x)^2 = 900$
The answers are not the same.

15. Management is probably using the modal salary whereas the union is using the mean salary.

17. Extreme numbers have a great effect on the mean.

19. At 68°F

Section 9.3

21. Range = 326, standard deviation = $\sqrt{11567.10938}$ or 107.55

23. Range = 52, standard deviation = $\sqrt{215.354}$ or 14.67

25. Mean = 82.4, standard deviation = $\sqrt{103.84}$ or 10.19

27. a) For Matty, range = 22,
standard deviation = $\sqrt{59.5}$ or 7.71
For Harriet, range = 31,
standard deviation = $\sqrt{100.5}$ or 10.02

b) Matty

29. No change. Subtracting a constant amount c from each number leaves the standard deviation unaffected.

31. The new mean will be increased by \$15 so that it will now be \$855. The standard deviation will remain the same.

Section 9.4

33.

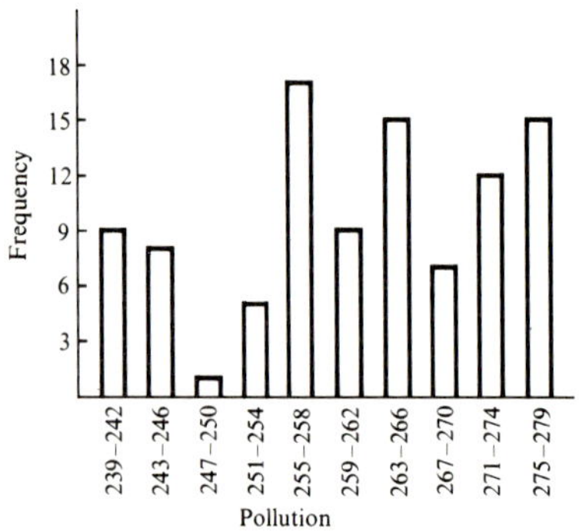

35.

Age	Frequency
15–18	8
19–20	4
21–22	4
23–24	2
25–26	3
27–28	4
29–30	3
31–32	6
33–34	4
35–36	2

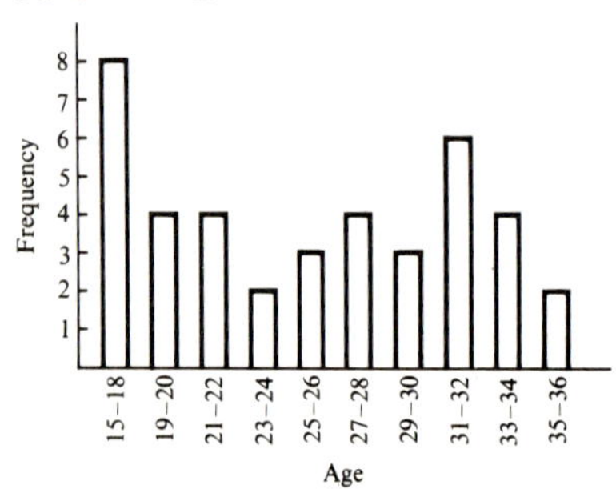

37.

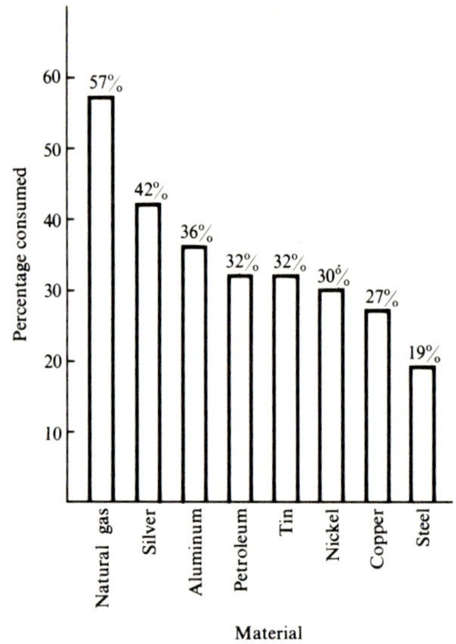

39.

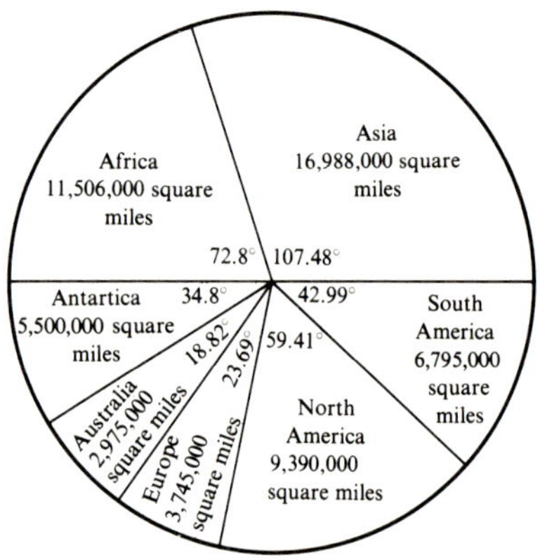

41.

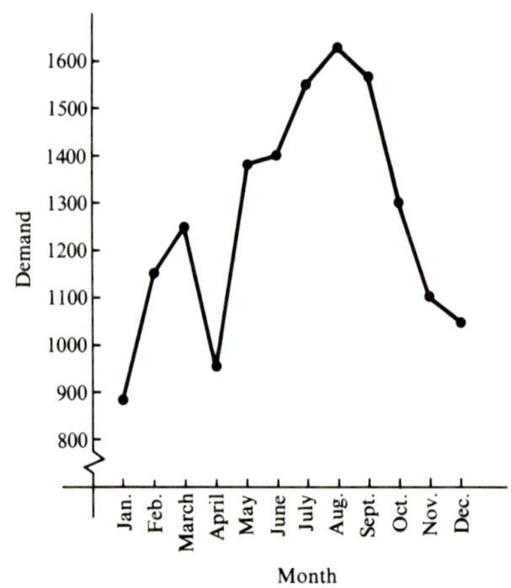

43.

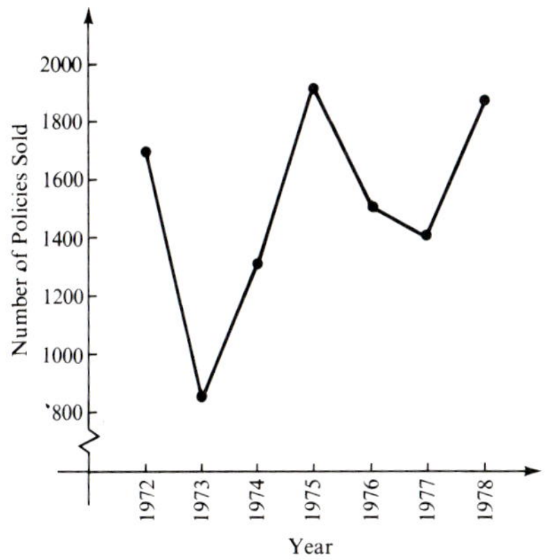

Section 9.5

45. a) 0.4608 b) 0.2224
c) 0.1894 d) 0.0427
e) 0.0146 f) 0.8452
g) 0.8587 h) 0.1343

47. a) 1 b) -2.3
c) 6 d) 0

49. a) 0.8164 b) 0.3707
c) 0.0099 d) 0.5899

51. 0.8577 **53.** 0.0314

Section 9.6

55.

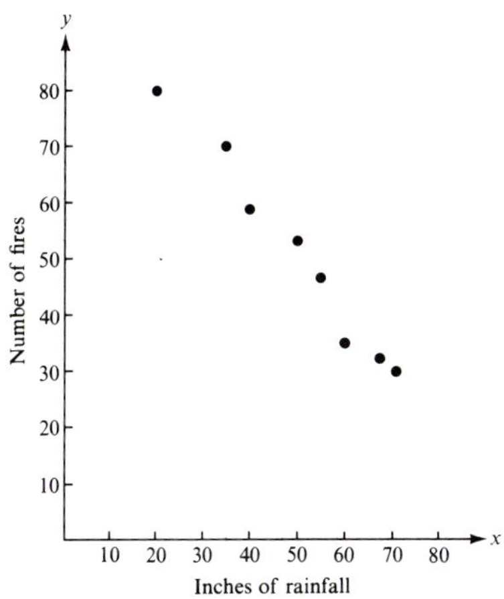

57.

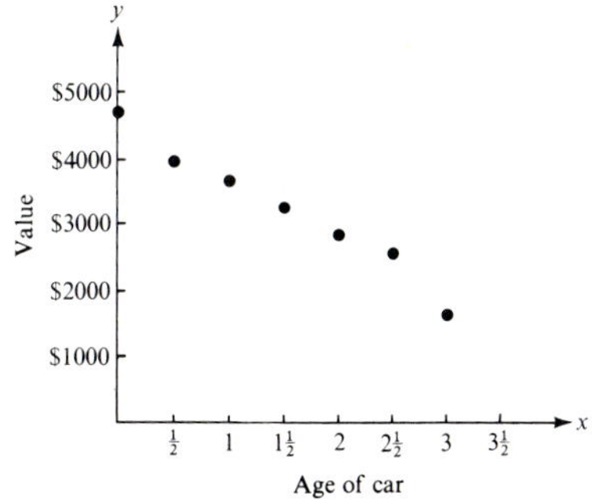

59.

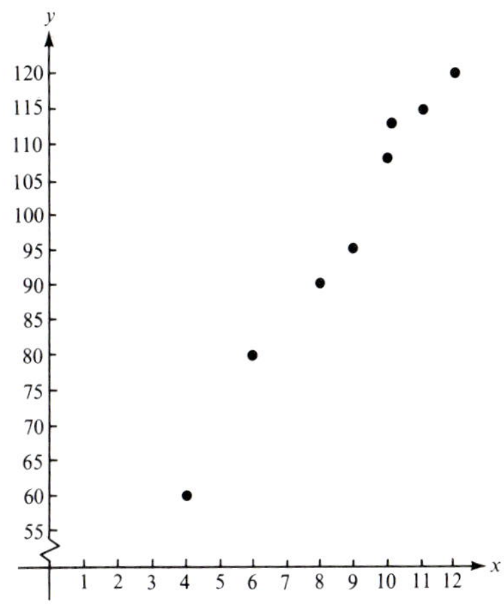

b) 0.9848 c) $y = 97.5 + 7.64(x - 8.75)$
d) 145.25

61. a) $y = 27.67 + .834(x - 82.5)$ b) 42.265

63. a) $y = 42.43 + 4.607(x - 3)$ b) 65.465

Mastery Tests

FORM A

1. (a) **2.** (d) **3.** (a) **4.** (d) **5.** (c)
6. (e) **7.** (b) **8.** (c) **9.** (c) **10.** (b)
11. (b) **12.** (a) **13.** (a)

FORM B

1. a) 1798 b) 8464
2. Probably the mode
3. 85.4 **4.** 150 **5.** (c)
6. Positive correlation

Chapter 10

Section 10.2

1. \$1417.50 **3.** \$920.00 **5.** 34.29%
7. \$3500 **9.** a) \$181.48 b) \$180.00 c) \$184.00
11. \$27.09 **13.** \$42.14 **15.** \$926.10
17. \$6749.29 **19.** 8.33% **21.** \$4538.94
23. 41.42% compounded annually
25. \$6422.41 assuming he makes his monthly payments on time.

Section 10.4

27. \$16,095.87 **29.** 9,014.68 **31.** \$1,862.40
33. \$7,380.18
35. 15%. (This assumes that the total cost to the school, with interest added, is \$23,134.72)
37. Neither. Each bank is offering the mortgage at an 8% yearly interest rate.

Section 10.5

39. \$37.66 **41.** \$13,432.80 **43.** \$99.64
45. \$255.22
47. a) A
b) Option A \$30,386.40; option B \$46,935; option C \$61,774.08
49. 17 payments

Mastery Tests

FORM A

1. Interest **2.** \$122.50 **3.** 365 **4.** \$7.55
5. \$14,190.68 **6.** 7.519% **7.** An annuity
8. Interest **9.** False **10.** True

FORM B

1. In 1980, it would be worth \$1,167,410,436.
2. 7.35 years **3.** \$1,418.52 **4.** \$2564.07
5. 96 months **6.** Lease now. **7.** \$946.53
8. 8.71% **9.** \$29,672.03 **10.** \$3532.27
11. \$8001.81 **12.** \$22,623.49
13. a) \$1079.03 b) \$12,528.96
14. \$4956.19 **15.** \$19,548.53

Chapter 11

Section 11.3

1. Yes **3.** No **5.** Yes **7.** Not regular
9. Regular **11.** Regular **13.** Not regular
15. Not regular **17.** Regular **19.** Regular
21. $\left(\frac{3}{5} \quad \frac{2}{5}\right)$ **23.** $\left(\frac{9}{16} \quad \frac{7}{16}\right)$ **25.** $\left(\frac{8}{21} \quad \frac{1}{3} \quad \frac{2}{7}\right)$
27. Absorbing **29.** Not absorbing
31. a) $\frac{4}{25}$ b) $\frac{84}{125}$ c) $\frac{164}{625}$

33. Two step $= \begin{pmatrix} \frac{7}{20} & \frac{13}{20} \\ \frac{13}{50} & \frac{37}{50} \end{pmatrix}$; three step $= \begin{pmatrix} \frac{61}{200} & \frac{139}{200} \\ \frac{139}{500} & \frac{361}{500} \end{pmatrix}$

35. Two step $= \begin{pmatrix} \frac{41}{150} & \frac{11}{75} & \frac{29}{50} \\ \frac{3}{10} & \frac{9}{40} & \frac{19}{40} \\ \frac{11}{40} & \frac{13}{60} & \frac{61}{120} \end{pmatrix}$;

three step $= \begin{pmatrix} \frac{427}{1500} & \frac{309}{1500} & \frac{191}{375} \\ \frac{659}{2400} & \frac{239}{1200} & \frac{421}{800} \\ \frac{1003}{3600} & \frac{701}{3600} & \frac{79}{150} \end{pmatrix}$

37. $\frac{6}{13}$

Section 11.4

39. $\begin{pmatrix} .52213304 & .24827556 & .2295914 \\ .4043699 & .41276205 & .1828605 \\ .1863276 & .12367996 & .68999244 \end{pmatrix}$

41. $\frac{7}{16}$ **43.** $\frac{17}{32}$

45. Transition matrix is

		Year 2: Movie goers	Year 2: Non-moviegoers
Year 1	Movie goers	.90	.10
	Non-moviegoers	.20	.80

47.

	$i=0$	$i=1$	$i=2$	$i=3$	$i=4$	$i=5$	$i=6$
$i=0$	1	0	0	0	0	0	0
$i=1$	.6	0	.4	0	0	0	0
$i=2$	0	.6	0	.4	0	0	0
$i=3$	0	0	.6	0	.4	0	0
$i=4$	0	0	0	.6	0	.4	0
$i=5$	0	0	0	0	.6	0	.4
$i=6$	0	0	0	0	0	0	1

49. Transition matrix is

Prob that they will switch to

		F	G	U
	F	.75	.15	.10
Now	G	.19	.69	.12
	U	.03	.17	.80

51. Transition matrix is

		Prob that outfit worn on Tuesday is		
		Red	Blue	Black
Outfit worn Monday	Red	0	0	0
	Blue	1	0	0
	Black	$\frac{1}{2}$	$\frac{1}{2}$	0

Mastery Tests

FORM A

1. True **2.** (a) **3.** (c) **4.** (d)
5. (a), (b), (c), and (d) **6.** State a_2 since $P_{22} = 1$
7. $(\frac{3}{11} \quad \frac{8}{11})$

8. $\begin{pmatrix} \frac{23}{54} & \frac{31}{54} \\ \frac{31}{72} & \frac{41}{72} \end{pmatrix}$

9. Positive numbers
10. 1 **11.** False **12.** Yes

FORM B

1. a)

		Prob that lion will next move to cell			
		I	II	III	IV
Lion now in cell	I	0	$\frac{1}{2}$	$\frac{1}{2}$	0
	II	$\frac{1}{2}$	0	0	$\frac{1}{2}$
	III	$\frac{1}{2}$	0	0	$\frac{1}{2}$
	IV	0	$\frac{1}{2}$	$\frac{1}{2}$	0

b) $\frac{1}{2}$
2. 0.175

3.

	$i=0$	$i=1$	$i=2$	$i=3$	$i=4$	$i=5$	$i=6$	$i=7$
$i=0$	1	0	0	0	0	0	0	0
$i=1$	$\frac{1}{2}$	0	$\frac{1}{2}$	0	0	0	0	0
$i=2$	0	$\frac{1}{2}$	0	$\frac{1}{2}$	0	0	0	0
$i=3$	0	0	$\frac{1}{2}$	0	$\frac{1}{2}$	0	0	0
$i=4$	0	0	0	$\frac{1}{2}$	0	$\frac{1}{2}$	0	0
$i=5$	0	0	0	0	$\frac{1}{2}$	0	$\frac{1}{2}$	0
$i=6$	0	0	0	0	0	$\frac{1}{2}$	0	$\frac{1}{2}$
$i=7$	0	0	0	0	0	0	0	1

4. a)

		Prob that next day she will order		
		Strawberry shortcake	Ice cream sundae	Jello
If one day she orders	Strawberry shortcake	0	$\frac{4}{7}$	$\frac{3}{7}$
	Ice cream sundae	$\frac{2}{3}$	0	$\frac{1}{3}$
	Jello	$\frac{1}{2}$	$\frac{1}{2}$	0

b) $\frac{5}{21}$

5. Transition matrix is

	K	N	L
K	0	$\frac{5}{8}$	$\frac{3}{8}$
N	$\frac{1}{2}$	0	$\frac{1}{2}$
L	$\frac{2}{7}$	$\frac{5}{7}$	0

6. a)

	1st base man	2nd base-man	Short-stop	3rd base-man
1st baseman	0	.8	.05	.15
2nd baseman	.6	0	.3	.1
Shortstop	.5	.3	0	.2
3rd baseman	.4	.35	.25	0

b) .13835

7. a)

		Tomorrow	
		Functioning	Not functioning
Today	Functioning	.9	.1
	Not Functioning	.8	.2

b) .8889

8. a)

		Tomorrow		
		N.Y.	N.J.	Conn.
Today	N.Y.	$\frac{1}{3}$	$\frac{1}{3}$	$\frac{1}{3}$
	N.J.	$\frac{2}{3}$	0	$\frac{1}{3}$
	Conn.	$\frac{1}{2}$	$\frac{1}{2}$	0

b) .4691

9. Transition matrix is

		Year 2	
		Prefer automatic	Prefer standard
Year 1	Prefer automatic	.85	.15
	Prefer standard	.30	.70

10. Transition matrix is

		Feb. 1978		
		Students	Working	Unemployed
	Students	.80	.17	.03
Feb. 1977	Working	.05	.75	.20
	Unemployed	.20	.68	.12

11. Transition matrix is

		Prob next semester will teach			
		Calculus	Algebra	Statistics	Finite math
	Calculus	.10	.40	.30	.20
Teaches now	Algebra	.25	.05	.44	.26
	Statistics	.20	.30	.10	.40
	Finite math	.15	.25	.35	.25

a) .2848

Index

82 83 84 85 9 8 7 6 5 4 3 2 1

Frequently Used Symbols

Symbol	Meaning
$\{a, b, c, d, \ldots\}$	set of elements
$\varnothing$	null set
$\in$	is an element of
$\notin$	is not an element of
$A \subset B$	A is a subset of B
A'	complement of set A
$A \cup B$	A union B
$A \cap B$	A intersect B
U	universal set
$y = f(x)$	y is a function of x
$n(A)$	the number of elements in set A
m	slope of a line
$n!$	n factorial
${}_nP_r$	the number of permutations of n things taken r at a time
${}_nC_r$	the number of combinations of n things taken r at a time
Prob(A)	the probability of event A
Prob($A \mid B$)	the probability of event A given that event B has occurred
σ	standard deviation
$\leq$	less than or equal to
$\geq$	greater than or equal to
$>$	greater than
$<$	less than
z	z-score or z-value
r	coefficient of linear correlation
P_{ij}	probability that an experiment now in state i will next move to state j